T0338265

COMPUTATIONAL INTELLIGENCE

COMPUTATIONAL INTELLIGENCE

SYNERGIES OF FUZZY LOGIC, NEURAL NETWORKS AND EVOLUTIONARY COMPUTING

Nazmul Siddique
University of Ulster, UK

Hojjat Adeli
The Ohio State University, USA

A John Wiley & Sons, Ltd., Publication

This edition first published 2013
© 2013 John Wiley & Sons, Ltd

Registered office

John Wiley & Sons Ltd, The Atrium, Southern Gate, Chichester, West Sussex, PO19 8SQ, United Kingdom

For details of our global editorial offices, for customer services and for information about how to apply for permission to reuse the copyright material in this book please see our website at www.wiley.com.

Library of Congress Cataloging-in-Publication Data

Siddique, N. H.
   Computational intelligence : synergies of fuzzy logic, neural networks, and evolutionary computing / Nazmul Siddique, Hojjat Adeli.
      pages cm
   Includes bibliographical references and index.
   ISBN 978-1-118-33784-4 (cloth)
   1. Computational intelligence.   I. Adeli, Hojjat, 1950– II. Title.
   Q342.S53 2013
   006.3–dc23

                                                                                          2012047736

A catalogue record for this book is available from the British Library

ISBN: 9781118337844

Typeset in 10/12pt Times by Aptara Inc., New Delhi, India
Printed and bound in Singapore by Markono Print Media Pte Ltd

To Kaniz, Oyndrilla, Opala and Orla
– Nazmul

To Nahid, Amir, Anahita, Cyrus and Mona
– Hojjat

Contents

Foreword

Computational Intelligence: Synergies of Fuzzy Logic, Neural Networks and Evolutionary Computing, or CIS for short, is a true *magnum opus*. Authored by Dr Nazmul Siddique and Professor Hojjat Adeli, CIS is a profoundly impressive work. It breaks new ground on many levels and in many directions. It contains a wealth of information which is new, and if not new, hard to find elsewhere.

In recent years, computational intelligence (CI) has been growing rapidly in visibility and importance. CIS's coverage of CI is very thorough, very authoritative, very insightful and very reader-friendly. CIS paves the way for making courses on CI a requirement in engineering curricula.

What is computational intelligence? A bit of history is in order. The core of CI is the conception, design and utilization of intelligent systems. The concept of an intelligent system began to crystallize in the 1980s, at a time when AI was undergoing an identity crisis, moving from logic to probability theory. There were many competing methodologies, among them traditional AI, fuzzy logic, neuro-computing and evolutionary computing. Each of these methodologies had a community, with each community claiming superiority over the others. In that climate, I came to the conclusion that what was needed was a unification of the methodologies which were competing with AI, gaining strength through unity. This was the genesis of the concept of soft computing (SC) as a partnership of fuzzy logic, neuro-computing, evolutionary computing and probabilistic computing. The guiding principle that underlies SC is that generally, superior performance can be achieved when the constituent methodologies of SC are employed in combination rather than in stand-alone mode. The Berkeley Initiative in Soft Computing (BISC) was launched in 1991 with very lukewarm support of my colleagues but strong backing from the Dean of the College of Engineering, David Hodges. Today, there are 20 journals with 'soft computing' in the title.

A few years after the launch of BISC, Jim Bezdek used his influence to create a Computational Intelligence Society within the IEEE. The concept of computational intelligence is closely related to the concept of soft computing. The principal difference between SC and CI is that SC is a partnership of fuzzy logic, neuro-computing, evolutionary computing and probabilistic computing, whereas CI is a partnership of fuzzy logic, neuro-computing and evolutionary computing. There is no competition between CI and SC, but with the backing of the IEEE, CI has been growing rapidly in visibility and acceptance. An important factor in the growth of CI is that today the principal concepts and techniques which are employed in the conception, design and utilization of intelligent systems are drawn, in large measure, from CI rather than from AI. What is worthy of note is that in 1995, at about the time when CI came into existence, Professors Adeli and Hung published a book entitled *Machine Learning –*

Neural Networks, Genetic Algorithms, and Fuzzy Systems, which in effect was the first treatise on CI. In retrospect, the importance of this seminal treatise is hard to exaggerate.

CIS is a remarkable work. The introductory chapter on CIS presents the authors' perception of what CI is and what it has to offer. The introductory chapter is followed by two chapters on the basics of fuzzy set theory and fuzzy logic. The authors' exposition is succinct, insightful and reader-friendly. I am highly impressed by their exposition of fuzzy logic and its applications.

The concepts of a linguistic variable, fuzzy if–then rules and fuzzy control received a great deal of attention, with the stress on applications. The authors' discussion of the concept of a linguistic variable reminds me of the hostile criticism of what I wrote at the time, 1973, about the concept of a linguistic variable. The criticism reflected a deep-seated tradition within science and engineering – the tradition of respect for numbers and disrespect for words. What my critics did not understand is that the use of words in place of numbers opens the door to exploitation of tolerance for imprecision, and thereby reduces cost and achieves simplicity. Today, almost all applications of fuzzy logic employ the concept of a linguistic variable. Several important applications of fuzzy logic, among them fuzzy control, are discussed in detail. As in other chapters of CIS, the exposition concludes with MATLAB® programs and references.

Following their masterly exposition of the basics of fuzzy logic, the authors turn to neuro-computing and neural systems. In the three decades since its debut, neuro-computing has become a highly important body of concepts and techniques, with wide-ranging applications in system identification, simulation and adaptation. A critical event in the evolution of neuro-computing was the invention of the backpropagation algorithm, originally due to Paul Werbos in the early 1970s and independently reinvented and developed by David Rumelhart in the early 1980s. The backpropagation algorithm opened the door to a wide variety of applications of neuro-computing. Many examples of such applications are described in CIS.

The exposition of neuro-computing is followed by an equally masterly exposition of evolutionary computing. Evolutionary computing is rooted in the pioneering work of John Holland on genetic algorithms, in combination with the seminal work of Larry Fogel. In essence, evolutionary computing is systematized random search. What is surprising is that systematized random search can be so effective in optimization, and especially in global maximization. In CIS, one finds insightful expositions of various non-standard approaches to optimization, including multi-objective optimization and Pareto optimization. But what is really fascinating is what remarkable results can be achieved through the employment of John Koza's genetic programming.

I noted earlier that the guiding principle of computational intelligence is that, in general, superior performance is achieved when fuzzy logic, neuro-computing and evolutionary computing are used in combination rather than in stand-alone mode. The last part of CIS is motivated by this guiding principle. There are very informative discussions of neuro-fuzzy systems, evolutionary fuzzy systems, evolutionary neural systems and evolutionary fuzzy neural systems. Much of the information in these chapters is hard to find elsewhere. There is much that is original to the authors.

In sum, CIS is a major contribution to the literature – it is authoritative, thorough, up-to-date, insightful and reader-friendly. CIS should be on the desk of anybody who is interested in the conception, design and utilization of intelligent systems. Professors Nazmul Siddique and Hojjat Adeli (and their publisher, John Wiley & Sons Ltd) deserve our compliments and loud applause.

Lotfi A. Zadeh
UC Berkeley
December 2012

Preface

Creating intelligent systems has been of interest to scientists for many years. In the early days of science, scientists developed systems which imitated the behaviour of living organisms. A famous example is Jacque De Vaucanson's mechanical duck from 1735, which could move its head, tail and wings as well as swallow food. This gave the illusion of intelligence and delighted and amused people at the time. The whole control mechanism was based on rotating cylinders, with gudgeons used in music boxes to control the timely execution of different behaviours. The behaviour was mechanistic as the duck always showed the same behaviour according to the mechanical control system used. To generate a new behaviour, the mechanical control system had to be changed. That means, to generate different behaviours in different situations and environments, the mechanical duck needed different control systems. Rather than using different control systems for different behaviours, scientists attempted to provide different behaviours with a single control system. This posed the challenge of developing adaptive and learning systems. In the beginning, the hope of success was based only on the belief that some general laws of adaptation should exist. The endeavour went through different stages of development, such as deterministic, stochastic and adaptive. In the happy days of determinism, various mathematical and analytical tools were developed to describe and analyse systems. These methodologies were successful, especially for linear systems. Difficulties arose as soon as nonlinear factors had to be considered. New tools had to be developed for nonlinear systems.

Uncertainty was an issue to be avoided at all costs and was not addressed until the late nineteenth century. Newtonian mechanics was replaced by statistical mechanics to describe uncertainty with the help of probability theory, developed by Thomas Bayes in the eighteenth century, which continued until the late twentieth century. The gradual evolution of probability theory for the expression of uncertainty was challenged by new theories of vagueness and fuzzy set theory, developed by Lotfi Zadeh, which came into being in the latter part of the twentieth century as a measure of uncertainty. Fuzzy systems theory has proved to be a powerful tool for the approximation of nonlinear and complex systems where traditional analytic functions or numeric relations are unable to manage.

The long-suffering stage is the time taken for model development of unknown systems, which cannot even be determined experimentally. This is evidenced by the emergence of new theories of adaptivity. The possibility of developing system models under incomplete and very little *a priori* information is based on adaptation and learning. That is, a system capable of adapting and learning is to be considered intelligent. Among the many interesting mathematical and non-traditional apparatuses for adaptation and learning, neural networks are the most widely used. Various learning algorithms have been developed since the 1960s.

The problem of developing systems to satisfy specific criteria appeared at some stage due to design, technology and development constraints. The problem of optimality then became one of the key issues in developing models or systems. In fact, the problem of optimality is a central issue in science, engineering, economy and everyday life. In deterministic or stochastic processes, the criterion of optimality (i.e., the functional) should be known explicitly *a priori* with sufficient information. The conditions of optimality only define local extrema. If the number of such extrema is large, the problem of finding the global extremum becomes a complex one. Various conventional mathematical and derivative-based optimization techniques have been developed over the past decades. Unfortunately, very often these methods cannot be applied to a wide range of problems since the functional (or the objective functions) are not analytically treatable or even not available in closed form. Further, many real-life optimization problems have constraints that either cannot be defined mathematically or are highly nonlinear implicit and discontinuous functions of design variables. These led researchers to seek stochastic methods such as evolutionary and bio-inspired optimization techniques that are capable of searching a high-dimensional space.

The inherent capability and appeal of such traditional approaches diminished as the complexity of systems grew. There arose a need for non-traditional approaches inspired by nature, such as human thinking, perception and reasoning, biological neural networks and evolution in nature.

In 1995 H. Adeli and S.L. Hung published a seminal book, *Machine Learning – Neural Networks, Genetic Algorithms, and Fuzzy Systems* (John Wiley & Sons) as the first treatise to present the three main fields of computational intelligence in a single book and demonstrate that through integration of the three emerging computing paradigms, intractable problems could be solved more effectively. Since then, research on computational intelligence has grown exponentially and the field of computational intelligence is now well established. That seminal book has inspired the current book. Computational intelligence schemes are presented in this book with the development of a suitable framework for fuzzy, neural and evolutionary computing, evolutionary/fuzzy systems, evolutionary neural systems, neuro-fuzzy systems and finally hybridization of the three basic paradigms. Applications to linear and nonlinear systems are discussed, with examples and MATLAB® exercises.

This book is designed for final-year undergraduate, postgraduate, research students and professionals. It is written at a comprehensible level for students who have some basic knowledge in calculus, differential equations and some exposure to optimization theory. Owing to the emphasis on systems and control, the book should be appropriate for electrical, control, computer, industrial and manufacturing engineering students as well as computer and information science students. With mathematical and programming references and applications in each chapter, the book is self-contained. It should also serve already practicing engineers and scientists who intend to study the field of computational intelligence and system science. In particular, it is assumed that the reader has no experience in fuzzy logic, neural networks or evolutionary computing.

The final goal of the authors is the adroit integration of three different computational intelligence technologies and problem-solving paradigms: fuzzy systems, neural networks and evolutionary computing. The book is organized in ten chapters. It includes three introductory chapters (Chapters 2, 4 and 6) on basic techniques of fuzzy logic, neural networks and evolutionary computing in order to introduce the reader to these three different computing paradigms. It then presents different applications of the three technologies in a wide range of application domains in Chapters 3, 5 and 7. Hybridization of the three technologies is an interesting feature, which has been presented in Chapters 8, 9 and 10. Most of the book covers applications in systems modelling, control and optimization.

There have been many texts, research monographs and edited volumes published since the 1990s. There are a few books that cover some topics on the combination of the three basic technologies. They are all referenced in each chapter of this book, which the reader may find useful in further reading for research. There is no single book covering all topics on fuzzy, neural networks or evolutionary computing or their combinations that is well suited for such a wide-ranging audience, especially undergraduate, postgraduate and research students. This book is an attempt to attract all groups, putting the emphasis on a combination of the three methodologies. The book has not been written for any specific course, however, it could be used for courses in computational intelligence, intelligent control, intelligent systems, fuzzy systems, neural computing, evolutionary computing and hybrid systems. As such, the book is appropriate for beginners in the field of computational intelligence. The book is also applicable as prescribed material for a final-year undergraduate course. The book is written based on the experience of many years following pedagogical features with illustrations, step-by-step algorithms, worked examples and MATLAB® code for real-world problems. The intention of the book is not to provide a thorough discussion of all computational intelligence paradigms and methods, but to give an overview of the most popular and frequently used methods.

Acknowledgements

It is necessary to thank a number of people who have helped in many ways (unknown to them) in the preparation of this book. First of all, the authors would like to thank all staff at the learning resource centre of the Magee Campus of the University of Ulster. Special thanks go to Mr Lewis Childs, who was very kind in finding the latest and rare literature from different sources across the UK. The initial material was developed and used for an MSc course at the School of Computing and Intelligent Systems, University of Ulster. Useful feedback was received from many postgraduate students: Dr Neil Glackin, Dr Leo Galway, Dr Patric Gormley, Dr Michael McBride, Dr Julie Wall, Mr Brian McAlister, Mr Jai Verdhan Singh, Dr Erich Michols and Dr Faraz Hasan among them. Thanks go to Dr Tom Lunney, who as the MSc course director communicated many helpful suggestions for improving the course material. Professor Robert John, as an external examiner of the MSc course, was very encouraging in developing the material as a book. Thanks to Professor Liam Maguire, Head of the School of Computing and Intelligent Systems, who was very supportive of the first author during the spring semester of 2012, allowing him to be dedicated full time to manuscript preparation.

The authors would like to thank many of their collaborators: Dr Bala Amavasai, Dr Richard Mitchell, Dr Michael O'Grady, Dr Mourad Oussalah, Dr John St. Quinton, Dr Osman Tokhi, Professor Alamgir Hossain, Professor Ali Hessami, Dr Takatoshi Okuno, Professor Akira Ikuta, Professor Hydeuki Takagi, Dr Filip Ponulak, Dr David Fogel and Professor Bernard Widrow. The authors would like to thank all the staff at John Wiley & Sons Ltd associated with the publication of this book, especially Tom Carter, Anne Hunt, Eric Willner and Genna Manaog for their support and help throughout the preparation of the manuscript and the production of this book.

The first author's eldest sister passed away during the preparation of the manuscript; she would have been happy to see the book published. The first author would like to thank his wife Kaniz for her love and patience during the entire endeavour of the book, without which it would never have been published, and his daughters Oyndrilla, Opala and Orla for making no complaints during this time.

1

Introduction to Computational Intelligence

Keep it simple:
As simple as possible,
But not simpler.
                    −Albert Einstein

1.1   Computational Intelligence

Much is unknown about intelligence and much will remain beyond human comprehension. The fundamental nature of intelligence is only poorly understood and even the definition of intelligence remains a subject of controversy. Considerable research is currently being devoted to the understanding and representation of intelligence. According to its dictionary definition, intelligence means the ability to comprehend, reason and learn. From this point of view, a definition of intelligence can be elicited whereby an intelligent system is capable of comprehending (with or without much *a priori* information) the environment or a process; reasoning about and identifying different environmental or process variables, their inter-relationship and influence on the environment or process; and learning about the environment or process, its disturbance and operating conditions. Other aspects of intelligence that describe human intelligence are creativity, skills, consciousness, intuition and emotion.

Traditional artificial intelligence (AI) has tried to simulate such intelligent behaviour in systems requiring exact and complete knowledge representation (Turing, 1950). Unfortunately, many real-world systems cannot be described exactly with complete knowledge. It has been demonstrated that the use of highly complex mathematical description can seriously inhibit the ability to develop system models. Furthermore, it is required to cope with significant unmodelled and unanticipated changes in the environment or process and in the model objectives. This will involve the use of advanced decision-making processes to generate actions so that a certain performance level is maintained even though there are drastic changes in the operating conditions. Thus, the dissatisfaction with conventional modelling techniques is growing with

Computational Intelligence: Synergies of Fuzzy Logic, Neural Networks and Evolutionary Computing, First Edition.
Nazmul Siddique and Hojjat Adeli.

the increasing complexity of dynamical systems, necessitating the use of more human exper-tise and knowledge in handling such processes. Computational intelligence techniques are thus a manifestation of the crucial time when human knowledge will become more and more important in system modelling and control as an alternative approach to classical mathemati-cal modelling, whose structure and consequent outputs in response to external commands are determined by experimental evidence (i.e., the observed input/output behaviour of the system or plant). The system is then a so-called intelligent system. Intelligent techniques are properly aimed at processes that are ill-defined, complex, nonlinear, time-varying and stochastic. Intel-ligent systems are not defined in terms of specific algorithms. They employ techniques that can sense and reason without much *a priori* knowledge about the environment and produce control actions in a flexible, adaptive and robust manner.

1.2   Paradigms of Computational Intelligence

Many attempts have been made by different authors and researchers to define the term com-putational intelligence (CI). Despite the widespread use of the term, there is no commonly accepted definition of CI. The term was first used in 1990 by the IEEE Neural Networks Coun-cil. Bezdek (1994) first proposed and defined the term CI. A system is called computationally intelligent if it deals with low-level data such as numerical data, has a pattern-recognition component and does not use knowledge in the AI sense, and additionally when it begins to exhibit computational adaptivity, fault tolerance, speed approaching human-like turnaround and error rates that approximate human performance (Bezdek, 1994). At the same time, the birth of CI is attributed to the IEEE World Congress on Computational Intelligence in 1994. Since then there has been much explanation published on the term CI. The IEEE Computa-tional Intelligence Society (formerly the IEEE Neural Networks Council) defines its subject of interest as neural networks (NN), fuzzy systems (FS) and evolutionary algorithms (EA) (Dote and Ovaska, 2001). Some authors argue that computational intelligence is a collection of heuristic algorithms encompassing techniques such as swarm intelligence, fractals, chaos theory, immune systems and artificial intelligence. There are also other approaches that satisfy the AI techniques. Marks (1993) clearly outlined the distinction between CI and AI, although both CI and AI seek similar goals. Based on three levels of analysis of system complexity, Bezdek (1994) argues that CI is a subset of AI.

Zadeh (1994, 1998) proposed a different view of machine intelligence, where he distin-guishes hard computing techniques based on AI from soft computing techniques based on CI. In hard computing, imprecision and uncertainty are undesirable features of a system whereas these are the foremost features in soft computing. Figure 1.1 shows the difference between AI and CI along with their alliance with hard computing (HC) and soft computing (SC). Zadeh defines soft computing as a consortium of methodologies that provide a foundation for design-ing intelligent systems. Some researchers also believe that SC is a large subset of CI (Eberhart and Shui, 2007). The remarkable features of these intelligent systems are their human-like capability to make decisions based on information with imprecision and uncertainty.

Fogel (1995a) views adaptation as the key feature of intelligence and delineates the technolo-gies of neural, fuzzy and evolutionary systems as the rubric of CI, denoting them as methods of computation that can be used to adapt solutions to new problems without relying on explicit human intervention. Adaptation is defined as the ability of a system to change or evolve its parameters or structure in order to better meet its goal. Eberhart and Shui (2007) believe that

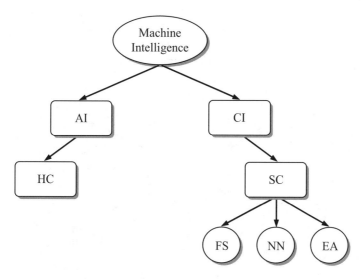

Figure 1.1   Difference between AI-HC and CI-SC

adaptation and self-organization play an important role in CI and argue that adaptation is central to CI, comprising the practical concept, paradigms, algorithms and implementations that facilitate intelligent behaviour. They argue further that CI and adaptation/self-organization are synonymous (Eberhart and Shui, 2007).

1.3   Approaches to Computational Intelligence

Central to computational intelligence is the construction of a process or system model (King, 1999; Konar, 2005), which is not amenable to mathematical or traditional modelling because:

(i)   the processes are too complex to represent mathematically;
(ii)   the process models are difficult and expensive to evaluate;
(iii)   there are uncertainties in process operation;
(iv)   the process is nonlinear, distributed, incomplete and stochastic in nature.

The system has the ability to learn and/or deal with new or unknown situations and is able to make predictions or decisions about future events. The term computational intelligence, as defined by Zadeh, is a combination of soft computing and numerical processing. The area of computational intelligence is in fact interdisciplinary and attempts to combine and extend theories and methods from other disciplines, including modern adaptive control, optimal control, learning theory, reinforcement learning, fuzzy logic, neural networks and evolutionary computation. Each discipline approaches computational intelligence from a different perspective, using different methodologies and toolsets towards a common goal. The inter-relationship between these disciplines is illustrated in Figure 1.2.

Computational intelligence uses experiential knowledge about the process that generally produces a model in terms of input/output behaviour. The question is how to model this human

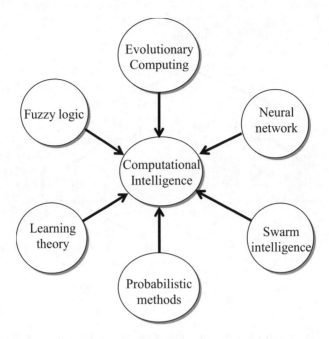

Figure 1.2   Periphery of computational intelligent methodologies

knowledge and represent it in such a manner as to be computationally efficient. Engelbrecht (2002) considers the following five basic approaches to computational intelligence:

  (i) Fuzzy logic,
 (ii) Neural networks,
(iii) Evolutionary computing,
 (iv) Learning theory,
  (v) Probabilistic methods,
 (vi) Swarm intelligence.

In this book, the three methodologies of fuzzy logic, neural networks and evolutionary computing and their synergies will be covered and all other methodologies (such as swarm intelligence, learning theory and probabilistic methods) will be addressed as supportive methods in computational intelligence.

1.3.1   Fuzzy Logic

It has been suggested by researchers that measurements, process modelling and control can never be exact for real and complex processes. Also, there are uncertainties such as incompleteness, randomness and ignorance of data in the process model. The seminal work of Zadeh introduced the concept of fuzzy logic to model human reasoning from imprecise and incomplete information by giving definitions to vague terms and allowing construction of a rule base (Zadeh, 1965, 1973). Fuzzy logic can incorporate human experiential knowledge and give it an engineering flavour to model and control such ill-defined systems with nonlinearity and uncertainty. The fuzzy logic methodology usually deals with reasoning and inference on a higher level, such as semantic or linguistic.

1.3.2 Neural Networks

Neurons are the fundamental building blocks of the biological brain. Neurons receive signals from neighbouring neurons through connections, process them in the cell body and transfer the results through a long fibre called an axon. An inhibiting unit at the end of the axon, called the synapse, controls the signal between neurons. The axon behaves like a signal-conducting device. An artificial neural network is an electrical analogue of the biological neural network. Neural networks originated from the work of Hebb in the 1940s and more recently the work of Hopfield, Rumellhart, Grossberg and Widrow in the 1980s has led to a resurgence of research interest in the field (Hebb, 1949; Grossberg, 1982; Hopfield, 1982; Rummelhart *et al.*, 1986; Widrow, 1987). Neural networks are biologically inspired, massively parallel and distributed information-processing systems. Neural networks are characterized by computational power, fault tolerance, learning from experiential data and generalization capability, and are essentially low-level computational algorithms that usually demonstrate good performance in processing numerical data. The learning takes place in different forms in neural networks, such as supervised, unsupervised, competitive and reinforcement learning. Research on neural network-based control systems has received considerable interest over the past several years, firstly because neural networks have been shown to be able to approximate any nonlinear function defined on a compact set of data to a specified accuracy and secondly because most control systems exhibit certain types of unknown nonlinearity, which suits neural networks as an appropriate control technology.

1.3.3 Evolutionary Computing

Evolutionary computing is the emulation of the process of natural selection in a search procedure based on the seminal work on evolutionary theory by Charles Darwin (Darwin, 1859). In nature, organisms have certain characteristics that influence their ability to survive in adverse environments and pass on to successive progeny with improved abilities. The genetic information of species can be coded into chromosomes that represent these characteristics. The species undergo reproduction and give birth to new offspring with features of capability to combat the adverse environment and survive. The process of natural selection ensures that the more fit individuals have the opportunity to mate most of the time, leading to the expectation that the offspring will have a similar or higher level of fitness. Evolutionary computation uses iterative progress and development in a population. This population is then selected in a guided random search using parallel processing to achieve the desired population of solutions. Such processes are often inspired by biological mechanisms of evolution. Nearly a century after Darwin's theory of evolution, Fraser (1957) was the first to conduct a simulation of genetic systems representing organisms by binary strings. Box (1957) proposed an evolutionary operation to optimize industrial production. Friedberg (1958) proposed an approach to evolve computer programs. The fundamental works of Lowrence Fogel (Fogel, 1962) in evolutionary programming, John Holland (Holland, 1962) in genetic algorithms, Ingo Rechenberg (Rechenberg, 1965) and Hans-Paul Schwefel (Schwefel, 1968) in evolutionary strategies have had significant influence on the development of evolutionary algorithms and computation as a general concept for problem solving and a powerful tool for optimization. Since the developmental years of the 1960s there have been significant contributions to the field by many people, including De Jong (1975), Goldberg (1989) and Fogel (1995b) to name a few. The 1990s saw another set of developments in evolutionary algorithms, for example Koza (1992) developed genetic programming, Reynolds (1994, 1999) developed cultural algorithms and Storn and Price (1997) developed differential evolution. Evolutionary algorithms have now found widespread application in almost all branches of science and engineering.

1.3.4 Learning Theory

Humans appear to be able to learn new concepts without much effort in a conventional sense. The mechanism of learning in humans is little known. In psychology, learning is the process of bringing together cognitive, emotional and environmental effects and experiences to acquire, enhance or change knowledge, skills, values and world views (Ormrod, 1995; Illeris, 2004). For any learning, it is also important how information is input, processed and stored. Learning theories provide explanations of such processes, and how exactly they occur (Vapnik, 1998).

Learning theories fall into three main philosophical frameworks: behaviourism, cognitive theories and constructivism. Behaviourism deals with the objectively observable aspects of learning. Cognitive theories look at how learning occurs in the brain. Constructivism views learning as a process in which the learner actively constructs or builds new ideas or concepts.

A new scientific discipline of machine learning (Samuel, 1959) has evolved based on the psychological learning theories. In machine learning, researchers use and apply four basic forms of learning. Supervised learning generates a function that maps inputs to desired outputs. Unsupervised learning models a set of input features and maps them to similar patterns, like clustering. Semi-supervised learning combines both labelled and unlabelled examples to generate an appropriate function or classifier. Reinforcement learning indicates how to act on a given observation from the environment. Every action has some impact on the environment, and the environment provides feedback in the form of rewards that guide the learning process. Learning mechanisms are an essential part of any intelligent system and hence are powerful tools for computational intelligence.

1.3.5 Probabilistic Methods

Probability theory has been viewed as the methodology of choice for dealing with uncertainty and imprecision. The probabilistic method involves considering an appropriate probability space over a wider family of structures, and proving that a sample point corresponding to the required structure has positive probability in this space. This method was introduced by Erdos and Spencer (1974) and has made major contributions in areas of mathematics and computer science such as combinatorics, functional analysis, number theory, topology, group theory, combinatorial geometry and theoretical computer science. Probabilistic behaviour or stochasticity (randomness) is also sometimes listed as an attribute of intelligent systems. A complex nonlinear dynamic system very often shows chaotic behaviour, that is, chaotic phenomena are features of complex dynamical systems (Grim, 1993). It is somewhat uncertain whether the attribute should be represented as randomness or chaos.

The term chaos refers to complicated dynamical behaviour. There is no uniform agreement as to the precise definition, but a significant body of literature uses the term to refer to systems of a particular type with a set of periodic points and an orbit which are dense in a closed invariant set Λ and these are very sensitive to initial conditions (Devaney, 1989). In principle, the future behaviour of a chaotic system is completely determined by the past, but in practice, any uncertainty in the choice of initial conditions grows exponentially with time. Chaotic behaviour has been observed in the laboratory in a variety of systems, such as electrical and electronic circuits, lasers, oscillating chemical reactions, fluid dynamics, mechanical and magneto-mechanical systems (Sumathi and Surekha, 2010). The dynamic behaviour of a chaotic system is predictable in the short term but impossible to predict in the long term. Chaos theory is essentially a recent extension of a larger field of mathematics which is part of complex nonlinear system dynamics. However, these theories seem to permeate many aspects of natural intelligent systems, from basic biology to behavioural intelligence, as well as most artificial intelligent processes and systems.

1.3.6   Swarm Intelligence

Swarm systems in nature are perhaps one of the most mesmerizing things to observe. A flock of birds twisting in the evening light, the V-shaped structure of migrating geese, winter birds hunting for food, the dancing of starlings in the evening light, ants marching to forage, the synchronized flashing of fireflies and mound building by termites are some of the fascinating examples of swarm systems. But how do they produce such well-choreographed collective behaviour without any central coordinator or leader? How do they communicate with each other? How does an ant which has found food tell other ants about the location of the food? How do the flocks of migrating geese maintain a V-shaped structure? How do fireflies know when to glow? Is there a central control or coordinator for the collective behaviours? Scientists and biologists have been researching for decades to answer some of these questions.

The collective behaviours of insects living in colonies (such as ants, bees, wasps and termites) have attracted researchers and naturalists for many years. Close observation of an insect colony shows that the whole colony is very organized, with every single insect having its own agenda. The seamless integration of all individual activities does not have any central control or any kind of supervision. Researchers are interested in this new way of achieving a form of collective intelligence, called swarm intelligence (SI) (Bonabeau *et al.*, 1999; Kennedy and Eberhart, 2001). SI is widely accepted as a computational intelligence technique based around the study of collective behaviour in decentralized and self-organized systems typically made up of a population of simple agents interacting locally with one another and with their environment (Kennedy and Eberhart, 2001; Garnier *et al.*, 2007). Although there is normally no centralized control structure dictating how individual agents should behave, local interactions between such agents often lead to the emergence of global behaviour. Examples of systems like this can be found in nature, including particle swarms, ant colonies, birds flocking, animals herding, fish schooling and bacterial foraging. Recently, biologists and computer scientists have studied how to model biological swarms to understand how such social insects interact, achieve goals and evolve.

Ants are social insects. They live in colonies and their behaviour is governed by the goal of colony survival rather than the survival of individuals. When searching for food, ants initially explore the surrounding area close to the nest in a random manner. While moving, ants leave a chemical pheromone trail on the ground. Ants can smell the pheromone. When choosing their way, they tend to choose, in probability, paths marked by strong pheromone concentrations. As soon as an ant finds a food source, it evaluates the quantity and the quality of the food and carries some of it back to the nest. During the return trip, the quantity of pheromone that an ant leaves on the ground may depend on the quantity and quality of the food. The pheromone trails will guide other ants to the food source. It has been shown (Deneubourg *et al.*, 1990; Dorigo and Stützle, 2004) that the indirect communication between the ants via pheromone trails enables them to find shortest paths between their nest and food sources.

Ant colonies or societies in general can be compared to distributed systems, which present a highly structured social organization in spite of simple individuals. The ant colonies can accomplish complex tasks far beyond their individual capabilities due to the structured organization of their society. The inspiring source of ant colony optimization (ACO) is the foraging behaviour of real ant colonies (Blum, 2005). Dorigo *et al.* (1996) were the first to propose a simple stochastic model that adequately describes the dynamics of the ants' foraging behaviour, and in particular, how ants can find shortest paths between food sources and their nest.

ACO is a meta-heuristic optimization algorithm that can be used to find approximate solutions to difficult combinatorial optimization problems and has been applied successfully to an impressive number of optimization problems. Applications of ACO include routing optimization in networks and vehicle routing, graph colouring, timetabling, scheduling and solving the quadratic assignment problem, the travelling salesman problem (Blum, 2005). Studies of the nest building of ants and bees have resulted in the development of clustering and structural optimization algorithms.

Flocking is seen as a feature of coherent manoeuvring of a group of individuals in space. This is a commonly observed phenomenon in some animal societies. Flocks of birds, herds of quadrupeds and schools of fish are often shown as fascinating examples of self-organized coordination (Camazine *et al.*, 2001). Natural flocks maintain two balanced behaviours: a desire to stay close to the flock and a desire to avoid collisions within the flock (Shaw, 1975). Joining a flock or staying with a flock seems to be the result of evolutionary pressure from several factors, such as protecting and defending from predators, improving the chances of survival of the (shared) gene pool from attacks by predators, profiting from a larger effective search for food, and advantages for social and mating activities (Shaw, 1962). Reynolds (1987) was the first to develop a model to mimic the flocking behaviour of birds, which he described as a general class of polarized, non-colliding, aggregate motion of a group of individuals. Such flocking behaviours were simulated using three simple rules: collision avoidance with flock mates, velocity matching with nearby flock mates, and flock centring to stay close to the flock. Flocking models have numerous applications. Some include the simulation of traffic patterns, such as the flow of cars on a motorway which has a flock-like motion, animating troop movement in real-time strategy games and in simulating mobile robot movement (Momen *et al.*, 2007; Turgut *et al.*, 2008).

One of the interesting features in the behaviour of fishes is the fish school. About half of all fish species are known to form fish schools at some stage in their lives. Fish can form loosely structured groups called shoals and highly organized structures called fish schools. Fish schools are seen as self-organized systems consisting of individual autonomous agents (Shaw, 1962). Fish schools also come in many different shapes: stationary swarms, predator-avoiding vacuoles and flash expansions, hourglasses and vortices, highly aligned cruising parabolas, herds and balls (Parrish *et al.*, 2002). A fish school can be of various sizes, for example, a herring school often exceeds 5000 individuals and spreads over 700 square metres

(Mackinson, 1999). Modelling the behaviour of fish schools has been a subject of research for a long time. Niwa (1996) studied the collective behaviour of fishes and proposed a model based on Newtonian dynamics which results in emergent patterns. Couzin *et al.* (2002) proposed an alternative model where each fish is considered as an autonomous agent interacting with its local neighbours and producing a complex pattern by following three simple rules: (i) move away from very near neighbours; (ii) follow the same direction as close neighbours; (iii) avoid becoming isolated. Following the rules, each individual fish can have three zones: repulsion, alignment and attraction. Individuals are attracted to neighbours over a larger range than they align with in the attraction zone. Individuals always move away from neighbours in the repulsion zone. If the radius of the alignment zone is increased, individuals would go from a loosely packed stationary swarm to a torus where individuals circle round their centre of mass and, finally, to a parallel group moving in a common direction.

Particle swarm optimization (PSO) was developed by Kennedy and Eberhart (1995) based on the social behaviour of swarms such as fish and birds in nature. PSO has similarities with evolutionary algorithms, but it is simpler in the sense that it does not apply any mutation or crossover operation, instead using real-number randomness and global communication among the swarming particles. Each particle, referring to an individual in the swarm representing a candidate solution to the optimization problem, is flown through the multidimensional search space, adjusting its position in the search space according to its own experience and that of its neighbouring particles. Particles make use of the best positions encountered and the best positions of their neighbours to position themselves towards an optimum solution. The performance of each particle is measured according to a predefined fitness function which is related to the problem being solved. Applications of PSO include function approximation, clustering, optimization of mechanical structures and solving systems of equations. There are now as many as about 20 different variants of PSO.

Rumours are a form of social communication. The way a rumour propagates within a population in society was first modelled by Daley and Kendal (1965) at the University of Cambridge. The spreading of a rumour often has severe consequences on the perception of celebrities, financial markets and even society (Nekovee *et al.*, 2007). Rumours can also be manipulated intentionally to disrupt competitor organizations. They can cause panic during wars and can create disaster in stock markets.

The flashing of fireflies in the summer sky in tropical regions has been one of the most hypnotic and wonderful experiences for explorers and naturalists for many years. There are about 2000 firefly species, and most fireflies produce short and rhythmic flashes. The flashing light can be seen as a signalling system and the true function of such signalling system is not really known yet. However, two fundamental functions of such flashes are to attract mating partners and to attract potential prey; flashing may also serve as a warning mechanism. The rhythm, rate and duration of flashing form part of the signal system that brings two fireflies together. For example, females respond to a male's unique pattern of flashing. This unique feature of fireflies can be formulated in such a way as to make it possible to formulate new optimization algorithms.

Yang (2009) proposed a new heuristic algorithm, called the firefly algorithm (FA), based on three idealized rules: (i) fireflies attract one another with flashing lights; (ii) the level of attractiveness is proportional to their brightness and a less bright firefly will move towards a brighter one, otherwise it will move randomly; (iii) the brightness of a firefly is determined by the landscape of its objective function. For a maximization problem, a population of fireflies is generated and the brightness is simply proportional to the value of the objective function (fitness value). FA has found many applications in engineering and multi-objective optimization problems (Yang, 2008, 2010).

Quite a number of cuckoo species engage in obligate brood parasitism by laying their eggs in the nests of other host birds of different species. If the host bird discovers the eggs are not its own, it either throws the eggs away or abandons the nest. Some cuckoo species are specialized in the mimicry of colour and pattern of eggs of their chosen host species, thus reducing the chances of their eggs being thrown out or abandoned. Yang and Deb (2010) developed a new meta-heuristic optimization algorithm, called cuckoo search (CS), which is based on the interesting breed behaviour of certain cuckoo species. There have been many applications of cuckoo search reported in the literature (Yang, 2008, 2010).

MacArthur and Wilson (1967) began working together on mathematical models of bio-geography in the 1960s. They were trying to develop mathematical models of biogeography that describe how species migrate from one island to another, how new species arise and how species become extinct. Since then biogeography has become a major area of research, which studies the geographical distribution of biological species. The concept of biogeography can be used to derive a new family of algorithms for optimization called biogeography-based optimization (BBO). BBO has been applied to benchmark functions and to a sensor-selection problem, providing performance on a par with other population-based methods (Simon, 2008).

Passino (2002) pointed out, in a seminal paper, how individual and groups of bacteria forage for nutrients and how to model this as a distributed optimization process; the natural foraging strategy can lead to optimization and the idea can be applied to solve real-world optimization problems. Based on this concept, Passino (2002) proposed an optimization technique known as the bacterial foraging optimization algorithm (BFOA). To date, BFOA has been applied successfully to real-world problems such as optimal controller design, harmonic estimation, transmission loss reduction, active power filter synthesis and learning of artificial neural networks (Das *et al.*, 2009).

Several new meta-heuristic optimization algorithms inspired by nature have been introduced in recent years. Among them are a galaxy-based search algorithm (Hosseini, 2011) and spiral dynamics-inspired optimization (Tamura and Yasuda, 2011).

EA and SI together form a broader class of search and optimization paradigm termed global search and optimization (GSO). The classification of the different algorithms and techniques of GSO is shown in Figure 1.3.

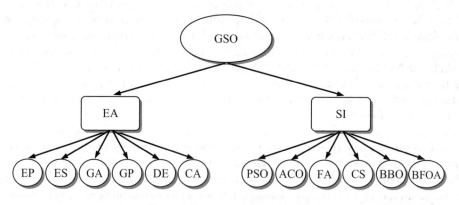

Figure 1.3 Classification of GSO

1.4 Synergies of Computational Intelligence Techniques

The synergistic combination of all the methodologies has a very rational basis for applications and designing intelligent systems. An individual method can be excellent in approximate reasoning and modelling uncertainty but may not be good at learning with experiential data or may not be good at adapting in an unknown environment. Thus, a combined approach with computational intelligence techniques and their implementation is of importance for overall performance, computation cost and convenience of application. This combination is called a hybrid intelligent system by many researchers. Zadeh (1994) thinks hybrid intelligent systems are definitely the way of the future.

Fuzzy logic is good at approximate reasoning but does not have any learning ability or adaptive capacity. Neural networks, on the other hand, have efficient mechanisms in learning from experiential data. Evolutionary algorithms enable a system to adapt behaviour or optimize structure. The synergistic combination of these methodologies can provide better computational models that will complement the limitations of any single method. Depending on the compatibility of the individual methodologies, the synergism can be classified into two types: strongly coupled and weakly coupled. In strongly coupled synergism, the individual methodologies are hybridized in such as way as to be inseparable and each individual methodology loses most of its identity in the combined structure. In weakly coupled synergism, each individual methodology plays its own part by upholding the structural identity and working towards a common goal.

The different forms of synergisms of fuzzy logic, neural networks and evolutionary algorithms are shown in Figure 1.4. The common forms of synergism of fuzzy systems and evolutionary algorithms include tuning, optimization and learning of membership functions,

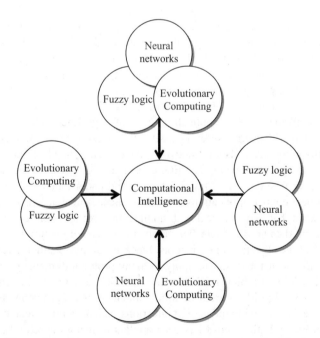

Figure 1.4 Synergies of computational intelligent methodologies

parameters and rule-based fuzzy systems using evolutionary algorithms. Another form of synergism is the control of different parameters of evolutionary algorithm by a fuzzy controller. Both forms of fuzzy evolutionary synergism are weakly coupled.

The common forms of weakly coupled synergisms of neural networks and evolutionary algorithms include training, designing, optimizing architecture and parameters of neural networks and feature selection, transformation and scaling of training data for neural networks using evolutionary algorithms. Also, neural networks are being used to control parameters of evolutionary algorithms. There is also a strongly coupled synergism between the two methodologies where the genetic operations are represented in the form of a neural network and the training epochs are meant to be the generations of evolution.

Synergisms of neural networks and fuzzy systems are the most common and have proved to be very powerful tools for system modelling and control. In a weakly coupled synergism, neural networks and fuzzy systems work independently towards a common goal, where neural networks assist fuzzy systems to acquire knowledge and rules, tuning or adjusting membership functions. In strongly coupled synergism, a fuzzy system is represented in the form of a neural network, which can learn from experiential data. The literature is rich in this type of synergism.

The final type of synergism is a combination of the three methodologies. The most common synergism is the training or optimizing structure of a hybrid neuro-fuzzy system using an evolutionary algorithm. A strongly coupled synergism of the three methodologies may not be possible. There are other types of synergisms possible between swarm intelligence, fuzzy systems, evolutionary algorithms and neural networks.

1.5 Applications of Computational Intelligence

The essence of systems based on computational intelligence is the process that interprets information and data of various natures. The other feature of computational intelligence is that where processing of information in algorithms becomes difficult. Developed theories have been quickly applied to various fields of computer science, engineering, data analysis and biomedicine. Each component methodology of the CI has its application areas. Certainly, more than one technology can be applied to the same application. For example, data clustering can be performed using neural networks and fuzzy logic but the difference would be in the accuracy of the performance. The application areas of neural networks can be categorized into five groups, such as data analysis and classification, associative memory, clustering, generation of patterns and control. Neural networks have been applied to analyse and classify medical data and images, for example, EEG, cancer data, etc. Neural networks have also been widely used for face detection, fraud detection and pattern analysis. The use of neural networks in nonlinear control applications is the most successful area. The inherent advantage of neural networks is that they can deal with nonlinearities of a system and model such systems when sufficient data are available. The application areas of evolutionary algorithms are optimization and multi-objective optimization. Since traditional mathematical optimization techniques are difficult or too costly to apply to many problem domains such as robot tract determination, scheduling problems, DNA analysis, optimization of large structural parameters, etc., evolutionary algorithms are becoming popular for these problems. Fuzzy logic and fuzzy systems have found a wide range of applications such as control, image processing and decision making. Fuzzy logic control has been applied to many household appliances such as washing machines, microwave ovens, toasters, vacuum cleaners, etc. One application of a fuzzy controller is well known: its implementation on a video camera to stabilize the image while holding the camera unsteadily. Fuzzy expert systems have been applied to many areas of medical diagnostics, scheduling, foreign exchange trading and business strategy selection.

1.6 Grand Challenges of Computational Intelligence

Though the CI techniques have been applied successfully to scientific, engineering, economic, business and industrial problems, CI seriously lacks efficient knowledge acquisition, representation and retrieval structures. The grand challenges for the CI community would be to propose more efficient knowledge representation and retrieval mechanisms. Feigenbaum (2003) thinks the grand challenge would be to build a large knowledge base by simply reading text and thus reducing the knowledge engineering effort by one order of magnitude. Some researchers argue that CI should be more human-centric, helping humans to formulate their goals and solve their problems, leading to personal fulfilment (Duch, 2007). A long-term goal for CI would be to create cognitive systems that can compete with humans in a large number of problems. A good part of CI research is concerned with low-level cognitive functions such as perception, object recognition, signal analysis, finding structure in data and association tasks. Despite great progress in CI, artificial systems designed to solve lower-level cognitive tasks are far behind simple natural systems. From this point of view, CI needs to focus on higher-level cognitive systems using symbolic knowledge representation. CI is more than the study of the design of intelligent systems; it includes all non-algorithmizable processes which humans can perform with various degrees of competence. In that sense, Goldberg and Harik (1996) see CI more as a way of thinking about problems rather than a solution to problems using specific techniques.

1.7 Overview of the Book

Chapter 2 describes the concepts of fuzzy logic, fuzzy sets and the description of fuzzy sets by membership functions, different types of membership functions and their features. To apply fuzzy logic one needs to understand the operations of fuzzy sets and fuzzy relations, which are discussed in the chapter with examples. The chapter also describes the interesting features of linguistic variables and hedges. The chapter shows the fascinating features of fuzzy if–then rules and inference mechanisms that will help in developing applications. The chapter also provides a set of worked examples.

Chapter 3 presents an investigation into different types of fuzzy systems, fuzzy modelling and fuzzy control methods and techniques in general. These include simple Mamdani-, Sugeno- and Tsukomoto-type fuzzy modelling and control techniques. A comparative study of the suitability of different methods for applications is made. The chapter also presents different types of fuzzy controllers, namely PD, PI and PID. Different approaches to rule reduction are investigated and analysed as well.

Chapter 4 presents an introduction to biological neurons, different models of neurons, activation functions and basics of neural networks. The chapter then introduces different feedforward architectures such as the multilayer perceptron, radial-basis function, regression networks, probabilistic, belief and stochastic networks and recurrent architectures such as Elman, Jordan and Hopfield networks. The chapter describes different learning algorithms of neural networks, such as supervised and unsupervised.

Chapter 5 presents neural systems with application to nonlinear systems. Different techniques of identification and modelling of nonlinear systems using neural networks have been discussed. Application of neural networks to control problems is a very popular and widely used technique. Different schemes of neuro-control, such as direct, indirect, backpropagation through time and inverse control, have been discussed. Neural networks have also been popular for predictive and adaptive control. Different schemes of predictive and adaptive neuro-control with applications have also been discussed. This chapter introduces different application developments with MATLAB® as well.

Chapter 6 presents evolutionary computing and algorithms. Basic to any evolutionary computing is the chromosome representation and genetic operators. This chapter describes different types of encoding scheme, selection and crossover and mutation operators. Finally, it introduces different evolutionary algorithms such as genetic algorithms, genetic programming, evolutionary programming, evolutionary strategies, differential evolution and cultural algorithms.

Chapter 7 presents an investigation into different evolutionary systems and their applications. Multi-objective optimization is one of the promising areas of application of evolutionary algorithms. This chapter also investigates co-evolution of populations and different symbiotic relationships between species. Another aspect in evolutionary algorithms is parallelism, where multiple populations work together with a common goal. This chapter presents an account of these techniques widely used in evolutionary computing.

Chapter 8 presents combinations of fuzzy systems and evolutionary computing. Different kinds of combination are possible, such as controlling parameters of evolutionary algorithms by fuzzy logic and optimizing parameters of a fuzzy system by evolutionary algorithms. This chapter presents the optimization of fuzzy systems, especially membership functions, rule-based or both using evolutionary algorithms and also highlights the fuzzy control of genetic operators in limited applications.

Chapter 9 presents combinations of evolutionary algorithms and neural networks. Mainly two types of combination, supportive and collaborative, between evolutionary algorithms and neural networks have been reported in the literature. In supportive combinations, one of the two technologies is the primary problem solver and the other plays a supporting role, such as setting up initial conditions or parameters. In collaborative combinations, both of the technologies act together as a problem solver. This chapter will explore these combinations in designing and training neural networks, learning control parameters, activation functions and setting up initial conditions.

Chapter 10 presents combinations of neural networks and fuzzy systems, the most important of which are cooperative and hybrid combinations. In cooperative combinations, fuzzy systems or neural networks are used to control parameters, initial conditions and/or structures of neural networks or fuzzy systems. This chapter covers the detailed description, architectures and use of possible combinations of these two technologies. In hybrid combinations, each of these technologies loses its identity and presents a new single system to address the problem at hand. The most successful and widely used hybrid system is the ANFIS. This chapter will introduce ANFIS and different variants as well as other hybrids.

1.8   MATLAB® Basics

MATLAB® is a high-level language for scientific and engineering computation. MATLAB® is an integrated software environment and provides numeric computation, data analysis, graphics visualization and system simulation. The integrated environment of MATLAB® is shown in

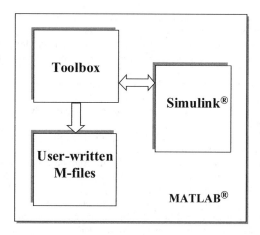

Figure 1.5   MATLAB® environment

Figure 1.5. The language, tools and built-in maths functions enable users to explore multiple approaches and reach faster solutions than with other programming languages. It provides tools for creating customized toolboxes or harnessing with other toolboxes such as fuzzy logic, direct search and genetic algorithms or neural network toolboxes. Applications are developed by writing M-files and running at the command prompt. Simulink® is a software package for modelling, simulating and analysing dynamical systems under MATLAB®. It supports linear and nonlinear systems, modelled in continuous time, sampled time or a hybrid of the two. Different parts of the system can have different rates. Simulink® provides a graphical user interface (GUI) for building models as block diagrams using click-and-drag mouse operations. Simulink® also includes a comprehensive block library of sinks, sources, linear and nonlinear components and connectors.

This section provides a brief introduction to different command-line functions of MATLAB®. A brief introduction to MATLAB®, different functions, control statements, writing M-files and plot functions are discussed, with examples in Appendix A.

References

Bezdek, J.C. (1994) What is computational intelligence? In *Computational Intelligence Imitating Life*, J.M. Zurada, R.J. Marks II and C.J. Robinson (eds), IEEE Press, New York, pp. 1–12.

Blum, C. (2005) Ant colony optimisation: introduction and recent trends, *Physics of Life Reviews*, 2, 353–373.

Bonabeau, E., Dorigo, M. and Theraulaz, G. (1999) *Swarm Intelligence: From Natural to Artificial Systems*, Oxford University Press, New York.

Box, G.E.P. (1957) Evolutionary operation: a method for increasing industrial productivity, *Applied Statistics*, 6(2), 81–101.

Camazine, S., Deneubourg, J.-L., Franks, N.R., Sneyd, J., Theraulaz, G. and Bonabeau, E. (2001) *Self-Organization in Biological Systems*, Princeton University Press, Princeton, NJ.

Couzin, I.D., Karause, J., James, R., Ruxton, G.D. and Franks, N.R. (2002) Collective memory and spatial sorting in animal groups, *Journal of Theoretical Biology*, 218, 1–11.

Daley, D.J. and Kendal, D.G. (1965) Stochastic rumours, *Journal of the Institute of Mathematics and its Applications*, 1, 42–55.

Darwin, C. (1859) *The Origin of Species by Means of Natural Selection or the Preservation of Favoured Races in the Struggle for Life*, Mentor, New York (reprint 1958).

Das, S., Dasgupta, S., Biswas, A., Abraham, A. and Konar, A. (2009) On the stability of the chemotactic dynamics in bacterial-foraging optimisation algorithm, *IEEE Transactions on Systems, Man and Cybernetics A*, 39(3), 670–679.

Dote, Y. and Ovaska, S.J. (2001) Industrial applications of soft computing: a review, *Proceedings of the IEEE*, 89(9), 1243–1265.

De Jong, K.A. (1975) *Analysis of the behaviour of a class of genetic adaptive systems*, PhD Thesis, University of Michigan, Ann Arbor, MI.

Deneubourg, J.-L., Aron, S., Goss, S. and Pasteels, J.-M. (1990) The self-organizing exploratory pattern of the Argentine ant, *Journal of Insect Behaviour*, 3, 159–168.

Devaney, R.L. (1989) *An Introduction to Chaotic Systems*, 2nd edn, Addison Wesley, New York.

Dorigo, M., Maniezzo, V. and Colorni, A. (1996) Ant system: optimization by a colony of cooperating agents, *IEEE Transactions on System, Man and Cybernetics B*, 26(1), 29–41.

Dorigo, M. and Stützle, T. (2004) *Ant Colony Optimization*, MIT Press, Cambridge.

Duch, W. (2007) Towards comprehensive foundations of computational intelligence. In *Challenges for Computational Intelligence*, W. Duch and J. Mandziuk (eds), Springer-Verlag, Berlin.

Eberhart, R.C. and Shui, Y. (2007) *Computational Intelligence – Concepts to Implementations*, Elsevier, Amsterdam.

Engelbrecht, A.P. (2002) *Computational Intelligence: An Introduction*, John Wiley & Sons, New York.

Erdos, P. and Spencer, J. (1974) *Probabilistic Methods in Combinatorics*, Academic Press, New York.

Feigenbaum, E.A. (2003) Some challenges and grand challenges for computational intelligence, *Journal of the ACM*, 50(1), 32–40.

Fogel, D.B. (1995a) Review of computational intelligence: imitating life, *IEEE Transactions on Neural Networks*, 6, 1562–1565.

Fogel, D.B. (1995b) *Evolutionary Computation – Toward a New Philosophy of Machine Intelligence*, IEEE Press, New York.

Fogel, L.J. (1962) Autonomous automata, *Industrial Research*, 4, 14–19.

Fraser, A.S. (1957) Simulation of genetic systems by automatic digital computers, I. Introduction, *Australian Journal of Biological Sciences*, 10, 484–491.

Friedberg, R.M. (1958) A learning machine: Part I, *IBM Journal of Research and Development*, 2(1), 2–13.

Garnier, S., Gautrais, J. and Theraulaz, G. (2007) The biological principles of swarm intelligence, *Swarm Intelligence*, 1, 3–31.

Goldberg, D.E. (1989) *Genetic Algorithms in Search, Optimization, and Machine Learning*, Addison Wesley, New York.

Goldberg, D.E. and Harik, G. (1996) A case study in abnormal CI: the design manufacturing and others anthropocentric systems, *International Journal of Computational Intelligence and Organisations*, 1, 78–93.

Grim, P. (1993) Self-reference and chaos in fuzzy logic, *IEEE Transactions on Fuzzy Systems*, 1(4), 237–253.

Grossberg, S. (1982) *Studies of Mind and Brain: Neural Principles of Learning, Perception, Development, Cognition, and Motor Control*, Reidell Press, Boston, MA.

Hebb, D.O. (1949) *The Organization of Behaviour: A Neuropsychological Theory*, John Wiley, New York.

Holland, J.H. (1962) Outline for a logical theory of adaptive systems, *Journal of ACM*, 3, 297–314.

Hopfield, J.J. (1982) Neural networks and physical systems with emergent collective computational abilities. *Proceedings of National Academy of Sciences*, 79, 2554–2558.

Hosseini, H.S. (2011) Principal component analysis by galaxy-based search algorithm: a novel metaheurisitc for continuous optimisation, *International Journal of Computational Science and Engineering*, 6(1&2), 132–140.

Illeris, K. (2004) *Three Dimensions of Learning*, Krieger Publishing, Malabar, FL.

Kennedy, J. and Eberhart, R.C. (2001) *Swarm Intelligence*, Morgan-Kaufmann, New York.

King, R.E. (1999) *Computational Intelligence in Control Engineering*, Marcel Dekker, New York.

Konar, A. (2005) *Computational Intelligence: Principles, Techniques and Applications*, Springer-Verlag, Berlin.

Koza, J.R. (1992) *Genetic Programming: On the Programming of Computers by Means of Natural Selection*, MIT Press, Cambridge, MA.

MacArthur, R. and Wilson, E. (1967) *The Theory of Biogeography*, Princeton University Press, Princeton, NJ.

Mackinson, S. (1999) Variation in structure and distribution of pre-spawning Pacific herring shoals in two regions of British Columbia, *Journal of Fish Biology*, 55, 972–989.

Marks, R. (1993) Intelligence: computational versus artificial, *IEEE Transactions on Neural Networks*, 4(5), 737–739.

Momen, S., Amavasai, B.P. and Siddique, N.H. (2007) Mixed species flocking for heterogenous robotic swarms. In *The International Conference on Computer as a Tool* (EUROCON 2007), Piscataway, NJ. IEEE Press, New York, pp. 2329–2336.

Nekovee, M., Moreno, Y., Bianconi, G. and Marsili, M. (2007) Theory of rumour spreading in complex social networks, *Journal of Physica A*, 374, 457–470.

Niwa, H.S. (1996) Newtonian dynamical approach to fish schooling, *Journal of Theoretical Biology*, 181, 47–63.

Ormrod, J.E. (1995) *Human Learning*, Prentice Hall, Englewood Cliffs, NJ.

Parrish, J.K., Viscido, S.V. and Grunbaum, D. (2002) Self-organized fish schools: an examination of emergent properties, *The Biological Bulletin*, 202, 296–305.

Passino, K.M. (2002) Biomimicry of bacterial foraging for distributed optimization and control, *IEEE Control System Magazine*, 22(3), 52–67.

Rechenberg, I. (1965) *Cybernetic Solution Path of an Experimental Problem*, Royal Aircraft Establishment, Library Translation No. 1122, Farnborough, UK.

Reynolds, C. (1987) Flocks, herds, and schools: a distributed behavioural model, *Computer Graphics*, 21(4), 25–34.

Reynolds, R.G. (1994) Introduction to cultural algorithms. In *Proceedings of the Third Annual Conference on Evolutionary Programming*, A.V. Sebald and L.J. Fogel (eds), World Scientific Press, Singapore, pp. 131–139.

Reynolds, R.G. (1999) *An Overview of Cultural Algorithms: Advances in Evolutionary Computation*, McGraw-Hill, New York.

Rumelhart, D.E., Hinton, G.E. and Williams, R.J. (1986) Learning representations by back-propagation errors, *Nature*, 323, 533–536.

Samuel, A.L. (1959) Some studies in machine learning using game checkers, *IBM Journal of Research and Development*, 3, 211–229.

Schwefel, H.-P. (1968) *Projekt MHD-Strausstrhlrohr: Experimentelle Optimierung einer Zweiphasenduese*, Teil I, Technischer Bericht 11.034/68, 35, AEG Forschungsinstitute, Berlin.

Shaw, E. (1962) The schooling of fishes, *Scientific American*, 206, 128–138.

Shaw, E. (1975) Fish in schools, *Natural History*, 84(8), 40–46.

Simon, D. (2008) Biogeography-based optimization, *IEEE Transactions on Evolutionary Computation*, 12(6), 702–713.

Storn, R. and Price, K. (1997) Differential evolution – a simple and efficient heuristic for global optimisation over continuous space, *Journal of Global Optimisation*, 11(4), 431–459.

Sumathi, S. and Surekha, P. (2010) *Computational Intelligence Paradigms – Theory and Applications Using MATLAB*, CRC Press/Taylor & Francis, Boca Raton, FL.

Tamura, K. and Yasuda, K. (2011) Primary study of spiral dynamics inspired optimisation, *IEEJ Transactions on Electrical and Electronic Engineering*, 6(S1), 98–100.

Turgut, A.E., Çelikkanat, H., Gökçe, F. and Sahin, E. (2008) Self-organized flocking in mobile robot swarms, *Swarm Intelligence*, 2, 97–120.

Turing, A.M. (1950) Computing machinery and intelligence, *Mind*, 59, 433–460.

Vapnik, V.N. (1998) *Statistical Learning Theory*, John Wiley & Sons, New York.

Widrow, B. (1987) Adaline and Madaline, *Proceedings of the IEEE First International Conference on Neural Networks*, 1, 145–157.

Yang, X.S. (2008) *Nature-Inspired Metaheuristic Algorithms*, Luniver Press, Bristol.

Yang, X.S. (2009) Firefly algorithms for multimodal optimization. In *Stochastic Algorithms: Foundations and Applications*, SAGA 2009, Lecture Notes in Computer Sciences, Vol. 5792, Springer-Verlag, Berlin, pp. 169–178.

Yang, X.S. (2010) *Engineering Optimisation: An Introduction with Metaheuristic Applications*, John Wiley & Sons, New York.

Yang, X.S. and Deb, S. (2010) Engineering optimisation by cuckoo search, *International Journal of Mathematical Modelling and Numerical Optimisation*, 1(4), 330–343.

Zadeh, L.A. (1965) Fuzzy sets, *Information and Control*, 8(3), 338–353.

Zadeh, L.A. (1973) Outline of a new approach to the analysis of complex systems and decision process, *IEEE Transactions on System, Man and Cybernetics*, 3, 28–44.

Zadeh, L.A. (1998) Roles of soft computing and fuzzy logic in the conception, design and deployment of information/intelligent systems. In *Computational Intelligence: Soft Computing and Fuzzy-Neuro Integration with Applications*, O. Kaynak, L.A. Zadeh, B. Tiirksen and I.J. Rudas (eds), Springer-Verlag, Berlin, pp. 10–37.

Zadeh, L.A. (1994) Fuzzy logic, neural networks and soft computing, *Communications of the ACM*, 37, 77–84.

2

Introduction to Fuzzy Logic

2.1 Introduction

In classical (Newtonian) mechanics, uncertainty was considered as undesirable and to be avoided by any means. In the late nineteenth century, researchers started to realize that no physical system exists without a certain amount of uncertainty. This is a phenomenon without which the description of a system or model is incomplete. A trend started then in science and engineering to incorporate uncertainty in system models. At this stage uncertainty was quantified with the help of probability theory, developed in the eighteenth century by Thomas Bayes (Price, 1763). The expression of uncertainty using probability theory was first challenged by Max Black (Black, 1937). He proposed a degree as a measure of vagueness. Vagueness can be used to describe a certain kind of uncertainty. For example, John is young. The proposition defined here is vague. He pointed out two main ideas: one is the nature and observability of vagueness and the other is the relevance of vagueness for logic. Black proposed vague sets defined by a membership curve. This was the first attempt to give a precise mathematical theory for sets where there is a membership curve.

There was another movement present in the philosophy, among logicians. The most basic assumptions of classical (or two-valued) propositional as well as first-order logic are the principles of bivalence and compositionality. The principle of bivalence is the assumption that each sentence is either true or false under any one of the interpretations, i.e., has exactly one of the truth values usually denoted numerically by 1 and 0. The problem of future contingencies was a source of many unresolved debates during the middle ages, continuing until the revival of the field of logic in the second half of the nineteenth century. In the second half of the nineteenth century, dissatisfaction with the principle of bivalence appeared (Gottwald, 2001). Charles Sanders Peirce laughed at the 'sheep and goat separators' who split the world into true and false. Around 1867, Peirce set up a triadic trichotomic semiotic as a new type of logic of universal nature. It necessarily derives from a general philosophical system, the doctrine of the continuum. All that exists is continuous and such a continuum governs knowledge and implies generality (Eisele, 1979).

Following the doctrine of the continuum, new interest in multi-valued logic began in the early twentieth century. The real starting phase of many-valued logic began in the 1920s and continued until 1930. The main driving force behind the development was the Polish

Computational Intelligence: Synergies of Fuzzy Logic, Neural Networks and Evolutionary Computing, First Edition.
Nazmul Siddique and Hojjat Adeli.
© 2013 John Wiley & Sons, Ltd. Published 2013 by John Wiley & Sons, Ltd.

School of Logic under the leadership of Jan Lukasiewicz (Lukasiewicz, 1930; Lukasiewicz and Tarski, 1930). Lukasiewicz proposed a formal model of many-valued logic, claiming that the three-valued and the infinite-valued case are of interest for applications. His first intention was to use a third truth value for 'possible'. In his three-valued proposal, 1 stands for true, 0 stands for false and $1/2$ stands for possible. This intended application to modal logic and the Lukasiewicz system did not work out well. At the same time, the American mathematician Post (1921) introduced a family of finitely valued systems. His interest was in the problem of functional completeness.

Following the movement of the Polish School of Logic, there were many theoretical developments in many-valued logic in the 1930s and 1940s. The work of Goedel (1932) and Jaskowski (1936) clarified the mutual relations between the intuitionist and many-valued logic. Goedel tried to understand intuitionist logic in terms of many truth degrees. The outcome was the family of Goedel systems. Also the result, namely, that intuitionist logic does not have a characteristic logical matrix with only finitely many truth degrees. Jaskowski (1936) constructed an infinite-valued characteristic matrix for intuitionist logic. It appeared that the truth degrees of the matrix do not have a suitable intuitive interpretation.

It was Albert Einstein who first pointed out that mathematical precision does not correspond to reality. His remarkable comment during the lecture on 'Geometrie und Erfahrung' clarified that so far as the laws of mathematics refer to reality, they are not certain. And so far as they are certain, they do not refer to reality (Einstein, 1921). This was a landmark assertion that physical variables cannot be measured to their equivalent mathematical exactness despite the availability of high-precision instruments. The tolerance and acceptance of imprecision and uncertainty was gradually mounting among scientists and engineers, realizing that precision and uncertainty incur costs for industry. Human thinking is also not stirred by numeric calculations, rather by approximate reasoning based on manipulation of imprecise information. Vagueness is consciously accepted in daily life to facilitate perception and communication. Bertrand Russell sees both vagueness and precision as features of language, they are not reality. He further argues that vagueness is clearly a matter of degree (Rolf, 1982). If all these are seen as a historical consequence, the events of multi-valued logic and imprecision about the real world have led to the development of fuzzy logic.

Beside these developments inside pure many-valued logic, Zadeh (1965) started an application-oriented approach towards formalization of vague notions by generalized set-theoretic means. His main argument was that as the complexity of a system increases, the ability to make precise and significant statements about its behaviour diminishes until a threshold is reached beyond which precision and significance become almost mutually exclusive characteristics (Zadeh, 1973). He introduced the concept of a fuzzy set in his seminal paper published in 1965 (Zadeh, 1965). As the *father of fuzzy logic*, he was instrumental in making fuzzy logic a major field of study to complement probability theory and its widespread use, with numerous applications.

2.2 Fuzzy Logic

Logic is a tool for reasoning propositions that can be manipulated with mathematical precepts. A proposition is a declarative or linguistic statement within a universe of discourse. For example, Elizabeth is tall. In classical logic a proposition is either true or false. That means the proposition 'Elizabeth is tall' can be either true or false. As another example, Figure 2.1

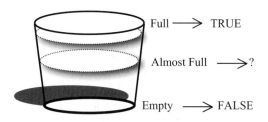

Figure 2.1 A real-world situation

shows a real-world situation where the glass is more than half full of water. The values true or false in classical two-valued logic cannot describe a situation like this. Fuzzy logic is a transition from absolute truth to partial truth. That is, from a variable x (True or False) to a linguistic variable 'Almost full', 'Very close to empty', etc. From this perspective, fuzzy logic can be seen as a reasoning formalism of humans where all truths are partial or approximate and any falseness is represented by partial truth.

2.3 Fuzzy Sets

A fuzzy set A in X is characterized by a membership function $\mu_A(x)$ which associates with each point in X a real number in the interval [0, 1], with the values of $\mu_A(x)$ at x representing the grade of membership of x in A. That is, the class of objects A belongs to X with a continuum of grades of membership $\mu_A(x)$. For example, a fuzzy set $A = \{x_1, x_2, x_3, x_4\}$ in X is characterized by the membership function $\mu_A(x)$ which maps each point x in X to real values 0.5, 1, 0.75 and 0.5. $\mu_A(x)$ represents the degree of membership of x in A and the mapping is only limited by $\mu_A(x) \in [0, 1]$. In classical set theory, the membership function can take only two values: 0 and 1, i.e., either $\mu_A(x)=1$ or $\mu_A(x) = 0$. In set-theoretic notation this is written as $\mu_A(x) \in \{0, 1\}$. A fuzzy set is an extension of a classical set. If X is the universe of discourse and its elements are denoted by x, then a fuzzy set A in X is defined as a set of ordered pairs

$$A = \{x, \mu_A(x) \,|\, x \in X\} \tag{2.1}$$

This mapping can be depicted pictorially, as shown in Figure 2.2.

In Figure 2.2, x_1, x_2, x_3 and x_4 have membership grades of 0.5, 1, 0.75 and 0.5, respectively, written as $\mu_A(x_1) = 0.5$, $\mu_A(x_2) = 1$, $\mu_A(x_3) = 0.75$ and $\mu_A(x_4) = 0.5$. A notational

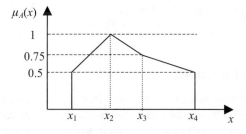

Figure 2.2 Fuzzy set

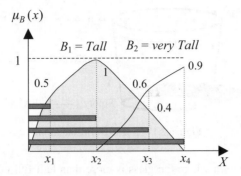

Figure 2.3 Set of tall boxes

convention of fuzzy sets for a discrete and finite universe of discourse X in practice is written as

$$A = \{\mu_A(x_1)/x_1 + \mu_A(x_2)/x_2 + \cdots + \mu_A(x_n)/x_n\} = \sum_{i=1}^{n} \mu_A(x_i)/x_i \qquad (2.2)$$

where '+' does not mean arithmetic addition or logical OR.

Example 2.1   Let $A = \{x_1, x_2, x_3, x_4\}$ in the universe of discourse X having membership values of 0.4, 1.0, 0.7 and 0.8, respectively. This fuzzy set can be written as

$$A = \{0.4/x_1 + 1.0/x_2 + 0.7/x_3 + 0.8/x_4\}$$

Example 2.2   Let $B_1 = \{x_1, x_2, x_3, x_4\}$ be a set of tall boxes and $B_2 = \{x_1, x_2, x_3, x_4\}$ be a set of very tall boxes in the universe of discourse X. The fuzzy sets for the tall and very tall boxes can be written as

$$B_1 = \{0.5/x_1 + 1.0/x_2 + 0.4/x_3 + 0/x_4\}$$
$$B_2 = \{0/x_1 + 0/x_2 + 0.6/x_3 + 0.9/x_4\}$$

The two fuzzy sets for tall and very tall boxes are shown graphically in Figure 2.3. It should be noted that box x_3 belongs to fuzzy set $B_1 = Tall$ with a grade of membership 0.4 and to $B_2 = very\ Tall$ with a grade of membership 0.6.

2.4 Membership Functions

Very often, real-world situations are not certain and cannot be described precisely. For example, the uncertainty in Example 2.2 is belonging to Tall or very Tall. The uncertainties of expressions like 'very nice', 'too small', 'high value' are called fuzziness. The function that characterizes the fuzziness of a fuzzy set A in X, which associates each point in X with a real number in

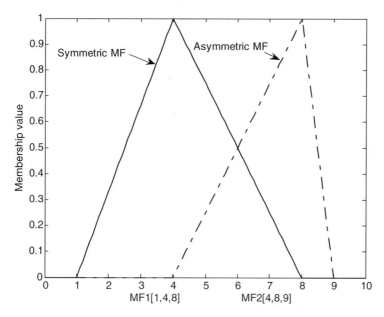

Figure 2.4 Triangular MF

the interval [0, 1], is called a membership function (MF). There is no strict rule for defining a membership function. The choice of membership function is usually problem-dependent and often determined heuristically and subjectively. Most widely used MFs in the fuzzy logic literature are triangular, trapezoidal, Gaussian and bell-shaped functions.

2.4.1 Triangular MF

A triangular MF is specified by three parameters $\{a, b, c\}$, shown in Figure 2.4 and defined as

$$\mu(x) = \max\left(\min\left(\frac{x-a}{b-a}, \frac{c-x}{c-b}\right), 0\right) \qquad (2.3)$$

The parameters $\{a, b, c\}$ with $a < b < c$ determine the x coordinates of the three corners of the underlying triangular MF. Triangular MFs can be asymmetric, depending on the relations $a \leq b$ and $b \leq c$. Figure 2.4 shows a symmetric and an asymmetric triangular MF.

2.4.2 Trapezoidal MF

A trapezoidal MF is specified by four parameters $\{a, b, c, d\}$, shown in Figure 2.5 and defined as

$$\mu(x) = \max\left(\min\left(\frac{x-a}{b-a}, 1, \frac{d-x}{d-c}\right), 0\right) \qquad (2.4)$$

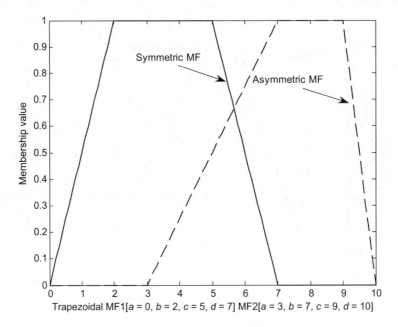

Figure 2.5 Trapezoidal MF

The parameters $\{a, b, c, d\}$ with $a < b < c < d$ determine the x coordinates of the four corners of the underlying trapezoidal MF. Trapezoidal MFs can be asymmetric, depending on the relations $a \leq b$ and $c \leq d$. Both triangular and trapezoidal MFs can be symmetric or asymmetric, which is seen as an advantage for some applications. Owing to their simple formulae and computational efficiency, both triangular and trapezoidal MFs have been used extensively, especially in online applications.

2.4.3 Gaussian MF

A Gaussian MF is specified by two parameters $\{m, \sigma\}$, shown in Figure 2.6 and defined as

$$\mu(x) = \exp\left[-\frac{1}{2}\left(\frac{x - m}{\sigma}\right)^2\right] \tag{2.5}$$

The parameters m and σ represent the centre and width of the Gaussian MF, respectively.

2.4.4 Bell-shaped MF

A bell-shaped MF is specified by three parameters $\{m, \sigma, a\}$, shown in Figure 2.7 and defined as

$$\mu(x) = \frac{1}{1 + \left|\dfrac{x - m}{\sigma}\right|^{2a}} \tag{2.6}$$

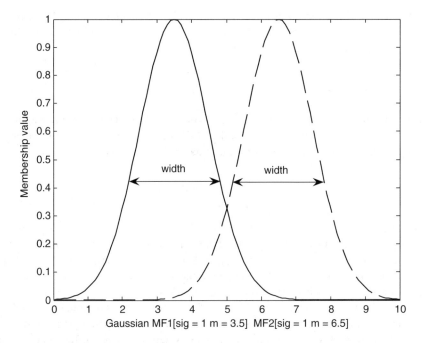

Figure 2.6   Gaussian MF

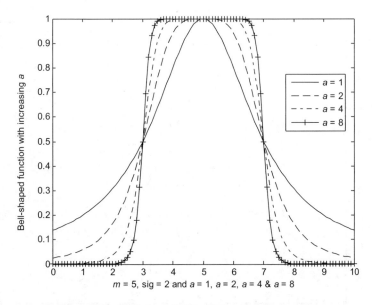

Figure 2.7   Bell-shaped MF with increasing value of a

The parameters m and σ represent the centre and width of the bell-shaped MF, respectively. Parameter a, usually positive, controls the slope of the MF as shown in Figure 2.7. The MF is narrower with increasing value of a.

2.4.5   Sigmoidal MF

Gaussian and bell-shaped MFs are smooth and symmetric MFs. There are many applications where asymmetric MFs are useful. A sigmoidal MF is asymmetric and is either open left or right (Jang *et al.*, 1997). A parameterized sigmoidal MF is defined by

$$\mu(x) = \frac{1}{1 + \exp\left[-a\left(x - c\right)\right]} \tag{2.7}$$

The parameter a controls the slope of the MF at the cross-point $x = c$. Two sigmoidal MFs are shown in Figure 2.8. One is right open and the other is left open. The sign of the parameter a determines the open-end direction of the sigmoidal MF. If a is positive, the MF will open to the right and if a is negative, the MF will open to the left. This property of the sigmoidal MF helps to define extreme positive or extreme negative MFs.

There is no general rule for choosing the type of MFs for a particular problem or application. It is rather application-dependent; the shape of MF depending on the parameters of the MF used, which greatly influences the performance of a fuzzy system. There are different approaches to construct membership functions, such as heuristic selection (the most widely used), clustering approach, C-means clustering approach, adaptive vector quantization and self-organizing map.

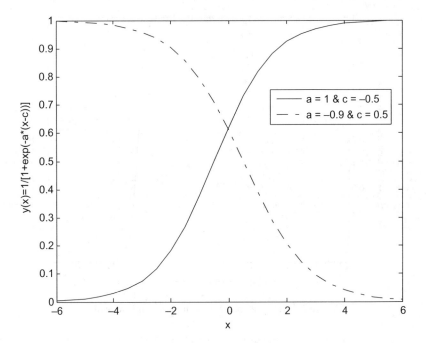

Figure 2.8   Two sigmoidal MFs

For a detailed description of these approaches, interested readers are directed to Chi *et al.* (1996). Some methods of tuning and optimization of MFs are discussed in Chapter 8.

2.5   Features of MFs

Several concepts used to define membership functions are discussed here: support set, core, singleton, crosspoints (or crossover points), peak point, symmetric or asymmetric membership function, left and right width.

2.5.1   Support

The support of a fuzzy set A is the set of all points $x \in X$ at which $\mu_A(x) > 0$. Assume A is a fuzzy subset of X. The support of A, denoted *Support*(A), is the crisp subset of X whose elements have nonzero membership grades in A, defined as

$$Support(A) = \{x \mid \mu_A(x) > 0 \text{ and } x \in X\} \tag{2.8}$$

2.5.2   Core

The core of a fuzzy set A, denoted $Core(A)$, is the crisp subset of X consisting of all elements with membership grade 1. This is defined as

$$Core(A) = \{x \mid \mu_A(x) = 1 \text{ and } x \in X\} \tag{2.9}$$

The support set *Support*(A) and core set $Core(A)$ of a trapezoidal MF are shown in Figure 2.9.

2.5.3   Fuzzy Singleton

A fuzzy set whose support is a single point in X at which $\mu_A(x) = 1$ is called a singleton. A singleton is shown in Figure 2.10.

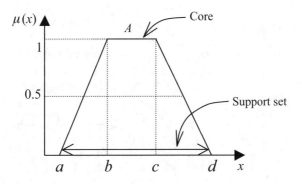

Figure 2.9   Support set and core of MF A

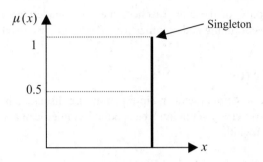

Figure 2.10 Singleton

2.5.4 *Crossover Point*

A crossover point of a fuzzy set A is a point $x \in X$ at which $\mu_A(x) = \alpha$ with $\alpha \in [0, 1]$:

$$Crossover(A) = \{x \mid \mu_A(x) = 0.5\} \qquad (2.10)$$

The crossover point of the membership function A with B is at 0.5. In other words, it is the overlap of two neighbouring membership functions. The overlap of the two triangular MFs A and B is shown in Figure 2.11. The crossover of A is the same as the crossover of B.

A triangular MF has three features by which it can be parameterized: the peak, left width and right width. These are the anchor points of the three corners of a triangular MF. At the peak point, point b of MF A and point c of MF B in Figure 2.11, the membership value is 1. The left width is the distance of the left anchor point from the peak point and the right width is the distance of the right anchor point from the peak point. The left and right widths for the MF B are shown in Figure 2.11. If the left and right widths are not equal the MF is asymmetric, otherwise it is symmetric. The MF B in Figure 2.11 is asymmetric triangular whereas the MF A is symmetric. Symmetric and asymmetric trapezoidal MFs are shown in Figure 2.5.

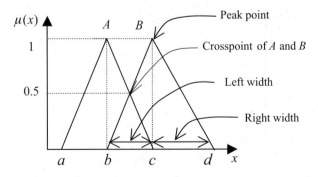

Figure 2.11 Crossover point, left and right width of MF

2.6 Operations on Fuzzy Sets

The membership function is the main component defining the basic fuzzy set operations. Zadeh and other researchers have given additional and alternative definitions for set-theoretic operations.

α**-cut of a fuzzy set:** The α-cut of a fuzzy set A, denoted A_α, is a subset of X consisting of all the elements in X defined by

$$A_\alpha = \{x \mid \mu_{A_\alpha}(x) \geq \alpha \text{ and } x \in X\} \tag{2.11}$$

This means that the fuzzy set A_α contains all elements with a membership of $\alpha \in [0, 1]$ and higher, called the α-cut of the membership function. The α-cut of a fuzzy set A is shown in Figure 2.12. At a resolution level of α, it will have support of A_α. The higher the value of α, the higher the confidence in the parameter.

Example 2.3 Let A be a fuzzy set in the universe of discourse X and $(x_1, x_2, x_3, x_4) \in X$ defined as follows:

$$A = \{0.3/x_1, 1/x_2, 0.5/x_3, 0.9/x_4, 1/x_5\}$$

A_α for $\alpha > 0.5$ is

$$A_{\alpha>0.5} = \{1/x_2, 0.9/x_4, 1/x_5\}$$

Union of fuzzy sets: The union of two fuzzy sets A and B with membership functions μ_A and μ_B, respectively, is a fuzzy set C, denoted $C = A \cup B$, with the membership function μ_C. There are two definitions for the union operation: the max membership function and the product rule, as defined in Equations (2.12) and (2.13):

$$\mu_C(x) = \max [\mu_A(x), \mu_B(x)] \tag{2.12}$$

$$\mu_C(x) = \mu_A(x) + \mu_B(x) - \mu_A(x)\mu_B(x) \tag{2.13}$$

where x is an element in the universe of discourse X.

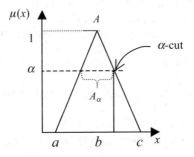

Figure 2.12 α-cut of the membership function

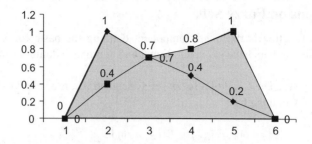

Figure 2.13   Union of fuzzy sets A and B using max operation

Example 2.4   Let A and B be two fuzzy sets in the universe of discourse X and $(x_1,$ $x_2, x_3, x_4) \in X$ defined as follows:

$$A = \{0/x_1 + 1/x_2 + 0.7/x_3 + 0.4/x_4 + 0.2/x_5 + 0/x_6\}$$
$$B = \{0/x_1 + 0.4/x_2 + 0.7/x_3 + 0.8/x_4 + 1/x_5 + 0/x_6\}$$

The union of fuzzy sets A and B using the max membership function is

$$C_{\max} = A \cup B = \{0/x_1 + 1/x_2 + 0.7/x_3 + 0.8/x_4 + 1/x_5 + 0/x_6\}$$

where $\mu_C(x_i)$ is calculated from $\max[\mu_A(x_i), \mu_B(x_i)]$ for $i = 1, 2, 3, \ldots, 6$. Alternatively, using the product rule it is

$$C_{\text{prod}} = A \cup B = \{0/x_1 + 1/x_2 + 0.91/x_3 + 0.88/x_4 + 1/x_5 + 0/x_6\}$$

where $\mu_C(x_i)$ is calculated using $[\mu_A(x_i) + \mu_B(x_i) - \mu_A(x_i) * \mu_B(x_i)]$ for $i = 1, 2, 3, \ldots, 6$.
The union operation of fuzzy sets A and B is shown in Figure 2.13.

Intersection of fuzzy sets: The intersection of two fuzzy sets A and B with membership functions μ_A and μ_B, respectively, is a fuzzy set C, denoted $C = A \cap B$, with membership function μ_C defined using the min membership function or the product rule as

$$\mu_C(x) = \min[\mu_A(x), \mu_B(x)] \tag{2.14}$$
$$\mu_C(x) = \mu_A(x) * \mu_B(x) \tag{2.15}$$

Example 2.5   Let A and B be two fuzzy sets in the universe of discourse X and $(x_1, x_2, x_3, x_4) \in X$ defined as in the previous example.
The intersection of fuzzy sets A and B using the min membership function is

$$C_{\min} = A \cap B = \{0/x_1 + 0.4/x_2 + 0.7/x_3 + 0.4/x_4 + 0.2/x_5 + 0/x_6\}$$

where $\mu_C(x_i)$ is calculated from $\mu_C(x) = \min[\mu_A(x), \mu_B(x)]$ for $i = 1, 2, 3, \ldots, 6$. Alternatively, using the product rule it is

$$C_{\text{prod}} = A \cap B = \{0/x_1 + 0.4/x_2 + 0.49/x_3 + 0.32/x_4 + 0.2/x_5 + 0/x_6\}$$

where $\mu_C(x_i)$ is calculated from $\mu_C(x) = \mu_A(x) * \mu_B(x)$ for $i = 1, 2, 3, \ldots, 6$.

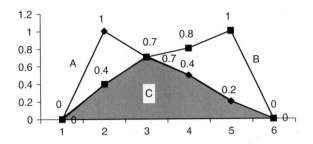

Figure 2.14   Intersection of fuzzy sets A and B using the min operation

The intersection operation of fuzzy sets A and B is shown in Figure 2.14.

Complement of fuzzy set: The complement of a fuzzy set A with membership function μ_A is a fuzzy set, denoted $\sim A$, with membership function $\mu_{\sim A}$ defined as

$$\mu_{\sim A}(x) = 1 - \mu_A(x) \tag{2.16}$$

Example 2.6   Let A be a fuzzy set in the universe of discourse X and $(x_1, x_2, x_3, x_4, x_5, x_6, x_7, x_8) \in X$ defined as follows:

$$A = \{1/x_1 + 1/x_2 + 0.9/x_3 + 0.8/x_4 + 0.7/x_5 + 0.3/x_6 + 0.1/x_7 + 0/x_8\}$$

The complement of fuzzy set A is $\sim A$:

$$\sim A = \{0/x_1 + 0/x_2 + 0.1/x_3 + 0.2/x_4 + 0.3/x_5 + 0.7/x_6 + 0.9/x_7 + 1/x_8\}$$

where $\mu_{\sim A}(x)$ is calculated from $[1 - \mu_A(x)]$ for $i = 1, 2, 3, \ldots, 8$.
The complement operation of fuzzy set A is shown in Figure 2.15.

Fuzzy subsets or containment: Let A and B be two fuzzy sets with membership functions μ_A and μ_B, respectively. A is a subset of B (or A is contained in B), written $A \subset B$, if and only if

$$\mu_A \leq \mu_B \; \forall x, x \in X \tag{2.17}$$

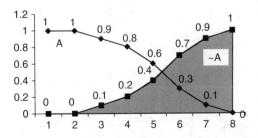

Figure 2.15   Complement of fuzzy set A

Equality of fuzzy sets: Two fuzzy sets A and B with membership functions μ_A and μ_B, respectively are equal, written $A = B$, if and only if

$$\mu_A = \mu_B \; \forall x, x \in X \tag{2.18}$$

Example 2.7   Let A and B be two fuzzy sets in the universe of discourse X and $(x_1, x_2, x_3, x_4) \in X$ defined as follows:

$$A = \{0/x_1 + 1/x_2 + 0.7/x_3 + 0.4/x_4 + 0.2/x_5 + 0/x_6\}$$
$$B = \{0/x_1 + 1/x_2 + 0.7/x_3 + 0.4/x_4 + 0.2/x_5 + 0/x_6\}$$

All the membership values of A are equal to those of B, i.e., $\mu_A = \mu_B$, therefore $A = B$.

Null (or empty) fuzzy set: A fuzzy set is null (or empty, denoted $\emptyset$) if and only if its membership function $\forall x \in X$ (for all elements in X) is identically zero on X. This is defined as

$$\mu_\emptyset(x) = \{x \mid \mu_\emptyset(x) = 0 \text{ and } \forall x \in X\} \tag{2.19}$$

Properties of fuzzy sets: Assume A, B and C are fuzzy sets of X. The following properties hold for union, intersection and fuzzy subsets.

(i) **Commutativity**

$$A \cup B = B \cup A$$
$$A \cap B = B \cap A$$

(ii) **Idempotency**

$$A \cup A = A$$
$$A \cap B = B \cap A$$

(iii) **Associativity**

$$A \cup (B \cup C) = (A \cup B) \cup C = A \cup B \cup C$$
$$A \cap (B \cap C) = (A \cap B) \cap C = A \cap B \cap C$$

(iv) **Distributivity**

$$A \cup (B \cap C) = (A \cup B) \cap (A \cup C)$$
$$A \cap (B \cup C) = (A \cap B) \cup (A \cap C)$$
$$A \cup \emptyset = A$$
$$A \cap \emptyset = \emptyset$$
$$A \cup X = X$$
$$A \cap X = A$$

(v) **Transitivity**

If $A \subset B$ then $B = A \cup B$ and $A = A \cap B$

If $A \subset B$ and $B \subset C$ then $A \subset C$

The following properties hold for complements of fuzzy subsets.

(vi) **De Morgan's Law**

$$\overline{A \cup B} = \bar{A} \cup \bar{B}$$
$$\overline{A \cap B} = \bar{A} \cap \bar{B}$$

A significant feature of fuzzy set that distinguishes them from classical sets is that $\bar{A} \cap \emptyset \neq \emptyset$ and $\bar{A} \cup A \neq X$.

2.7   Linguistic Variables

A linguistic variable is a variable whose values are words or sentences, used as labels of fuzzy subsets (Zadeh, 1975a,b, 1976a). Such linguistic variables serve as a means of approximate characterization of systems which cannot be described precisely by numerical values or other traditional quantitative terms. For example, speed is a linguistic variable if its values are slow, medium, fast, not slow, very fast, not very slow, etc. In this case, fast is a linguistic value of speed and is imprecise compared with an exact numeric value such as 'speed is 77 mph'. The relation between a numerical variable $s = 77$ and the linguistic variables slow, medium and fast is illustrated graphically in Figure 2.16.

In general, a linguistic variable is characterized by a quintuple $\{X, T, U, G, M\}$, where X is the name of the variable (e.g., Speed), T denotes the term set of X (i.e., the set of names of linguistic labels of X over a universe of discourse U: slow, medium, fast, etc.), G is the syntactic rule or grammar for generating names and M is the semantic rule for associating with each X its meaning, $M(X) \subseteq U$ (Zadeh, 1975a).

2.7.1   Features of Linguistic Variables

Theoretically, the term set $T(X)$ is infinite but in practical applications, $T(X)$ is defined with a small number of terms so that each element of $T(X)$ defines a mapping between each element and the function $M(X)$, which associates a meaning with each term in the term set. Let the term set of the linguistic variable Speed be {slow, medium, fast} within the universe of discourse $U = [0, 120]$. The term set can be expressed as

$$T(Speed) = \{slow, medium, fast\}$$

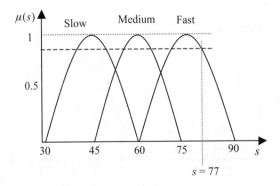

Figure 2.16   Relation between linguistic and numerical variables

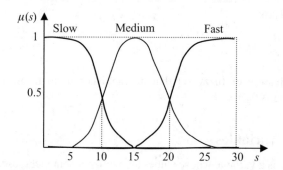

Figure 2.17   Three fuzzy sets for speed – Slow, Medium and Fast

The semantic rule of linguistic variables can be expressed using context-free grammar. For example

$$T = \{slow, very\ slow, very\ very\ slow, \ldots\}$$

Using context-free grammar the above expression can be written as

$$T \rightarrow slow$$
$$T \rightarrow very\ T$$

Here, 'very' is called a linguistic hedge, which is used to derive new linguistic variables. Linguistic hedges will be discussed further in the next section.

A linguistic variable can be a word or sentence and such natural language expressions are fuzzy, e.g., Slow OR Medium, Medium AND Fast. Figure 2.17 shows three MFs: Slow, Medium and Fast.

The linguistic variable 'Slow OR Medium' is shown graphically in Figure 2.18. It is the shaded area representing the union of the membership functions 'Slow or Medium'.

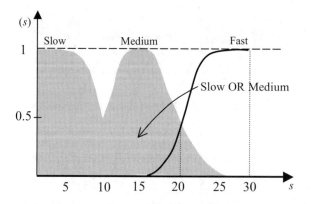

Figure 2.18   Expression for 'Slow OR Medium'

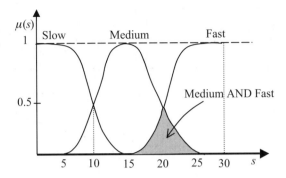

Figure 2.19 Expression for 'Medium AND Fast'

The linguistic variable 'Medium AND Fast' is shown graphically in Figure 2.19. It is the shaded area representing the intersection of the membership functions Medium AND Fast.

In the above examples, OR and AND are connectives, which play an important role in the description of linguistic variables. It can be seen from Figures 2.18 and 2.19 that they are used to derive new linguistic variables from the term sets. The role of connectives will be discussed further in the use of linguistic hedges presented in the next section.

2.8 Linguistic Hedges

The purpose of the hedges is to generate a larger set of values for a linguistic variable from a small collection of primary terms. Hedges are realized on primary terms through the processes

- Intensification or concentration,
- Dilation and
- Fuzzification.

This can be represented as a quadruple $\{H, M, T, C\}$, where H is the set of hedges, M is the marker, T is the set of primary terms (e.g., slow, medium, fast, etc.) and C is the set of connectives. Parentheses are used as markers in the definition of linguistic variables to separate the term set from the hedge, e.g., Very (Small). Figure 2.20 depicts the format of the use of the different term sets, hedges and connectives for defining linguistic variables.

For example, Big but Not Very (Big). Here 'Big' is a primary term set, 'but' is a connective (which means AND in this case) and 'Very' is a hedge. 'Not' is a complement operation on the term set. Parentheses '( )' are used as a marker.

Example 2.8 The hedge 'Very' is a concentration (or intensification) operation and performed by squaring the membership values of the primary fuzzy set. The operation is shown for two primary fuzzy sets Small and U in the example below.

$$\text{Very (Small)} = \text{Small}^2 = [\mu_{\text{Small}}]^2$$
$$\text{Very (Very } (U)) = (\text{Very } ([\mu_U]))^2 = ([\mu_U]^2)^2 = [\mu_U]^4$$

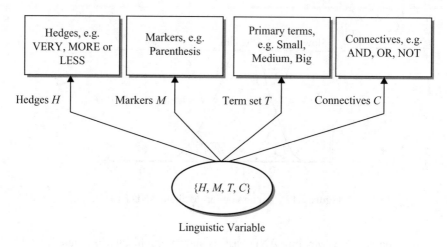

Linguistic Variable

Figure 2.20   Linguistic variables and hedges

Example 2.9   Consider the fuzzy set A of short pencils defined by

$$A = \left\{ \frac{0.20}{p_1} + \frac{0.5}{p_2} + \frac{1}{p_3} + \frac{1}{p_4} + \frac{0.9}{p_5} \right\}$$

Then the fuzzy set for very short pencils can be expressed by the use of a hedge on the fuzzy set A:

$$Very\,(A) = [\mu_A]^2 = \left\{ \frac{0.04}{p_1} + \frac{0.25}{p_2} + \frac{1}{p_3} + \frac{1}{p_4} + \frac{0.81}{p_5} \right\}$$

The linguistic hedge 'More or less' is a dilation operation defined as More or less $(A) = A^{1/2}$. The fuzzy set for more or less short pencils can be expressed by the following:

$$More\ or\ less\,(A) = [\mu_A]^{1/2} = \left\{ \frac{0.45}{p_1} + \frac{0.71}{p_2} + \frac{1}{p_3} + \frac{1}{p_4} + \frac{0.95}{p_5} \right\}$$

The application of the linguistic hedges 'Very' and 'More or less' is demonstrated through the concentration (or intensification) and dilation process as shown in Figure 2.21.

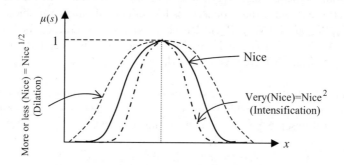

Figure 2.21   Dilation and concentration (or intensification)

Table 2.1   Hedges and their meaning

Hedge	Meaning
About, around, near, roughly	Approximates a scalar
Above, more than	Restricts a fuzzy region
Almost, definitely, precisely	Contrasts intensification
Below, less than	Restricts a fuzzy region
Generally, usually	Contrasts diffusion
Neighbouring, close to	Approximates narrowly
Not	Negation or complement
Quite, rather, somewhat	Dilutes a fuzzy region
Very, extremely	Intensifies a fuzzy region

A linguistic variable can be used with more than one hedge, for example

Almost very fast but generally below 100 km/hr.

Close to 100 m but not very high.

Not more than about zero.

'But' is a connective here which is equivalent to AND. The equivalent versions of the above linguistic variables with markers can be expressed as

Almost (Very (Fast)) AND Generally (below 100 km/hr).

Close (100 m) AND Not (Very (High)).

Not (More than (About zero)).

The operation of multiple hedges can result in the same primary fuzzy set. For example, the operation of the hedges 'More or less very nice' is represented by the following expression. It can be seen that the operation of the hedges on the primary term (fuzzy) set 'Nice' resulted in the same primary fuzzy set:

$$\text{More or less (Very (Nice))} = \text{More or less (Nice}^2) = (\text{Nice}^2)^{1/2} = \text{Nice}$$

Some widely used hedges and their meanings are given in Table 2.1.

Linguistic variables and hedges allow us to construct mathematical models for expressions of natural language. These models can then be used to write process rules and computer programs and simulate real-world processes and behaviour.

2.9   Fuzzy Relations

Having described the operations on fuzzy sets, we need to look at how we can represent linguistic statements mathematically. In fact, many application problems are described using fuzzy relations.

The concept of a relation has a natural extension to fuzzy sets and plays an important role in the theory of such sets and their applications. A fuzzy relation R from the fuzzy set A in X to the fuzzy set B in Y is a fuzzy set defined by the Cartesian product $A \times B$ in the Cartesian product space $X \times Y$. R is characterized by the membership function expressing various degrees of strength of relations:

$$R = A \times B = \sum \mu_R(x, y)/(x, y) = \sum \min(\mu_A(x), \mu_B(y)) \qquad (2.20)$$

$$R = A \times B = \sum \mu_R(x, y)/(x, y) = \sum \mu_A(x)^* \mu_B(y) \qquad (2.21)$$

In Equations (2.20) and (2.21) the sum does not mean a mathematical summation operation, it means all possible combinations of all elements.

R is also called the relational matrix. The Cartesian product is implemented in the same fashion, as is the cross product of two vectors. For example, fuzzy set A with 4 elements (a column vector of dimension 4×1) and fuzzy set B with 5 elements (a row vector of dimension 1×5) will provide the resulting fuzzy relation R which is represented by a matrix of dimension 4×5.

Example 2.10 Let A and B be two fuzzy sets defined by

$$A = \{1/1 + 0.8/2 + 0.6/3 + 0.5/4\}$$
$$B = \{0.5/1 + 1/2 + 0.3/3 + 0/4\}$$

The fuzzy relation (i.e., the Cartesian product of A and B using the min operation) will be

$$\mathbf{R = A \times B} = \begin{bmatrix} \{1, .5\} & \{1, 1\} & \{1, .3\} & \{1, 0\} \\ \{.8, .5\} & \{.8, 1\} & \{.8, .3\} & \{.8, 0\} \\ \{.6, .5\} & \{.6, 1\} & \{.6, .3\} & \{.6, 0\} \\ \{.5, .5\} & \{.5, 1\} & \{.5, .3\} & \{.5, 0\} \end{bmatrix} = \begin{bmatrix} 0.5 & 1 & 0.3 & 0 \\ 0.5 & 0.8 & 0.3 & 0 \\ 0.5 & 0.6 & 0.3 & 0 \\ 0.5 & 0.5 & 0.3 & 0 \end{bmatrix}$$

The fuzzy relation using the product operation will be

$$\mathbf{R = A \times B} = \begin{bmatrix} \{1, .5\} & \{1, 1\} & \{1, .3\} & \{1, 0\} \\ \{.8, .5\} & \{.8, 1\} & \{.8, .3\} & \{.8, 0\} \\ \{.6, .5\} & \{.6, 1\} & \{.6, .3\} & \{.6, 0\} \\ \{.5, .5\} & \{.5, 1\} & \{.5, .3\} & \{.5, 0\} \end{bmatrix} = \begin{bmatrix} 0.5 & 1 & 0.3 & 0 \\ 0.4 & 0.8 & 0.24 & 0 \\ 0.3 & 0.6 & 0.18 & 0 \\ 0.25 & 0.5 & 0.15 & 0 \end{bmatrix}$$

2.9.1 Compositional Rule of Inference

If R is a fuzzy relation in $X \times Y$ and A is a fuzzy set in X then the fuzzy set B in Y is given by

$$B = A \circ R \qquad (2.22)$$

B is inferred from A using the relation matrix R which defines the mapping between X and Y and the operation '$\circ$' is defined as the max/min operation.

Example 2.11   Let A be a fuzzy set defined by

$$A = \{0.9/1 + 0.4/2 + 0/3\}$$

with the fuzzy relation R given by the following relational matrix:

$$\mathbf{R} = \mathbf{A} \times \mathbf{B} = \begin{bmatrix} 1 & 0.8 & 0.1 \\ 0.8 & 0.6 & 0.3 \\ 0.6 & 0.3 & 0.1 \end{bmatrix}$$

Then the fuzzy output B in Y using the max/min operation will be

$$\mathbf{B} = \mathbf{A} \circ \mathbf{R} = \begin{bmatrix} \dfrac{0.9}{1} & \dfrac{0.4}{2} & \dfrac{0}{3} \end{bmatrix} \circ \begin{bmatrix} 1 & 0.8 & 0.1 \\ 0.8 & 0.6 & 0.3 \\ 0.6 & 0.3 & 0.1 \end{bmatrix}$$

$$\mathbf{B} = \begin{bmatrix} \{0.9, 1\} & \{0.9, 0.8\} & \{0.9, 0.1\} \\ \{0.4, 0.8\} & \{0.4, 0.6\} & \{0.4, 0.3\} \\ \{0, 0.6\} & \{0, 0.3\} & \{0, 0.1\} \end{bmatrix}$$

Taking the minimum values row-wise, we obtain

$$\mathbf{B} = \begin{bmatrix} 0.9 & 0.8 & 0.1 \\ 0.4 & 0.4 & 0.3 \\ 0 & 0 & 0 \end{bmatrix}$$

Taking the maximum values column-wise, we obtain the fuzzy set B from the compositional relation:

$$\mathbf{B} = \begin{bmatrix} 0.9 & 0.8 & 0.3 \end{bmatrix}$$

2.10   Fuzzy If–Then Rules

Fuzzy sets and their operations are the subjects and verbs of fuzzy logic. If–Then rule statements are used to formulate the conditional statements that comprise fuzzy logic. A single fuzzy If–Then rule assumes the form

$$\text{If } < \text{fuzzy proposition} > \text{ Then } < \text{fuzzy proposition} > \tag{2.23}$$

For example,

$$\text{If } < x \text{ is } A_1 > \text{ Then } < y \text{ is } B_2 >$$

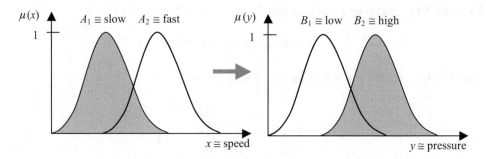

Figure 2.22   If–Then rule

where A_1 and B_2 are linguistic variables defined by fuzzy sets on the ranges (i.e., the universe of discourse) X and Y, respectively. The If part of the rule 'x is A_1' is called the antecedent or premise and the Then part of the rule 'y is B_2' is called the consequent. In other words, the conditional statement can be expressed in mathematical form:

$$\text{If } A_1 \text{ Then } B_2 \text{ or } A_1 \rightarrow B_2 \qquad (2.24)$$

Example 2.12   The speed and pressure of a steam engine can be expressed with the following linguistic conditional statement:

If *Speed* is *Slow* Then *Pressure* should be *High*

Graphically, this statement is represented in Figure 2.22.

2.10.1   Rule Forms

In general, three forms exist for any linguistic variables:

(i) Assignment statement
     e.g., x is *not large* AND *not very small*.
(ii) Conditional statement
     e.g., IF x is *very big* THEN y is *medium*.
(iii) Unconditional statement
     e.g., set pressure *high*.

2.10.2   Compound Rules

A linguistic statement expressed by a human might involve compound rule structures. By using basic properties and operations defined for fuzzy sets, any compound rule structure may be decomposed and reduced to a number of simple canonical rules.

Conjunctive antecedents: A multiple conjunctive antecedent can have the following form:

$$\text{IF } x \text{ is } A_1 \text{ AND } x \text{ is } A_2 \ldots \text{ AND } x \text{ is } A_n \text{ THEN } y \text{ is } B_S \qquad (2.25)$$

Equation (2.25) can be rewritten as

$$\text{IF } x \text{ is } A_S \text{ THEN } y \text{ is } B_S \qquad (2.26)$$

where $A_S = A_1 \cap A_2 \cap A_3 \cap \cdots \cap A_n$ and A_S is expressed by means of a membership function based on the definition of fuzzy intersection operation as

$$\mu_{A_S}(x) = \min\left[\mu_{A_1}(x), \mu_{A_2}(x), \ldots, \mu_{A_n}(x)\right] \qquad (2.27)$$

Disjunctive antecedents: Similarly, a multiple disjunctive antecedent can have the following form:

$$\text{IF } x \text{ is } A_1 \text{ OR } x \text{ is } A_2 \ldots \text{ OR } x \text{ is } A_n \text{ THEN } y \text{ is } B_S \qquad (2.28)$$

Equation (2.28) can be rewritten as

$$\text{IF } x \text{ is } A_S \text{ THEN } y \text{ is } B_S \qquad (2.29)$$

where $A_S = A_1 \cup A_2 \cup A_3 \cup \cdots \cup A_n$ and A_S is expressed by means of a membership function based on the definition of fuzzy union operation as

$$\mu_{A_S}(x) = \max\left[\mu_{A_1}(x), \mu_{A_2}(x), \ldots, \mu_{A_n}(x)\right] \qquad (2.30)$$

2.10.3   Aggregation of Rules

Most rule-based systems have more than one rule. The process of obtaining the overall consequent from the individual consequents contributed by each rule is the aggregation of rules. In the case of a system of rules that must be jointly satisfied, the rules are connected by AND connectives. The aggregated output y is found by fuzzy intersection of the entire individual rule consequent y_i, where $i = 1, 2, 3, \ldots, r$:

$$y = y_1 \text{ AND } y_2 \text{ AND } \cdots \text{ AND } y_r \qquad (2.31)$$

or

$$y = y_1 \cap y_2 \cap \cdots \cap y_r$$

The output is defined by means of a membership function based on the definition of fuzzy intersection operation as

$$\mu_y(y) = \min\left[\mu_{y_1}(y), \mu_{y_2}(y), \ldots, \mu_{y_r}(y)\right] \text{ for } y \in Y \qquad (2.32)$$

For the case of a disjunctive system of rules where at least one rule must be satisfied, the rules are connected by OR connectives. The aggregated output y is found by fuzzy union of all the individual rule consequents y_i, where $i = 1, 2, 3, \ldots, r$:

$$y = y_1 \ OR \ y_2 \ OR \ \cdots OR \ y_r \qquad (2.33)$$

or

$$y = y_1 \cup y_2 \cup \cdots \cup y_r$$

The output is defined by means of a membership function based on the definition of fuzzy union operation as

$$\mu_y(y) = \max \left[\mu_{y_1}(y), \mu_{y_2}(y), \ldots, \mu_{y_r}(y) \right] \text{ for } y \in Y \qquad (2.34)$$

Example 2.13 Let us consider a fuzzy system with two inputs x_1 and x_2 (antecedents) and a single output y (consequent). Inputs x_1 and x_2 have three linguistic variables small, medium and big with a triangular membership function. Output y has two linguistic variables small and big with a triangular membership function as shown in Figure 2.23. The rule base consists of the following two rules:

Rule 1: IF x_1 is *small* AND x_2 is *medium* THEN y is *big*

Rule 2: IF x_1 is *medium* AND x_2 is *big* THEN y is *small*

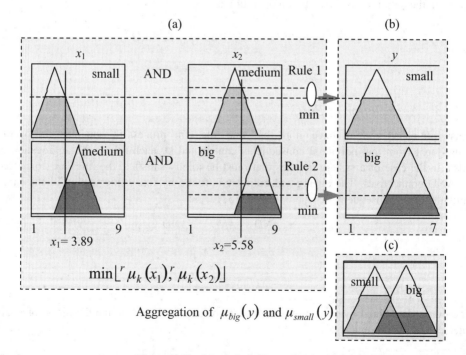

Figure 2.23 Max/min inference method

The inputs $x_1 = 3.89$ and $x_2 = 5.58$ are crisp values for which the membership values $\mu_k(x_1)$ and $\mu_k(x_2)$ (k denotes the MFs small, medium or big) are calculated for triangular membership functions. The aggregated outputs for r rules are given by

$$r1 : \mu_{big}(y) = \max \lfloor \min \lfloor \mu_{small}(x_1), \mu_{medium}(x_2) \rfloor \rfloor$$
$$r2 : \mu_{small}(y) = \max \lfloor \min \lfloor \mu_{medium}(x_1), \mu_{big}(x_2) \rfloor \rfloor$$

In this example, $r = 1, 2$. The minimum membership values of $\lfloor \mu_{small}(x_1), \mu_{medium}(x_2) \rfloor$ and $\lfloor \mu_{medium}(x_1), \mu_{big}(x_2) \rfloor$ for the antecedents are calculated and propagate through to the consequent part. This operation is shown in Figure 2.23(a). The membership function for the consequent of each rule is then truncated by taking the maximum values, i.e. $\max \lfloor \min \lfloor \mu_{small}(x_1), \mu_{medium}(x_2) \rfloor \rfloor$ and $\max \lfloor \min \lfloor \mu_{medium}(x_1), \mu_{big}(x_2) \rfloor \rfloor$ are computed, which is shown in Figure 2.23(b). The truncated membership functions for each rule, i.e. $\mu_{big}(y)$ and $\mu_{small}(y)$ are aggregated using the graphical equivalent of either conjunctive or disjunctive rules. The aggregation operation max results in an aggregated membership function comprising the outer envelope of the individual truncated membership forms from each rule. This operation is shown in Figure 2.23(c).

It has been demonstrated in Figure 2.23 that any numeric value (or crisp value) has to be converted into a fuzzy input and then a conclusion can be drawn using the rule of inference on consequent fuzzy sets. There are three distinct steps in the process. They are described in the following sections.

2.11   Fuzzification

The process that allows converting a numeric value (or crisp value) into a fuzzy input is called fuzzification. There are two methods of fuzzification.

- Singleton fuzzification: This maps a real value $x_i \in X$ into a fuzzy singleton A_{x_i} which has membership value 1 at $x = x_i$ and 0 at all other points in X. This is expressed as

$$\mu_{A_{x_i}}(x) = \begin{cases} 1 & \text{if} \quad x = x_i \\ 0 & \text{otherwise} \end{cases} \tag{2.35}$$

Singleton fuzzification greatly simplifies computation but is generally used in implementations where there is no noise. There is no widespread use of singleton fuzzification in fuzzy systems and applications.

- A_{x_i} is fuzzy: This maps a real $x_i \in X$ into a fuzzy set A_{x_i} in X described by a membership function:

$$\mu_{A_{x_i}}(x) = \begin{cases} 1 & \text{if} \quad x = x_i \\ [0, 1] & \text{decreases from 1 as } x \text{ moves from } x_i \end{cases} \tag{2.36}$$

In other words, fuzzification actually provides a membership grade of a real (or crisp) value $x_i \in X$ as its belongingness to a fuzzy set A_{x_i}. The fuzzy set can be described by various

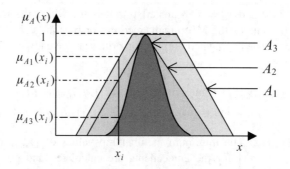

Figure 2.24 Fuzzification in different types of MFs

membership functions discussed in Section 2.4. Figure 2.24 shows the fuzzification of $x_i \in X$ using three different types of membership functions: trapezoidal (A_1), triangular (A_2) and Gaussian (A_3). It demonstrates that $x = x_i \in X$ has a different fuzzified value, i.e. membership grade, depending on the type (i.e., shape) of the membership function. The membership grades are $\mu_{A_1}(x_i)$ for trapezoidal MF, $\mu_{A_2}(x_i)$ for triangular MF and $\mu_{A_3}(x_i)$ for Gaussian MF.

Example 2.14 Let A be a fuzzy set defined by the bell-shaped MF (with centre $m = 5$, width $\sigma = 1$ and shape parameter $a = 1$ in Equation (2.6)) as follows:

$$\mu_A(x) = \frac{1}{1 + \left| \dfrac{x - 5}{1} \right|^2}$$

The fuzzification of the value $x_i = 6$ will yield the grade of membership as

$$\mu_A(x_i) = \frac{1}{1 + \left| \dfrac{6 - 5}{1} \right|^2} = 0.5$$

It is obvious that the shape of the MF plays an important role in fuzzification and in any subsequent process. Fuzzification using well-defined MFs can suppress noise in the inputs of a fuzzy system (Wang, 1997).

2.12 Defuzzification

Defuzzification is the reverse process of fuzzification. Mathematically, the defuzzification of a fuzzy set is the process of conversion of a fuzzy quantity into a crisp value. This is necessary when a crisp value is to be provided from a fuzzy system to the user. For example, if we develop a fuzzy system for blood pressure control, we will probably want to tell the user what blood pressure is expected to be in the next time instant.

Fuzzy control engineers have many different ways of defuzzifying. However, there are quite simple methods in use. It is intuitive that fuzzification and defuzzification should be reversible.

That is, if we fuzzify a number into a fuzzy set and immediately defuzzify it, we should be able to get the same number back again.

There are many defuzzification methods available in the literature. Very often standard defuzzification methods fail in some applications. It is, therefore, important to select the appropriate defuzzification method for a particular application. Unfortunately, there is no standard rule for selecting a particular defuzzification method for an application. The choice of the most appropriate method depends on the application. A good study on the selection of appropriate defuzzification methods has been reported by Runker (1997). In the next few sections, some widely used methods of defuzzification are presented.

Max-membership method: Also known as the height method, the max-membership method is both simple and quick. This method takes the peak value of each fuzzy set and builds the weighted sum of these peak values. This method is given by the algebraic expression in Equation (2.38).

$$x^* = \frac{\sum\limits_{k=1}^{m} c_k . h_k}{\sum\limits_{k=1}^{n} h_k} \tag{2.37}$$

Defuzzification using the max-membership method is shown in Figure 2.25(a). c_k is the peak value of the fuzzy sets and h_k is the height of the clipped fuzzy sets, as shown in the figure.

Centre of gravity method: Also referred to as centre of area or centroid method in the literature. This is the most widely used defuzzification method. The centre of area method finds the centroid of the area under the membership function. In the continuous case it is given by the expression in Equation (2.39).

$$x^* = \frac{\int \mu_c(x) . x dx}{\int \mu_c(x) dx} \tag{2.38}$$

and for a discrete universe with m quantization levels in the output it is given by

$$x^* = \frac{\sum\limits_{i=1}^{m} \mu_c(x_i) . x_i}{\sum\limits_{i=1}^{m} \mu_c(x_i)} \tag{2.39}$$

Figure 2.25(b) shows this operation in a graphical way. The value x^* is the centroid of the area, which is the defuzzified value of the combined overlapped consequent fuzzy sets of the rule. Some numerical aspects of the centre of area method of defuzzification are reported by Patel and Mohan (2002).

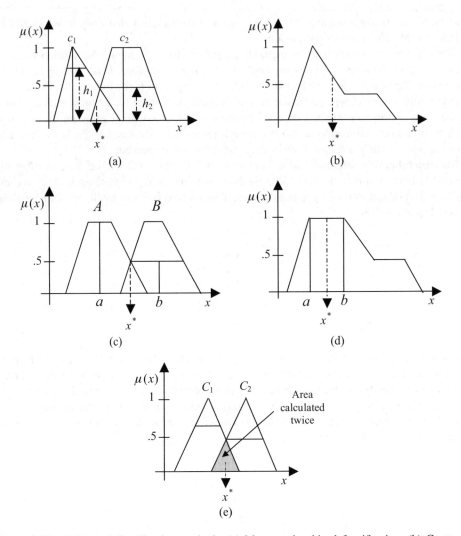

Figure 2.25 Different defuzzification methods. (a) Max-membership defuzzification; (b) Centre of gravity defuzzification; (c) Weighted average defuzzification; (d) Mean-max defuzzification; (e) Centre of sums defuzzification

Weighted average method: This method is suitable for symmetric membership functions. It is given by the algebraic expression

$$x^* = \frac{\sum \mu_c(x).x'}{\sum \mu_c(x)} \tag{2.40}$$

where Σ denotes an algebraic sum. This is shown in Figure 2.25(c). In the figure, there are two trapezoidal membership functions A and B. $\mu_A(x) = 1$ is the weight for $x' = a$ and

$\mu_B(x) = 0.5$ is the weight for $x' = b$. The defuzzified value of the two clipped trapezoidal MFs can be calculated using Equation (2.41) as

$$x^* = \frac{\{a(1) + b(0.5)\}}{(1 + 0.5)}$$

Mean-max membership: Also known as the middle of maxima method, a single defuzzified output is generated by the mean or average of all local maxima defined by

$$x^* = \frac{\sum\limits_{i=1}^{N} \mu_{\max}(x_i)}{N} \qquad (2.41)$$

where $\max \mu(x_i)$ is the maximum membership value and N is the number of times the membership function reaches the maximum support value. Figure 2.25(d) shows the two maxima a and b. The defuzzified output x^* is calculated from the mean of the two values as follows:

$$x^* = \frac{(a + b)}{2}$$

Centre of sums: This process involves the algebraic sum of individual output fuzzy sets instead of computing the union of the two fuzzy sets. Since it calculates the area of individual fuzzy sets, the method is faster than the centre of gravity method. The defuzzified value x^* is formally given in the discrete case by the expression

$$x^* = \frac{\sum\limits_{i=1}^{m} x_i \cdot \sum\limits_{k=1}^{n} \mu_k(x_i)}{\sum\limits_{i=1}^{m} \cdot \sum\limits_{k=1}^{n} \mu_k(x_i)} \qquad (2.42)$$

One drawback of this method is that the overlapping area is added twice. There are two fuzzy sets C_1 and C_2 in Figure 2.25(e) and the defuzzified value for the centre of sums method is shown in the figure. The shaded area is the overlapped area of the fuzzy sets C_1 and C_2, which is calculated twice.

Finally, Figure 2.26 shows the entire process of fuzzification, aggregation and defuzzification in a fuzzy system. Figure 2.26(a) shows the fuzzification of the input MFs, Figure 2.26(c) shows the aggregation of the output MFs and Figure 2.26(d) shows the defuzzification process using the centre of gravity method.

There have been various studies reported on defuzzification methods in the literature (Driankov *et al.*, 1993). An empirical study of the performance of defuzzification methods applied to different fuzzy controllers has been reported by Lancaster and Wierman (2003). They investigated standard methods such as the true centre of gravity, fast centre of gravity and mean of maxima and found these methods have some advantages over the other methods.

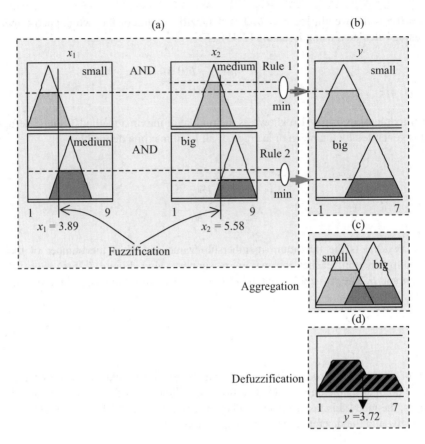

Figure 2.26   Fuzzification, aggregation and defuzzification

They also developed some new methods, such as the plateau average, weighted plateau average, sparus, capitis and clivosus. These methods are not discussed further here, but interested readers are referred to Lancaster and Wierman (2003) and Van Broekhoven and De Baets (2004). A very comprehensive review of defuzzification methods can also be found in Van Leekwijck and Kerre (1999) and Roychowdhury and Pedrycz (2001). A comparative analysis of different defuzzification methods is given in Driankov *et al.* (1993). Good theoretical analyses of defuzzification processes and problems have been reported in Yager and Filev (1994) and Kickert and Mamdani (1978).

2.13   Inference Mechanism

Inference is the process of formulating a nonlinear mapping from a given input space to output space. The mapping then provides a basis from which decisions can be made. The process of fuzzy inference involves all the membership functions, operators and if–then rules.

There are three types of fuzzy inference mechanism, which have been widely employed in various fuzzy systems and applications. The differences between these three fuzzy inferences,

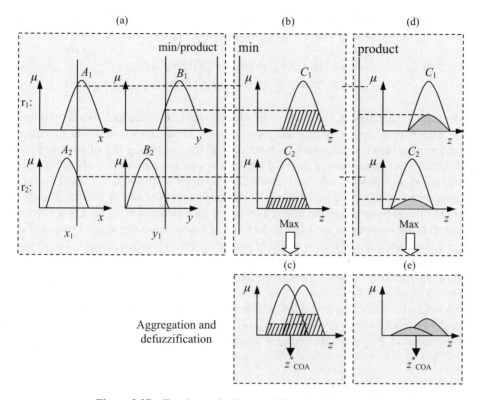

Figure 2.27 Two-input single-output Mamdani fuzzy model

also called fuzzy models, mainly lie in the consequent parts of their fuzzy rules, aggregations and defuzzification procedures. These fuzzy inferences are:

1. Mamdani fuzzy inference,
2. Sugeno fuzzy inference,
3. Tsukamoto fuzzy inference.

2.13.1 Mamdani Fuzzy Inference

The Mamdani-type fuzzy inference was first proposed as an attempt to control a steam engine and boiler using a set of linguistic control rules obtained from an experienced human operator (Mamdani and Assilian, 1974). Figure 2.27 illustrates a Mamdani-type fuzzy inference model. The system consists of two inputs x and y (antecedents) and a single output z (consequent). Each input x, y and output z has two MFs: $\{A_1, A_2\}$, $\{B_1, B_2\}$ and $\{C_1, C_2\}$, respectively. A typical rule in a Mamdani-type fuzzy model is described by a collection of R rules of the form

$$k : \text{If } x \text{ is } A_i^k \text{ and } y \text{ is } B_j^k \text{ then } z \text{ is } C_l^k \tag{2.43}$$

where $k = 1, 2, \ldots, R$, $i = 1, 2, \ldots, N$, $j = 1, 2, \ldots, M$ and $l = 1, 2, \ldots, L$. N, M and L are the numbers of membership functions for inputs and output, respectively. The maximum number of rules in the Mamdani-type fuzzy system here is $R \subset N \times M$. There are only two rules used to demonstrate the inferencing mechanism in Figure 2.27. These are:

$$r1 : \text{IF } x \text{ is } A_1 \text{ AND } y \text{ is } B_1 \text{ THEN } z \text{ is } C_1$$
$$r2 : \text{IF } x \text{ is } A_2 \text{ AND } y \text{ is } B_2 \text{ THEN } z \text{ is } C_2$$

In Mamdani's fuzzy model, crisp values are used as inputs. For example, two crisp values x_1 and y_1 are measured for the inputs x and y, respectively. Figure 2.27(a) shows the fuzzification and inferencing using a minimum or product rule for computing the firing strengths for rules with ANDed antecedent. Figure 2.27(b,d) shows the consequent part of each rule using max/min and max/product rule, respectively. An analogy of Example 2.13 will be helpful for understanding this process. Max/min is the most common rule of composition. In the max/min rule of composition the inferred output of each rule is a fuzzy set chosen from the minimum firing strength. In the max/product rule of composition the inferred output of each rule is a fuzzy set scaled down by its firing strength via the algebraic product. The truncated membership functions for each rule, i.e., $\mu_{C_1}(z)$ and $\mu_{C_2}(z)$ in this case, are aggregated. The aggregation operation is shown for both the max/min and max/product rule of composition in Figure 2.27(c,e).

In Mamdani's fuzzy model, defuzzification (see Section 2.12 for different defuzzification methods) is carried out to convert a fuzzy set to a crisp value. Figure 2.27(c,e) shows the defuzzified value z^*_{COA} using the centre of area method.

2.13.2   Sugeno Fuzzy Inference

The Sugeno fuzzy inference, also known as the TSK fuzzy model, was proposed by Takagi, Sugeno and Kang (Takagi and Sugeno, 1985; Sugeno and Kang, 1988) in an effort to develop a systematic approach to generate fuzzy rules from a given input/output data set. A typical rule in the Sugeno-type fuzzy model for two-input single-output is described by a collection of rules of the form

$$k : \text{If } x \text{ is } A_i^k \text{ and } y \text{ is } B_j^K \text{ then } z^k = f(x, y) \tag{2.44}$$

where x and y are the inputs and z is the output, A_i and B_j are fuzzy MFs for the inputs in the antecedent part, $z = f(x, y)$ is a crisp function in the consequent part $k = 1, 2, \ldots, R$, $i = 1, 2, \ldots, N$, $j = 1, 2, \ldots, M$. N and M are the numbers of membership functions for inputs and R is the maximum number of rules. Figure 2.28 illustrates a two-input single-output Sugeno fuzzy model. In this example, each input x and y has two MFs $\{A_1, A_2\}$ and $\{B_1, B_2\}$ while $\{z_1, z_2\}$ are consequent functions. There are only two rules used to demonstrate the Sugeno-type inferencing mechanism in Figure 2.28. These are:

$$r1 : \text{IF } x \text{ is } A_1 \text{ AND } y \text{ is } B_1 \text{ THEN } z_1 = a_1 x + b_1 y + c_1$$
$$r2 : \text{IF } x \text{ is } A_2 \text{ AND } y \text{ is } B_2 \text{ THEN } z_2 = a_2 x + b_2 y + c_2$$

Usually, $z = f(x, y)$ is polynomial in the input variables x and y but it can be any function as long as it can appropriately describe the output of the model within the fuzzy region specified by the antecedent of the rule. $\{a_1, b_1, c_1\}$ and $\{a_2, b_2, c_2\}$ are the parameters of the

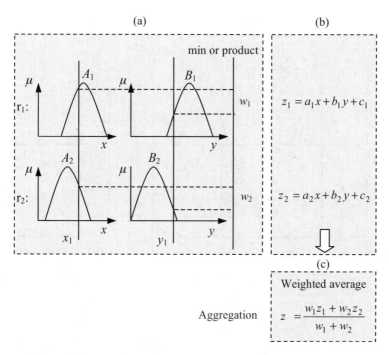

Figure 2.28 Two-input single-output first-order Sugeno fuzzy model

polynomial function $z = f(x, y)$. When $z = f(x, y)$ is a first-order polynomial, the resulting fuzzy inference system is called a first-order Sugeno fuzzy model as proposed by Takagi and Sugeno (1985) and Sugeno and Kang (1988). In the Sugeno-type fuzzy model in Figure 2.28, two measured crisp values x_1 and y_1 are used for the inputs x and y, respectively. Figure 2.28(a) shows the fuzzification and inferencing using the minimum or product rule for computing the firing strengths w_1 and w_2 for the rules with ANDed antecedent. The firing strength is calculated using the minimum or product rule as

$$w_r = \min(\mu_{A_i}, \mu_{B_j}) \text{ or } w_r = \text{prod}(\mu_{A_i}, \mu_{B_j}) \text{ for } r = 1, 2 \qquad (2.45)$$

When $z = f(x, y)$ is a constant, it is called a zero-order Sugeno fuzzy model. The rules of a zero-order Sugeno fuzzy model are as follows:

$$\text{r1 : IF } x \text{ is } A_1 \text{ AND } y \text{ is } B_1 \text{ THEN } z_1 = c_1$$
$$\text{r2 : IF } x \text{ is } A_2 \text{ AND } y \text{ is } B_2 \text{ THEN } z_2 = c_2$$

where c_1 and c_2 are constant values. This can be considered as a special case of the Mamdani fuzzy model, in which the consequent of each rule is specified by a fuzzy singleton or by a pre-defuzzified value of the consequent fuzzy set of a Mamdani-type fuzzy system or a special case of the Tsukamoto fuzzy model (described in Section 2.13.3) in which the consequent of each rule is specified by an MF of a step function.

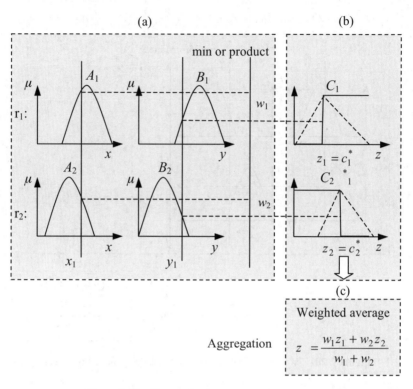

Figure 2.29   Zero-order Sugeno-type fuzzy system

The two special cases for the consequent MFs, triangular and step function, of the zero-order Sugeno system are shown in Figure 2.29. c_1^* and c_2^* are the defuzzified values of the MFs. The output of the zero-order Sugeno model is a smooth function of its input variables when there is a good overlap between the neighbouring MFs in the antecedent part. In other words, the overlap of the MFs in the consequent of a Mamdani model does not have a decisive effect on the smoothness of the output. It is the overlap of the antecedent MFs that determines the smoothness of the resulting input/output behaviour of the fuzzy system (Jang, 1993; Jang and Sun, 1993).

Once the parameters $\{a_k, b_k, c_k\}$, $k = 1, 2, \ldots, R$ are known, the consequent z_k are calculated for each rule using a first-order polynomial. Figure 2.28(b) shows the consequent part of the Sugeno-type system where z_1 and z_2 are computed. An analogy of the Mamdani-type system described in the previous section will be helpful for understanding this process.

The overall output of a Sugeno fuzzy model is obtained via the weighted average of the crisp outputs z_k, thus avoiding the time-consuming process of defuzzification required by a Mamdani model using the centre of gravity method. The weights are the firing strengths of each rule calculated in Equation (2.45). The weighted average defuzzification is computed by

$$z = \frac{w_1 z_1 + w_2 z_2}{w_1 + w_2}$$

Figure 2.28(c) illustrates the aggregation and final defuzzified value for the Sugeno-type system. In practice, the weighted average operator is sometimes replaced with the weighted sum operator defined by

$$z = w_1 z_1 + w_2 z_2$$

The weighted sum helps reduce further computation, especially in the training of a fuzzy inference system. However, this simplification could lead to the loss of MF linguistic meaning unless the sum of firing strengths is close to unity (Jang *et al.*, 1997).

2.13.3 Tsukamoto Fuzzy Inference

In the Tsukamoto fuzzy inference, the consequent of each fuzzy if–then rule is represented by a monotonic MF (Tsukamoto, 1979). A typical rule in the Tsukamoto-type fuzzy model for two-inputs single-output is described by a rule set of the form

$$k : \text{If } x \text{ is } A_i^k \text{ and } y \text{ is } B_j^k \text{ then } z \text{ is } C_l^k \qquad (2.46)$$

The system consists of two inputs x and y (antecedents) and a single output z (consequent), where $k = 1, 2, \ldots, R$, $i = 1, 2, \ldots, N$, $j = 1, 2, \ldots, M$ and $l = 1, 2, \ldots, L$. N, M and L are the numbers of membership functions for inputs and output, respectively. The maximum number of rules of the Tsukamoto-type fuzzy system is $R \subset N \times M$. Figure 2.30(a) illustrates

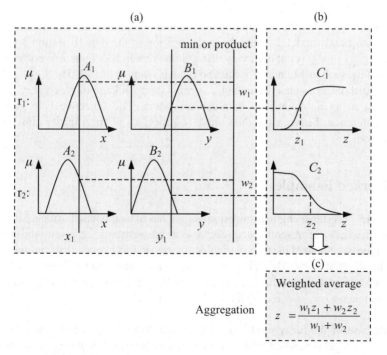

Figure 2.30 Two-input single-output Tsukamoto fuzzy model

a two-input single-output Tsukamoto fuzzy model. In this case, each input x and y has two MFs $\{A_1, A_2\}$ and $\{B_1, B_2\}$ while $\{C_1, C_2\}$ are consequent monotonic functions. There are only two rules used to demonstrate the inferencing mechanism in Figure 2.30(a). These are:

$$r1 : \text{IF } x \text{ is } A_1 \text{ AND } y \text{ is } B_1 \text{ THEN } z \text{ is } C_1$$
$$r2 : \text{IF } x \text{ is } A_2 \text{ AND } y \text{ is } B_2 \text{ THEN } z \text{ is } C_2$$

In a Tsukamoto-type fuzzy model, measured crisp values x_1 and y_1 are used for fuzzification. Fuzzification and inferencing using the minimum or product rule for computing the firing strengths w_r, $r = 1, 2$ for the rules with ANDed antecedent are shown in Figure 2.30(a). The firing strength is calculated using the minimum or product rule as

$$w_r = \Gamma\left(\mu_{A_i}, \mu_{B_j}\right) \text{ for } r = 1, 2 \tag{2.47}$$

$\Gamma(.)$ is a minimum or product operation in Equation (2.47). If a Tsukamoto fuzzy model consists of R rules with firing strengths $w_1, w_2, \ldots, w_R$ then the defuzzified outputs will be $z_1, z_2, \ldots, z_R$ (one z value for each rule), as shown in Figure 2.30(b). As shown in the figure, it thus makes the computation of defuzzification simple. The overall output is taken as the weighted average of each rule's output. Since each rule infers a crisp output, the Tsukamoto fuzzy model aggregates each rule's output by the method of weighted averages. The aggregated output of the fuzzy system will be

$$z = \frac{w_1 z_1 + w_2 z_2 + \cdots + w_R z_R}{w_1 + w_2 + \cdots + w_R} \tag{2.48}$$

Figure 2.30(c) illustrates the aggregation and defuzzification of the consequent outputs of the Tsukamoto fuzzy model. Despite the simplification of the defuzzification procedure, the Tsukamoto fuzzy model is not used very often. Some researchers think it is not as transparent as other models, such as Mamdani or Sugeno models (Jang *et al.*, 1997).

In the next three examples, Mamdani-, Sugeno- and Tsukamoto-type fuzzy models will be explained using a hypothetical simulation system. The objective here is to demonstrate the differences between the three models, processes of fuzzification, inferencing and defuzzification.

2.14   Worked Examples

Example 2.15   A steam engine simulation system has to be modelled using a Mamdani-type fuzzy system, where x represents speed, y represents pressure and z represents throttle position. No units of measurement of the three variables are used in this hypothetical simulation system. The membership functions (MF) for speed x, pressure y and throttle position z, defined within the same universe of discourse [0, 15], are shown in Figure 2.31(a–c). For each of the variables, the MFs are taken to be low (L), medium (M) and high (H).

The rule base of the fuzzy model consists of nine rules, as given in Table 2.2. A single iteration of graphical simulation of the steam engine is described in this example for the initial values $x = 9$ and $y = 10$.

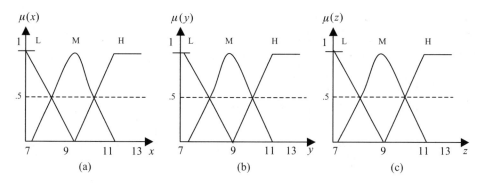

Figure 2.31   MFs for x, y and z. (a) MFs for speed; (b) MFs for pressure; (c) MFs for throttle position

For initial values $x = 9$ and $y = 10$, only two rules are fired. These are:

$$R1 : IF\ x \text{ is } M \text{ and } y \text{ is } M \text{ THEN } z \text{ is } \mathbf{M}$$
$$R2 : IF\ x \text{ is } M \text{ and } y \text{ is } H \text{ THEN } z \text{ is } \mathbf{H}$$

Figure 2.32(a) shows the fuzzification of the two input values using the antecedent MFs shaded in grey and inferencing using the min rule in the antecedent part, which yields the output MF in the consequent part (shaded area) for rule 1. Similarly, Figure 2.32(b) shows the fuzzification and inferencing using the min rule that yielded the output MF (shaded area) for rule 2. The two consequent MFs (i.e., shaded M and H) are aggregated and defuzzified using the centre of gravity method illustrated in Figure 2.32(c). This gives the throttle position of 9.8.

Example 2.16   The same steam engine simulation system has been developed using a Takagi–Sugeno fuzzy model, where x represents speed, y represents pressure and z represents throttle position. Two membership functions for speed x and pressure y, defined within the same universe of discourse [0, 15], are shown in Figure 2.33(a–b). For variable x, the MFs are taken to be $\{A_1, A_2\}$ and for variable y, the MFs are taken to be $\{B_1, B_2\}$. The throttle position z is defined by the four first-order polynomial functions below:

$$z_1 = 3x + 2y + 1$$
$$z_2 = x + 3y + 1$$
$$z_3 = x + 2y$$
$$z_4 = 2x + 5$$

Table 2.2   Rule base for a Mamdani-type fuzzy model

	y		
x	L	M	H
L	H	M	L
M	H	**M**	**H**
H	H	M	L

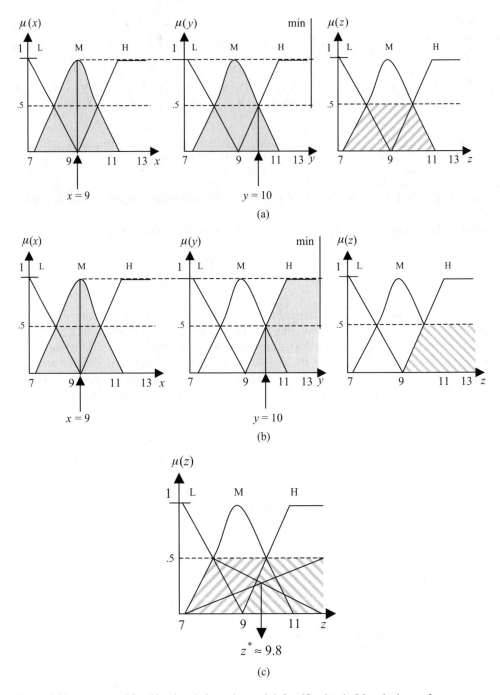

Figure 2.32 Process of fuzzification, inferencing and defuzzification in Mamdani-type fuzzy system. (a) Rule 1: IF x is M and y is M THEN z is M; (b) Rule 2: IF x is M and y is H THEN z is H; (c) Aggregated and defuzzified output using COG method

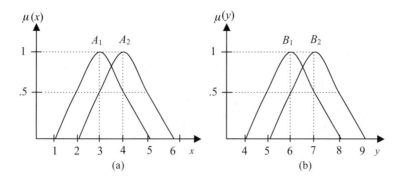

Figure 2.33 MFs for inputs x and y. (a) MFs for x; (b) MFs for y

The rule base of the fuzzy model consists of four rules, as given in Table 2.3.

A single iteration of graphical simulation of the Sugeno-type fuzzy model of the steam engine is described in this example for initial values $x = 3$ and $y = 7$. For initial values $x = 3$ and $y = 7$, all four rules are fired. Each fired rule is shown in Figure 2.34(a–d).

Figure 2.34(a) shows the fuzzification of the two input values using the antecedent MFs shaded in grey and the inferencing using the min rule in the antecedent part, which yields the firing strength $w_1 = 0.5$. The polynomial function z_1 of the consequent part of rule 1 is evaluated using the input values $x = 3$ and $y = 7$, which yields $z_1 = 24$ and is shown in the shaded area on the right. In the same way, the firing strengths w_2, w_3 and w_4 and the consequent polynomial functions z_2, z_3 and z_4 are computed for rules 2, 3 and 4. Figure 2.34(b–d) shows the firing strengths $w_2 = 1$, $w_3 = 0.5$ and $w_4 = 0.5$ and the values of the consequent polynomial functions $z_2 = 25$, $z_3 = 17$ and $z_4 = 11$ for the rules 2, 3 and 4, respectively.

The consequent polynomial function values $z_1 = 24$, $z_2 = 25$, $z_3 = 17$ and $z_4 = 11$ for all four rules are aggregated by appropriate weights. The weights are the firing strengths calculated for each rule. The final throttle position is calculated using the weighted average of defuzzification shown below. It gives a throttle position of 20.4.

$$z = \frac{w_1 z_1 + w_2 z_2 + w_3 z_3 + w_4 z_4}{w_1 + w_2 + w_3 + w_4}$$

$$z = \frac{0.5 * 24 + 1 * 25 + 0.5 * 17 + 0.5 * 11}{0.5 + 1 + 0.5 + 0.5}$$

$$z = \frac{12 + 25 + 8.5 + 5.5}{2.5}$$

$$z = \frac{51.0}{2.5} = 20.4$$

Table 2.3 Rule base for the Takagi–Sugeno-type fuzzy model

		y	
x		B_1	B_2
A_1		z_1	z_2
A_2		z_3	z_4

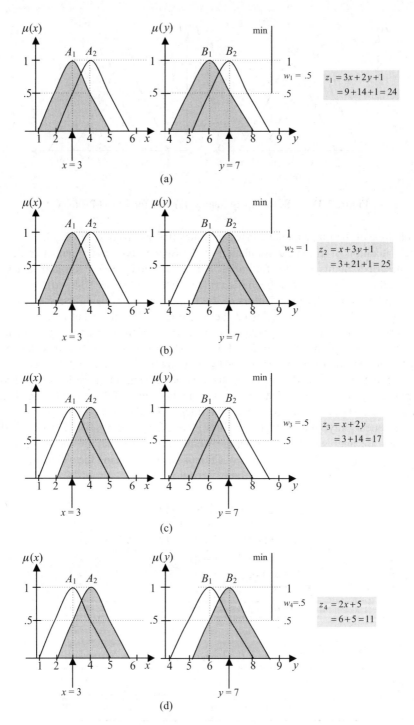

Figure 2.34   Fired rules in Sugeno-type fuzzy system. (a) Rule 1: IF x is A_1 and y is B_1 THEN z is $z_1 = 3x + 2y + 1$; (b) Rule 2: IF x is A_1 and y is B_2 THEN $z_2 = x + 3y + 1$; (c) Rule 3: IF x is A_2 and y is B_1 THEN $z_3 = x + 2y$; (d) Rule 4: IF x is A_2 and y is B_2 THEN $z_4 = 2x + 5$

Table 2.4 Rule base for the Tsukamoto-type fuzzy model

	y	
x	B_1	B_2
A_1	C_1	C_2
A_2	C_2	C_3

It is to be noted that if the product rule is used for calculating the firing strengths, a different throttle position would have been reached.

Example 2.17   The same steam engine simulation system has been developed using a Tsukamoto-type fuzzy model, where x represents speed, y represents pressure and z represents throttle position. Two membership functions for speed x, pressure y and three membership functions for throttle position z are defined within the same universe of discourse [0, 15]. For variable x, the MFs are taken to be $\{A_1, A_2\}$. For variable y, the MFs are taken to be $\{B_1, B_2\}$. For variable z, the MFs are taken to be $\{C_1, C_2, C_3\}$. There are four rules that describe the relationship between the inputs and output of the fuzzy model. Table 2.4 describes the rule base of the system.

The MFs for speed (x) and pressure (y) are described by the membership functions shown in Figure 2.35(a–b). The throttle positions are defined by three monotone membership functions C_1, C_2 and C_3 as shown in Figure 2.35(c).

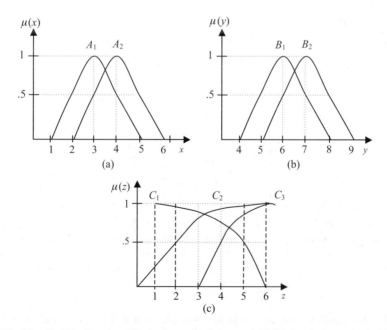

Figure 2.35   Membership functions for the inputs and output. (a) MFs for the input x; (b) MFs for the input y; (c) MFs for the output z

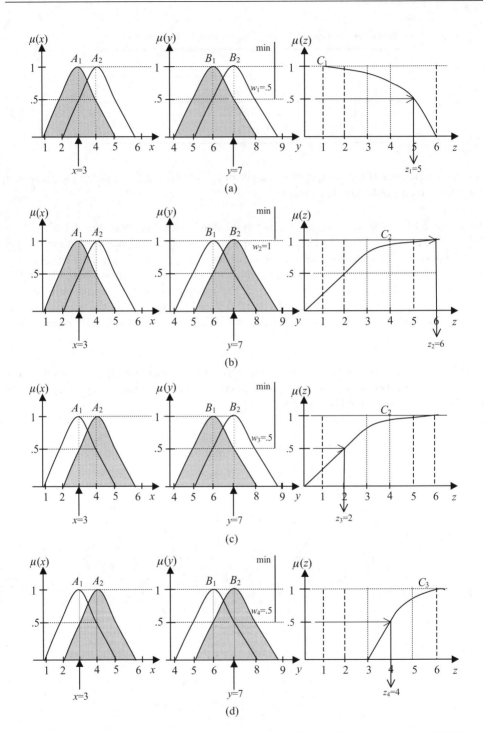

Figure 2.36   Fired rules in Tsukamoto-type fuzzy system. (a) Rule 1: IF x is A_1 and y is B_1 THEN z is C_1; (b) Rule 2: IF x is A_1 and y is B_2 THEN z is C_2; (c) Rule 3: IF x is A_2 and y is B_1 THEN z is C_2; (d) Rule 4: IF x is A_2 and y is B_2 THEN z is C_3

It is required to compute the throttle position z at speed $x = 3$ and pressure $y = 7$ using the Tsukamoto fuzzy model described above. All fired rules are shown graphically in Figure 2.36(a–d).

Figure 2.36(a) shows the fuzzification of the two input values $x = 3$ and $y = 7$ using the antecedent MFs shaded in grey and the inferencing using the product rule in the antecedent part of the rule, which yields the firing strength $w_1 = 0.5$. Using this membership grade of 0.5, it then evaluates the consequent membership function C_1 to produce the defuzzified value of $z_1 = 5$. This is shown in the consequent part for rule 1 in the figure. In the same way, the firing strengths w_2, w_3 and w_4 are calculated and the defuzzified values of the consequent monotone functions z_2, z_3 and z_4 are computed for rules 2, 3 and 4. Figure 2.36(b–d) shows the firing strengths $w_2 = 1$, $w_3 = 0.5$ and $w_4 = 0.5$ and the values of the consequent functions $z_2 = 6$, $z_3 = 2$ and $z_4 = 4$ for rules 2, 3 and 4, respectively.

The consequent values $z_1 = 5$, $z_2 = 6$, $z_3 = 2$ and $z_4 = 4$ for all four rules are aggregated by appropriate weights. The weights are the firing strengths calculated for each rule. The final throttle position is calculated using the weighted average of defuzzification according to Equation (2.48) shown below. It gives a throttle position of 4.6.

$$z = \frac{w_1 z_1 + w_2 z_2 + w_3 z_3 + w_4 z_4}{w_1 + w_2 + w_3 + w_4}$$

$$z = \frac{0.5 * 5 + 1 * 6 + 0.5 * 2 + 0.5 * 4}{0.5 + 1 + 0.5 + 0.5}$$

$$z = \frac{2.5 + 6 + 1 + 2}{2.5} = \frac{11.5}{2.5} = 4.6$$

It is to be noted that if the minimum rule of inference is used for calculating the firing strengths, the same firing strengths will result in the same throttle position.

2.15 MATLAB® Programs

The Fuzzy Logic Toolbox in MATLAB® provides tools to create and edit fuzzy membership functions, define fuzzy variables, create fuzzy inference systems and create rule bases within the MATLAB® platform. Fuzzy systems can also be integrated with Simulink® for simulation. The toolbox provides three categories of tools:

- Command-line functions,
- GUI interface tools and
- Simulink® blocks.

In this chapter, only command-line functions are used. Few MATLAB® programs are provided to illustrate construction of membership functions, fuzzy inferencing, creating rule bases, defuzzification and simulation of fuzzy inference systems described in this chapter. Details of codes with associated descriptions and plots are given in Appendix B.

References

Black, M. (1937) Vagueness: an exercise in logical analysis, *International Journal of General Systems*, 17, 107–128.

Chi, Z., Yan, H. and Phan, T. (1996) *Fuzzy Algorithms: With Applications to Image Processing and Pattern Recognition*, World Scientific, Singapore.

Driankov, D., Hellendoorn, H. and Reinfrank, M. (1993) *An Introduction to Fuzzy Control*, Springer-Verlag, Heidelberg.

Einstein, A. (1921) Geometrie und Erfahrung, Offentlische Sitzung 27. Januar, 1921, Sitzungsbericte der Koeniglische Presse, Akademie der Wissenschaften, Berlin, pp. 123–130.

Eisele, C. (1979) *Studies in the Scientific and Mathematical Philosophy of Charles S. Peirce*, R.M. Martin (ed.), Mouton, The Hague.

Goedel, K. (1932) Zum Intuitionistischen Aussagenkalkuel, *Anzeiger Akademie der Wissenschaften Wien, Math.-naturwissenschaft Klasse*, Vol. 69, pp. 65–66.

Gottwald, S. (2001) *A Treatise on Many-Valued Logics*, Studies in Logic and Computation, Vol. 9, Research Studies Press, Baldock.

Jang, J.-S.R. (1993) ANFIS: adaptive-network-based fuzzy inference systems, *IEEE Transactions on Systems, Man and Cybernetics*, 23(3), 665–685.

Jang, J.-S.R. and Sun, C.-T. (1993) Functional equivalence between radial basis function networks and fuzzy inference system, *IEEE Transactions on Neural Networks*, 4, 156–159.

Jang, J.-S.R., Sun, C.-T. and Mizutani, E. (1997) *Neuro-Fuzzy and Soft Computing: A Computational Approach to Learning and Machine Intelligence*, Prentice-Hall, Upper Saddle River, NJ.

Jaskowski, S. (1936) Recherches sur le Systeme de la Logique Intuitioniste, Actes du Congres Internationale de Philosophie Scientifique, Paris, Vol. 6, pp. 58–61 [English translation: *Studia Logica*, 34, 1975, pp. 117–120].

Kickert, W.J.M. and Mamdani, E.H. (1978) Analysis of fuzzy logic controller, *Fuzzy Sets and Systems*, 1, 29–44.

Lancaster, S.S. and Wierman, M.J. (2003) Empirical study of defuzzification. In *Proceedings of the International Conference of the North American Fuzzy Information Processing Society*, pp. 121–126.

Lukasiewicz, J. (1930) Philosophische Bemerkungen zu Mehrwertigen Systemen des Aussagenkalkuels, *Comptes Rendus Sieances Societe des Sciences et Lettres Varsovie*, cl. III, 23, 51–77.

Lukasiewicz, J. and Tarski, A. (1930) Untersuchungen ueber den Aussagenkalkuel, *Comptes Rendus Sieances Societe des Sciences et Lettres Varsovie*, cl. III, 23, 30–50.

Mamdani, E.H. and Assilian, S. (1974) Application of fuzzy algorithms for control of simple dynamic plant, *Proceedings of IEEI*, 121, 1585–1588.

Patel, A. and Mohan, B. (2002) Some numerical aspects of centre of area defuzzification method, *Fuzzy Sets and Systems*, 132, 401–409.

Post, E.L. (1921) Introduction to a general theory of elementary propositions, *American Journal of Mathematics*, 43, 163–185.

Price, R. (1763) An essay towards solving a problem in the doctrine of chances by the late Rev. Mr. Bayes, *Philosophical Transactions of the Royal Society of London*, pp. 370–418.

Rolf, B. (1982) Russell's Theses on Vagueness, *History and Philosophy of Logic*, Vol. 3, pp. 68–83.

Roychowdhury, S. and Pedrycz, W. (2001) A survey of defuzzification strategies, *International Journal of Intelligent Systems*, 16, 679–695.

Runker, T.A. (1997) Selection of appropriate deffuzification methods using application specific properties, *IEEE Transactions on Fuzzy Systems*, 5(1), 72–79.

Sugeno, M. and Kang, G.T. (1988) Structure identification of fuzzy model, *Fuzzy Sets and Systems*, 28, 15–33.

Takagi, T. and Sugeno, M. (1985) Fuzzy identification of systems and its application to modeling and control, *IEEE Transactions on Systems, Man and Cybernetics*, 15, 116–132.

Tsukamoto, Y. (1979) An approach to fuzzy reasoning method. In *Advances in Fuzzy Set Theory and Applications*, M.M. Gupta, R.K. Ragade and R. Yager (eds), North-Holland, Amsterdam, pp. 137–149.

Van Broekhoven, E. and De Baets, B. (2004) A comparison of three methods for computing the centre of gravity defuzzification. In *Proceedings of the 2004 IEEE International Conference on Fuzzy Systems*, pp. 1537–1542.

Van Leekwijck, W. and Kerre, E. (1999) Defuzzification: criteria and classification, *Fuzzy Sets and Systems*, 108(2), 159–178.

Wang, L.-X. (1997) *A Course in Fuzzy Systems and Control*, Prentice-Hall International, Upper Saddle River, NJ.

Yager, R.R. and Filev, D.P. (1994) *Essential of Fuzzy Modelling and Control*, John Wiley & Sons, New York.

Zadeh, L.A. (1965) Fuzzy sets, *Information and Control*, 8, 338–353.

Zadeh, L.A. (1968) Fuzzy algorithms, *Information and Control*, 12, 94–102.

Zadeh, L.A. (1972) A fuzzy-set-theoretic interpretation of linguistic hedges, *Journal of Cybernetics*, 2, 4–34.

Zadeh, L.A. (1973) Outline of a new approach to the analysis of complex systems and decision process, *IEEE Transactions on System, Man and Cybernetics*, 3, 28–44.

Zadeh, L.A. (1975a) The concept of linguistic variable and its application to approximate reasoning – I, *Information Sciences*, 8, 199–249.

Zadeh, L.A. (1975b) The concept of linguistic variable and its application to approximate reasoning – II, *Information Sciences*, 8, 301–357.

Zadeh, L.A. (1976a) The concept of linguistic variable and its application to approximate reasoning – III, *Information Sciences*, 9, 43–80.

Zadeh, L.A. (1976b) The linguistic approach and its application to decision analysis. In *Directions in Large Scale Systems*, Y.C. Ho and S.K. Mitter (eds), Plenum Press, New York, pp. 339–370.

Zadeh, L.A. (1999) From computing with numbers to computing with words – from manipulation of measurements to manipulation of perceptions, *IEEE Transactions on Circuits and Systems – I: Fundamental Theory and Applications*, 45(1), 105–119.

3

Fuzzy Systems and Applications

3.1 Introduction

The fundamental limitation of traditional mathematics, its tools and techniques is that they cannot cope with humanistic or biological systems (Zadeh, 1999). Some examples of such humanistic systems are economic systems, biological systems, social systems, political systems and, more generally, man-made systems of various types. In other words, the conventional quantitative approaches of system analysis and modelling are intrinsically unsuited for dealing with humanistic systems or any system whose complexity is comparable to that of humanistic systems. Thus, to deal with such systems realistically, we need approaches that are not obsessive about precision and rigorous mathematical formalisms. The alternative approach to traditional notions of systems is based on Zadeh's fuzzy sets and linguistic variables, which bear an approximate relation to primary data. Fuzzy systems are those whose inputs and outputs are described by fuzzy variables and fuzzy relations. The seminal ideas of fuzzy systems can be found in the early papers of Zadeh (1968, 1971, 1972, 1973). Since then, fuzzy logic has found applications in system identification (Zadeh, 1994; Espinosa *et al.*, 2005), modelling (Yager and Filev, 1994; Zadeh, 1994; Shin and Xu, 2009), control (Zadeh, 1994; Wang, 1997; Kovacic and Bogdan, 2006), clustering (Hoeppner *et al.*, 1997; Oliveira and Pedrycz, 2007), image processing (Bezdek *et al.*, 1999) and many others. This chapter will present the foundations of fuzzy systems, the manner in which fuzzy logic, linguistic variables, fuzzy inferencing and fuzzy theory are applied to the formulation and solution of real-world problems, along with examples drawn from current research in the field of computational intelligence.

There are two justifications for fuzzy systems. Firstly, the real world is too complicated and complex. It has been demonstrated that the use of highly complex mathematical description of systems can seriously inhibit the ability to develop system models. Furthermore, it is required to cope with significant unmodelled and unanticipated changes in the operating environment. Fuzziness was introduced to obtain a reasonable model of such systems. Secondly, dissatisfaction is growing with conventional approaches and human knowledge is becoming more and more important in modelling and control of real-world systems. The fuzzy system concept can formulate knowledge in a systematic manner and give it an engineering flavour. A fuzzy system can combine the available information, such as the knowledge of the human expert, the mathematical descriptions and sensory measurements, effectively. The combination of such

---

Computational Intelligence: Synergies of Fuzzy Logic, Neural Networks and Evolutionary Computing, First Edition.
Nazmul Siddique and Hojjat Adeli.
© 2013 John Wiley & Sons, Ltd. Published 2013 by John Wiley & Sons, Ltd.

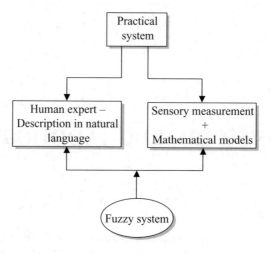

Figure 3.1 Example of a practical system

information is illustrated in Figure 3.1. The primary goal of this chapter is to introduce fuzzy systems, modelling and control based on the foundation of the previous chapter.

3.2 Fuzzy System

Let $u(t)$, $y(t)$ and $x(t)$ denote the input, output and state of a system S at time t, respectively, where $t = -1, 0, 1, 2, 3, \ldots$ The state equations of the system S can be defined as

$$x(t + 1) = f(x(t), u(t)) \tag{3.1}$$

$$y(t) = g(x(t), u(t)) \tag{3.2}$$

where f and g are mappings defined by $f : X \times U \rightarrow X$ and $g : X \times U \rightarrow Y$, respectively. The system S is a fuzzy system if $u(t)$ or $y(t)$ or $x(t)$ or any combination of them ranges over fuzzy sets. Similarly, even if the state $x(t)$ is described in terms of adjectives or linguistic hedges such as light, heavy, more or less light, very heavy, etc., then the system S is still a fuzzy system. If, for any given $\{x(t), u(t)\}$, $X(t + 1)(x(t), u(t))$ and $Y(t + 1)(x(t), u(t))$ denote the sets of values of $x(t + 1)$ and $y(t)$, respectively, then the state equation of the fuzzy system S defined above can be rewritten as

$$X(t + 1) = F(x(t), u(t)) \tag{3.3}$$

$$Y(t) = G(x(t), u(t)) \tag{3.4}$$

Extending the classical definition of a system model, we arrive at fuzzy models using the apparatus of fuzzy set theory. Thus, the system model described by Equations (3.3) and (3.4) is to be called a fuzzy system model.

3.3   Fuzzy Modelling

The general purpose of a model is to describe the functioning of a system in terms of input/output behaviour. Traditional techniques of system modelling have significant limitations. In many cases it is difficult to describe the system behaviour by a set of mathematical equations when the system is nonlinear and partially known or unknown. Moreover, there are many uncertainties and unpredictable dynamics that do not allow the system model to be described mathematically. Such uncertainties and unpredictable behaviour in complicated and ill-defined systems can be modelled using the linguistic approach as a model of human thinking, which introduced fuzziness into systems theory (Zadeh, 1965, 1973). Therefore, fuzzy system modelling is an important issue while control of such systems is of concern. There are many interpretations of fuzzy system modelling. For instance, a fuzzy set is a fuzzy model of a human concept. In this study, a fuzzy system model is understood as an approach to form a system model using a descriptive language based on fuzzy logic with fuzzy predicates. In other words, fuzzy models consist of linguistic explanations of system behaviour. Apart from fuzzy control, there are many studies on fuzzy modelling. These are divided into two groups. The first group deals with a fuzzy model of the system itself or a fuzzy model for simulation (Tong, 1980; Pedrycz, 1984; Filev, 1991; Pedrycz and Gomide, 1998). The second group deals with fuzzy modelling of a plant for control (Czogala and Pedrycz, 1981; Takagi and Sugeno, 1985; Lygeros, 1996). Just as in modern control theory, a fuzzy controller can be designed based on a fuzzy model of a plant if the fuzzy model can be identified (Sugeno and Yasukawa, 1993; Gilachet and Foulloy, 1995) with appropriate structure and variables.

In fuzzy system modelling, the identification method used is very important (Emami *et al.*, 1998). Identification for fuzzy modelling involves two distinct aspects:

- Structure identification and
- Parameter identification.

In general, structure identification is a difficult and extremely ill-defined process and not readily amenable to automated techniques. The problem of parameter identification is closely related to the estimation of the membership functions and parameters of the membership functions, or alternatively the fuzzy relation associated with the fuzzy model.

3.3.1   Structure Identification

Generally, structure identification constitutes two problems.

- To find input/output variables from a number of input/output candidates by a heuristic method based on experience and/or common sense.
- To find input/output relations in the form of if–then rules.

In a fuzzy model, the structure identification of this kind is stated in a different way. A fuzzy model consists of a number of if–then rules. The number of rules, n, in a fuzzy model corresponds to the order of the model in a conventional method (Jang, 1994). There are two

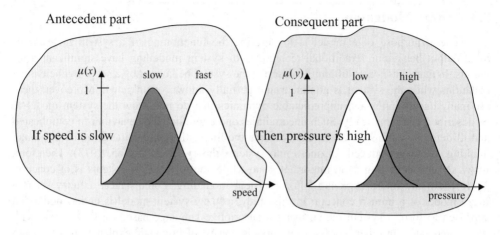

Figure 3.2 Antecedent and consequent parts of the if–then rule

parts in an if–then rule. Figure 3.2 shows an example of an if–then rule. The 'if' part of the rule is called the antecedent and the 'then' part is called the consequent.

The antecedent of a fuzzy rule defines a local fuzzy region, while the consequent describes the behaviour within the region via various constituents. The consequent constituents can be consequent membership functions (e.g., MFs in a Mamdani or Tsukamoto fuzzy model), a constant value (e.g., in a zero-order Sugeno model) or a linear equation (e.g., in a first-order Sugeno model). Different consequent constituents result in different fuzzy inference systems, but their antecedents are always the same and determine the model order by partitioning the input space.

For example, let us consider the fuzzy model with the following rules:

$$\text{If } x_1 \text{ is } A_1, \text{ then } y \text{ is } C_1$$
$$\text{If } x_1 \text{ is } A_2 \text{ and } x_2 \text{ is } B_1, \text{ then } y \text{ is } C_2$$
$$\text{If } x_1 \text{ is } A_2 \text{ and } x_2 \text{ is } B_2, \text{ then } y \text{ is } C_3$$

The inputs x_1 and x_2 in the above model are partitioned into three subspaces and the number of rules corresponds to the number of subspaces. Different ways of partitioning the two-dimensional input space are shown in Figure 3.3.

For example, let us consider the fuzzy model with three inputs and a single output defined by the following rules:

$$\text{If } x_1 \text{ is } A_2 \text{ and } x_2 \text{ is } B_1 \text{ and } x_3 \text{ is } C_1, \text{ then } z \text{ is } D_2$$
$$\text{If } x_1 \text{ is } A_2 \text{ and } x_2 \text{ is } B_2 \text{ and } x_3 \text{ is } C_2, \text{ then } z \text{ is } D_3$$

The inputs x_1, x_2 and x_3 of the above model are partitioned into three subspaces and the number of rules corresponds to the number of subspaces. Different ways of partitioning the three-dimensional input space are shown in Figure 3.4.

The problem in hand is combinatorial in nature and hence a heuristic rule can be applied to find an optimal partition together with a criterion.

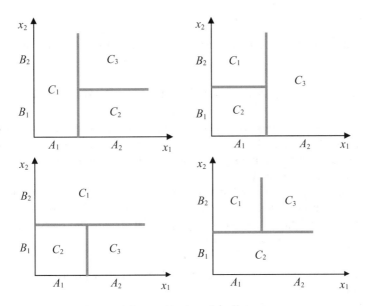

Figure 3.3 Partitioning of the input space

Conceptually, fuzzy modelling can be pursued in two stages, which are not totally disjoint. The first stage is the identification of the structure, which includes the following tasks:

- selection of relevant inputs/outputs;
- choice of a specific type of fuzzy inference system, e.g. Mamdani, Sugeno or Tsukamoto;
- determining the number of linguistic terms associated with each input and output variable;
- generating a set of fuzzy if–then rules.

To accomplish these tasks, the designer mainly relies on knowledge such as common sense and simple physical laws of the system, information provided by human experts or operators or simply by trial and error. After the first stage of fuzzy modelling, a rule base is obtained that describes the behaviour of the system in terms of linguistic variables.

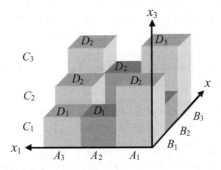

Figure 3.4 Three-dimensional input space partitioning

3.3.2   Parameter Identification

In ordinary system identification, parameters are the coefficients in a functional system model. For example

$$y(k+1) = a_0 x(k) + a_1 x(k-1) + a_2 x(k-2) + b_0 y(k) + b_1 y(k-1) \qquad (3.5)$$

where a_0, a_1, a_2, b_0, b_1 are parameters of the system model to be estimated. In fuzzy modelling, the parameters are those in the membership functions of the fuzzy sets. In fuzzy modelling, parameter identification is merely an optimization problem with a conventional criterion or objective function, such as output error.

Determination of the meaning of linguistic variables is referred to as identification of the deep structure. The deep structure determines the membership functions of each linguistic variable (the coefficient of each rule's output polynomial in a Sugeno-type fuzzy model). The identification of the deep structure includes:

- choosing an appropriate family of parameterized MFs (e.g., triangular, trapezoidal, Gaussian or bell-shaped);
- applying heuristic selection or interviewing human operators to determine the parameters of the MFs;
- refining the parameters of the MFs using suitable optimization techniques.

The first two tasks above can be achieved with a human expert, but for the third task an input/output data set will be required or a tuning approach should be adopted. Various parameter identification and optimization techniques can be used, such as least-square methods, derivative-based methods and derivative-free methods (e.g., evolutionary algorithms). Some of these are discussed in Chapter 8.

3.3.3   Construction of Parameterized Membership Functions

Choosing membership functions is the first and essential step in parameter identification of fuzzy modelling. A convenient and concise way to construct an MF is to parameterize it and then express the MF mathematically in terms of parameters. For example, the MF shown in Figure 3.5(a) may be a good representation of the input space and may look attractive, but it will be difficult to express mathematically in terms of few parameters, which leads to complications in model development. The full envelop of the input space in Figure 3.5(a) can be described by parametric MFs such as the trapezoidal function, as shown in Figure 3.5(b).

A trapezoidal MF is specified by four parameters $\{a, b, c, d\}$ as shown in Figure 3.6(a) and defined by

$$\mu(x) = \max\left(\min\left(\frac{x-a}{b-a}, 1, \frac{d-x}{d-c}\right), 0\right) \qquad (3.6)$$

The parameters $\{a, b, c, d\}$ with $a < b < c < d$ determine the x coordinates of the four corners of the underlying trapezoidal MF.

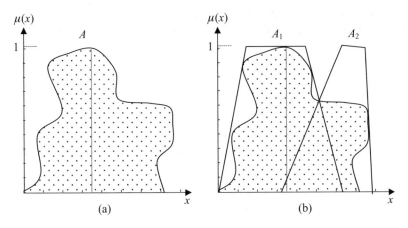

Figure 3.5 Hypothetical MF and parametric MF. (a) MF representing envelop of input space; (b) Parametric representation of MFs

A triangular MF is specified by three parameters $\{a, b, c\}$ as shown in Figure 3.6(b) and defined by

$$\mu(x) = \max\left(\min\left(\frac{x-a}{b-a}, \frac{c-x}{c-b}\right), 0\right) \tag{3.7}$$

The parameters $\{a, b, c\}$ with $a < b < c$ determine the x coordinates of the three corners of the underlying triangular MF. Owing to the simple formulae and computational efficiency, both triangular and trapezoidal MFs have been popular and used extensively, especially in real-time applications.

A Gaussian MF is specified by two parameters $\{m, \sigma\}$ and defined by

$$\mu(x) = \exp\left[-\frac{1}{2}\left(\frac{x-m}{\sigma}\right)^2\right] \tag{3.8}$$

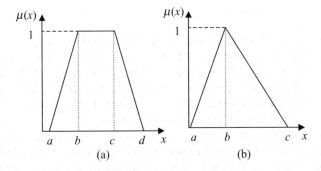

Figure 3.6 Parameters of MFs. (a) Trapezoidal MF; (b) Triangular MF

where the parameters m and σ represent the centre and width of the Gaussian MF, respectively. A bell-shaped MF is specified by three parameters $\{m, \sigma, a\}$ and defined by

$$\mu(x) = \frac{1}{1 + \left| \dfrac{x - m}{\sigma} \right|^{2a}} \tag{3.9}$$

The parameters m and σ represent the centre and width of the bell-shaped MF, respectively. Parameter a, usually positive, controls the slope of the MF at the crossover point. A sigmoidal MF has two parameters and is defined by

$$\mu(x) = \frac{1}{1 + \exp\left[-a\left(x - c\right)\right]} \tag{3.10}$$

where c is the centre and the parameter a controls the slope of the MF. The sign of the parameter a determines the open-end direction of the MF.

Gaussian and bell-shaped MFs achieve smoothness but they are unable to specify asymmetric MFs, which are important in certain applications. A sigmoidal MF, which is either open left or right, can be used in such applications. There are other mixed types of membership functions in use, such as difference sigmoidal, product sigmoidal, Π-shaped, Z-shaped and S-shaped. A detailed description of the different types of parameterized MFs is provided in Section 2.4 of Chapter 2.

The shape of the MFs eventually represents the fuzziness of the variables that describe the fuzzy system. Therefore, the shape of the MFs, their number and distribution influence the performance of fuzzy systems and controllers (Kovacic and Bogdan, 2006). Many techniques have been proposed which reflect the actual data distribution and using learning algorithms where some input/output data are available. To further enhance system performance, the generated membership functions can be tuned using techniques such as gradient descent with neural networks or optimized using evolutionary algorithms for instance. The literature is quite rich on this topic. Some of the techniques will be discussed later in this book. The different approaches to construct and tune parameterized membership functions in use can be categorized into groups such as (i) heuristic selection, (ii) clustering approach, (iii) neural networks and (iv) evolutionary algorithms. A detailed description of these approaches can be found in Chi *et al.* (1996) and Ross (2004).

(i) Heuristic selection

Heuristic selection of parameters of membership functions is widely used and practised in fuzzy modelling and applications. This involves common-sense knowledge, application of physical laws or general information about the system. For example, Figure 3.7 shows various shapes in the universe of temperature. The parameters are chosen heuristically if the MFs are referred to the ranges of human comfort. If the MFs are defined for a steam engine, we get a different range of values as parameters.

If data are available, parameters can be chosen from the data distribution heuristically. For example, Figure 3.8(a) shows a fairly linear distribution of available data. Inspecting the data, two centres $\{c_1, c_2\}$ can be chosen heuristically and define the two membership functions S and B as shown in Figure 3.8(b).

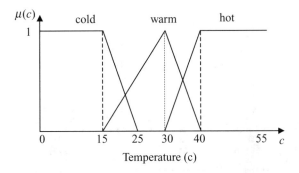

Figure 3.7   Heuristic selection of parameters

The data distribution will not necessarily be linear as in Figure 3.8. It can be nonlinear or scattered over the whole input space. An example of a scattered distribution is shown in Figure 3.9. In such cases, heuristic selection of cluster centres may not work well. An appropriate method for estimation of cluster centres is required (Tung and Quek, 2002).

(ii) Clustering approach

A clustering algorithm can be applied to estimate the actual data distribution and the resulting clusters can be used to produce the membership functions, which will interpret the data better and also be useful in producing a concise representation of a system's behaviour by identifying natural groupings of data from a large data set. For example, two parameters, namely the centre and the width, are required to define Gaussian or bell-shaped MFs. Clustering techniques can be used to determine the initial locations of the centres and the width can be calculated from the distance between the two centres. In the case of triangular MFs, the same centre can be used and the left and right width can be calculated from the previous and next centres. Figure 3.9 shows the data distribution of two variables, say the error and change of error. Three cluster centres $\{c_1, c_2, c_3\}$ are determined using a clustering technique and the widths are determined from the distance between centres.

Clustering performs partitioning of data into an appropriate number of subsets, i.e. partitioning of the input space accordingly helps choose the parameters of the MFs. Although for some applications users can determine the number of clusters K in terms

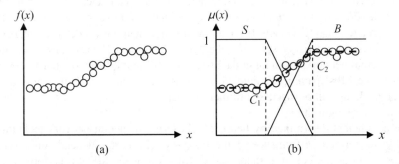

Figure 3.8   Construction of MF from data. (a) Data distribution; (b) Parameters of MFs

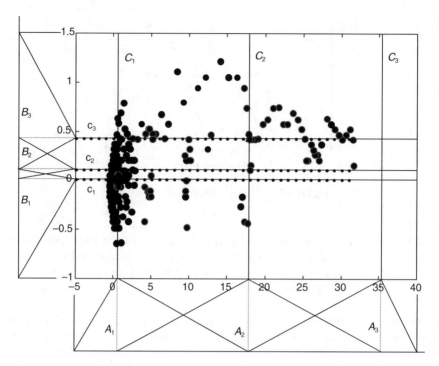

Figure 3.9   Determining MF parameters using clustering

of their expertise, under some circumstances the value of K is unknown and needs to be estimated exclusively from the data itself. Most of the clustering algorithms ask for K to be provided as an input parameter, and it is obvious that the quality of the resulting clusters is largely dependent on the estimation of K. A division with too many clusters complicates interpretation and analysis of the data, while too few clusters cause loss of information, misleading decisions or degraded performance. There are different algorithms to determine clusters, such as the C-means algorithm (Bezdek, 1981), the Gath–Geva algorithm (Gath and Geva, 1989) and the Gustafson–Kessel algorithm (Gustafson and Kessel, 1979).

(iii) Neural networks

Neural networks have also been used to determine the parameters of MFs when data are available. For example, adaptive vector quantization (Dickenson and Kosko, 1993) or self-organized map (Kohonen, 1990) are well-suited methods for determining parameters such as cluster centres or spread of the MFs. Takagi and Hayashi (1991) created MFs using neural networks from input data sets. Mostly, neural network approaches are used to tune the MFs' parameters defined by clustering techniques (Espinosa *et al.*, 2005). Some methods will be discussed further in Chapter 10.

(iv) Evolutionary algorithms

The parameters determined by heuristic, clustering or neural network techniques may not yield optimal performance of the fuzzy system as this depends on several other issues, such as the position, shape and distribution of the MFs. Therefore, a tuning or optimization of the parameters is necessary. Evolutionary algorithms are derivative-free optimization techniques and are popular methods for optimizing parameters of MFs. Different approaches and applications of evolutionary algorithms to fuzzy systems are discussed in greater detail in Chapter 8.

A good account of discussions on constructing MFs using gradient descent, clustering and gradient descent and evolutionary strategies can be found in Espinosa *et al.* (2005).

3.4 Fuzzy Control

Most of the classical design methodologies such as Nyquist, Bode, state-space and optimal control are based on assumptions that the process is linear and stationary and hence is represented by a finite-dimensional constant-coefficient linear model. These methods do not suit complex systems well because few of these represent uncertainty and incompleteness in process knowledge or complexity in design. But the fact is that the real world is too complex. In particular, many industrial processes are highly nonlinear and complex. As the complexity of a system increases, quantitative analysis and precision become difficult. However, many processes that are nonlinear, uncertain, incomplete or non-stationary have subtle and interactive exchanges with the operating environment and are controlled successfully by skilled human operators. Rather than mathematically modelling the process, the human operator models the process in a heuristic or experiential manner. It is evident that human knowledge is becoming more and more important in control system design. This experiential perspective in controller design requires the acquisition of heuristic and qualitative, rather than quantitative, knowledge or expertise from the human operator. During the past several years, fuzzy control has emerged as one of the most active and powerful areas for research in the application of such complex and real-world systems using fuzzy set theory (Zadeh, 1965, 1994). The control of complex nonlinear systems has been approached in recent years using fuzzy logic techniques. A fuzzy logic controller (FLC) has the basic configuration illustrated in Figure 3.10.

3.4.1 Fuzzification

In fuzzy control applications, the observed data are usually crisp. Since the data manipulation in an FLC is based on fuzzy sets, fuzzification is necessary. Therefore, fuzzification is defined as a mapping from an observed input space to fuzzy sets in a certain input universe of

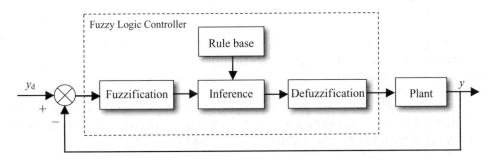

Figure 3.10   Configuration of a fuzzy logic controller

discourse. This process consists of associating to each fuzzy set a membership function. These functions can be thought of as maps from the real numbers to the interval $\mu = [0, 1]$. Fuzzification consists of associating a fuzzy vector with the quantity x by passing x through all the membership functions p_i providing grade of membership functions $\mu_i(x), i = 1, 2, \ldots n$:

$$F : R \rightarrow \mu(x) \tag{3.11}$$

$$x \rightarrow \begin{bmatrix} p_1 \\ \vdots \\ p_n \end{bmatrix} = \begin{bmatrix} \mu_1(x) \\ \vdots \\ \mu_n(x) \end{bmatrix} \tag{3.12}$$

3.4.2   Inference Mechanism

Inference is the process of formulating a nonlinear mapping from a given input space to an output space. The mapping then provides a basis from which decisions can be taken. The process of fuzzy inference involves all the membership functions, fuzzy logic operators and if–then rules.

There are three basic types of fuzzy inference, which have been widely employed in various control applications. Larsen's product rule model is a variant of Mamdani-type model. The differences between these four fuzzy inferences, also called fuzzy models, lie in the consequents of their fuzzy rules, aggregations and defuzzification procedures. These fuzzy models are:

  (i) Mamdani-type fuzzy model;
 (ii) Sugeno-type fuzzy model;
(iii) Tsukamoto-type fuzzy model;
(iv) Larsen's product rule model.

  (i) Mamdani-type fuzzy model
    The Mamdani-type fuzzy model was first proposed as an attempt to control a steam engine and boiler using a set of linguistic control rules obtained from an experienced human operator (Mamdani and Assilian, 1974). Figure 3.11 illustrates a two-input single-output Mamdani-type fuzzy model.
    A typical rule in a Mamdani-type fuzzy model with two-input single-output has the form

If x is A and y is B then z is C

    In Mamdani's fuzzy model, crisp values are used as inputs and defuzzification (see Chapter 2) is used to convert a fuzzy set to a crisp value.
 (ii) Sugeno-type fuzzy model
    The Sugeno-type fuzzy model, also known as the TSK fuzzy model, was proposed by Takagi, Sugeno and Kang (Takagi and Sugeno, 1985; Sugeno and Kang, 1988) in an effort to develop a systematic approach to generate fuzzy rules from a given input/output data set. A typical fuzzy rule in a Sugeno fuzzy model has the form

If x is A and y is B then $z = f(x, y)$

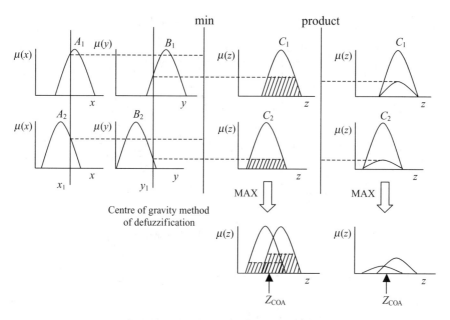

Figure 3.11 Mamdani-type fuzzy model

Usually, $f(x, y)$ is polynomial in the input variables x and y but it can be any function as long as it can appropriately describe the output of the model within the fuzzy region specified by the antecedent of the rule (Wang and Langari, 1995; see also Chapter 2). Figure 3.12 illustrates a two-input single-output Sugeno fuzzy model.

(iii) Tsukamoto-type fuzzy model

In the Tsukamoto-type fuzzy model, the consequent of each fuzzy if–then rule is represented by a fuzzy set with a monotonic MF (Tsukamoto, 1979). As a result, the inferred

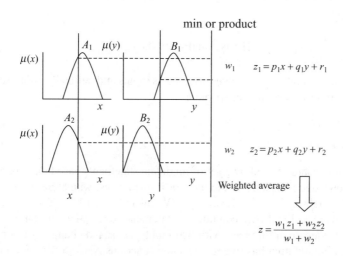

Figure 3.12 Sugeno-type fuzzy model

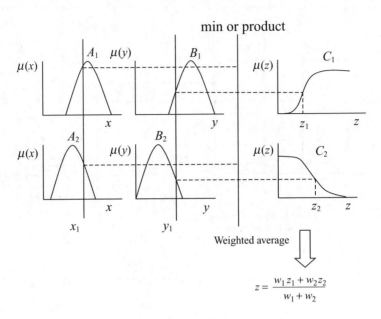

Figure 3.13   Tsukamoto-type fuzzy model

output of each rule is defined as a crisp value included in the rule's firing strength. The overall output is taken as the weighted average of each rule's output (see also Chapter 2). Figure 3.13 illustrates a two-input single-output Tsukamoto fuzzy model.

(iv) Larsen's product rule

Larsen's product rule model is similar to the Mamdani-type fuzzy modelling. The only difference is that it takes the product of the consequent membership functions as shown in Figure 3.14, which is an illustration of a two-input single-output Mamdani-type fuzzy model (Larsen, 1980; Figueiredo *et al.*, 1993). A typical rule with two inputs and single output has the form

$$\text{If } x \text{ is } A \text{ and } y \text{ is } B \text{ then } z \text{ is } C$$

In Mamdani's fuzzy model, crisp values are used as inputs and defuzzification (see Chapter 2) is used to convert a fuzzy set to a crisp value.

3.4.3   Rule Base

A fuzzy system is characterized by a set of linguistic statements based on expert knowledge. The expert knowledge is usually in the form of if–then rules, which are easily implemented by fuzzy conditional statements in fuzzy logic (Wong and Lin, 1997). The collection of fuzzy rules that are expressed as fuzzy conditional statements forms the rule base or the rule set of an FLC. For example, a rule base with two inputs, error and change of error, is shown in Table 3.1. Each input/output has five membership functions NB, NS, ZO, PS and PB, where PB=positive big, PS=positive small, ZO=zero, NS=negative small and NB=negative big.

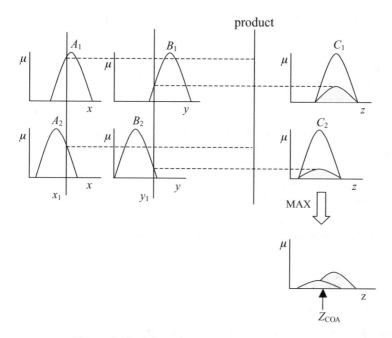

Figure 3.14 Inferencing with Larsen's product rule

The design parameters of the rule base include:

- Choice of process state and control output variables;
- Choice of the content of the rule antecedent and the rule consequent;
- Choice of term sets for the process state and control output variables;
- Derivation of the set of rules.

If one has made the choice to design a P-, PD-, PI-, or PID-like fuzzy logic controller this already implies the choice of process state and control output variables, as well as the content of the rule antecedent and rule consequent for each of the rules. The process state

Table 3.1 FLC rule base with error and change of error

	Change of error				
Error	NB	NS	ZO	PS	PB
NB	PB	PB	PB	PS	ZO
NS	PB	PS	PS	ZO	NS
ZO	PS	ZO	ZO	ZO	NS
PS	PS	ZO	NS	NS	NB
PB	ZO	NS	NB	NB	NB

variables representing the contents of the rule antecedent ('if' part of the rule) are selected as follows:

- Error, denoted by e;
- Change of error, denoted by Δe;
- Sum of error, denoted by Σe.

The control output (process input) variables representing the content of the rule consequent ('then' part of the rule) are selected as follows:

- Control output, denoted by u;
- Change of control output, denoted by Δu.

By analogy with the conventional controller, these are defined as

- $e(k) = y_d - y(k)$
- $\Delta e(k) = e(k) - e(k-1)$
- $\displaystyle\sum_{k=1}^{n} e(k) = \sum_{k=1}^{n-1} e(k) + e(k)$
- $\Delta u(k) = u(k) - u(k-1)$ or $u(k) = u(k-1) + \Delta u(k)$

where y_d is the desired output or set point, y is the process output, k is the sampling time and n is the maximum sample number.

3.4.4 Defuzzification

Basically, defuzzification is a mapping from a space of fuzzy control actions defined over an output universe of discourse into a space of non-fuzzy (crisp) control actions. In a sense this is the inverse of fuzzification, even though mathematically the maps need not be inverses of one another. In general, defuzzification can be viewed as a mapping DF from a fuzzy vector μ with n fuzzy sets to a real number:

$$DF : \mu \to R \tag{3.13}$$

In general, there are different methods for defuzzifying a fuzzy set A defined over the universe of discourse Z. A detailed description of the defuzzification methods is given in Chapter 2.

Considering the demand for low computation time in real-time applications, the following defuzzification methods are widely used:

(i) centre of area Z_{COA} is the centre of gravity of the aggregated area of the output MFs;
(ii) bisector of area Z_{BOA} is a vertical line that divides the area into two equal areas;
(iii) mean of maximum Z_{MOM} is the average of the maximizing z at which the membership function reaches a maximum μ^*;
(iv) smallest of maximum Z_{SOM} is the minimum in terms of magnitude of the maximizing z;
(v) largest of maximum Z_{LOM} is the maximum in terms of magnitude of the maximizing z.

The calculation needed to carry out these defuzzification operations is still time-consuming. Researchers are trying to find new approaches to minimize the time involved. One example of such an attempt is the zero-order Sugeno-type fuzzy system, where a pre-defuzzified constant value is used in the consequent part of the rule. An overview of defuzzification methods and the lack of a systematic approach to the defuzzification problem can be found in Lee (1990) and Yager and Filev (1994).

3.5 Design of Fuzzy Controller

Let $x = (x_1, \ldots, x_n)$ be a vector of process state variables, y the process output variable and u the process input variable or control variable. The conventional closed-loop model, when linearized around the set point, is given by

$$x(k + 1) = A \cdot x(k) + b^T \cdot u(k) \tag{3.14}$$

$$y(k) = c^T \cdot x(k) \tag{3.15}$$

$$u(k) = k \cdot y(k) \tag{3.16}$$

where A is the process matrix, b and c are vectors and k is a scalar. The state equations can be written as

$$x(k + 1) = A \cdot x(k) + b^T \cdot u(k) \tag{3.17}$$

$$u(k) = k \cdot c^T \cdot x(k) \tag{3.18}$$

The fuzzy counterpart of the above model can be described as follows. Let the linguistic variable x_i (e.g., error, change of error, etc.) have the term set X_i (e.g., NB, NS, etc.) and the membership function for X_i be denoted by $\widetilde{X}_i$. Thus, the linguistically defined process state vector is denoted by $\widetilde{X} = (\widetilde{X}_1, \ldots, \widetilde{X}_n)$. Similarly, u takes linguistically defined values U with membership functions $\widetilde{U}$. Thus, the fuzzy model of the closed-loop system can be described as

$$\widetilde{X}(k + 1) = \left[\widetilde{X}(k) \times \widetilde{U}(k)\right] \circ \widetilde{A} \tag{3.19}$$

$$\widetilde{U}(k) = \widetilde{X}(k) \circ \widetilde{K} \tag{3.20}$$

where $\widetilde{A}$ is a fuzzy relation on $X \times U \times X$, $\circ$ is the composition operation and $\widetilde{K}$ is the controller which is a fuzzy relation on $X \times U$ representing the meaning of a set of if–then rules of the form

$$\text{If } x_1 \text{ is } X_i \text{ and } \ldots x_n \text{ is } X_j \text{ then } u \text{ is } U_k \tag{3.21}$$

$\widetilde{A}$ can be obtained in explicit form by on/off-line identification or $\widetilde{A}$ is the fuzzy relation giving the overall meaning of a set of production rules.

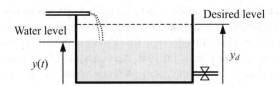

Figure 3.15   Water level in a tank

3.5.1   Input/Output Selection

Suppose the fuzzy controller has to control the water level of a tank as shown in Figure 3.15. Rather than going for development of a mathematical model of the system with available states, a fuzzy model using the available states, namely the error e, change of error Δe, sum of error Σe and valve position u at each discrete time step during the control process, can be developed.

The states of water level and state of valve can be measured directly from the system, whereas the error e, change of error Δe and sum of error Σe can be derived from these states as follows:

$$e = y_d - y \tag{3.22}$$

$$\Delta e = e(k) - e(k-1) \tag{3.23}$$

$$\sum e(k) = \sum e(k-1) + e(k) \tag{3.24}$$

where y is the measured water level and y_d is the desired water level.

3.5.2   Choice of Membership Functions

Since Lotfi Zadeh introduced fuzzy sets, the main difficulties have been with the meaning and measurement of membership functions as well as their extraction, modification and adaptation to dynamically changing conditions. There is no general rule for choice of membership functions, and this mainly depends on the problem domain. In general, use of narrower membership functions results in a faster response but causes larger oscillations, overshoot and settling time.

Gaussian and bell-shaped membership functions involve calculation of exponential terms and use substantial processing time. Trapezoidal membership functions have four parameters and can burden the optimization procedure. Triangular membership functions are the best choice and used for simplicity.

3.5.3   Creation of Rule Base

The fuzzy rules R must be completed and covered by fuzzy partitioning the input space. Figure 3.16 shows an input space partitioning for two-input single-output systems.

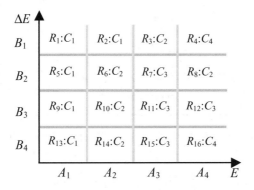

Figure 3.16 Fuzzy input space partitioning

For example, the error and change of error and valve position of a PD-like FLC can be divided into four partitions (i.e., partitioned into four fuzzy sets) as:

$$\text{error } E = \{A1,\ A2,\ A3,\ A4\}$$
$$\text{change of error } \Delta E = \{B_1,\ B_2,\ B_3,\ B_4\}$$
$$\text{valve position } U = \{C_1, C_2, C_3, C_4\}$$

where E, ΔE and U are the universe of discourse for error, change of error and valve position, respectively. The nth rule for the two-input single-output system is

$$R_n : \text{IF } (e \text{ is } A_i) \text{ and } (\Delta e \text{ is } B_j) \text{ THEN } (u \text{ is } C_k)$$

where R_n, $n = 1, 2, \ldots, 16$, is the nth fuzzy rule. A_i, B_j and C_k, $i = 1, 2, \ldots, 4$, $j = 1, 2, \ldots, 4$ and $k = 1, 2, \ldots, 4$, are primary fuzzy sets. There are 16 rules obtained from this uniform partitioning. Initially, fuzzy rules are based on input/output data and these rules are refined through trial and error.

3.5.4 Types of Fuzzy Controller

A fuzzy controller can be constructed using e, Δe and Σe as inputs and control input u as output depending on the type of controller, e.g. PD, PI or PID type.

P-like FLC: The equation for a conventional proportional (P)-like controller is given as

$$u = k_p \cdot e(k) \tag{3.25}$$

where k_p is the proportional gain coefficient. The rule for a P-like controller is given in symbolic form as

$$\text{If } e \text{ is } A_i \text{ then } u \text{ is } B_j \tag{3.26}$$

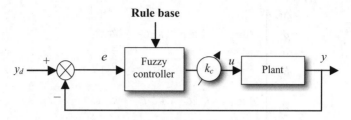

Figure 3.17   Block diagram of a P-type FLC with error

where A_i and B_j, $i, j = 1, 2, \ldots, n$, are the linguistic variables. Figure 3.17 shows the block diagram of a P-type single-input single-output fuzzy controller for a plant. The function of the control output for such a single-input single-output (SISO) system is then a curve, as shown in Figure 3.18 for $n = 4$.

PD-like FLC: A conventional proportional differential (PD)-like FLC can be developed by using an error and change of error model as

$$u = k_p.e + k_d.\Delta e \tag{3.27}$$

where k_p and k_d are the proportional and differential gain coefficients and e is the error, Δe is the change of error. In this type of FLC, it is assumed that no mathematical model for the system is available except two states, namely, the error and change of error. Only output y is measured from the system and the error and change of error are derived. The error and change of error are defined as

$$e(k) = y_d - y(k) \tag{3.28}$$

$$\Delta e(k) = e(k) - e(k - 1) \tag{3.29}$$

where y_d is the desired output and $y(k)$ is the actual output. Figure 3.19 shows the block diagram of a PD-like FLC with error and change of error as inputs. The PD-like

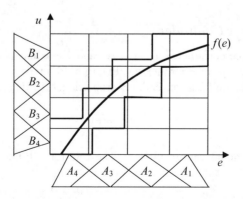

Figure 3.18   Function of control output for SISO systems

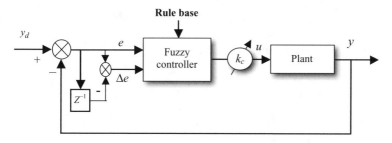

Figure 3.19   Block diagram of PD-like FLC with error and change of error

FLC consists of rules of the form

$$\text{If } e \text{ is } A_i \text{ and } \Delta e \text{ is } B_j \text{ then } u \text{ is } C_k \qquad (3.30)$$

where A_i, B_j and C_k are the linguistic variables and $i = 1, \ldots, n_1$, $j = 1, \ldots, n_2$ and $k = 1, \ldots, m$.

The control surface of a two-input single-output (MISO) system is shown in Figure 3.20, where X and Y represent inputs and Z represents the controller output. For a PD-type controller, X represents error and Y represents change of error. For a PI-type controller, X represents error and Y represents sum of error.

Example 3.1: PD-like FLC with error and change of error   A simple PD-like FLC is developed for a manipulator. A schematic representation of the flexible-link manipulator system considered here is shown in Figure 3.21, where X_oOY_o and XOY represent the stationary and moving coordinates, respectively and τ represents the applied torque at the hub. E, I, ρ, V, I_h and M_P represent the Young's modulus, area moment of inertia, mass density per unit volume, cross-sectional area, hub inertia and payload of the manipulator, respectively. In this example, the motion of the manipulator is confined to the X_oOY_o plane.

In a PD-type FLC, it is assumed that no mathematical model for the flexible link is available except two states, namely, the hub angle error and change of error. Only the hub angle θ is

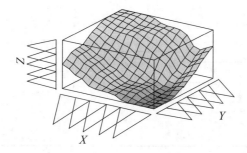

Figure 3.20   Control surface of a two-input single-output system

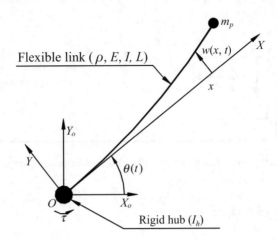

Figure 3.21 Schematic representation of the single-link flexible manipulator

measured from the system and the error and change of error are derived from θ. The hub angle error and change of error are defined as

$$e(k) = \theta_d - \theta(k) \tag{3.31}$$
$$\Delta e(k) = e(k) - e(k-1) \tag{3.32}$$

where θ_d is the desired hub angle, e is the error and Δe is the change in angle error. Figure 3.22 shows a block diagram of the PD-like FLC with error and change of error as inputs.

Triangular membership functions are chosen for inputs and output. The membership functions for angle error, change of angle error and torque input are shown in Figure 3.23. The universe of discourse for the angle error and change in angle error are chosen as [−36, +36] degrees and [−25, +25] respectively. The universe of discourse of the output (i.e., input torque) is chosen as [−3, +3] volts. To construct a rule base, the angle error, change of angle error and torque input are partitioned into five primary fuzzy sets as

hub angle error $E = \{NB, NS, ZO, PS, PB\}$

change of angle error $C = \{NB, NS, ZO, PS, PB\}$

torque $U = \{NB, NS, ZO, PS, PB\}$

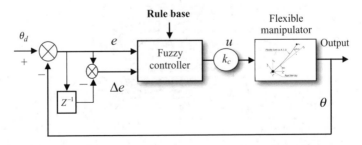

Figure 3.22 PD-like FLC with angle error and change of angle error

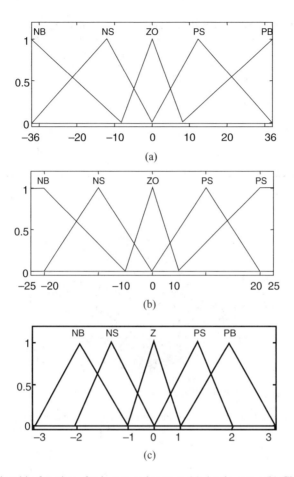

Figure 3.23 Membership functions for inputs and output. (a) Angle error; (b) Change of angle error; (c) Torque input

where E, C and U are the universes of discourse for hub angle error, change of angle error, and torque input, respectively. The nth rule of the rule base for the FLC, with error and change of error as inputs, is

$$R_n : \text{IF } (e \text{ is } E_i) \text{ and } (\Delta e \text{ is } C_j) \text{ THEN } (u \text{ is } U_k)$$

where R_n, $n = 1, 2, \ldots, N_{max}$ is the nth fuzzy rule, E_i, C_j and U_k, for $i, j, k = 1, 2, \ldots, 5$ are the primary fuzzy sets. The rule base is shown in Table 3.2.

The membership functions defined in Figure 3.23 and the rule base defined in Table 3.2 form the control surface of the controller, which is shown in Figure 3.24. The controller is applied to the single-link manipulator described above. The response of the manipulator system is shown in Figure 3.25. For a demanded angle of 36 degrees, it reached a maximum overshoot of 50 degrees. The PD-type FLC shows rapid response at transient state, i.e., a rise time of 17 time units and a settling time of 44 time units. The performance of the PD-type FLC is

Table 3.2 FLC rule base with angle error and change of angle error

	Change of error				
Error	NB	NS	ZO	PS	PB
NB	PB	PB	PB	PS	ZO
NS	PB	PS	PS	ZO	NS
ZO	PS	ZO	ZO	ZO	NS
PS	PS	ZO	NS	NS	NB
PB	ZO	NS	NB	NB	NB

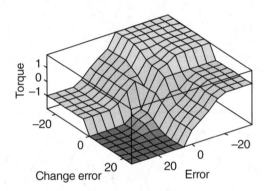

Figure 3.24 Control surface of the controller with hub angle error and change of hub angle error

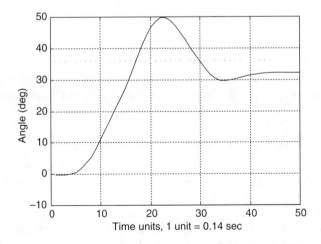

Figure 3.25 Hub angle with FLC with hub angle error and change of hub angle error

very promising in respect of rise time, maximum overshoot and settling time but it shows a significant amount of steady-state error of 2.56 degrees. Steady state error is a characteristic feature of any PD-type controller.

PI-like FLC: A conventional proportional-integral (PI)-like controller is described as

$$u = k_P e + k_I \int e \, dt \qquad (3.33)$$

where k_P and k_I are the proportional and integral gain coefficients. Taking the derivative with respect to time of Equation (3.30) yields

$$\dot{u} = k_P \cdot \dot{e} + k_I \cdot e \qquad (3.34)$$

which can be rewritten as

$$\Delta u = k_P \cdot \Delta e + k_I \cdot e \qquad (3.35)$$

This yields an incremental PI-like controller equation. The PI-like FLC rule base accordingly consists of rules of the form

$$\text{If } e \text{ is } A_i \text{ and } \Delta e \text{ is } B_j \text{ then } \Delta u \text{ is } C_k$$

In this case, to obtain the value of the control output $u(k)$, the change of control output $\Delta u(k)$ is added to $u(k-1)$ such that

$$u(k) = \Delta u(k) + u(k-1) \qquad (3.36)$$

Another way to express the PI-like controller is the absolute integral PI-like controller:

$$u = k_P \cdot e + k_I \cdot \Sigma e \qquad (3.37)$$

where Σe is the sum of error, k_p and k_I are the proportional and integral gain coefficients. The absolute PI-like FLC consists of rules of the form

$$\text{If } e \text{ is } A_i \text{ and } \Sigma e \text{ is } B_j \text{ then } u \text{ is } C_k \qquad (3.38)$$

where A_i, B_j and C_k are the linguistic variables. A block diagram of the absolute PI-type FLC is shown in Figure 3.26. In this type, the output is measured from the system and the error and sum of error are derived.

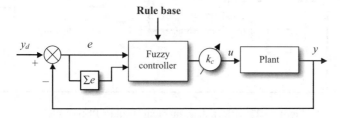

Figure 3.26   Block diagram of a PI-type FLC with error and sum of error

Example 3.2: PI-type FLC with error and sum of error   It is well known that the PI-type FLC has good performance at steady state, like the traditional PI-type controllers. That is, the PI-like FLC reduces steady-state error, but yields penalized rise time and settling time (Chao and Teng, 1997). The PI-type controllers give inevitable overshoot when attempting to reduce the rise time, especially when a system of order higher than one is under consideration (Lee, 1993). These undesirable characteristics of fuzzy PI controllers are caused by integral operation of the controller, even though the integrator is introduced to overcome the problem of steady-state error.

The incremental PI-like FLC for the flexible-link manipulator will be like that described in Equation (3.32). The inputs are the same as a PD-like FLC with error and change of error, except the control input is incremented at each time. Actually, the rules of the fuzzy controller are designed with a phase plane in mind, in which the fuzzy controllers drive a system into the so-called sliding mode. The tracking boundaries in the phase plane, however, are related not to the incremental control input but to the control input itself, which is calculated by Equation (3.33). To select the maximum variation of the incremental control input Δu giving satisfactory rise time and maximum overshoot is not so easy as in the case where the control input itself is to be determined (Lee, 1993). One natural approach to overcome such a difficult situation is to adopt the rate of change of error. Such a controller may be called a PID fuzzy controller, and will be addressed later. Furthermore, a primary objective of this example is to demonstrate the performance of the PD- and PI-type FLCs with different inputs such as error (e), change of error (Δe) and sum of error (Σe), and hence this type of controller is not discussed further in this chapter. Rather, an absolute PI-type controller is presented. In an absolute PI-type FLC, the error and sum of error are used as inputs and it is described by Equation (3.34).

A block diagram of the absolute PI-type FLC is shown in Figure 3.27. In this type of controller, the angle is measured from the system and the sum of angle error is derived from the angle error. Triangular membership functions are chosen for inputs and output. The membership functions for hub angle error, sum of hub angle error and torque input are shown in Figure 3.28(a–c). The universes of discourse for the hub angle error and sum of hub angle error are chosen as [−36, +36] degrees and [−150, +150] degrees respectively. The universe of discourse of the output is chosen as [−3, +3] volts.

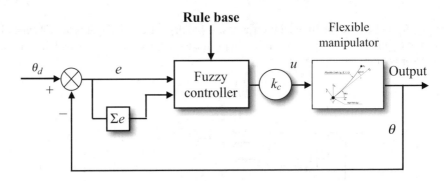

Figure 3.27   Block diagram of an absolute PI-type FLC

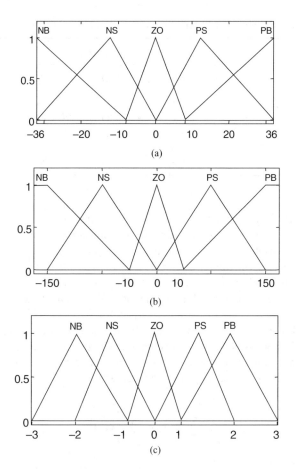

Figure 3.28 Membership functions for inputs and output. (a) Angle error; (b) Sum of angle error; (c) Torque input

To construct a rule base, the angle error, sum of angle error and torque input are partitioned into five primary fuzzy sets as

$$\text{angle error } E = \{\text{NB, NS, ZO, PS, PB}\}$$
$$\text{sum of angle } S = \{\text{NB, NS, ZO, PS, PB}\}$$
$$\text{torque } U = \{\text{NB, NS, ZO, PS, PB}\}$$

where E, S and U are the universes of discourse for hub angle error, sum of hub angle error and torque input, respectively. The nth rule of the rule base for this PI-type FLC is

$$R_n : \text{IF } (e \text{ is } E_i) \text{ and } (s \text{ is } S_j) \text{ THEN } (u \text{ is } U_k)$$

where R_n, $n = 1, 2, \ldots, N_{\max}$ is the nth fuzzy rule, E_i, S_j and U_k, for $i, j, k = 1, 2, \ldots, 5$ are the primary fuzzy sets. The rule base for the PI-type controller is shown in Table 3.3.

Table 3.3 Rule base for PI-type FLC with error and sum of error

			Sum of error		
Error	NB	NS	ZO	PS	PB
NB	PB	PB	PB	PS	ZO
NS	PB	PS	ZO	ZO	NS
ZO	PS	ZO	ZO	ZO	NS
PS	PS	ZO	ZO	NS	NB
PB	ZO	NS	NB	NB	NB

A difficulty arises from deciding on the number of time units to go back in calculating the sum in Equation (3.34). Even the literature on conventional control theory tends to be somewhat vague on this point, and many texts use an indefinite integral type of notation when representing the integral term, though obviously it is not to be taken literally. The reason for this vagueness may be that traditionally, in conventional control, the integral term is approximated by analogue circuitry, and the integral limits cannot easily be stated precisely anyway (Lewis, 1997).

Experience with the system suggests using 10 time units to indicate recent tendencies in the error, and experimentation demonstrates that this works very well. It is also convenient to work with an average rather than a sum so that the base value can easily be compared with the current error. Thus, the $\sum e$ base value is calculated as

$$\sum e(1) = \sum_{k=-8}^{1} e(k) \tag{3.39}$$

The control surface of the controller with angle error and sum of angle error is shown in Figure 3.29.

The controller was implemented on the flexible-link manipulator. The response of the absolute PI-type FLC for the flexible-link manipulator is shown in Figure 3.30. It can be seen that the response has a very good performance for a demanded hub angle of 36 degrees with a small steady-state error of –0.34 degrees. It has a rise time of 12 time units, which is less than the rise time of the PD-like FLCs in Example 3.1 and a larger overshoot of 66.45 degrees with

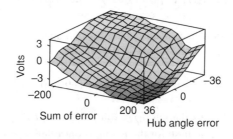

Figure 3.29 Control surface of the controller with error and sum of error

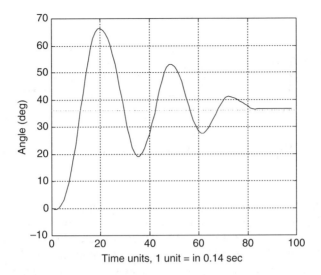

Figure 3.30   Angle error with PI-type FLC

an excessive oscillation around the set point. The oscillations caused a prolonged settling time of 85 time units.

PID-like FLC: A further option to obtain a better performance in respect of rise time, settling time, overshoot and steady-state error is to develop a proportional-integral-derivative (PID)-like FLC. The basic idea of a PID controller is to choose the control law by considering the error e, change of error Δe and integral of error Σe (or $\int_0^t edt$). The PID-type fuzzy controller is described by

$$u_{\text{PID}} = k_P \cdot e + k_d \cdot \Delta e + k_I \cdot \int_0^t e \cdot dt \qquad (3.40)$$

By replacing the integral of error term $\int_0^t edt$ with the sum of error term Σe, the PID-type fuzzy controller in discrete time is described by

$$u_{\text{PID}} = k_P \cdot e + k_d \cdot \Delta e + k_I \cdot \Sigma e \qquad (3.41)$$

The fuzzy control rule corresponding to the PID controller (as shown in Figure 3.31) has the form

$$\text{If } e \text{ is } A_i \text{ and } \Delta e \text{ is } B_j \text{ and } \Sigma e \text{ is } C_k \text{ then } u \text{ is } D_l \qquad (3.42)$$

where $i = 1, \ldots, n_1, j = 1, \ldots, n_2, k = 1, \ldots, n_3$ and $l = 1, \ldots, m$. Theoretically, the number of rules to cover all possible input combinations and variations for a three-term fuzzy controller is $n_1 \times n_2 \times n_3$, where n_1, n_2 and n_3 are the number of linguistic labels of the three input variables.

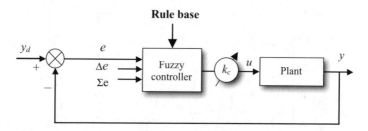

Figure 3.31   PID-type FLC with error, change of error and sum of error

Generally, a PD-type two-term fuzzy controller cannot eliminate steady-state error whereas a PI-type two-term fuzzy controller can eliminate steady-state error but it has a slower response due to the integral term in the control variable. These characteristics have been demonstrated in the example PD- and PI-type FLCs in earlier sections for the flexible-link manipulator. In order to meet the design criteria of fast rise time, minimum overshoot, shorter settling time and zero steady-state error, a further option is to develop a PID-type FLC which enables fast rise time, smaller overshoot and settling time from the PD part and minimum steady-state error from the PI part of the PID controller. The generic fuzzy PID controller is a four-dimensional (three-input single-output) fuzzy system. The basic idea of a PID controller is to choose the control law by considering the error e, change of error Δe and integral of error or sum of error Σe, thus giving the controller defined in Equation (3.37).

Theoretically, the number of rules to cover all possible input variations for a three-term fuzzy controller is $n_1 \times n_2 \times n_3$, where n_1, n_2 and n_3 are the number of linguistic terms of the three input variables. In particular, if $n_1 = n_2 = n_3 = 5$, then the number of rules $R = 5 \times 5 \times 5 = 125$. In practical applications the design and implementation of such a huge rule base is a tedious task, and it will take a substantial amount of memory space and reasoning time. Because of a long reasoning time, the response of such a generic PID-type FLC will be too slow and hence not suitable for applications where a fast response is desired, for example the flexible-link manipulator system discussed in two examples.

A variety of approaches have been proposed to overcome the problems of a generic PID-type fuzzy controller in Tzafestas and Papanikolopoulos (1990) and Brehm (1994). The inherent feature of a PD-type fuzzy controller is that it has fast rise time, less overshoot and steady-state error. So Kwok et al. (1990, 1991) have considered a novel means of decomposing a PID controller into a fuzzy PD controller in parallel with various types of fuzzy gains, fuzzy integrators, fuzzy I controller and deterministic integral control to minimize the steady-state error. The various PID configurations are shown in Figure 3.32. For a process whose steady-state gain is known or can be measured easily as k_p, then integral action is not necessary. This combination of fuzzy PD with steady-state gain control is shown in Figure 3.32(a). The output of the integral action is defined as $u_I = \frac{r}{k_p}$, where r is the set point or desired output. If the proportional gain k_p is not known, then an integral action is necessary. The integral action is implemented by placing a conventional integral controller in parallel with the fuzzy PD controller. The implementation of the fuzzy PD with integral action controller is shown in Figure 3.32(b). In this case, the output of the integral action is defined as $u_I = k_I \Sigma e$, where k_I is the integral gain to be determined by trial and error. Some researchers argue that the two implementations in Figure 3.32(a,b) are not true fuzzy PID controllers as they apply

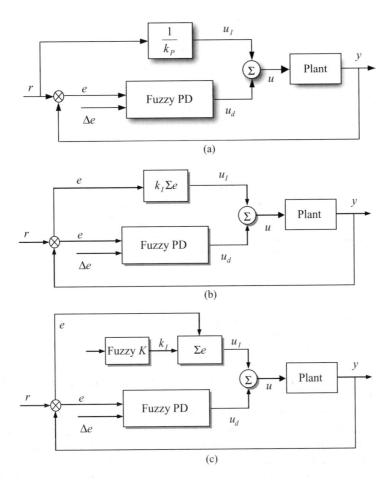

Figure 3.32 Different PID-type FLC configurations. (a) Fuzzy PD with steady-state gain control;
(b) Fuzzy PD with integral action control; (c) Fuzzy PD with fuzzified k_I

deterministic control for the integral part. As a remedy they suggest fuzzifying the integral
gain k_I. The implementation of the fuzzy PD with fuzzified integral gain controller is shown
in Figure 3.32(c). The output of the integral action is defined as $u_I = fuzzy(k_I)\Sigma e$. A detailed
description of these kinds of decompositions can be found in Harris *et al.* (1993).

A typical method for rule reduction in a fuzzy PID-type controller is to divide the three-term
PID controller into two separate fuzzy PD and fuzzy PI parts (Kwok *et al.*, 1990; Zhang and
Mizumoto, 1994; Chen and Linkens, 1998). The parallel combination of PD- and PI-type
fuzzy controllers is shown in Figure 3.33(a). This combination of PD and PI controllers with
n linguistic labels in each input variable requires only $n \times n + n \times n = 2n^2$ rules (e.g., for
$n = 5$ there will be $5 \times 5 + 5 \times 5 = 50$ rules), which is significantly smaller than the n^3
rules (e.g., $5 \times 5 \times 5 = 125$) required by a generic PID controller. A further possibility is to
combine a fuzzy I part with a fuzzy PD part, as shown in Figure 3.33(b). This combination
of PD and I controllers will require only $n \times n + n = n^2 + n$ rules (e.g., for $n = 5$ there will
be $5 \times 5 + 5 = 30$ rules), which is much smaller than 50 rules. This is the number of rules

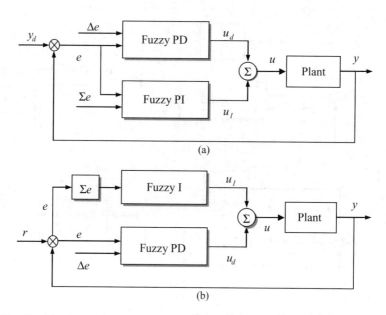

(a)

(b)

Figure 3.33 Combination of fuzzy PD and fuzzy PI. (a) Combination of fuzzy PD and fuzzy PI; (b) Combination of fuzzy PD and fuzzy I

processed during execution of the controller, consuming a significant amount of processing time and memory space.

Siddique (2002) proposed a further rule reduction by employing a switching PD/PI-type FLC where the fuzzy controller is switched from PD- to absolute PI-type after a certain period of time. Only one rule base consisting of $n \times n$ rules for each type of controller is executed at a time, and thus the number of executed rules in a controller will be reduced to only 25 rules for 5 linguistic labels for each input variable. The state variables used in a PD/PI-type FLC are the same as in the PD-type and PI-type FLCs described earlier in this section. The functional block diagram of the switching PD/PI-type controller is shown in Figure 3.34.

Determination of a switching point is important and can result in poor performance if chosen inappropriately. It is obvious that if the controller is switched at the point of maximum overshoot of the PD-like FLC, it can yield the best performance. But surprisingly, it does

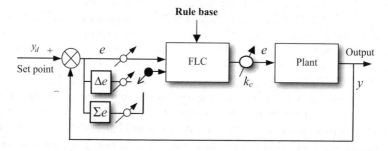

Figure 3.34 Block diagram of a PD/PI-type FLC

not give a good result. Experimental investigations show that a switching point just before or after the maximum overshoot gives a better result than at the point of maximum overshoot, suggesting a trial-and-error method to find the switching point around the point of maximum overshoot.

The above approaches may be useful in reducing the number of rules for a three-input single-output PID fuzzy controller. A systematic study of the fuzzy PID controller has been reported in Hu *et al.* (2001) and Siddique (2002). However, control problems are complicated in the industry, where multiple variables need to be controlled. The situation becomes more complicated when interaction effects exist among the cross-coupled input/output variables of the system. In such situations, it is difficult to design fuzzy if–then rules for cross-coupled input/output variables. The simplest way to design a fuzzy controller for such a problem is to decompose the problem with respect to inputs, outputs or objectives and design fuzzy controller modules for each task or subsystem. It has been demonstrated that the combined performance of the modular fuzzy controllers is comparable to a single monolithic fuzzy controller (Syljak, 1991; Chi, 1995; Chi *et al.*, 1997; Shin and Xu, 2009).

3.6 Modular Fuzzy Controller

A generic problem with FLCs is that the number of rules grows exponentially with the number of input/output variables and linguistic terms for each variable. For a complete rule base with input variables $\{X_i | i = 1, \ldots, n\}$ with linguistic terms $\{A_{ij} | j = 1, \ldots, m_i\}$ and output variables $\{Y_k | k = 1, \ldots, l\}$ with linguistic terms $\{B_{kj} | j = 1, \ldots, p_k\}$, the number of rules will be

$$R = \prod_{i=1}^{n} m_i \tag{3.43}$$

The rules have the form

$$\text{If } (X_1 \text{ is } A_{11}) \text{ and } \ldots \text{ and } (X_n \text{ is } A_{nm}) \text{ then } (Y_1 \text{ is } B_{11}) \text{ and } \ldots \text{ and } (Y_l \text{ is } B_{lp}) \tag{3.44}$$

This large number of rules complicates the design of an FLC, because for each of the R different premises the expert must provide a combination of term sets for the output variables, which is nearly impossible for a human expert to guess. It is possible to omit a set of rules if it could be guaranteed that a certain combination of input/output variables will never occur during control of the dynamic system. A modular structure of FLCs with minimum number of input/output variables can reduce the number of rules R.

For large-scale and complex systems, the reduction in computation and design complexity remains a challenge of intelligent control systems. Hierarchical and modular methodologies have gained wide popularity because of their simplicity in design and robustness. There are several approaches to decomposing a system into modules, such as the decentralized approach, time-scale decomposition, hierarchical system and workspace decomposition (Syljak, 1991; Chi *et al.*, 1997). For control problems with multiple objectives of different priority, a sub-controller with a subset of input/output variables can be designed for each objective. Furthermore, antecedents can be decomposed into single input modules. Each fuzzy module is designed to handle one specific input affiliated with one of the decoupled antecedents $\{X_i | i = 1, \ldots, n\}$ and produces a crisp action $\{Y_k | k = 1, \ldots, l\}$. For example, an

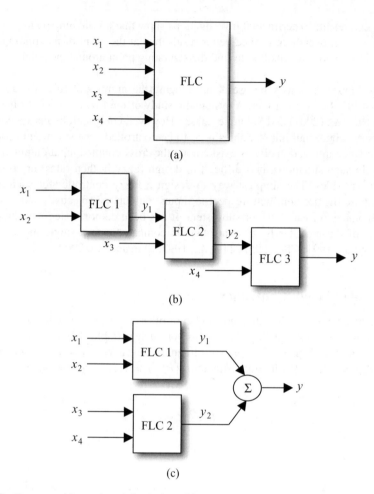

Figure 3.35 Decomposition and modular design of fuzzy controller. (a) Classical monolithic controller; (b) Hierarchical combination of modular FLCs; (c) Parallel combination of modular FLCs

FLC with four inputs and a single output, as shown in Figure 3.35(a), can be modularized in two ways. That is, the FLC can be a hierarchical or cascade combination of modules and a parallel combination of modules. Figure 3.35(b) shows the hierarchical modular architecture and Figure 3.35(c) shows the parallel modular architecture.

The total possible number of fuzzy rules that can be generated for the rule base is L^k, where k is the number of inputs and L is the number of fuzzy linguistic terms or MFs. Compared with the modular FLC design, each input represents one fuzzy control module. The total number of rules for each module is determined by the number L of MFs. Thus, the total number of fuzzy rules for all k modules is kL. This clearly shows a significant reduction in the number of fuzzy rules from L^k to kL, as well as savings in computation time.

Ahmad *et al.* (2010a,b) developed a parallel modular fuzzy controller (MFC) for a two-wheeled wheelchair. The generic architecture of the parallel modular architecture is shown in Figure 3.36. The objective of the MFC is to achieve a zero-degree upright position of the

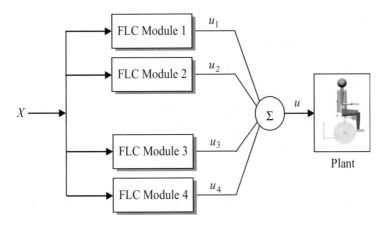

Figure 3.36   Modular FLC for control of a wheelchair

wheelchair on two wheels. The objectives of the four fuzzy controllers are to lift and stabilize the two links of the wheelchair. They achieved a significant rule reduction with simplified fuzzy controller structure and satisfactory performance.

3.7   MATLAB® Programs

Fuzzy inference systems have been applied successfully to various fields such as control, data classification, computer vision and decision systems. Because of their wide areas of application, fuzzy inference systems are also well known by different names, such as fuzzy rule-based systems, fuzzy expert systems, fuzzy modelling, fuzzy associative memory and fuzzy logic controllers. All of these names eventually fall under the umbrella of fuzzy systems.

Many industrial processes are highly nonlinear and complex. As the complexity of a system increases, quantitative analysis and controller design become difficult. Rather than mathematically modelling the process, the human operator models the process in a heuristic or experiential manner. This experiential perspective in controller design requires the acquisition of heuristic and qualitative, rather than quantitative, knowledge or expertise from the human operator. During the past several years, fuzzy logic has emerged as one of the most active and powerful areas for research in the application of control systems design. Fuzzy systems and applications can be developed using the Fuzzy Logic Toolbox in MATLAB®. The toolbox provides three categories of tools:

- Command-line functions,
- GUI interface tools and
- Simulink® blocks.

In Chapter 2, command-line functions are discussed. With the help of these command-line functions, examples of control applications are presented in Appendix B. In this chapter, some fuzzy system applications are presented. A brief introduction to GUI interface tools and Simulink® blocks is also given. Details of codes with associated descriptions and plots are given in Appendix C.

References

Ahmad, S., Siddique, N.H. and Tokhi, M.O. (2010a) A modular fuzzy control approach for two-wheeled wheelchair, *Journal of Intelligent and Robotic Systems*, 64(3&4), 401–426.

Ahmad, S., Tokhi, M.O. and Siddique, N.H. (2010b) Modular fuzzy control with input shaping technique for transformation of two-wheeled wheelchair to four-wheeled mode, *2010 IEEE Symposium on Industrial Electronics and Applications* (ISIEA 2010), 3–5 October, Penang, Malaysia, pp. 562–566.

Bezdek, J.C. (1981) *Pattern Recognition with Fuzzy Objective Function Algorithms*, Plenum Press, New York.

Bezdek, J.C., Keller, J., Krisnapuram, R. and Pal, N.R. (1999) *Fuzzy Models and Algorithms for Pattern Recognition and Image Processing*, Kluwer Academic, Dordrecht.

Brehm, T. (1994) Hybrid fuzzy logic PID controller, *Proceedings of 3rd IEEE Conference on Fuzzy Systems*, Vol. 3, pp. 1682–1687.

Chao, C.-T. and Teng, C.-C. (1997) A PD-like self-tuning fuzzy controller without steady-state error, *Fuzzy Sets and Systems*, 87, 141–154.

Chen, M. and Linkens, D.A. (1998) A hybrid neuro-fuzzy PID controller, *Fuzzy Sets and Systems*, 99, 27–36.

Chi, C.W. (1995) *Modular intelligent control*, Master's Thesis, Department of Electrical Engineering, Santa Clara University, USA.

Chi, Z., Yan, H. and Phan, T. (1996) *Fuzzy Algorithms: With Applications to Image Processing and Pattern Recognition*, World Scientific, Singapore.

Chi, C.W., His, S.T.C. and Tseng, H.C. (1997) Adaptive aggregation of modular fuzzy control. *IEEE International Conference on Systems, Man and Cybernetics*, Vol. 2, pp. 1329–1334.

Czogala, E. and Pedrycz, W. (1981) On identification in fuzzy systems and its application in control problems, *Fuzzy Sets and Systems*, 6, 73–83.

Dickenson, J. and Kosko, B. (1993) Fuzzy function learning with covariance ellipsoids, *Proceedings of IEEE International Conference on Neural Networks*, San Francisco, CA, Vol. 3, pp. 1162–1167.

Emami, M.R., Turksen, I.B. and Goldberg, A.A. (1998) Development of a systematic methodology of fuzzy logic modelling, *IEEE Transactions on Fuzzy Systems*, 6(3), 346–361.

Espinosa, J., Vandewalle, J. and Wertz, V. (2005) *Fuzzy Logic, Identification and Predictive Control*, Springer-Verlag, London.

Figueiredo, M., Gomide, F., Rocha, A. and Yager, R. (1993) Comparison of Yager's level set method for fuzzy logic control with Mamdani's and Larsen's method, *IEEE Transactions on Fuzzy Systems*, 1(2), 156–159.

Filev, D. (1991) Fuzzy modelling of complex systems, *International Journal of Approximate Reasoning*, 5, 281–290.

Gath, I. and Geva, A.B. (1989) Unsupervised optimal fuzzy clustering, *IEEE Transactions on Pattern Analysis and Machine Intelligence*, 7, 773–781.

Gilachet, S. and Foulloy, L. (1995) Fuzzy controllers: synthesis and equivalences, *IEEE Transactions on Fuzzy Systems*, 3(2), 140–148.

Gustafson, D.E. and Kessel, W.C. (1979) Fuzzy clustering with fuzzy covariance matrix, *Proceedings of IEEE CDC*, San Diego, CA, pp. 761–766.

Harris, C.J., Moore, C.G. and Brown, M. (1993) *Intelligent Control: Aspects of Fuzzy Logic and Neural Nets*, World Scientific, Singapore.

Hoeppner, F., Klawonn, F. and Kruse, R. (1997) *Fuzzy-Clusteranalyse: Verfahren fuer die Bilderkennung, Klassifizierung und Datenanalyse*, Vieweg Verlag, Braunschweig.

Hu, B.-G., Mann, G.K.I. and Gosine, R.G. (2001) A systematic study of fuzzy PID controllers – function-based evaluation approach, *IEEE Transactions on Fuzzy Systems*, 9(5), 699–712.

Jang, J.-S.R. (1994) Structure determination in fuzzy modelling: a fuzzy CART approach, *Proceedings of International Conference on Fuzzy Systems*, Orlando, FL, pp. 380–385.

Klir, G.J. and Yuan, B. (1995) *Fuzzy Sets and Fuzzy Logic: Theory and Applications*, Prentice-Hall, Upper Saddle River, NJ.

Kohonen, T. (1990) The self-organising map, *Proceedings of the IEEE*, 78(9), 1464–1480.

Kovacic, Z. and Bogdan, S. (2006) *Fuzzy Controller Design: Theory and Application*, CRC Press, Boca Raton, FL.

Kwok, D.P., Tam, D., Li, C.K. and Wang, P. (1990) Linguistic PID controllers, *Proceedings of 11th IFAC World Congress*, Tallin, USSR, pp. 192–197.

Kwok, D.P., Tam, D. and Li, C.K. (1991) Analysis and design of fuzzy PID control systems, *Proceedings of IEE Control '91 Conference*, Herriot Watt University, Edinburgh, pp. 955–960.

Larsen, P.M. (1980) Industrial applications of fuzzy logic control, *International Journal of Man-Machine Studies*, 12(1), 3–10.

Lee, C.C. (1990) Fuzzy logic in control systems: fuzzy logic controller – Part II, *IEEE Transactions on Systems, Man and Cybernetics*, 20, 419–435.

Lee, J. (1993) On methods for improving performance of PI-type fuzzy logic controllers, *IEEE Transactions on Fuzzy Systems*, 1(1), 298–301.

Lewis, H.W. (1997) *The Foundation of Fuzzy Control*, Plenum Press, New York.

Lygeros, J. (1996) A formal approach to fuzzy modelling, *Proceedings of American Control Conference*, pp. 3740–3744.

Mamdani, E.H. and Assilian, S. (1974) Application of fuzzy algorithms for control of simple dynamic plant, *Proceedings of IEE*, 121, 1585–1588.

Oliveira, J.V. De and Pedrycz, W. (2007) *Advances in Fuzzy Clustering and its Applications*, John Wiley & Sons, Chichester, UK.

Pedrycz, W. (1984) An identification algorithm in fuzzy relational systems, *Fuzzy Sets and Systems*, 13, 153–167.

Pedrycz, W. and Gomide, F. (1998) *An Introduction to Fuzzy Sets: Analysis and Design*, MIT Press, Cambridge, MA.

Ross, T.J. (2004) *Fuzzy Logic with Engineering Applications*, 2nd edn, John Wiley & Sons, New York.

Shin, Y.C. and Xu, C. (2009) *Intelligent Systems: Modelling, Optimisation and Control*, CRC Press, Boca Raton, FL.

Siddique, N.H. (2002) *Intelligent control of flexible-link manipulator systems*, PhD Thesis, Department of Automatic Control and Systems Engineering, The University of Sheffield, UK.

Sugeno, M. and Kang, G.T. (1988) Structure identification of fuzzy model, *Fuzzy Sets and Systems*, 28, 15–33.

Sugeno, M. and Yasukawa, T. (1993) A fuzzy-logic-based approach to qualitative modelling, *IEEE Transactions on Fuzzy Systems*, 1(1), 7–31.

Syljak, D. (1991) *Decentralised Control of Complex Systems*, Academic Press, New York.

Takagi, H. and Hayashi, I. (1991) NN-driven reasoning, *International Journal of Approximate Reasoning*, 5, 191–212.

Takagi, T. and Sugeno, M. (1985) Fuzzy identification of systems and its application to modeling and control, *IEEE Transactions on Systems, Man and Cybernetics*, 15, 116–132.

Tong, R.M. (1980) The evaluation of fuzzy models derived from experimental data, *Fuzzy Sets and Systems*, 4, 1–12.

Tsukamoto, Y. (1979) An approach to fuzzy reasoning method. In *Advances in Fuzzy Set Theory and Applications*, M.M. Gupta, R.K. Ragade and R. Yager (eds), North-Holland, Amsterdam, pp. 137–149.

Tung, W.L. and Quek, C. (2002) DIC: a novel discrete incremental clustering technique for derivation of fuzzy membership functions, *PRICAI: Trends in Artificial Intelligence*, Lecture Notes in Computer Science, Vol. 2417, pp. 485–491.

Tzafestas, S. and Papanikolopoulos, N.P. (1990) Incremental fuzzy expert PID control, *IEEE Transactions on Industrial Electronics*, 37, 365–371.

Wang, L.-X. (1997) *Adaptive Fuzzy Systems and Control: Design and Stability Analysis*, Prentice-Hall, Englewood Cliffs, NJ.

Wang, L. and Langari, R. (1995) Building Sugeno-type models using fuzzy discretization and orthogonal parameter estimation techniques, *IEEE Transactions on Fuzzy Systems*, 3(4), 454–458.

Wong, C.-C. and Lin, N.-S. (1997) Rule extraction for fuzzy modelling, *Fuzzy Sets and Systems*, 88, 23–30.

Yager, R.R. and Filev, D. (1994) *Essentials of Fuzzy Modelling and Control*, John Wiley & Sons, Chichester, UK.

Zadeh, L.A. (1965) Fuzzy sets, *Information and Control*, 8, 338–353.

Zadeh, L.A. (1968) Fuzzy algorithms, *Information and Control*, 12, 94–102.

Zadeh, L.A. (1971) Towards a theory of fuzzy systems. In *Aspects of Networks and Systems Theory*, R.E. Kalman and R.N. DeClairis (eds), Holt, Rinehart & Winston, New York, pp. 469–490.

Zadeh, L.A. (1972) A fuzzy-set-theoretic interpretation of linguistic hedges, *Journal of Cybernetics*, 2, 4–34.

Zadeh, L.A. (1973) Outline of a new approach to the analysis of complex systems and decision process, *IEEE Transactions on Systems, Man and Cybernetics*, 3, 28–44.

Zadeh, L.A. (1994) The role of fuzzy logic in modeling, identification and control, *Modelling, Identification and Control*, 15(3), 191–203.

Zadeh, L.A. (1999) From computing with numbers to computing with words – from manipulation of measurements to manipulation of perceptions, *IEEE Transactions on Circuits and Systems – I: Fundamental Theory and Applications*, 45(1), 105–119.

Zhang, Z.M. and Mizumoto, M. (1994) On rule self-generating for fuzzy control, *International Journal of Intelligent Systems*, 9(12), 1047–1057.

4

Neural Networks

4.1 Introduction

The study of the human brain is hundreds of years old. Advances in brain research promise an initial understanding of the mechanism of cognitive process in the brain. This shows that the brain stores information as patterns. Some of these patterns are very complicated, for example the ability to recognize individual faces from different angles. This process of storing information as patterns, utilizing those patterns, and then solving problems encompasses a new field in computing, which does not utilize traditional programming. This involves the creation of massively parallel networks and the training of those networks to solve specific tasks.

The exact workings of the human brain are still a mystery. Yet, some aspects of this amazing processor are known. In particular, the most basic element of the human brain is a specific type of cell, which provides us with our abilities to remember, think and apply previous experiences to our every action. These cells, all approximately 100 billion of them, are known as neurons. Each of these neurons can connect with up to 200,000 other neurons, although 1,000 to 10,000 is typical. The individual neurons convey information via a host of electrochemical pathways. Together, these neurons and their connections form a process which is not binary, not stable and not synchronous. This building block of the human brain has a few general capabilities. Basically, a biological neuron receives inputs from other sources, combines them in some way, performs a generally nonlinear operation on the result and then outputs the final result. Figure 4.1 shows a biological neuron and Figure 4.2 shows the four main functional parts and their relationships in a neuron. Recent experimental data has provided further evidence that biological neurons are structurally more complex than the simplistic explanation above.

The first model of artificial neural networks came in 1943 when Warren McCulloch, a neurophysiologist and Walter Pitts, a young mathematician outlined the first formal model of an elementary computing neuron (McCulloch and Pitts, 1943). McCulloch and Pitts' artificial

Computational Intelligence: Synergies of Fuzzy Logic, Neural Networks and Evolutionary Computing, First Edition.
Nazmul Siddique and Hojjat Adeli.
© 2013 John Wiley & Sons, Ltd. Published 2013 by John Wiley & Sons, Ltd.

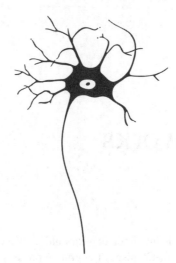

Figure 4.1   A biological neuron (motor neuron) – close to a schematic neuron

neuron model is shown in Figure 4.3. They modelled a simple neural network with electrical circuits. The firing rule for this model is defined as

$$O = \begin{cases} 1 & \text{if } \sum_{i=1}^{n} w_i x_i \geq T \\ 0 & \text{if } \sum_{i=1}^{n} w_i x_i < T \end{cases} \qquad (4.1)$$

where $i = 1, 2, \ldots, n$ and T is a threshold value.

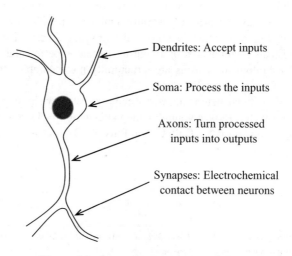

Figure 4.2   Four parts of a biological neuron

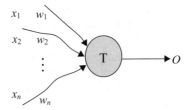

Figure 4.3 McCulloch and Pitts' neuron model

Reinforcing this concept of neurons, Donald Hebb first proposed a learning scheme for updating a neuron's connections that we now refer to as the Hebbian learning rule (Hebb, 1949). He pointed out that information can be stored in connections and postulated the learning technique. As computers advanced from their infancy in the 1950s, it became possible to begin to model the rudiments of these theories concerning human thought.

During the 1950s, neuron-like elements called perceptrons were invented by Frank Rosenblatt, a neurobiologist at Cornell University. A simple perceptron model is shown in Figure 4.4. A single-layer perceptron was found to be useful in classifying a continuously valued set of inputs into one of two classes (Rosenblatt, 1958). The perceptron computes a weighted sum of the inputs, subtracts a threshold and passes one of two possible values out as a result. The firing rule for this model is defined as

$$
O = \begin{cases} 1 & \text{if } \sum_{i=1}^{n} w_i x_i - b \geq T \\ 0 & \text{if } \sum_{i=1}^{n} w_i x_i - b < T \end{cases}
\tag{4.2}
$$

where $i = 1, 2, \ldots, n$ and b is called bias.

In 1959, Bernard Widrow and Marcian Hoff of Stanford University developed models they called ADALINE (ADAptive LINear Elements) and MADALINE (Multiple ADALINE) (Widrow and Hoff, 1960, 1962). These formed the so-called Widrow–Hoff learning rule and the first neural network to be applied to a real-world problem. The rule minimized the sum of squares error during training involving pattern classification. It is an adaptive filter, which eliminates echoes on phone lines.

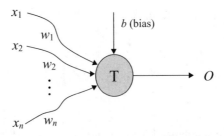

Figure 4.4 Rosenblatt's perceptron model

Unfortunately, the perceptron is limited and was proven as such in Marvin Minsky and Seymour Papert's book *Perceptrons* in 1969 (Minsky and Papert, 1969). Disappointment set in as promises were unfilled. The challenge was not answered until the mid-1980s. In 1982 several events caused a renewed interest in neural network research. John Hopfield of Caltech presented a paper to the National Academy of Sciences. Hopfield's approach was not to simply model brains but to create useful devices. With clarity and mathematical analysis, he showed how such networks could work and what they could do (Hopfield, 1982). Although the mathematical framework for the new training scheme of layered networks was discovered in 1974 by Paul Werbos (Werbos, 1974), it unfortunately went unnoticed at that time.

Another realization of the field came from the publication of two volumes on parallel distributed processing, edited by McClelland and Rumelhart (McClelland and Rumelhart, 1986). The new learning rule and other concepts introduced in this work have removed one of the most essential network training barriers. The publication opened a new era for the once underestimated computing potential of layered networks.

4.2 Artificial Neuron Model

McCulloch and Pitts proposed a mathematical model of the neurons and showed how neural-like networks could be computed. The weight vector W contains the weights connecting the various parts of the network. The term 'weight' is used in neural network terminology and is a means of expressing the strength of the connection between any two neurons (i.e., the weight of information flowing from neuron to neuron in the neural network). However, the model makes use of several drastic simplifications. It allows binary 0, 1 states only, operates under a discrete-time assumption and assumes synchronous operations of all neurons in a larger network. Weights and threshold are fixed in the model and no interaction among network neurons takes place except for signal flow. It is probably not desirable to stretch the analogy too far.

Every neuron model consists of a processing element with synaptic input connections and a single output. The first stage is a process where the inputs $x_1, x_2, \ldots, x_n$ multiplied by their respective weights $w_1, w_2, \ldots, w_n$ are summed by the neuron. The resulting summation process may be shown as

$$net = (w_1 \cdot x_1 + w_2 \cdot x_2 + \cdots + w_n \cdot x_n) \tag{4.3}$$

It can be written in vector notation form as

$$net = \left(\sum_{i=1}^{n} w_i \cdot x_i \right) \tag{4.4}$$

where w is the weight vector defined as $w = [w_1, w_2, \ldots, w_n]^T$ and x is the input vector defined as $x = [x_1, x_2, \ldots, x_n]$.

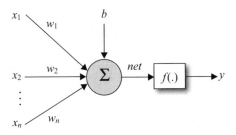

Figure 4.5 Perceptron model

A threshold value b, called the bias, plays an important role for some neuron models and needs to be mentioned explicitly as a separate neuron model parameter. Then Equation (4.4) looks like

$$net = \left(\sum_{i=1}^{n} w_i \cdot x_i \right) + b \tag{4.5}$$

In order to allow for varying input conditions and their effect on the output, it is usually necessary to include a nonlinear activation function $f(.)$ in the neuron arrangement. This is so that adequate levels of amplification may be used where necessary for small input signals, which avoids the risk of driving the output to unacceptable limits. Such a perceptron model is shown in Figure 4.5. The output of the neuron is now expressed in the form

$$y = f(net) \tag{4.6}$$

4.3 Activation Functions

There are a number of types of commonly used activation functions $f(.)$, such as the step function, linear function, ramp function and sigmoid functions. The activation functions are selected specific to the applications. Some common types of activation functions are shown in Figure 4.6.

Linear function: The effect of this function is to multiply the output by a constant factor, such as

$$y = f(net) = K \cdot net \tag{4.7}$$

Step function: The output of a step function is limited to only -1 and $+1$ depending on the value of the input signal (i.e., the value of y):

$$y = f(net) = \begin{cases} +1 & \text{if} \quad net > 0 \\ -1 & \text{if} \quad net < 0 \end{cases} \tag{4.8}$$

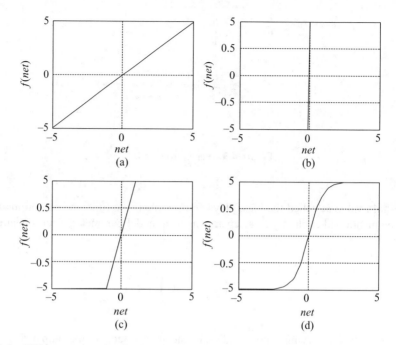

Figure 4.6   Different types of activation function. (a) Linear function; (b) Step function; (c) Ramp function; (d) Tansigmoidal function

Ramp function: The effect of the ramp function is to behave as a linear function between the upper and lower limits and once these limits are reached to behave as a step function. The output is limited within a max and min value and linear within this limit, such as

$$y = f(net) = \begin{cases} \text{max} & \text{if } net > upper\ limit \\ K \cdot net & \text{if } upper\ limit > net > lower\ limit \\ \text{min} & \text{if } net < lower\ limit \end{cases} \qquad (4.9)$$

Tansigmoid function: The tansigmoid function is an S-shaped curve. A number of mathematical expressions may be used to define an S-shaped curve. The most commonly used function is expressed as

$$y = f(net) = \frac{1 - e^{-net}}{1 + e^{-net}} \qquad (4.10)$$

This function is easy to differentiate and sometimes enables a simplification to be made in the neural network formulation.

4.4   Network Architecture

Two or more neurons can be combined in a layer to form a network and a network architecture can contain one or more such layers. As research efforts continue, new and extended definitions

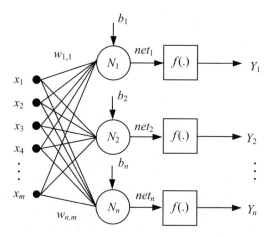

Figure 4.7   Single-layer feedforward network

may develop, but this definition is sufficient for the introductory study of artificial neural architectures and algorithms at this stage. Two basic types of networks are distinguished, considering the connectivity of the neurons in a network:

- Feedforward network and
- Recurrent network or feedback network.

4.4.1   Feedforward Networks

In a feedforward network, only forward connectivity of the neurons is considered. Figure 4.7 shows a single-layer feedforward network. The inputs to the network are the input vector

$$x = \begin{bmatrix} x_1 \\ x_2 \\ \vdots \\ x_m \end{bmatrix} \qquad (4.11)$$

The weights of the network are defined by the weight matrix

$$W = \begin{bmatrix} w_{1,1} & w_{1,2} & \cdots & w_{1,m} \\ w_{2,1} & w_{2,2} & \cdots & \\ \vdots & \vdots & \vdots & \vdots \\ w_{n,1} & w_{n,2} & \cdots & w_{n,m} \end{bmatrix} \qquad (4.12)$$

and the biases are defined by the bias vector

$$b = \begin{bmatrix} b_1 \\ b_2 \\ \vdots \\ b_n \end{bmatrix} \tag{4.13}$$

The output Y of the network can be written in vector form as

$$Y = f(W \cdot x + b) \tag{4.14}$$

The information-processing ability of a neural network depends on its topology (Yao, 1993). The selection of network architecture or topology is largely determined by the application and the number of neurons, connections and choice of transfer functions are fixed during the design. In the following sections, seven different types of feedforward neural network architectures will be investigated for the suitability of different applications:

   (i) Multilayer perceptron networks,
  (ii) Radial basis function networks,
 (iii) Generalized regression neural networks,
 (iv) Probabilistic neural networks,
  (v) Belief networks,
 (vi) Hamming networks and
(vii) Stochastic networks.

4.4.1.1 Multilayer Perceptron Networks

A network with several layers of perceptrons is an MLP network. Each layer has a weight matrix W, a bias vector b and output vector Y as shown in Figure 4.8.

The outputs of the first hidden layer are defined as $^1Y = {}^1f({}^1W \cdot X + {}^1b)$ and the output is defined as $^2Y = {}^2f({}^2W \cdot {}^1Y + {}^2b)$, where $f(.)$ is the chosen activation function.

In principle, the MLP can be employed in any sort of model, linear or nonlinear, and any sort of network, single-layer or multilayer. However, MLP networks have traditionally been associated with sigmoid, tansigmoid functions in a multiple-layer network (mostly three-layer), such as that shown in Figure 4.8. Each of the m components of the input vector $x = \{x_1, x_2, \ldots, x_m\}$ feeds forward to n neurons with sigmoid function defined as $f(x) = \frac{1}{1+e^{-x}}$ or tansigmoid or hyperbolic tangent function defined as $f(x) = \frac{1-e^{-x}}{1+e^{-x}}$, whose outputs are linearly combined with weights $w = \{w_1, w_2, \ldots, w_n\}$ into the network output $f(x)$. These are three of the most commonly used activation functions in MLP networks. They are popular because of the advantage of providing nonzero derivatives with respect to input signals, and they exhibit smoothness and show asymptotic properties. It is customary to use a linear activation function for the output of MLP networks for approximation of a continuous function.

There is no exact rule for determining the number of hidden layers and neurons in the hidden layer. In general, an MLP with one hidden layer will need at least $(P - 1)$ hidden

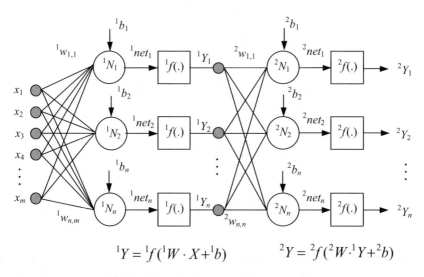

$${}^1Y = {}^1f({}^1W \cdot X + {}^1b) \qquad {}^2Y = {}^2f({}^2W \cdot {}^1Y + {}^2b)$$

Figure 4.8 Multilayer perceptron network

neurons to classify P patterns (Huang and Huang, 1991; Choi *et al.*, 2001). The hidden layer may memorize the input patterns rather than learning the features when the number of hidden-layer neurons exceeds the number of training patterns. If a single neuron memorizes an input pattern, the network will be at risk of failure of that neuron. Therefore, such memorization should be prevented and this can be done by varying the input patterns during training. In other words, the same pattern should not be used more than once for training. Applying a little noise to input patterns will make the training more robust. There are some common rules, such as that a neuron in the first hidden layer forms a hyperplane, which can approximate the boundaries between pattern classes in the pattern space. A neuron in the second hidden layer forms a hyper-region (i.e., convex areas bounded by hyperplanes). A neuron in the third hidden layer defines an area. Therefore, a three-layer network is able to solve a wide range of classification and approximation problems (Jain *et al.*, 1996; Rutkowski, 2005).

The number of neurons in the hidden layer has a decisive impact on the network operation and performance. It is not a critical parameter as the training time does not vary significantly for similar sized hidden layers. A large number of neurons in the hidden layer will make the training process lengthy. If the number of training samples is smaller than the size of the network, it may overtrain the network and lose generalization capability (Tsoukalas and Uhrig, 1997; Rutkowski, 2005).

4.4.1.2   Radial Basis Function Networks

Radial basis function (RBF) networks consist of receptive field units (hidden units). The activation of the receptive field units is defined by a special class of functions, whose response decreases (or increases) monotonically with distance from a central point chosen arbitrarily. The centre and shape of the radial function are the parameters of the radial basis function.

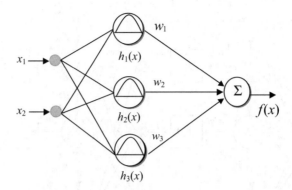

Figure 4.9 Schematic diagram of RBF network

Figure 4.9 illustrates a schematic diagram of an RBF network with three receptive fields. The activation function of the ith receptive field, also called the hidden unit, is defined as

$$h_i(x) = R_i \left(\frac{\|x - c_i\|}{\sigma_i} \right)$$
(4.15)

where x is the multidimensional input vector, the c_i are the centres of the basis functions, the σ_i are the radii of the basis functions and $R_i(.)$ is the ith radial basis function. $R_i(.)$ has a single maximum at the centre. The advantage of the radial basis function network is that there are no connection weights between the input layer and the hidden layer.

Typically, $R_i(.)$ is a Gaussian function defined in Equation (4.16) or a sigmoidal function defined in Equation (4.17):

$$R_i(x) = \exp \left(-\frac{\|x - c_i\|^2}{2\sigma_i^2} \right)$$
(4.16)

$$R_i(x) = \frac{1}{1 + \exp \left(-\frac{\|x - c_i\|^2}{2\sigma_i^2} \right)}$$
(4.17)

Mathematical background

Formally, the method of RBF is a technique for nonlinear discrimination and for multivariate interpolation in high-dimensional spaces (Powell, 1985). RBF method can be seen as an extension of spline functions of one variable to several variables. The problem here is, given a set of m input vectors $x_i \in R^n$, $i = 1, 2, \ldots, m$ and a set of real numbers $y_i \in R$, $i = 1, 2, \ldots, m$ to find a function $f : R^n \to R$ that satisfies the linear spline interpolation condition

$$y_i = f(x_i), i = 1, 2, \ldots, m$$
(4.18)

A set of n nonlinear arbitrary basis functions $\phi_i \|x - \xi_i\|$ is introduced with $\xi_i \in R^n$ such that a strict interpolation or mapping from $f : R^n \rightarrow R$ is implemented according to

$$f(x) = \sum_{i=1}^{n} \lambda_i \phi_i \|x - \xi_i\| \qquad (4.19)$$

where $x \in R^n$ is the input vector, $\phi_i \|\cdot\|$ is the radial basis function, $\lambda_i : i = 1, 2, \ldots, n$ are the weights and $\xi_i, i = 1, 2, \ldots, n$ are the knots. To ensure continuity at the knots, an approach is to introduce polynomial approximations which have higher derivatives at the knots. Commonly, an approximation which has first two derivatives at the knots is introduced. This then forms a basis for the space of well-known cubic splines. A different definition of cubic splines can be found in Powell (1981):

$$f(x) = \sum_{i=1}^{n+2} \lambda_i B_i \|x - \xi_i\| \qquad (4.20)$$

For the RBF models, the knots ξ_i are initially renamed as an equivalent number of centres c_i. Again because continuity is required at the interjoints of the segments, an approximation of the general form is as follows:

$$f(x) = \sum_{i=1}^{n} \lambda_i \phi_i \|x - c_i\| \qquad (4.21)$$

The measure of distance $\|.\|$ is taken as Euclidean norm and $\phi(.)$ is a fixed radially symmetric function. However, unlike the one-dimensional B-splines, the radial basis function for any x is very dependent on the set of input/output data. Choices of $\phi(.)$ that yield a good approximation include: linear $\phi(r) = r$, cubic $\phi(r) = r^3$, thin-plate spline $\phi(r) = r^2 \log(r)$, Gaussian $\phi(r) = e^{-r^2/\beta^2}$, multiquadratic $\phi(r) = (r^2 + k^2)^{1/2}$, inverse multiquadratic $\phi(r) = (r^2 + k^2)^{-1/2}$ and shifted logarithm $\phi(r) = \log(r^2 + k^2)$ with $r = (\|x - c_j\|)$ and β defined as the width of the locally tuned function (Moody, 1989), while k describes the sharpness of the hyperbola cone used in the radial basis function introduced by Hardy (1990). Both are positive constants. The width or shape is controlled by the additional parameters in the case of Gaussian, multiquadratic, inverse multiquadratic and shifted logarithmic RBFs. It can clearly be seen that all RBFs indicate a change of contributions in the approximation as they move radially away from the centre. This simple observation seems to agree with other theoretical investigations that the choice of $\phi(.)$ is non-crucial to the performance of the approximation problem (Moody and Darken, 1989; Moody, 1992).

RBF networks
In principle, RBFs can be employed in any sort of model, linear or nonlinear, and any sort of network, single-layer or multilayer. However, since Broomhead and Lowe's seminal paper (Broomhead and Lowe, 1988), RBF networks have traditionally been associated with radial functions in a single-layer network such as shown in Figure 4.9. Each of m components of the input vector $x = \{x_1, x_2, \ldots, x_m\}$ feeds forward to n basis functions $h(x) = \{h_1(x), h_2(x), \ldots, h_n(x)\}$ whose outputs are linearly combined with weights

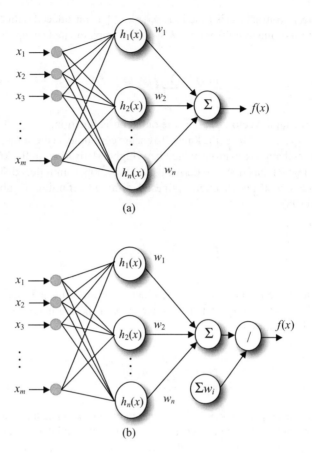

(a)

(b)

Figure 4.10   RBF network architecture. (a) Simple RBF network; (b) Weighted average RBF network

$w = \{w_1, w_2, \ldots, w_n\}$ into the network output $f(x)$. The output of the RBF can be computed in two ways. In the simpler way, the output is calculated as the weighted sum of the output as shown in Figure 4.10(a). A second way is to calculate the weighted average of the output as shown in Figure 4.10(b). An RBF network is nonlinear if the parameters of the basis functions, namely centre and radius, are moving or changing. In order to avoid nonlinear learning, RBF centres can be selected from the training data by some mechanism, and this effectively determines the hidden layer.

The mechanism employed should be simple and effective, and selected centres should suitably sample the input domain. The output layer of the RBF network is linear in the parameters, and its weights can be determined using learning laws such as the least square method (LSM). Thus the hidden layer performs a fixed nonlinear transformation with no adjustable parameters, and it maps the input space onto a new space. The output layer then implements a linear combiner on this new space and the only adjustable parameters are the weights of this linear combiner. Learning is equivalent to performing a linear optimization of the weight space to minimize the total output errors for a given input vector.

The RBF network effectively constructs a linear function space, which depends on the positions of the known data points according to an arbitrary distance measure. Alternatively, it can be seen as a method to perform a hypersurface reconstruction (Poggio and Girosi, 1990). This approach has been widely accepted to yield good multivariable approximation where other techniques fail (Franke, 1982). Good convergence properties RBF have been discussed by Jackson (1988).

4.4.1.3   General Regression Neural Network

The general regression neural network (GRNN) is Donald Specht's term (Specht, 1991) for Nadaraya–Watson kernel regression (Nadaraya, 1964; Watson, 1964), also reinvented in the neural network literature by Schioeler and Hartmann (1992). Kernels are also called Parzen windows. GRNNs can be thought of as normalized RBF networks in which there is a hidden unit centred at every training case. These RBF units are called kernels and are usually probability density functions such as the Gaussian (Wand and Jones, 1995). The hidden-to-output weights are just the target values, so the output is simply a weighted average of the target values of training cases close to the given input case. The only weights that need to be learned are the widths of the RBF units. These widths (often a single width is used) are called smoothing parameters or bandwidths, and are usually chosen by cross-validation or more esoteric methods that are not well known in the neural network literature.

A GRNN is a universal approximator for smooth functions, so it should be able to solve any smooth function approximation problem given enough data (Hocking, 1976, 1983). The main drawback of GRNN is that, like kernel methods in general, they suffer badly from the curse of dimensionality. A GRNN cannot ignore irrelevant inputs without major modifications to the basic algorithm. So, a GRNN is not likely to be the top choice if there are more than five or six non-redundant inputs (Caudill, 1993).

GRNNs are used to decide the problem of regression to find the most probable value of a random variable Z in any point of space X. Such estimation is based only on values of Z at some finite points of this space. We make the assumption that we have an equally normally (Gaussian) distributed random variable on the whole space. Thus, due to the principle of maximum likelihood, by the target function for minimization we understand the mean squared error (MSE). The GRNN is a nonparametric estimator. This means that to decide the problem there is no *a priori* given model; instead, it relies on the fact that the value of the function Z is connected with the coordinates of space X only by means of the probability density function (PDF). If the PDF is known, we can easily define the conditional mean of Z given X. This is called the regression of Z on X, given by

$$Z = \frac{\int\limits_{-\infty}^{+\infty} Z p(X, Z) \, dZ}{\int\limits_{-\infty}^{+\infty} p(X, Z) \, dZ} \tag{4.22}$$

where $X = (x, y) \triangleq$ space coordinates of data points, $Z(X) = Z(x, y) \triangleq$ measured (function) values and $p(X, Z) \triangleq$ function of conditional probability.

When $p(X, Z)$ is not known, it has to be estimated from a sample of observations. $p(X, Z)$ can be estimated by nonparametric kernel estimates proposed by Parzen (1962) for n sample observations with a p-dimensional vector variable:

$$\hat{p}(X, Z) = \frac{1}{\sqrt{(2\pi)^{p+1}} \sigma^{p+1}} \cdot \frac{1}{n} \sum_{i=1}^{n} \exp\left(-\frac{D_i^2}{2\sigma^2}\right) \cdot \exp\left(-\frac{(Z - Z^i)^2}{2\sigma^2}\right) \qquad (4.23)$$

$$D_i^2 = \left(X - X^i\right)^T \left(X - X^i\right) \qquad (4.24)$$

The estimate assigns a sample probability of width σ for each sample $\langle X^i, Z^i \rangle$. Substituting the conditional probability $p(X, Z)$ in Equation (4.23) into the conditional value in Equation (4.22) yields

$$\hat{Z}(X) = \frac{\sum\limits_{i=1}^{n} Z^i \exp\left(-\frac{D_i^2}{2\sigma^2}\right)}{\sum\limits_{i=1}^{n} \exp\left(-\frac{D_i^2}{2\sigma^2}\right)} \qquad (4.25)$$

This can be written in a simplified way as

$$\hat{Z}(X) = \frac{\sum\limits_{i=1}^{n} Z^i w_i}{\sum\limits_{i=1}^{n} w_i} \qquad (4.26)$$

where $w_i = \exp\left(\frac{-D_i^2}{2\sigma^2}\right)$ and w_i is treated as the weight of the ith data point (centre) for estimation of value $\hat{Z}$ at an estimated point with coordinate X, with distance D_i from this point to the centre and σ the only free (adaptive) parameter.

The denominator in Equations (4.25) and (4.26) is used as a normalized constant (common for all centres). The expression in Equation (4.26), also known as the Nadaraya–Watson kernel regression estimator, is at the heart of GRNN. The only free (adaptive) parameter σ defines the bandwidth of the Gaussian kernel. Advanced GRNN realizations may use an anisotropic parameter σ (different values along different directions).

Architecture of GRNNs

GRNNs comprise two layers of artificial neurons. The first layer consists of radial basis neurons, whose transfer function is a Gaussian with spreading factor σ. The first layer weights are simply the transpose of input vectors from the training set. A Euclidean distance is calculated between an input vector and these weights, which are then rescaled by the spreading factor. The radial basis output is then the exponential of the negatively weighted distance having the form

$$f(x) = \exp\left[-\frac{dist(x, w)^2}{\sigma^2}\right] = \exp\left[-\frac{\|x - w\|^2}{\sigma^2}\right] \qquad (4.27)$$

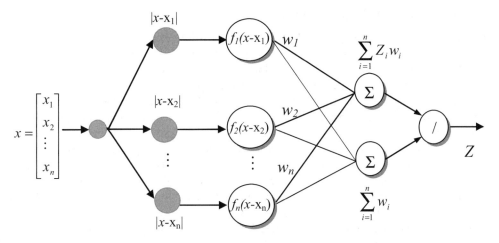

Figure 4.11 Architecture of GRNN

Therefore, if a neuron weight is equal to the input vector, the distance between the two is 0 giving an output of 1. This type of neuron gives an output characterizing the closeness between input vectors and weight vectors. The weight matrix size is defined by the size of the training data set (m parameters, n data points), while the number of neurons is the number of input vectors (n).

The second layer consists of neurons with a linear transfer function. Weights in this layer can be viewed as the slope (m) while the bias vector (b) is the y-intercept. Each is solved to minimize the sum-squared error (SSE) between the output of the first layer (x) and the desired output (y).

As the spreading factor σ increases, the radial basis function decreases in width. The network will respond with the target vector associated with the nearest design input vector. As the spreading factor σ becomes smaller, the radial basis function increases in width. Several neurons may then respond to an input vector. This is because the network does a weighted average of corresponding target vectors. As the radial basis function gets wider and wider, more neurons contribute to the average resulting in a smoother model function. The architecture of the GRNN described above is shown in Figure 4.11.

One advantage of the GRNN approach is its simplicity. The adjustment of the smoothing parameter σ is sufficient for determining the network. When the underlying parent distribution is not known, it is not possible to compute an optimum σ for a given number of observations n. It is therefore necessary to find σ on an empirical basis. Drawbacks of the GRNN can be the large network size and computational cost associated with developing this network for a large set of input test vectors. To adequately characterize the training data, it is typical that the number of neurons equals the number of training vectors. As the number of training vectors increases, the network size and computational load increase. Therefore, implementing a GRNN mainly involves reduction of the number of training vectors. One notable implicit feature of the GRNN solution is that the solution is not in compact or closed form. The function is essentially a look-up table of the network coefficients with matrix size equal to the training set. A detailed analysis of the normalization of the input and selection of the smoothing parameter can be found in Specht (1991).

4.4.1.4 Probabilistic Neural Network

The probabilistic neural network (PNN) is based on well-established statistical principles rather than heuristic approaches. Heuristic approaches usually involve making many small modifications to the system parameters, which gradually improves the system performance. The multilayer perceptron (MLP) neural network is typical of the heuristic approach and is associated with long training times with no guarantee of achieving a suitable solution within a reasonable training time. Specht introduced a three-layer, feedforward, nominal one-pass training algorithm called the PNN, derived from Bayes' decision strategy and nonparametric kernel-based estimators of probability density functions (Specht, 1990a,b). Consequently, it is guaranteed to approach the Bayes' optimal decision surface as the number of training samples increases and the Parzen or Parzen-like probability density function (PDF) (Parzen, 1962) estimator bandwidth approaches zero asymptotically, provided that the class of PDFs is smooth and continuous. One common PNN method is to use sums of spherical Gaussian functions centred at each training vector to estimate the class of PDFs. The spherical Gaussian basis is a Parzen PDF estimator and can be used to implement the PNN according to the following equation:

$$
f_i(x) = \frac{1}{(2\pi)^{\frac{p}{2}} \sigma^p} \frac{1}{M} \sum_{j=1}^{M} \exp\left[\frac{-(x - x_{ij})^T (x - x_{ij})}{2\sigma^2}\right] \tag{4.28}
$$

where i indicates the class number, j indicates the pattern number, x_{ij} is the jth training (or weight) vector from class i, x is the test vector, M is the number of test vectors in class i, p is the dimension of the pattern vector x, σ is the smoothing factor and $f_i(x)$ is the sum of multivariate Gaussian distributions centred at each of the training samples. Functions that are centred at each training vector are to estimate the class of PDFs.

PNN architecture
The PNN network is simply a parallel three-layer feedforward architecture that implements the PDF estimators for each class from their representative training samples (Watanabe and Fukumizu, 1998). The equation for $f_i(x)$, as defined by Equation (4.28), is used as the basis for the PNN. It can be written as a Bayes' decision function $D_i(x)$, as defined by Equation (4.24), if both vectors x and x_{ij} are normalized to unit length and if it is assumed that the number of representative sample vectors M_i (in each class) is in proportion to their *a priori* probability of occurrence. If the input vector x is normalized to unit length, then $\|x\|^2 = x.x = \|x_{ij}\|^2 = x_{ij}.x_{ij} = 1$. It can be shown that

$$
\exp\left[\frac{-(x - x_{ij})^T (x - x_{ij})}{2\sigma^2}\right] = \exp\left[\frac{(Z_{ij} - 1)}{\sigma^2}\right]
$$

The pattern units each form a dot product of the input pattern vector x with a weight vector x_{ij} (i.e., $Z_{ij} = x.x_{ij}$) and then performs a nonlinear operation on Z_{ij} before putting its activation level to the summation unit. PNN uses the nonlinear operation $\exp[\frac{(Z_{ij}-1)}{\sigma^2}]$ instead of a sigmoid activation function commonly used in backpropagation networks. A detailed derivation of the above equation can be found in Zaknich (2003). Normalization of vector x is strictly required to justify the underlying theory. However, in practice the vector normalization is often not necessary since it is still possible to form a satisfactory classification system without it. Often the relative vector magnitudes contain relevant discriminating information and it may be best not to normalize vectors for best performance. At other times vector normalization provides scale invariance, which may be important in some problems.

Another way to see the need for this normalization is to note that the underlying PDF is to be estimated with a basis function that has the same width in each direction. The factor $\frac{1}{M}$ is proportional to the reciprocal of the *a priori* probability of occurrence f_i and hence can be removed from Equation (4.28). Thus, Equation (4.28) becomes

$$f_i(x) = \frac{1}{(2\pi)^{\frac{p}{2}}\sigma^p} \sum_{j=1}^{M_i} \exp \frac{(Z_{ij} - 1)}{\sigma^2} \tag{4.29}$$

The factor $\frac{1}{(2\pi)^{\frac{p}{2}}\sigma^p}$ in Equations (4.28) and (4.29) is the same constant for all classes and can be ignored. This leaves the final equation for $f_i(x)$ as

$$f_i(x) = \sum_{j=1}^{M_i} \exp \frac{(Z_{ij} - 1)}{\sigma^2} \tag{4.30}$$

The decision function in Equation (4.30) is commonly implemented in the PNN architecture because it only involves a simple dot product, represented by $Z_{ij} = x.x_{ij}$, and an exponential activation function. The activation function is not limited to being an exponential or Gaussian, it can be chosen from a number of different types. Thus, the architecture of the PNN can be simplified as shown in Figure 4.12.

The input units are the distribution points for the vector elements and they supply the same values to each of the pattern units for each class, as shown in Figure 4.12. Each pattern unit represents a training vector x_{ij} and performs the operation of the exponential term $[\exp \frac{(Z_{ij}-1)}{\sigma^2}]$.

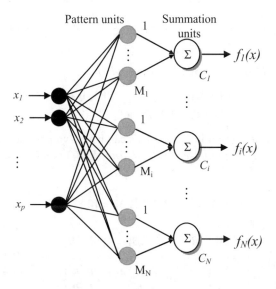

Figure 4.12   Architecture of PNN

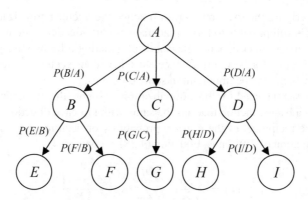

Figure 4.13   Tree of a belief network

These are summed for each class C_i to produce $f_i(x)$. Finally, the highest $f_i(x)$ value is chosen to determine the class decision for the unknown vector x. The smoothing factor σ has the same value throughout the network and it is the only adjustment made for optimizing the network. In practical problems it is not difficult to find a good value of σ by trial and error. The training of a PNN is fast, easy and typically requires only a few passes, but the most annoying disadvantage is that all training vectors must be stored and used – requiring a large memory space.

4.4.1.5   Belief Networks

A belief network is described by a directed acyclic graph, where nodes represent the events and connections to nodes represent cause-and-effect relationships between the nodes. A belief network is shown in Figure 4.13. The node A may have a number of possible values denoted by the probability distribution of A as $\{P(A_1), P(A_2), \ldots, P(A_n)\}$. For any two nodes A and B, a conditional probability matrix $[P(B/A)]$ represents the directed link from node A to node B if there exists a dependence $A \rightarrow B$ as shown in Figure 4.13. For example, the conditional probabilities $[P(B/A)], [P(C/A)]$ and $[P(D/A)]$ represent the dependencies $A \rightarrow B, A \rightarrow C$ and $A \rightarrow D$, respectively.

Given the probability distribution of $P(A)$, the probability distribution of the event B can be computed as

$$
\begin{aligned}
P(B) &= [P(B_1), P(B_2), \ldots, P(B_m)] \\
&= [P(A_1), P(A_2), \ldots, P(A_n)] * [P(B/A)] \\
&= [P(A)] * [P(B/A)]
\end{aligned} \tag{4.31}
$$

Each event in the directed graph in Figure 4.13 can have two possible values, true or false. That is, $P(A_1) = true$, $P(A_2) = false$, $P(B_1) = true$, $P(B_2) = false$. Let $P(A)$ and $P(B/A)$ be defined as

$$
P(A) = [P(A_1)P(A_2)] = [0.8 \quad 0.2]
$$

$$
P(B/A) = \begin{vmatrix} A \cdot B & B_1 & B_2 \\ \hline A_1 & P(B_1|A_1) & P(B_2|A_1) \\ A_2 & P(B_1|A_2) & P(B_2|A_2) \end{vmatrix} = \begin{vmatrix} 0.7 & 0.3 \\ 0.4 & 0.6 \end{vmatrix} \tag{4.32}
$$

The probability distribution of B can be computed according to Equation (4.31) as

$$P(B) = [P(A)] * [P(B/A)]$$

$$= \begin{bmatrix} 0.8 & 0.2 \end{bmatrix} \begin{bmatrix} 0.7 & 0.3 \\ 0.4 & 0.6 \end{bmatrix} \qquad (4.33)$$

$$= \begin{bmatrix} 0.64 & 0.36 \end{bmatrix}$$

That means $P(B_1) = 0.64$ and $P(B_2) = 0.36$. Similarly, the probability distributions of $C, D,$ E, F, G, H and I can be calculated with known conditional probabilities $[P(B/A)], [P(C/A)],$ $[P(D/A)], [P(E/B)], [P(F/B)], [P(G/C)], [P(H/D)]$ and $[P(I/D)]$. The interesting feature of the belief network is that the joint probability $P(A, B, C, D, E, F, G, H, I)$, denoted $P(Z)$ in the equation, can be calculated using the network defined by

$$P(Z) = P(A/B).P(A/C).P(A/D).P(B/E, F)P(C/G).P(D/H, I) \qquad (4.34)$$

If E and F are independent and H and I are independent, then Equation (4.34) becomes

$$P(Z) = P(A/B).P(A/C).P(A/D).P(B/E).P(B/F).P(C/G).P(D/H).P(D/I) \qquad (4.35)$$

Pearl (1987) proposed schemes for propagating beliefs in Bayesian networks. Such networks can be used as a causal reasoning tool and have found many applications (Neal, 1992; Haykin, 1999; Konar, 2005).

4.4.1.6 Hamming Network

The Hamming network (HN) is a two-layer feedforward neural network for classification of binary bipolar n-tuple input vectors using minimum Hamming distance denoted as D_H (Hamming, 1986). The first layer is the input layer for the n-tuple input vectors. The second layer (also called the memory layer) stores p memory patterns. A p-class Hamming network has p output neurons in this layer. The strongest response of a neuron is indicative of the minimum Hamming distance between the stored pattern and the input vector. The Hamming network for an n-tuple binary vector is shown in Figure 4.14. Let $x = [x_1, x_2, \ldots, x_n]$ be the n-tuple input vector, λ be the p-class of patterns (prototype vectors) $\lambda = [\lambda_1, \lambda_2, \ldots, \lambda_p]$, $\lambda_m \in \{-1, +1\}$ (bipolar binary element); $\lambda_m, m = 1, 2, \ldots, p$ is a pattern stored in the mth neuron in the network. The input pattern x is an n-dimensional vector of ± 1 s, a randomly generated version of one of the memory (stored) patterns. Each memory neuron is connected to all n neurons of the input layer. A memory pattern λ_m is stored in the network by letting the values of the connections between the memory neuron m and the input-layer neuron $i, i = 1, 2, \ldots, n$ (i.e., the weight vector) be $W_m = [w_{m1}, w_{m2}, \ldots, w_{mn}]$ where $m = 1, 2, \ldots, p$.

A Hamming network computes the Hamming distance between the input vector x and the memory pattern λ_m stored in the network and selects the memory which has the smallest Hamming distance, i.e., min $D_H(x, \lambda_m)$ for $m = 1, 2, \ldots, p$. The mth output of the Hamming network will be 1 for an input vector x if and only if $x = \lambda_m$. This would require the weights to be $W_m = \lambda_m$. Then, the network outputs are $x^T \lambda_1, x^T \lambda_2, \ldots, x^T \lambda_p$. The scalar product $x^T \lambda_m$ of two bipolar binary n-tuple vectors is equal to the difference between the total number of similar bit positions and the total number of different bit positions. The Hamming distance

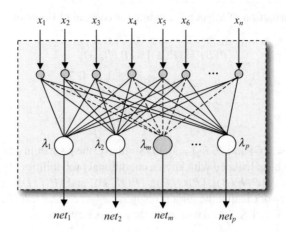

Figure 4.14 Hamming network for n-tuple binary vector classifier

$D_H(x, \lambda_m)$ is the number of bit positions that differ. For an n-bit input vector, the scalar product $x^T \lambda_m$ can be written as

$$x^T \lambda_m = [n - D_H(x, \lambda_m)] - D_H(x, \lambda_m) = n - 2D_H(x, \lambda_m)$$
$$\frac{1}{2} x^T \lambda_m = \frac{n}{2} - D_H(x, \lambda_m) \tag{4.36}$$

The Hamming distance can be derived from Equation (4.36) as

$$D_H(x, \lambda_m) = \frac{1}{2} x^T \lambda_m + \frac{n}{2} \tag{4.37}$$

where $\frac{n}{2}$ is seen as a fixed bias. From the earlier assumption that the weights should be $W_m = \lambda_m$, the weights of the Hamming network can be created by encoding the class vector prototypes as rows:

$$W_H = \frac{1}{2} \begin{bmatrix} \lambda_1^{(1)} & \lambda_1^{(2)} & \cdots & \lambda_1^{(n)} \\ \lambda_2^{(1)} & \lambda_2^{(2)} & \cdots & \lambda_2^{(n)} \\ \vdots & \vdots & \ddots & \vdots \\ \lambda_p^{(1)} & \lambda_p^{(2)} & \cdots & \lambda_p^{(n)} \end{bmatrix} \tag{4.38}$$

The factor $\frac{1}{2}$ is used for scaling purposes. Thus, the output of the mth neuron of the Hamming network is the Hamming distance $D_H(x, \lambda_m)$, that is

$$net_m = \frac{1}{2} x^T \lambda_m + \frac{n}{2} \tag{4.39}$$

Using the identity in Equation (4.36), net_m can be expressed as

$$net_m = n - D_H(x, \lambda_m) \tag{4.40}$$

Applying any activation function will simply need scaling of the neurons' output $f(net_m)$. A perfect match of input vector to class m will result in $D_H(x, \lambda_m) = 0$, giving the output $f(net_m) = 1$. The classification by a Hamming network is performed in a feedforward and instantaneous manner. Therefore, Hamming networks have found many applications (Zurada, 1992; Meilijson et al., 1998; Siddique et al., 2010).

4.4.1.7 Stochastic Networks (or Machines)

Stochastic networks model statistical behaviours using the principles of statistical mechanics. The network is shown some distribution of patterns to learn the internal model. The network is then capable of generating the same distribution when a set of input patterns is presented to the network. The laws of thermodynamics, concept of entropy, Gibbs distribution and Shannon's information theory are used as tools for the statistical mechanics. Examples of stochastic neural networks are the Boltzmann machine, Cauchy machine, and Helmholtz machine.

Boltzmann machine

The Boltzmann network (or machine) was probably the first multilayer learning machine inspired by statistical mechanics (Hilton, 1989). The network has two layers: the layer of visible units and the layer of hidden units. The network consists of stochastic neurons, uses bidirectional and symmetric connections between neurons, and neurons in the same layer have no connections between them. Weights are the same in both directions. Primarily a Boltzmann machine learns a neural network that can correctly model input patterns according to the Boltzmann distribution (Hilton, 1989; Anderson and Titterington, 1998; Haykin, 2009). The firing rule of the network may be expressed by the following:

$$P_j = \frac{1}{1 + \exp(-\Delta E_j/T)} \tag{4.41}$$

where $\Delta E_j = net_j = $ total inputs received by neuron j, T is the temperature of the network. Slow cooling rate, for example $T(n + 1) = T(0) * [1/\log(1 + n)], n \geq 1$, is used in Boltzmann machine.

Cauchy machine

The Cauchy machine is similar to the Boltzmann machine, where different temperatures and cooling rate patterns are used. Faster cooling rate such as $T(n + 1) = T(0)/(1 + T)$ is used in Cauchy machine. Cauchy distribution is characterised by longer tails than Boltzmann distribution. It increases the probability that larger changes will be made to speed up convergence. Thus, the presence of a few huge jumps enables faster escape from local minima. To move out of local minima, both allow error to increase under some conditions (Tsoukalas and Uhrig, 1997). Cauchy machine represents a possible solution to the local minima problem encountered with virtually every other neural network.

Helmholtz machine

The Helmholtz machine is a statistical inference engine. A recognition model is used to infer a probability distribution over the underlying causes from the input. A generative model is used to train the recognition model. The wake-sleep learning is a way of training the Helmholtz machine (Dayan *et al.*, 1995).

4.5   Learning in Neural Networks

Learning in a network is a procedure for modifying the weights and biases of a network, also referred to as a training algorithm, to force a network to yield a particular response to a specific input. Many learning rules are in use. Most of these rules are some sort of variation of the well-known rules. Research into different learning functions continues as new ideas routinely show up in the literature. Some researchers have the modelling of biological learning as their main objective. Others are experimenting with adaptations of their perceptions of how nature handles learning. Learning is certainly more complex than the simplifications represented by the learning rules used. Two different types of learning rules can be distinguished:

- Learning with supervision,
- Learning without supervision.

4.5.1   Supervised Learning

A supervised learning rule is provided with a set of input/output data (also called training data) of proper network behaviour. As the inputs are applied to the network, the network outputs are compared to the target outputs. The learning rule is then used to adjust the weights and biases of the network in order to move the network outputs closer to the targets. Supervised learning is illustrated in Figure 4.15.

The learning method tries to minimize the current errors of all processing elements. This global error reduction is created over time by continuously modifying the weights until an acceptable error goal is reached. Training consists of presenting input and output data to the

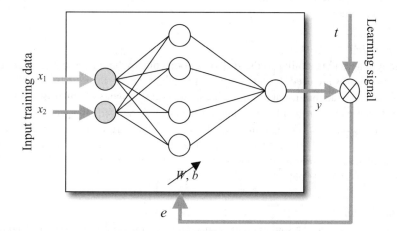

Figure 4.15   Supervised learning

network. This data is often referred to as the training set. That is, for each input set provided to the system, the corresponding desired output set is provided as well:

$$\{^1(x_1, x_2),^1 (t_1)\}, \{^2(x_1, x_2),^2 (t_1)\}, \ldots, \{^N(x_1, x_2),^N (t_1)\}$$ (4.42)

This training can consume a lot of time. In prototype systems, with inadequate processing power, learning can take days and even weeks. The rules that belong to supervised learning are:

- Widrow–Hoff rule,
- Gradient descent,
- Delta rule,
- Backpropagation rule,
- Cohen–Grossberg learning rule, and
- Adaptive conjugate gradient model of Adeli and Hung.

4.5.1.1   Widrow–Hoff Learning Algorithm

The Widrow–Hoff learning rule (Widrow and Hoff, 1960) is applicable for supervised training of neural networks. It is independent of the activation function of the neurons used since it minimizes the squared error between the desired output d_i and a neuron's actual output value. The Widrow–Hoff learning rule is shown diagrammatically in Figure 4.16. The weight vector increment under this learning rule is

$$\Delta w_i = \eta(d_i - o_i)x$$ (4.43)

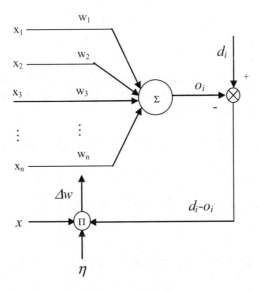

Figure 4.16   Widrow–Hoff learning

where η is the learning constant and x is the input vector. The output o_i is defined as

$$o_i = f(net_i) = net_i \qquad (4.44)$$

where net_i is defined as

$$net_i = \sum w_i^t x \qquad (4.45)$$

This rule can be considered a special case of the delta learning rule.

Example 4.1　Assume the neural network with a single neuron as shown below:

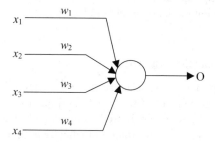

having initial weight vector w^1 and three input vectors:

$$w^1 = \begin{bmatrix} 1 \\ -1 \\ 0 \\ 0.5 \end{bmatrix}, \; x^1 = \begin{bmatrix} 1 \\ -2 \\ 1.5 \\ 0 \end{bmatrix}, \; x^2 = \begin{bmatrix} 1 \\ -0.5 \\ -2 \\ -1.5 \end{bmatrix} \text{ and } x^3 = \begin{bmatrix} 0 \\ 1 \\ -1 \\ 1.5 \end{bmatrix}$$

The network needs to be trained with a learning rate $\eta = 1$. The desired output is $d = [1 \quad -1 \quad 0]$. The activation function is defined as $f(net) = net$. Compute the weight vector after the first iteration of Widrow–Hoff learning.

Solution　In the first iteration, net^1 is calculated according to Equation (4.34):

$$net^1 = w^1 x^1 = \begin{bmatrix} 1 & -1 & 0 & .5 \end{bmatrix} \begin{bmatrix} 1 \\ -2 \\ 1.5 \\ 0 \end{bmatrix} = 3$$

The output is defined according to Equation (4.44) in the Widrow–Hoff learning rule:

$$o^1 = f(net^1) = net^1 = 3$$

The weight update is

$$\Delta w^1 = \eta(d^1 - o^1)x^1 = 1 * (1 - 3) \begin{bmatrix} 1 \\ -2 \\ 1.5 \\ 0 \end{bmatrix} = \begin{bmatrix} -2 \\ 4 \\ -3 \\ 0 \end{bmatrix}$$

Finally, the weight after the first iteration is

$$w^2 = w^1 + \Delta w^1 = \begin{bmatrix} 1 \\ -1 \\ 0 \\ 0.5 \end{bmatrix} + \begin{bmatrix} -2 \\ 4 \\ -3 \\ 0 \end{bmatrix} = \begin{bmatrix} -1 \\ 3 \\ -3 \\ 0.5 \end{bmatrix}$$

4.5.1.2 Gradient Descent Rule

This rule is similar to the Delta rule in that the derivative of the transfer function is still used to modify the delta error before it is applied to the connection weights. Here, however, an additional proportional constant tied to the learning rate is added to the final modifying factor acting upon the weight. This rule is commonly used, even though the convergence to a stable point is very slow. It has been shown that different learning rates for different layers of a network help the learning process converge faster. Among the gradient descent learning algorithms, backpropagation is the most popular and is an extension of the perceptrons to a multilayered neural network. There are a number of variations on the basic algorithm based on other optimization techniques, such as conjugate gradient and Newton methods.

Delta learning rule

This rule is a variation of Hebb's rule and based on the simple idea of continuously modifying the weights of the input connections to reduce the difference (the delta) between the desired output and the actual output of the network. It changes the weights in such a way as to minimize the mean squared error of the network. The delta error in the output layer is transformed by the derivative of the transfer function and this error is backpropagated into previous layers one layer at a time. The process of backpropagating the network errors continues until the first layer is reached. When using the delta rule, it is important to ensure that the input data set is well randomized. An ordered or structured presentation of the training data set can lead to a network which cannot converge to the desired accuracy, meaning that the network is incapable of learning the problem.

The delta learning rule is applicable for supervised training of neural networks and valid for a continuous activation function. It minimizes the squared error between the desired output y_d and the actual output, calculating the gradient vector with respect to w_i of the squared error defined as

$$E = \frac{1}{2}(y_d - o_i)^2 = \frac{1}{2}e^2 \tag{4.46}$$

where the output o_i is defined as

$$o_i = f(net_i) = f\left(w_i^t x\right) = w_i^t x \tag{4.47}$$

Here $f(.)$ is the activation function, which is continuous for delta learning. The minimization of the error requires the weight vector changes to be in the negative gradient direction, so we it is defined as

$$\Delta w_i = -\eta \nabla E \tag{4.48}$$

where η is the learning rate (constant). ∇E is defined as

$$\nabla E = -(y_d - o_i)f'(net)x \tag{4.49}$$

Here, $f'(.)$ is the derivative of the activation function. The components of the gradient vectors are

$$\frac{\partial E}{\partial w_{ij}} = -(y_d - o_i)f'(net)x_j, \quad j = 1, 2, \ldots, n \tag{4.50}$$

From Equations (4.48) and (4.49), the weight update for delta learning becomes

$$\Delta w_i = \eta(y_d - o_i)f'(net_i)x = \eta e_i f'(net)x \tag{4.51}$$

The process of delta learning is shown in Figure 4.17.

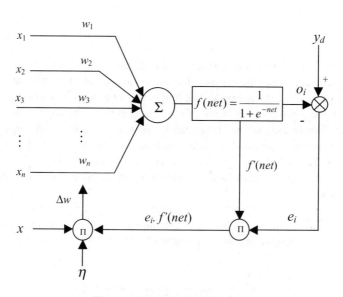

Figure 4.17   Delta learning rule

Example 4.2   Assume the neural network with a single neuron as shown below:

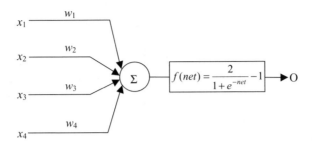

having initial weight vector w^1 and three input vectors:

$$w^1 = \begin{bmatrix} 1 \\ -1 \\ 0 \\ 0.5 \end{bmatrix}, x^1 = \begin{bmatrix} 1 \\ -2 \\ 0 \\ -1 \end{bmatrix}, x^2 = \begin{bmatrix} 0 \\ 1.5 \\ -0.5 \\ -1 \end{bmatrix} \text{ and } x^3 = \begin{bmatrix} -1 \\ 1 \\ 0.5 \\ -1 \end{bmatrix}$$

The desired responses for x^1, x^2 and x^3 are $d^1 = -1$, $d^2 = -1$ and $d^3 = 1$, respectively. The network needs to be trained with a learning constant $c = 1$. The activation function is defined as

$$f(net) = \frac{2}{1 + e^{-net}} - 1$$

The delta learning rule requires the value of $f'(.)$ to be computed in each step. For this purpose the following derivative is given:

$$f'(net) = \frac{1}{2}(1 - o^2)$$

Compute the weight update vector after the second iteration of delta learning.

Solution   In the first iteration, net^1 is calculated using the input vector x^1 and initial weight vector w^1

$$net^1 = w^1 x^1 = \begin{bmatrix} 1 & -1 & 0 & .5 \end{bmatrix} \begin{bmatrix} 1 \\ -2 \\ 0 \\ -1 \end{bmatrix} = 2.5$$

Output of the net is calculated using the activation function and derivative of the activation function is evaluated

$$o^1 = f(net^1) = \frac{2}{1 + \exp(net^1)} - 1 = 0.848$$

$$f'(net^1) = \frac{1}{2}\left[1 - (o^1)^2\right] = 0.140$$

w^2 is computed using the weight update rule in Equation (4.49) and the initial weight vector w^1

$$w^2 = \eta \left(d_1 - o^1 \right) f' \left(net^1 \right) x^1 + w^1$$

$$= 0.1 \left(-1 - 0.848 \right) * 0.140 * \begin{bmatrix} 1 \\ -2 \\ 0 \\ -1 \end{bmatrix} + \begin{bmatrix} 1 \\ -1 \\ 0 \\ 0.5 \end{bmatrix} = -0.0259 * \begin{bmatrix} 1 \\ -2 \\ 0 \\ -1 \end{bmatrix} + \begin{bmatrix} 1 \\ -1 \\ 0 \\ 0.5 \end{bmatrix}$$

$$= \begin{bmatrix} 0.974 \\ -0.948 \\ 0 \\ 0.487 \end{bmatrix}$$

In the second iteration, net^2 is calculated using the input vector x^2 and the weight vector w^2

$$net^2 = w^2 x^2 = \begin{bmatrix} 0.974 & -0.948 & 0 & 0.487 \end{bmatrix} \begin{bmatrix} 0 \\ 1.5 \\ -0.5 \\ -1 \end{bmatrix} = -1.948$$

Output of the net is calculated using the activation function and derivative of the activation function is evaluated

$$o^2 = f(net^2) = \frac{2}{1 + \exp(net^2)} - 1 = -0.75$$

$$f'(net^2) = \frac{1}{2} \left[1 - (o^2)^2 \right] = 0.218$$

Weight update after second iteration is:

$$\Delta w^2 = \eta \left(d_2 - o^2 \right) f' \left(net^2 \right) x^2$$

$$= 0.1 * (-1 + 0.75) * 0.218 * \begin{bmatrix} 0 \\ 1.5 \\ -0.5 \\ -1 \end{bmatrix} = \begin{bmatrix} 0 \\ -0.008 \\ 0.002 \\ 0.008 \end{bmatrix}$$

4.5.1.3 Generalized Delta Learning Rule

The delta learning rule can be generalized and applied to any feedforward layered network. The architecture of a two-layer network is considered in this case and shown in Figure 4.18.

The delta learning rule can now be applied to adjust the hidden layer weights (^1W) and output layer weights (^2W) of the two-layered network. This generalized delta learning rule is error backpropagation learning – the algorithm is shown as a block diagram in Figure 4.19 and will be further explained in the next section.

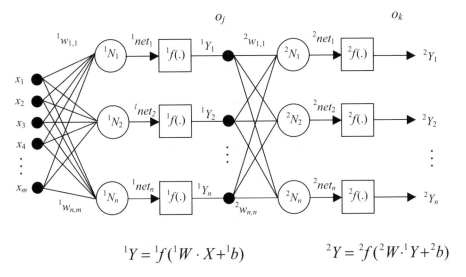

$${}^{1}Y = {}^{1}f({}^{1}W \cdot X + {}^{1}b) \qquad\qquad {}^{2}Y = {}^{2}f({}^{2}W \cdot {}^{1}Y + {}^{2}b)$$

Figure 4.18   Two-layer feedforward network

4.5.1.4   Backpropagation Learning Algorithm

Standard backpropagation is a gradient descent algorithm. The term 'backpropagation' refers to the manner in which the gradient is computed for nonlinear multiplayer networks. For example, we consider a three-layered NN shown in Figure 4.20.

Thus, the output layer is defined as

$$O_k = f(net_k) \tag{4.52}$$

$$net_k = \sum_j W_{kj} O_j + \theta_k \tag{4.53}$$

and the hidden layer is defined as

$$O_j = f(net_j) \tag{4.54}$$

$$net_j = \sum_i W_{ji} O_i + \theta_j \tag{4.55}$$

where $f(net)$ is given by

$$f(net) = \frac{1}{1 + \exp(-net)} \tag{4.56}$$

The learning procedure involves the presentation of a set of pairs of input/output patterns. The net propagates the pattern inputs to outputs to produce its own output patterns and then compares this with the desired output. The difference is called the error. If there is error, it is backpropagated to have the weights and biases changed. If there is no error, learning stops.

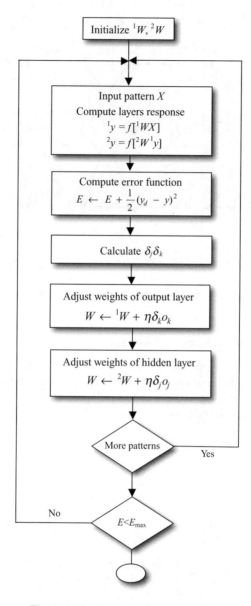

Figure 4.19   Error backpropagation rule

Derivation of the backpropagation algorithm

The backpropagation algorithm is based on the gradient descent method, in that the error function is defined by Equation (4.46). The principle of gradient descent method is shown in Figure 4.21, where the error function is compared with a rolling ball going down a valley.

$$E = \frac{1}{2} \sum e^2 = \frac{1}{2} \sum_k (t_k - O_k)^2 \qquad (4.57)$$

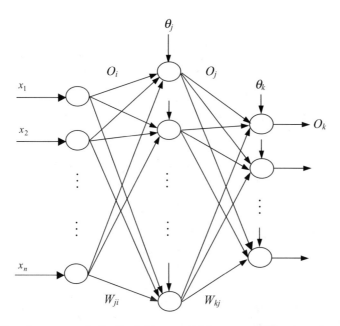

Figure 4.20 Three-layer network: $O_k, O_j, O_i \cong$ outputs of the output, hidden, input layers, respectively; $w_{kj} \cong$ connection weight from hidden layer j to output layer k; $w_{ji} \cong$ connection weight from input layer i to hidden layer j

Calculation of the output layer weight change. According to the steepest descent (gradient descent) method:

$$\Delta w_{kj} = -\eta \frac{\partial E}{\partial w_{kj}} \qquad (4.58)$$

where η is the learning rate and $\eta > 0$, Δw_{kj} is the weight change and $\Delta w_{kj} = w_{kj}^{new} - w_{kj}^{old}$.

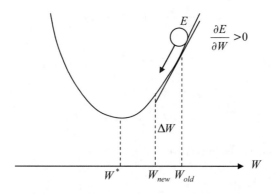

Figure 4.21 Principle of gradient descent method

Using the chain rule, we get

$$\frac{\partial E}{\partial w_{kj}} = \frac{\partial E}{\partial net_k} \cdot \frac{\partial net_k}{\partial w_{kj}} = -\delta_k \cdot \frac{\partial net_k}{\partial w_{kj}} \tag{4.59}$$

where $\delta_k = -\frac{\partial E}{\partial net_k}$, termed the generalized error signal.

Now, we want to derive the term $\frac{\partial net_k}{\partial w_{kj}}$:

$$\frac{\partial net_k}{\partial w_{kj}} = \frac{\partial \left(\sum_j w_{kj} O_j + \theta_k \right)}{\partial w_{kj}} = O_j \tag{4.60}$$

To compute the term δ_k, we apply the chain rule

$$\delta_k = -\frac{\partial E}{\partial net_k} = -\frac{\partial E}{\partial O_k} \frac{\partial O_k}{\partial net_k} \tag{4.61}$$

$$\frac{\partial E}{\partial O_k} = \frac{\frac{1}{2} \sum_k (t_k - O_k)^2}{\partial O_k} = -(t_k - O_k) \tag{4.62}$$

$$\frac{\partial O_k}{\partial net_k} = f'(net_k) \tag{4.63}$$

Note. $f'(x)$ denotes the derivative of $f(x)$ with respect to x and can easily be derived as follows:

$$f'(x) = \frac{e^{-x}}{(1 + e^{-x})^2} = \frac{1}{(1 + e^{-x})} \left(1 - \frac{1}{1 + e^{-x}} \right) \tag{4.64}$$

$$f'(x) = f(x)(1 - f(x))$$

Hence,

$$\frac{\partial O_k}{\partial net_k} = O_k (1 - O_k) \tag{4.65}$$

Thus, we have

$$\Delta w_{kj} = \eta \delta_k O_j \tag{4.66}$$

$$\delta_k = O_k (1 - O_k)(t_k - O_k)$$

Similarly, we can calculate the bias change

$$\Delta \theta_k = -\eta \frac{\partial E}{\partial \theta_k} = \eta \left(-\frac{\partial E}{\partial net_k} \frac{\partial net_k}{\partial \theta_k} \right) = \eta \delta_k \frac{\partial \left(\sum_j w_{kj} O_j + \theta_k \right)}{\partial \theta_k} = \eta \delta_k \tag{4.67}$$

Calculation of the hidden-layer weight change. According to the gradient descent method:

$$\Delta w_{ji} = -\eta \frac{\partial E}{\partial w_{ji}} \tag{4.68}$$

Using the chain rule we get

$$\frac{\partial E}{\partial w_{ji}} = \frac{\partial E}{\partial net_j} \cdot \frac{\partial net_j}{\partial w_{ji}} = -\delta_j \cdot \frac{\partial net_j}{\partial w_{ji}} \tag{4.69}$$

Now, we get

$$\frac{\partial net_j}{\partial w_{ji}} = O_i \tag{4.70}$$

Using the chain rule, we get

$$\delta_j = -\frac{\partial E}{\partial net_j} = -\sum_k \frac{\partial E}{\partial net_k} \frac{\partial net_k}{\partial O_j} \frac{\partial O_j}{\partial net_j} \tag{4.71}$$

$$\delta_j = \sum_k \delta_k w_{kj} f'(net_j) \tag{4.72}$$

$$\delta_j = O_j (1 - O_j) \sum_k \delta_k w_{kj} \tag{4.73}$$

Thus, we have

$$\Delta w_{ji} = \eta \delta_j O_i \tag{4.74}$$

and the bias change

$$\Delta \theta_j = -\eta \frac{\partial E}{\partial \theta_j} = \eta \delta_j \tag{4.75}$$

Backpropagation algorithm
1. Initialize w_{kj}, w_{ji}, θ_k and θ_j and set learning rate η.
2. Propagate inputs to network and calculate O_j, O_k.
3. Calculate δ_k by the formula

$$\delta_k = O_k (1 - O_k)(t_k - O_k)$$

4. Calculate change of weights and biases by

$$\Delta w_{kj} = \eta \delta_k O_j$$
$$\Delta \theta_k = \eta \delta_k$$

5. Calculate δ_j by the formula

$$\delta_j = O_j (1 - O_j) \sum_k \delta_k w_{kj}$$

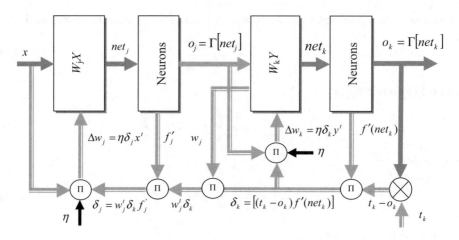

Figure 4.22   Backpropagation algorithm

6. Calculate change of weights and biases by

$$\Delta w_{ji} = \eta \delta_j O_i$$
$$\Delta \theta_j = \eta \delta_j$$

7. Calculate new weights and biases

$$w_{ji}(t+1) = w_{ji}(t) + \Delta w_{ji}$$
$$\theta_j(t+1) = \theta_j(t) + \Delta \theta_j$$
$$w_{kj}(t+1) = w_{kj}(t) + \Delta w_{kj}$$
$$\theta_k(t+1) = \theta_k(t) + \Delta \theta_k$$

8. Set $t \leftarrow t + 1$ and go to step 2.

The backpropagation algorithm is illustrated in Figure 4.22.

Problems with backpropagation learning

Backpropagation is based on the gradient descent algorithm to find the minimum error. We seek global minima, which are sometimes surrounded by many local minima or plateaux. Very often, backpropagation is stuck in a local minimum or plateau. This is shown in Figure 4.23. A momentum term α is added to the learning rule when backpropagation is stuck in a local minimum. Likewise, an acceleration term β is added to the learning rule when backpropagation is stuck in a plateau. The modified learning rules for the output and hidden layer are given by Equations (4.76) and (4.77):

$$\Delta w_{kj}(t) = -\eta \frac{\partial E}{\partial w_{kj}} + \alpha \Delta w_{kj}(t-1) + \beta \Delta w_{kj}(t-2) \tag{4.76}$$

$$\Delta w_{ji}(t) = -\eta \frac{\partial E}{\partial w_{ji}} + \alpha \Delta w_{ji}(t-1) + \beta \Delta w_{kj}(t-2) \tag{4.77}$$

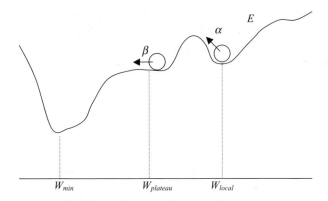

Figure 4.23 Momentum and acceleration

The values of the momentum term α and acceleration term β are chosen arbitrarily. Too big values will cause the algorithm to jump over the global minimum to another local minimum. Too small values will slow down the learning speed.

4.5.1.5 Cohen–Grossberg Learning Rule

Cohen–Grossberg learning comes from the psychological model of Pavlovian learning developed from the dog and food experiment. In Cohen–Grossberg learning, for two associated neurons:

- activity must increase with an external stimulus;
- activity must decrease when there is no stimulus;
- learning must respond to stimuli from neurons in the network.

Neuron j receives signals $y_i(t)$ from neurons i, $i = 1, 2, \ldots, n$ as shown in Figure 4.24. The activity $y_j(t)$ of neuron j can be represented by the differential equation

$$\frac{y_j(t)}{dt} = -\alpha y_j(t) + I_0(t) + \beta \sum_{i=1}^{n} w_{ij}(t) y_i(t) \tag{4.78}$$

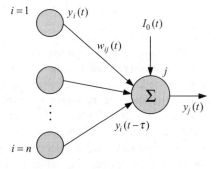

Figure 4.24 Receiving neuron in Cohen–Grossberg learning

where $y_i(t)$ is the activity of the ith neuron, $y_j(t)$ is the activity of the jth neuron, $I_0(t)$ is the external stimulus, $w_{ij}(t)$ is the weight between the ith and jth neurons, and α and β are the forgetting and learning constants, respectively. Two processes involved in the learning law are Hebbian learning and forgetting. The weight update rule is expressed by introducing an explicit forgetting factor, a threshold on the incoming activity term and a transmission delay:

$$\frac{w_{ij}(t)}{dt} = -Fw_{ij}(t) + Gy_j(t)\,|y_i(t-\tau) - T| \qquad (4.79)$$

where τ is the average transmission time from neuron i to neuron j, T is the threshold, G is the gain or learning constant and F is the forgetting factor ($F \leq 0.01$). An appropriate Cohen–Grossberg learning law would be to substitute the activities with the first derivative of the activities, i.e.

$$\frac{w_{ij}(t)}{dt} = -Fw_{ij}(t) + G\frac{y_j(t)}{dt}\left|\frac{y_i(t-\tau)}{dt} - T\right| \qquad (4.80)$$

Equation (4.80) is a version of the differential Hebbian learning (Tsoukalas and Uhrig, 1997).

4.5.1.6   Adaptive Conjugate Gradient Model of Adeli and Hung

The conjugate gradient method is an effective modification of the steepest descent method that was first proposed by Fletcher and Reeves (1964) in order to overcome the shortcomings of the backpropagation algorithm. Powell (1986) proposed a more robust approximate line search algorithm for convergence. The problem of arbitrary trial-and-error selection of the learning and momentum ratios encountered in the momentum backpropagation algorithm is circumvented in the adaptive conjugate gradient model proposed by Adeli and Hung (1994). Instead of constant learning and momentum ratios, the step length in the inexact line search is adapted during the learning process. The new adaptive algorithm provides a more solid mathematical foundation for neural network learning. The algorithm has been applied to the domain of image recognition and demonstrated a superior convergence property compared with backpropagation with momentum.

4.5.2   Unsupervised Learning

Unsupervised learning refers to learning without supervision, i.e., the target response is not known. That means no external signals are used to adjust network weights. Instead, they monitor internal representation for coding input patterns and learn to respond to different input patterns with different parts of the network. Since no information is available as to correctness or incorrectness of response, learning looks for features, regularities or trends present in the input patterns, and makes adaptations according to the function of the network. Even without being told whether it's right or wrong, the network still has some information on how to organize itself to respond to frequently occurring patterns. This information is built into the network topology during learning. Thus, unsupervised learning is usually suitable for data

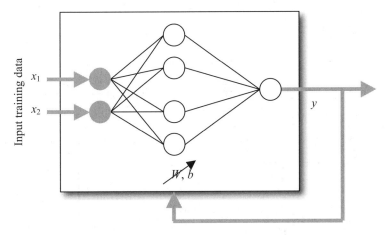

Figure 4.25   Unsupervised learning

clustering, feature extraction, categorization of persistent features in signals, classification and similarity measures.

An unsupervised learning algorithm might emphasize cooperation among clusters of processing elements. In such a scheme, the clusters would work together. Competition between processing elements could also form a basis for learning. Training of competitive clusters could amplify the responses of specific groups to specific stimuli. As such, it would associate those groups with each other and with a specific appropriate response. Normally, when competition for learning is in effect, only the weights belonging to the winning processing element will be updated.

Currently, this learning method is limited to networks known as self-organizing maps. At present, unsupervised learning is not well understood and is still the subject of research. An unsupervised learning scheme is illustrated in Figure 4.25.

The rules that belong to unsupervised learning are

- Hebb's rule and
- Kohonen's rule.

4.5.2.1   Hebbian Learning Rule

The first and undoubtedly the best known learning rule was introduced by Donald Hebb in his book *The Organization of Behavior* in 1949. The basic rule is: 'If a neuron receives an input from another neuron and if both are highly active (mathematically have the same sign), the weight between the neurons should be strengthened'. The Hebbian learning rule represents purely feedforward and unsupervised learning. The rule implements the interpretation of the classic statement – when an axon of cell A is near enough to excite cell B and persistently takes place in firing it, some growth or metabolic change takes place in one or both cells such that A's efficiency is increased. The rule states that if the cross-product of output and input, or correlation term $o_i x_j$ is positive, this results in an increase of weight w_{ji}, otherwise the weight decreases.

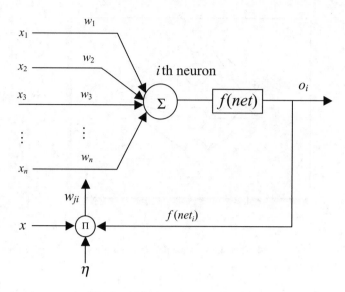

Figure 4.26 Hebbian learning rule

The weight update Δw in Hebbian learning becomes

$$\Delta w_{ji} = \eta o_i x_j = \eta f(net_i) x_j \tag{4.81}$$

where

$$o_i = f(net_i) \tag{4.82}$$

$$net_i = \sum w_{ji} x_j \tag{4.83}$$

with $i, j = 1, 2, 3, \ldots$ Here $f(.)$ is the activation function, which can be continuous or non-continuous. Hebbian learning is shown in Figure 4.26.

Example 4.3   Assume the neural network with a single bipolar binary neuron, as shown below:

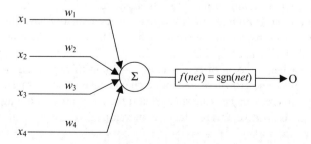

having initial weight vector w^1 and three input vectors x^1, x^2 and x^3:

$$w^1 = \begin{bmatrix} 1 \\ -1 \\ 0 \\ 0.5 \end{bmatrix}, x^1 = \begin{bmatrix} 1 \\ -2 \\ 1.5 \\ 0 \end{bmatrix}, x^2 = \begin{bmatrix} 1 \\ -0.5 \\ -2 \\ -1.5 \end{bmatrix} \text{ and } x^3 = \begin{bmatrix} 0 \\ 1 \\ -1 \\ 1.5 \end{bmatrix}$$

The network needs to be trained with a learning rate $\eta = 1$. The activation function is defined as

$$f(net) = \begin{cases} +1 & net \geq 0 \\ -1 & net < 0 \end{cases}$$

Compute the weight vector after the second iteration of Hebbian learning.

Solution   First iteration: net^1 is calculated using the input vector x^1 and initial weight vector w^1

$$net^1 = w^{1^T} x^1 = \begin{bmatrix} 1 & -1 & 0 & .5 \end{bmatrix} \begin{bmatrix} 1 \\ -2 \\ 1.5 \\ 0 \end{bmatrix} = 3.0$$

Since $net > 0$, o^1 will be $o^1 = f(net^1) = +1$

$$\Delta w^1 = \eta f\left(net^1\right) x^1 = 1 * (+1) * \begin{bmatrix} 1 \\ -2 \\ 1.5 \\ 0 \end{bmatrix} = \begin{bmatrix} 1 \\ -2 \\ 1.5 \\ 0 \end{bmatrix}$$

$$w^2 = w^1 + \Delta w^1 = \begin{bmatrix} 1 \\ -1 \\ 0 \\ 0.5 \end{bmatrix} + \begin{bmatrix} 1 \\ -2 \\ 1.5 \\ 0 \end{bmatrix} = \begin{bmatrix} 2 \\ -3 \\ 1.5 \\ 0.5 \end{bmatrix}$$

Second iteration: net^2 is calculated using input vector x^2 and weight vector w^2

$$net^2 = w^{2^T} x^2 = \begin{bmatrix} 2 & -3 & 1.5 & 0.5 \end{bmatrix} \begin{bmatrix} 1 \\ -0.5 \\ -2 \\ -1.5 \end{bmatrix} = -0.25$$

Since $net < 0$, o^2 will be $o^2 = f(net^2) = -1$

$$\Delta w^2 = \eta f\left(net^2\right) x^2 = 1 * -1 * \begin{bmatrix} 1 \\ -0.5 \\ -2 \\ -1.5 \end{bmatrix} = \begin{bmatrix} -1 \\ 0.5 \\ 2 \\ 1.5 \end{bmatrix}$$

$$w^3 = w^2 + \Delta w^2 = \begin{bmatrix} 2 \\ -3 \\ 1.5 \\ 0.5 \end{bmatrix} + \begin{bmatrix} -1 \\ 0.5 \\ 2 \\ 1.5 \end{bmatrix} = \begin{bmatrix} 1 \\ -2.5 \\ 3.5 \\ 2.0 \end{bmatrix}$$

The weight vector after the second iteration of Hebbian learning is $w^3 = \begin{bmatrix} 1 \\ -2.5 \\ 3.5 \\ 2.0 \end{bmatrix}$

4.5.2.2   Kohonen Learning

Kohonen reported very interesting and useful results from his research on self-organizing maps used for pattern recognition tasks (Kohonen, 1988). Kohonen's learning rule was inspired by learning in biological systems. In this learning system, the processing elements compete for the opportunity to learn, or update their weights. The processing element with the largest output is declared the winner and has the capability of inhibiting its competitors as well as exciting its neighbours. Only the winner is permitted an output, and only the winner plus its neighbours are allowed to adjust their connection weights. Further, the size of the neighbourhood can vary during the training period. The usual paradigm is to start with a larger definition of the neighbourhood, and narrow it as training progresses. The winning neuron or element is defined as the one that has the closest match to the input pattern. The result is that the wining neuron is more likely to win the competition the next time a similar vector is presented, and less likely to when a very different input vector is presented. As more and more inputs are presented, each neuron in the layer closest to a group of input vectors soon adjusts its weight towards those input vectors. Eventually, if there are enough neurons, every cluster of similar input vectors will have a neuron that outputs 1 when a vector of the cluster is presented, while outputting 0 at other times. Thus, the Kohonen network learns to categorize the input vectors it sees. In other words, networks model the distribution of the inputs. This is good for statistical or topological modelling of the data and is sometimes referred to as self-organizing maps. For a better understanding of the mechanism, see Zurada (1992).

Three types of learning will be considered:

- Competitive learning,
- Self-organizing maps, and
- Learning vector quantization or adaptive vector quantization.

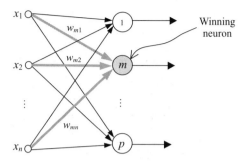

Figure 4.27   Winner-takes-all learning

Competitive learning

There are three basic elements to a competitive learning rule (Rumelhart and Zipser, 1985):

- A set of neurons that are all the same except for some randomly distributed synaptic weights, and which respond differently to a given set of input patterns.
- There is a limit imposed on the strength of each neuron.
- There is a mechanism in place that permits the neurons to compete for the right to respond to a given subset of inputs, such that only one output neuron, or only one neuron per group, is active at a time. The neuron that wins the competition is called a winner-takes-all neuron.

Typically, the winner-takes-all learning rule is used for learning statistical properties of inputs. The learning is based on the premise that one of the neurons in the layer, say the mth, has maximum response due to input x, as shown in Figure 4.27. The neuron is declared the winner. The winner selection is based on the following criterion of maximum activation among all p neurons participating in a competition:

$$w_m = \max_{i=1,2,\dots,p} \{w_i x\} \tag{4.84}$$

This criterion corresponds to finding the weight vector that is closest to the input x. As a result of this winning event, neuron m with weight vector $w_m = [\, w_{m1}\; w_{m2} \cdots w_{mn}\,]^T$ is the only neuron to adjust its weights.

The weight increment is computed as follows:

$$\Delta w_m = \alpha(x - w_m) \tag{4.85}$$

or the individual weight adjustment becomes

$$\Delta w_{mj} = \alpha(x_j - w_{mj}) \quad \text{for} \quad j = 1, 2, \dots, n \tag{4.86}$$

where $\alpha > 0$ is a small learning constant (rate), typically decreasing as learning progresses.

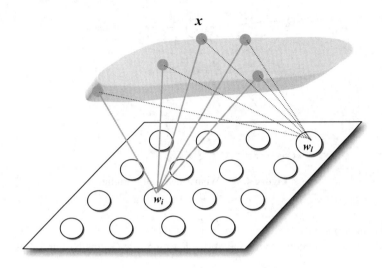

Figure 4.28   Two-dimensional SOM array

Self-organizing maps (SOM)

The self-organizing phenomenon is first demonstrated in an abstract system without reference to any biological structure or signal types. Consider Figure 4.28. The SOM defines a mapping from the input data $x \in \Re^n$ onto a two-dimensional array of nodes. With every node i is associated a parametric reference vector $w_i = [w_{i1}, w_{i2}, \ldots, w_{in}] \in \Re^n$. The lattice type of the array can be defined to be rectangular, hexagonal or even irregular. Hexagonal is effective for visual display. In the simplest case, an input vector $x = [x_1, x_2, \ldots, x_n] \in \Re^n$ is connected to all neurons in parallel via variable scalar weights w_{ij} (weight between ith neuron and jth input), which are in general different for different neurons.

In an abstract scheme it may be imagined that the input x, by means of some parallel computing mechanisms, is compared with all the w_i and the location of the best match in some metric is defined as the location of the response. The exact magnitude of the response need not be determined: the input is simply mapped onto this location, like a set of decoders. Vector x may be compared with all w_i in any metric, the smallest of the Euclidean distance $\|x - w_i\|$ can be made to define the best-matching node, denoted by c:

$$c = \|x - w_i\| = \min_i \{\|x - w_i\|\} \tag{4.87}$$

The Euclidean distance $\|x - w_i\|$ is defined by

$$\|x - w_i\| = \sqrt{\sum_{j=1}^{n} (x_j - w_{ij})^2} \tag{4.88}$$

where $x = [x_1, x_2, \ldots, x_n]$ and $w_i = [w_{i1}, w_{i2}, \ldots, w_{in}]$.

During learning, those nodes that are topographically close in the array up to a certain geometric distance will activate each other to learn something from the same input x. This

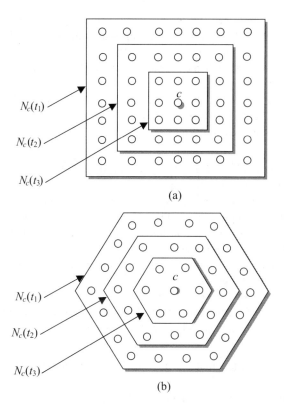

Figure 4.29 Different topological neighbourhoods in SOM. (a) Grid-type topological neighbourhood; (b) Hexagonal topological neighbourhood

will result in a local relaxation or smoothing effect on the weight vectors of neurons in the neighbourhood, which in continued learning leads to global ordering. The learning process will eventually converge, whereby the initial values of $w_i(0)$ can be arbitrary (e.g., random):

$$w_i(t + 1) = w_i(t) + h_{ci}(t)\,[x(t) - w_i(t)] \tag{4.89}$$

where $t = 0, 1, 2, 3, \ldots$

$h_{ci}(t)$ is the so-called neighbourhood function, a smoothing kernel defined over the lattice points. For convergence, it is necessary that $h_{ci}(t) \to 0$ when $t \to \infty$. In the literature, two simple choices for $h_{ci}(t)$ occur frequently. The simpler of them refers to a neighbourhood set of array points around node c, as shown in Figure 4.29. Let their index set be denoted by N_c and $N_c = N_c(t)$, whereby $h_{ci}(t) = \alpha(t)$. The value of $\alpha(t)$ is then identified with a learning rate factor $0 < \alpha(t) < 1$. Both $\alpha(t)$ and $N_c(t)$ are usually decreasing monotonically in time during the ordering process.

The algorithm presented here for preliminary simulations is only representative of many alternative forms. If the SOM network is not very large (say a few hundred nodes at most), selection of the process parameters is not very crucial. Special caution is required in the choice of the size of $N_c = N_c(t)$. If the neighbourhood is too small to start with, the map will not be ordered globally. Instead, various kinds of mosaic-like parcellations of the map are seen, between which the ordering direction changes discontinuously. This phenomenon can be avoided by starting with a fairly wide $N_c = N_c(0)$ and letting it shrink with time. The initial radius can even be more than half the diameter of the network. During the first 1000 steps or so, when the proper ordering takes place and $\alpha = \alpha(t)$ is fairly large, the radius of N_c can shrink linearly to one unit. During the fine-adjustment phase, N_c can still contain the nearest neighbours of cell c. The choice of $\alpha = \alpha(t)$ is critical during learning. Kohonen suggested some criteria for choosing values of $\alpha(t)$:

- $\alpha(t)$ should start with a value close to unity for approximately the first 1000 steps.
- $\alpha(t)$ should decrease monotonically thereafter.
- An accurate timing function is not important: $\alpha = \alpha(t)$ can be linear, exponential or inversely proportional to t. For example, $\alpha(t) = 0.9\left(1 - \frac{t}{1000}\right)$ may be reasonable.
- Ordering of the w_i occurs during this initial period, while the remaining steps are only needed for fine adjustment of the map.
- $\alpha = \alpha(t)$ should attain small values (e.g., on the order of 0.2 or less) over a long period during the fine-adjustment period.
- For a very large map, selection of an optimal $\alpha(t)$ law may be crucial for convergence.
- A rule of thumb is that, for good statistical accuracy, the number of steps in the fine-adjustment phase must be at least 500 times the number of network units.

For a detailed analogy and other variants of SOM, readers are directed to the book by Kohonen (1995).

The SOM algorithm can be summarized as follows:

1. Initialize weights $w_i(0)$. $w_i(0)$ can be set to equal the first M samples.
2. Calculate the Euclidean distance of neuron D_j from the inputs:

$$D_j = \sqrt{\sum_{j=1}^{n}\left(x_j - w_{ij}\right)^2} \tag{4.90}$$

3. Determine the winner neuron with weight vector w_c. w_c becomes the centre of a group of weight vectors that lie within a distance D from w_c.
4. Adjust all such neurons in the neighbourhood by

$$w_i(t+1) = w_i(t) + \alpha_t\left[x - w_i(t)\right]$$

where $0 < \alpha_t < 1$.

As the network trains, gradually reduce the values of D and α.

Example 4.4 Calculate the weight update of the two competing neurons shown in Figure 4.30 using a SOM with $\alpha = 0.5$.

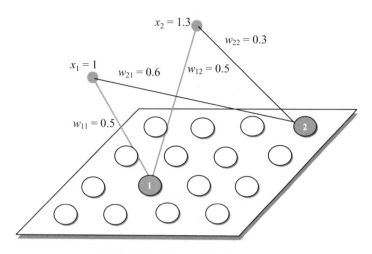

Figure 4.30   Two competing neurons

Solution   The initial weights are provided. So the next step is to calculate Euclidean distance of neuron 1 and 2 from x_1 and x_2 using Equation (4.88). Euclidean distance of neuron 1 from x_1 and x_2 is:

$$D_1 = \sqrt{[x_1 - w_{11}]^2 + [x_2 - w_{12}]^2} = \sqrt{[1 - 0.5]^2 + [1.3 - 0.5]^2}$$
$$= \sqrt{0.5^2 + 0.8^2} = \sqrt{0.89} = 0.943$$

Euclidean distance of neuron 2 from x_1 and x_2 is:

$$D_2 = \sqrt{[x_1 - w_{21}]^2 + [x_2 - w_{22}]^2} = \sqrt{[1 - 0.6]^2 + [1.3 - 0.3]^2}$$
$$= \sqrt{0.4^2 + 1^2} = \sqrt{1.16} = 1.077$$

Determine the winning neuron by finding the minimum distance $D_1 = \min\{D_1, D_2\}$. Therefore, the winning neuron is 1. The weights to neuron 1 are now allowed to update. The new weights are

$$w_{11}(t + 1) = w_{11}(t) + \alpha_t\,[x_1 - w_{11}(t)] = 0.5 + 0.5(1 - 0.5) = 0.5 + 0.25 = 0.75$$

$$w_{12}(t + 1) = w_{12}(t) + \alpha_t\,[x_2 - w_{12}(t)] = 0.5 + 0.5(1.3 - 0.5) = 0.5 + 0.40 = 0.9$$

Learning continues until the maximum epoch is reached or the distance becomes an acceptable minimum.

Learning vector quantization (LVQ) or adaptive vector quantization (AVQ)
Competitive or unsupervised learning systems have been a very rich research topic (Rumelhart and Zipser, 1985; Grossberg, 1987; Kohonen, 1990). They can be used as a mechanism for

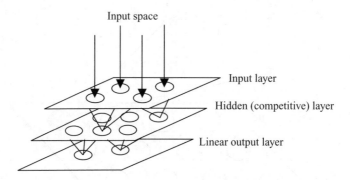

Figure 4.31   LVQ network

adaptive or learning vector quantization, in which the system adaptively quantizes the pattern space by discovering a set of representative prototypes. Therefore, the trained system can carry out the task of classification or recognition (Kohonen, 1990).

There exist a number of variations of LVQ with different network configurations or algorithms. However, the basic idea is quite similar (i.e., to categorize vector-valued stochastic data into different groups by employing some metric measures with a winner-selection criterion).

An LVQ network has two fully connected layers, an input layer with n units and a competitive layer with M units, and a second linear layer, shown in Figure 4.31. The competitive layer learns to classify the input in much the same way as the competitive learning. The linear layer transforms the competitive layer's classes into a target classification defined by the user. Classes learned by the competitive layer are referred to as subclasses and classes of the linear layer as target classes. Both the competitive and linear layers have one neuron per (sub or target) class. Thus, the competitive layer can learn up to S^1 subclasses and is combined with the linear layer to form the S^2 target class; S^1 is always larger than S^2.

- The input layer receives the incoming vector $x(t) \in R^n$ and forwards it to the competitive layer through the weights $w_j(t) \in R^n$.
- $\|x(t) - w_j(t)\| \le \varepsilon$, a very small value, means $w_j(t)$ is close to $x(t)$; the jth unit is the winning unit.
- According to the winner-takes-all activation function, I_j of all units is set to 0 except for the winning unit j, where I_j is set to 1.
- The winner's weight is updated by adding a scaled difference $\|x(t) - w_j(t)\|$ to $w_j(t)$, i.e.

$$w(t+1) = w(t) + \alpha[x(t) - w_j(t)] \qquad (4.91)$$

$\alpha \stackrel{\Delta}{=}$ scaling factor

The LVQ algorithm can be summarized as follows:

1. Initialize weights $w_j(0)$; $w_j(0)$ can be set to equal the first M samples, i.e. $w_j(0) = x_j, j = 1, 2, 3, \ldots, M$.
2. Calculate the distance metric $\|x(t) - w_j(t)\|$.

3. Determine the winner among M units:

$$\|x(t) - w_j(t)\| = \min_j \|x(t) - w_j(t)\|$$

4. Update the weight vectors:

$$w_j(t+1) = \begin{cases} w_j(t) + \alpha_t[x(t) - w_j(t)] & \text{if } j = I \\ w_j(t) & \text{else} \end{cases}$$

where $0 < \alpha_t < 1$.

4.6   Recurrent Neural Networks

Recurrent networks are of considerable research interest. A recurrent network is obtained from the feedforward network by connecting the neuron's output to their inputs. Such a recurrent or feedback network is depicted in Figure 4.32. The essence of closing a feedback loop is to enable control of output through outputs. Such control is especially meaningful if the present output $O(t)$ controls the output at the following instant $O(t + \Delta)$. The time delay Δ has a symbolic meaning – it is an analogy to the refractory period of an elementary biological neuron model. Using the notation introduced for feedforward networks, the mapping of $O(t)$ and $O(t + \Delta)$ can now be written as

$$O(t + \Delta) = \Gamma(W O(t)) \tag{4.92}$$

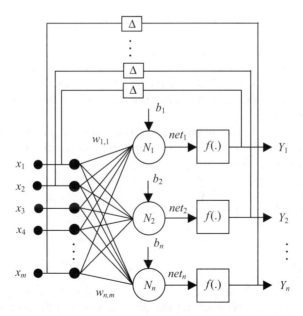

Figure 4.32   Single-layer recurrent network

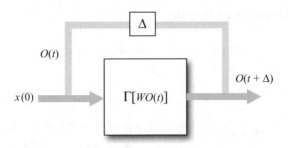

Figure 4.33   Block diagram of mapping in Equation (4.91)

This is represented by the block diagram in Figure 4.33. The input $x(0)$ is only required to initialize this network so that $O(0) = x(0)$. The input is then removed and the system remains autonomous for $t > 0$. Thus, a special case of this feedback configuration is considered such that $x(t) = x(0)$ and no input is provided to the network thereafter, or for $t > 0$. For a discrete-time neural network we can write

$$O(k + 1) = \Gamma(WO(k)) \quad \text{for } k = 1, 2, 3, \ldots \qquad (4.93)$$

The network in Figure 4.33 is called recurrent since its response at the $(k + 1)$th instant depends on the entire history of the network starting at $k = 0$. Indeed, Equation (4.92) can be shown as a series of nested solutions:

$$
\begin{aligned}
O(1) &= \Gamma\,[Wx(0)] \\
O(2) &= \Gamma\,[W\Gamma\,[Wx(0)]] \\
&\;\;\vdots \\
O(k + 1) &= \Gamma\,[W\Gamma\,[\cdots \Gamma\,[Wx(0)] \cdots]]
\end{aligned}
\qquad (4.94)
$$

The network begins state transitions once it is initialized at instant 0 with $x(0)$, and it goes through state transitions $O(k)$ for $k = 1, 2, 3, \ldots$ until it possibly finds an equilibrium state. This equilibrium state is often called an attractor. The sequence of states is generally non-deterministic and in addition, there are often many equilibrium states that can potentially be reached by the network.

Different variants of recurrent network have been discussed and their adaptive capacity has been investigated by Siddique and Amavasai (2012). However, three types of recurrent network are in wide use in the research community:

- Elman networks,
- Jordan networks and
- Hopfield networks.

4.6.1   Elman Networks

Elman networks are three-layer backpropagation networks with the addition of a feedback connection from the output of the hidden layer to its input. This feedback path allows Elman networks to learn to recognize and generate temporal patterns as well as spatial patterns (Elman, 1990).

Let's consider a multilayer perceptron with input units $i_1, \ldots, i_m$ and output units $j_1, \ldots, j_n$ computing a function $f : \mathfrak{R}^m \to \mathfrak{R}^n$ with $f(x_1, \ldots, x_m) = (o_{j1}, \ldots, o_{jm})$. For each unit j, the output o_j is defined recursively by

$$o_j = \begin{cases} x_k & \text{for unit } i_k \\ \sigma\left(\sum_i w_{ji} o_i + \theta_j\right) & \text{otherwise} \end{cases} \tag{4.95}$$

where the sum is taken over all units i in the previous layer, w_{ji} are the weights, θ_j is the bias of unit j and σ is an activation function.

A function $f : (\mathfrak{R}^m)^* \to \mathfrak{R}^l$ is computed by an Elman network if there exists a number $n \in N$, two functions $g : \mathfrak{R}^{m+n} \to \mathfrak{R}^n$ and $h : \mathfrak{R}^n \to \mathfrak{R}^l$ computed by multilayer perceptrons without hidden layers, and a context vector $x \in \mathfrak{R}$ with $f : h \circ \tilde{g}_x$. The last n input units of the function g are called context units in both cases.

The Elman architecture is more specific in that Elman networks have an extra layer of neurons that copy the current activations in the hidden-layer neurons, and after delaying these values for one time unit, feed them back as additional inputs into the hidden layer neurons. The architecture of an Elman network is shown in Figure 4.34. The architecture shown depicts the original Elman network with three layers of neurons. The first layer consists of two different groups of neurons. These are the group of external input neurons and the group of internal input neurons, also called context units. The inputs to the context units are the outputs of the hidden-layer neurons. The outputs of all the context units and the external input neurons are fed to the hidden neurons. Context units are also known as memory units as they store the previous output of the hidden neurons. In the Elman network architecture, the feedback is assumed to be unity (i.e., the feedback weights are not trainable).

Although, theoretically, an Elman network with all feedback connections from the hidden layer to the context layer set to 1 can represent an arbitrary nth-order system, where n is the number of context units, it cannot be trained using the standard backpropagation algorithm. By introducing self-feedback connections to the context units of the original Elman network

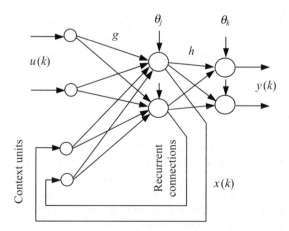

Figure 4.34   Architecture of Elman network

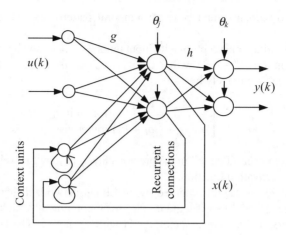

Figure 4.35   Structure of modified Elman network

and thereby increasing its dynamic memory capacity, it is possible to apply the standard BP algorithm to teach the network that task. The modified Elman network is shown in Figure 4.35. The idea of introducing self-feedback connections for the context units is borrowed from the Jordan network. The values of the self-connection weights are fixed between 0 and 1 before the start of training.

Training algorithm
At each epoch, the entire input sequence is presented to the network and its outputs are calculated and compared with the target sequence to generate an error sequence. For each time step, the error is backpropagated to find gradients of errors for each weight and bias. This gradient is actually an approximation since the contributions of weights and biases to errors via the recurrent connections are ignored and a user-chosen value for the gradient is used to update the weights with the backpropagation training function.

Owing to the use of approximation for the error gradient, Elman networks may not be as reliable as some other kinds of network as they are less capable of finding the most appropriate weights for the hidden neurons. For an Elman network to be good at learning a problem it needs more neurons in the hidden layer than are actually required for a solution by some other kinds of network. Therefore, the training of the Elman network should start with a fair number of hidden neurons.

4.6.2   Jordan Networks

Jordan neural networks are like modified Elman networks and have three layers, with the main feedback connections taken from the output layer to the context layer. The structure of a Jordan network is shown in Figure 4.36. It has been shown theoretically that a Jordan network is not capable of representing arbitrary dynamic systems. Therefore, by adding feedback connections from the hidden layer to the context layer, similar to the case of an Elman network, a modified Jordan network is proposed. The Jordan network can be trained using the standard BP algorithm to model different dynamic systems. The modified Jordan network is shown in Figure 4.37.

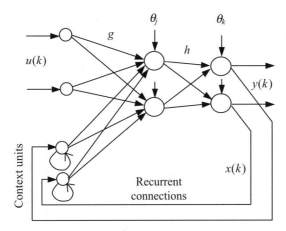

Figure 4.36   Structure of original Jordan network

As with the modified Elman network, the values of the feedback connection weights have to be fixed by the user if the standard BP algorithm is employed.

4.6.3   Hopfield Networks

A Hopfield network is a recurrent network proposed by Hopfield in 1982, which possesses auto-associative properties. It is a fully connected network, except for connection to itself (Hopfield, 1982). The network is used to store one or more stable target vectors. These stable vectors can be viewed as memories that the network recalls when provided with similar vectors. A classic paper in this field is that of Li *et al.* (1989). Following the postulates of Hopfield, the single-layer feedback neural network is assumed as shown in Figure 4.38. It consists of

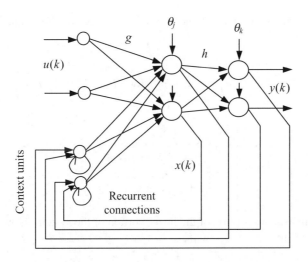

Figure 4.37   Structure of modified Jordan network

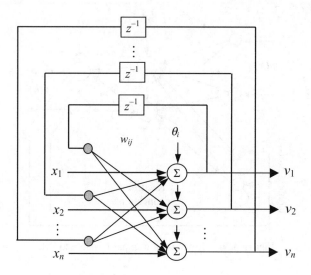

Figure 4.38   Architecture of Hopfield network

n neurons having threshold values θ_i. The feedback input to the ith neuron is equal to the weighted sum of neuron outputs v_j, where $j = 1, 2, \ldots, n$. Denoting w_{ij} as the weight value connecting the output of the jth neuron with the input of the ith neuron, we can express the total input net_i of the ith neuron as

$$net_i = \sum_{j=1}^{n} w_{ij} v_j + x_i - \theta_i \quad \text{for} \quad j \neq i, \ i = 1, 2, \ldots, n \quad (4.96)$$

where x_i is the external input to the ith neuron. Using vector notation, Equation (4.95) can be written as

$$net_i = w_i^T v + x_i - \theta_i \quad \text{for} \quad i = 1, 2, \ldots, n \quad (4.97)$$

where $w_i = \begin{bmatrix} w_{i1} \\ w_{i2} \\ \vdots \\ w_{in} \end{bmatrix}$ and $v = \begin{bmatrix} v_1 \\ v_2 \\ \vdots \\ v_n \end{bmatrix}$.

The complete matrix description of the linear portion of the system shown is given by

$$net = Wv + x - \theta \quad (4.98)$$

where $net = \begin{bmatrix} net_1 \\ net_2 \\ \vdots \\ net_n \end{bmatrix}, x = \begin{bmatrix} x_1 \\ x_2 \\ \vdots \\ x_n \end{bmatrix}$ and $\theta = \begin{bmatrix} \theta_1 \\ \theta_2 \\ \vdots \\ \theta_n \end{bmatrix}$. Matrix W, called the connectivity matrix,

is an $n \times n$ matrix defined as

$$W = \begin{bmatrix} 0 & w_{12} & \cdots & w_{1n} \\ w_{21} & 0 & \cdots & w_{2n} \\ \vdots & \vdots & \cdots & \vdots \\ w_{n1} & w_{n2} & \cdots & 0 \end{bmatrix} \tag{4.99}$$

The matrix of synaptic weights W in this model is symmetric (i.e., $w_{ij} = w_{ji}$) and the diagonal elements are zero (i.e., $w_{ij} = 0$ for $i = j$) which means that no connection exists from any neuron back to itself.

The updated algorithm for a discrete-time recurrent network is then obtained as follows:

$$v^{k+1} = \Gamma(Wv^k + x - \theta) \quad \text{for} \quad k = 0, 1, \ldots \tag{4.100}$$

where Γ is the activation function defined as sgn($\cdot$) which operates on every scalar row of the bracketed matrix.

It is to be noted that a Hopfield network may be operated in a continuous mode or discrete mode, depending on the neuron model used. The continuous mode is based on an additive model (Haykin, 2009). On the other hand, the discrete mode is based on the McCulloch–Pitts model described previously. The discrete Hopfield network has attracted a great deal of attention in the literature as a content-addressable memory (Haykin, 2009).

4.7   MATLAB® Programs

The Neural Network Toolbox provides tools to define architecture, initialize, train and simulate networks within the MATLAB® platform (Demuth and Beale, 2000). The toolbox contains three categories of tools:

- Command-line functions,
- GUI interface tools and
- Simulink® blocks.

The neural network systems can also be integrated for simulation with a Simulink® toolbox. The first category of tools is made up of functions (M-files) that can be called from the command line. In this chapter, only command-line functions will be used. A description of the different functions to define the network architecture and training of networks, and MATLAB® codes for the examples, are presented in Appendix D.

References

Adeli, H. and Hung, S.-L. (1994) An adaptive conjugate gradient learning algorithm for effective training of multilayer neural networks, *Applied Mathematics and Computation*, 62(1), 81–102.

Anderson, N.H. and Titterington, D.M. (1998) Boltzmann machines: statistical associations and algorithms for training. In *Neural Network Systems Techniques and Applications: Implementation Techniques*, C.T. Leondes (ed.), Academic Press, New York, Vol. 3, pp. 51–89.

Broomhead, D.S. and Lowe, D. (1988) Multivariable functional interpolation and adaptive networks, *Complex Systems*, 2, 321–355.

Caudill, M. (1993) GRNN and bear it, *AI Expert*, 8(5), 28–33.

Choi, S., Ko, K. and Hong, D. (2001) A multilayer feedforward neural network having N/4 nodes in two hidden layers, *Proceedings of the IEEE International Joint Conference on Neural Networks*, Washington, DC, Vol. 3, pp. 1675–1680.

Dayan, P., Hinton, G.E., Neal, R.M. and Zemel, R.S. (1995) The Helmholtz machine, *Neural Computing*, 7(5), 1022–1037.

Demuth, H. and Beale, M. (2000) *Neural Network Toolbox for use with Matlab*. User's Guide, Version 4. The Math Works Inc.

Elman, J.L. (1990) Finding structure in time, *Cognitive Science*, 14, 179–211.

Fletcher, R. and Reeves, R.M. (1964) Function minimisation by conjugate gradients, *The Computer Journal*, 7(2), 149–160.

Franke, R. (1982) Scattered data interpolation: tests of some methods, *Mathematics of Computations*, 38, 181–200.

Grossberg, S. (1987) Competitive learning: from interactive activation to adaptive resonance, *Cognitive Science*, 11, 23–26.

Haerdle, W. (1990) *Applied Nonparametric Regression*, Cambridge University Press, Cambridge.

Hamming, R. (1986) *Coding and Information Theory*, Prentice-Hall, Englewood Cliffs, NJ.

Hardy, R.L. (1990) Theory and applications of the multiquadratic biharmonic method, *Computers Mathematics with Applications*, 19, 163–208.

Haykin, S. (1999) *Neural Networks – A Comprehensive Foundation*, Prentice-Hall, Upper Saddle River, NJ.

Haykin, S. (2009) *Neural Networks and Learning Machines*, 3rd edn, Pearson Education, Oxford.

Hebb, D.O. (1949) *The Organization of Behavior: A Neuropsychological Theory*, John Wiley, New York.

Hilton, G.E. (1989) Deterministic Boltzmann machine learning performs steepest descent in weight space, *Neural Computation*, 1, 143–150.

Hocking, R.R. (1976) The analysis and selection of variables in linear regression, *Biometrics*, 32, 1–49.

Hocking, R.R. (1983) Developments in linear regression methodology, *Technometrics*, 12(3), 219–249.

Hopfield, J.J. (1982) Neural networks and physical systems with emergent collective computational abilities, *Proceedings of National Academy of Sciences*, 79, 2554–2558.

Huang, S.-C. and Huang, Y.-F. (1991) Bounds on the number of hidden neurons in multilayer perceptrons, *IEEE Transactions on Neural Networks*, 2(1), 47–55.

Jackson, I.R.H. (1988) *Radial basis function methods for multivariable approximation*, PhD Thesis, DAMTP, University of Cambridge.

Jain, A.K., Mao, J. and Mohiudding, K.M. (1996) Artificial neural networks: a tutorial, *Computer*, 29(3), 31–44.

Kohonen, T. (1988) *Self-Organisation and Associative Memory*, 3rd edn, Springer-Verlag, New York.

Kohonen, T. (1990) The self-organising map, *Proceedings of the IEEE*, 78(9), 1464–1480.

Kohonen, T. (1995) *Self-Organizing Maps*, Springer Series in Information Sciences, Vol. 30, Springer-Verlag, Berlin.

Konar, A. (2005) *Computational Intelligence – Principles, Techniques and Applications*, Springer-Verlag, Berlin.

Li, J., Michel, A.N. and Porod, W. (1989) Analysis and synthesis of a class of neural networks: linear systems operating on a closed hypercube, *IEEE Transactions on Circuits and Systems*, 36(11), 1405–1422.

McClelland, W.S. and Rumelhart, D.E. (1986) *Parallel Distributed Processing*, MIT Press, Cambridge, MA and the PDP Research Group.

McCulloch, W.S. and Pitts, W.H. (1943) A logical calculus of the ideas imminent in nervous activity, *Bulletin of Mathematical Biophysics*, 5, 115–133.

Meilijson, I., Ruppin, E. and Sipper, M. (1998) Fast computing in Hamming and Hopfield networks. In *Neural Network Systems Techniques and Applications: Algorithms and Architectures*, C.T. Leondes (ed.), Academic Press, New York, Vol. 1, pp. 123–154.

Minsky, M. and Papert, S. (1969) *Perceptrons*, MIT Press, Cambridge, MA.

Moody, J.E. (1989) Fast learning in multi-resolution hierarchies. In *Advances in Neural Information Processing Systems*, D.S. Touretzky (ed.), Morgan Kaufmann, San Mateo, CA.

Moody, J.E. (1992) The effective number of parameters: an analysis of generalisation and regularisation in nonlinear learning systems. In *Neural Information Processing Systems*, J.E. Moody, S.J. Hanson and R.P. Lippmann (eds), Morgan Kaufmann, San Mateo, CA, Vol. 4, pp. 847–854.

Moody, J.E. and Darken, C.J. (1989) Fast learning in networks of locally-tuned processing units, *Neural Computing*, 1, 281–294.

Nadaraya, E.A. (1964) On estimating regression, *Theory of Probability and Applications*, 10, 186–190.

Neal, R.M. (1992) Connectionist learning of belief networks, *Artificial Intelligence*, 56, 71–113.

Parzen, E. (1962) On estimation of a probability density function and mode, *Annals of Mathematics and Statistics*, 33, 1065–1076.

Pearl, J. (1987) Distributed revision of composite beliefs, *Artificial Intelligence*, 29, 241–288.

Poggio, T. and Girosi, F. (1990) Networks for approximation and learning, *Proceedings of IEEE*, 78(9).

Powell, M.J.D. (1981) *Approximation Theory and Methods*, Cambridge University Press, Cambridge.

Powell, M.J.D. (1985) Radial basis functions for multivariable interpolation: a review, *IMA Conference Algorithm for Approximation of Functions and Data*, RMCS Shrivenham.

Powell, M.J.D. (1986) Convergence properties of algorithms for nonlinear optimisation, *SIAM Review*, 28(4), 487–500.

Rosenblatt, F. (1958) The perceptron: a probabilistic model for information storage and organisation in the brain, *Psychology Review*, 65, 386–408.

Rumelhart, D.E. and Zipser, D. (1985) Feature discovery by competitive learning, *Cognitive Science*, 9(1), 75–112.

Rutkowski, L. (2005) *New Soft Computing Techniques for System Modelling, Pattern Classification and Image Processing*, Springer-Verlag, Berlin.

Schioeler, H. and Hartmann, U. (1992) Mapping neural network derived from the Parzen window estimator, *Neural Networks*, 5, 903–909.

Siddique, N.H. and Amavasai, B.P. (2012) An investigation into adaptive capacity of recurrent neural networks. In *Innovations in Intelligent Machines – 3, SCI 442, Systems*, I. Jordanov and L.C. Jain (eds), Springer-Verlag, Berlin, pp. 119–138.

Siddique, N.H., Condell, J.V., McGinnity, T.M., Gatsoulis, Y. and Kerr, E. (2010) Hierarchical architecture for incremental learning in mobile robotics. In *11th Conference Towards Autonomous Robotic Systems* (TAROS 2010), T. Belpaeme, G. Bugmann, C. Melhuish and M. Witkowski (eds), University of Plymouth, pp. 271–277.

Specht, D.F. (1990a) Probabilistic neural networks and the polynomial ADALINE as complementary techniques for classification, *IEEE Transactions on Neural Networks*, 1(1), 111–121.

Specht, D.F. (1990b) Probabilistic neural networks, *Neural Networks, International Neural Network Society*, 3, 109–118.

Specht, D.F. (1991) A general regression neural network, *IEEE Transactions on Neural Networks*, 2(6), 568–576.

Tsoukalas, L.H. and Uhrig, R.E. (1997) *Fuzzy and Neural Approaches in Engineering*, John Wiley & Sons, New York.

Wand, M.P. and Jones, M.C. (1995) *Kernel Smoothing*, Chapman & Hall, London.

Watanabe, S. and Fukumizu, K. (1998) Probabilistic design. In *Neural Network Systems Techniques and Applications: Algorithms and Architectures*, C.T. Leondes (ed.), Academic Press, New York, Vol. 1, pp. 181–229.

Watson, G.S. (1964) Smooth regression analysis, *Sankhy*, A26, 359–372.

Werbos, P.J. (1974) Beyond regression: new tools for prediction and analysis in the behavioural sciences, Doctoral Dissertation, Applied Mathematics, Harvard University.

Widrow, B. and Hoff, M.E. (1960) *Adaptive Switching Circuits*, IRE Western Electric Show and Convention Record, part 4, pp. 96–104.

Widrow, B. and Hoff, M.E. (1962) Associative storage and retrieval of digital information in networks of adaptive neurons, *Biological Prototypes and Synthetic Systems*, 1, 16

Yao, X. (1993) Evolutionary artificial neural networks, *International Journal of Neural Systems*, 4(3), 203–222.

Zaknich, A. (2003) *Neural Networks for Intelligent Signal Processing*, World Scientific, Singapore.

Zurada, J.M. (1992) *Introduction to Artificial Neural Systems*, PWS Publishing Company, Boston, MA.

5

Neural Systems and Applications

5.1 Introduction

Neural networks (NNs) are meant to interact with the natural environment, and information about the latter is usually collected from the real world through very noisy but redundant sensory signals. On the other hand, in the control of effectors or actuators, one often has to coordinate many mutually dependent and redundant signals. In both cases, neural networks can be used to implement a great number of implicitly and/or poorly defined transformations between variables (Rosenblatt, 1958; Minsky and Papert, 1959; Widrow and Hoff, 1960; Werbos, 1974; McClelland and Rumelhart, 1986). Many applications have been found in simple mathematical methods, such as fitting nonlinear functional expansions into experimental data, pattern recognition and data clustering. The 1988 DARPA Neural Network Study lists various applications of NNs, beginning in about 1984 with an adaptive channel equalizer. This device, which is an outstanding commercial success, is a single neuron network used in long-distance telephone systems to stabilize voice signals. NNs have been applied in many other fields since the DARPA report. Some of the applications mentioned in the literature are as follows:

- Aerospace – high-performance autopilot, flight path simulation, aircraft control systems, fault detection system.
- Automotive – automobile automatic guidance system.
- Banking, Financial and Business – cheque and other document reading, credit application evaluation, credit card activity.
- Defence – weapon steering, target tracking, object discrimination, facial recognition.
- Industrial, Manufacturing and Electronics – control, process identification, machine diagnosis, quality inspection.
- Medical – cancer cell analysis, EEG and ECG signal analysis, optimization of transplant times.
- Speech – speech recognition, compression, text-to-speech synthesis.
- Telecommunication – image and data compression, speech processing, real-time translation of spoken language.

Computational Intelligence: Synergies of Fuzzy Logic, Neural Networks and Evolutionary Computing, First Edition.
Nazmul Siddique and Hojjat Adeli.
© 2013 John Wiley & Sons, Ltd. Published 2013 by John Wiley & Sons, Ltd.

These applications fall mainly into two categories (Miller *et al.*, 1990; Kim and Lewis, 1998):

- Prediction and control applications;
- Data analysis and processing.

In this chapter, we will address some popular network architectures for modelling, identification, prediction and control applications that have been implemented using neural networks.

5.2   System Identification and Control

5.2.1   System Description

Mathematical systems theory is a fairly old but powerful scientific discipline which deals with the analysis and synthesis of linear and nonlinear dynamic systems. In general, a system with concentrated parameters can be described by ordinary differential equations. It is now well established to describe a dynamic system by a set of differential equations or difference equations. For example, an n-input m-output system is described by differential equations of the form

$$\dot{x}(t) = \Phi\left[x(t), u(t)\right]$$
$$y(t) = \Psi\left[x(t)\right]$$

(5.1)

where $u(t) = [u_1(t), u_2(t), \ldots, u_n(t)]^T$, $x(t) = [x_1(t), x_2(t), \ldots, x_P(t)]^T$ and $y(t) = [y_1(t), y_2(t), \ldots, y_m(t)]^T$ are the input vector, the state vector and the output vector, respectively. $\Phi : R^P \times R^n \to R^P$ and $\Psi : R^n \to R^m$ are static nonlinear mapping functions. In systems theory, the input/output state representation of a system is described by Equation (5.1). The difference equation version of the system in Equation (5.1) can be described by

$$x(k+1) = \Phi\left[x(k), u(k)\right]$$
$$y(k) = \Psi\left[x(k)\right]$$

(5.2)

where $x(.)$, $u(.)$ and $y(.)$ are discrete time sequences. If the system described by Equation (5.2) is assumed to be linear and time-invariant, the system can be expressed as

$$x(k+1) = Ax(k) + Bu(k)$$
$$y(k) = Cx(k)$$

(5.3)

where $\frac{\partial \Phi(x,u)}{\partial x} = A$, $\frac{\partial \Phi(x,u)}{\partial u} = B$ and $\frac{\partial \Psi(x)}{\partial x} = C$. The system in Equation (5.3) is called a state-space description, where A, B and C are stability, observability and controllability matrices. The system is thus parameterized by the three matrices and is called a linear time-invariant (LTI) system when A, B and C are known.

5.2.2   System Identification

A dynamical mathematical model is a mathematical description of the dynamic behaviour of a system in either the time or frequency domain. System identification is a general mathematical procedure to build dynamical models from measured input/output data. That means the system

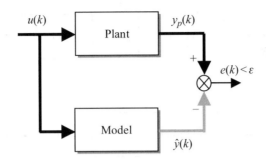

Figure 5.1   System (or plant or process) identification

identification needs to deal with analysis, determination of order, determination of parameters and estimation of parameters of the dynamic system. When the system's functions Φ and Ψ described in Equation (5.2) or the matrices A, B and C in Equation (5.3) are unknown, then it is the problem of system identification. In other words, system identification is to find the mapping of the functions $\Phi : X \times U \to X$ and $\Psi : X \to Y$ for $\forall x \in X, \forall u \in U$ and $\forall y \in Y$ or estimate the values of the parameters A, B and C of the system model described in Equation (5.3). The system identification problem is then transformed to constructing a suitable model of the system from input/output data which, when subjected to the same input $u(k)$, produces an output $\hat{y}(k)$ such that $\|y_P(k) - \hat{y}(k)\| < \varepsilon$ for some desired $\varepsilon > 0$ and $\|.\|$ is a suitably defined norm. The general system identification process is shown in Figure 5.1, where $e(k) = y_p(k) - \hat{y}(k)$ and the norm $\|y_P(k) - \hat{y}(k)\|$ is absolute error or squared error.

One of the aims of this chapter is to demonstrate the use of neural networks for system identification. Therefore, a general discussion on system identification is beyond the scope of this chapter. Readers more interested in systems identification are directed to the widely accepted book by Lennart Ljung (Ljung, 1999).

5.2.3   System Control

Control theory is an interdisciplinary branch of engineering and mathematics that deals with the behaviour of dynamical systems. The main objectives of control theory are to deal with the analysis and synthesis of dynamical systems where certain variables are to be controlled within prescribed limits so that the output follows a reference signal. If the functions Φ and Ψ in Equation (5.2) are known, the problem is to design a controller which generates the desired control input $u(k)$. There are a number of frequency and time domain techniques that are available for synthesis of controllers for linear systems of the form described in Equation (5.3) when A, B and C are known. Unfortunately, there are no known methods available for nonlinear systems, even when $\Phi(.)$ and $\Psi(.)$ are known.

A block diagram of a feedback control system is illustrated in Figure 5.2. The desired output of a system is called the reference. When one or more output variables of a system need to follow a certain reference $y_r(k)$ over time, a controller is needed to manipulate the control inputs $u(k)$ to the system to obtain the desired effect on the output $y_p(k)$ of the system. There are basically two approaches to NN control, namely, non-adaptive and adaptive approaches. For non-adaptive NN control of a plant, an accurate approximation of the dynamics of the plant has to be obtained first. This is carried out offline by training the NN using an input/output

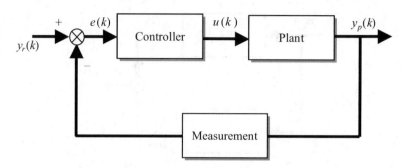

Figure 5.2   Block diagram of feedback control system

data set. The approximation can be used to develop the appropriate control strategy. The non-adaptive approach works well for many systems (Narendra and Parthasarathy, 1990; Nguyen and Widrow, 1990; Lee *et al.*, 1992). The non-adaptive approach has no mechanism to handle changes in system parameters or in the environment (Ge *et al.*, 1998). Moreover, the problem of controlling a system can be difficult for two reasons: firstly, the system can be described with a known structure but the parameters are unknown and secondly, the system model may contain unknown dynamics or uncertainty. In such situations, an adaptive approach is chosen. In adaptive control, a mechanism is devised to estimate or model uncertainty of the system by adjusting the control parameters. Two distinct approaches to adaptive control are widely in use: (i) direct control and (ii) indirect control. In direct control, the parameters of the controller are adjusted directly by minimizing some performance criterion. A block diagram of direct adaptive control is shown in Figure 5.3, where $y_r(k)$ is the reference signal, $y_p(k)$ is the plant output, $u(k)$ is the control signal to the system (or plant) and $e(k)$ is the output error used for adjustment of the controller parameters.

In indirect control, an approximator (often referred to as an identifier in the adaptive control literature) is used to estimate the unknown parameters (or measures of uncertainty) of the system assuming that the estimates are the true parameters of the system. The approximator is then adjusted using some performance index. A block diagram of indirect adaptive control is shown in Figure 5.4, where $y_p(k)$ is the plant output, $\hat{y}(k)$ is the estimate from the approximator, the same control signal $u(k)$ is employed for the system and approximator and $e(k)$ is the output error between the plant output and the approximator output, which is used for adjustment of the approximator's parameters.

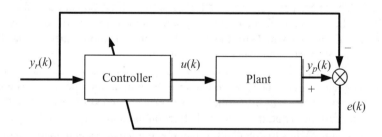

Figure 5.3   Block diagram of direct adaptive control system

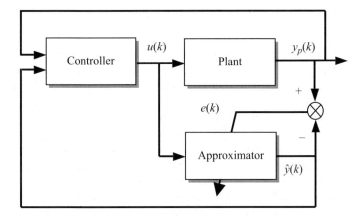

Figure 5.4 Block diagram of indirect adaptive control system

The primary interest of this chapter lies in the identification and control of unknown nonlinear dynamical systems using neural networks. Therefore, there is no general discussion on control theories of linear and nonlinear control techniques. In the next few sections, the different techniques of adaptive control based on neural networks will be explored.

5.3 Neural Networks for Control

Traditional control methodologies are mainly based on linear systems theory, while real systems are nonlinear in nature and have unmodelled dynamics, immeasurable noise, uncertainty, multi-loops, etc. which create problems for engineers in trying to design control algorithms (Kawato *et al.*, 1987, 1988; Ungar, 1996). It is, therefore, very challenging for engineers to design an efficient control algorithm. From the designer's point of view, the specifications for control algorithms should be simple enough to be implemented and understood with properties such as learning ability, robustness and nonlinearity. One of the reasons that neural networks have become very popular in control applications is that they satisfy some of these criteria for design and implementation. From a practical point of view, the inherent massive parallelism and fast adaptability of neural network implementations are additional advantages. Powerful learning algorithms, variety of architectures and the ability to train the neural networks from input/output functions and/or experiential data make neural networks the preferred technology for many applications. Neural networks provide simpler solutions to complex control problems. The success of the backpropagation algorithm to train multilayered networks led to an explosion in the application of neural networks for control purposes. The use of neural networks in control applications – including process control, robotics, manufacturing and aerospace applications, among others – has recently experienced rapid growth. The basic objective of neuro-control is to provide the appropriate input signal to a given physical system (process or plant) to yield its desired response. There are typically two steps involved when using neural networks for control:

- System (plant or process) identification and
- Control design.

5.3.1   System Identification for Control Design

The key concept of system identification is the process of determining a dynamic model for unknown systems. A brief description of system identification is given in Section 5.2.2. The identified model can be used subsequently for control purposes. System identification consists of two main steps: the first step is to choose an appropriate parametric model and the second step is to adjust the parameters of the model according to some adaptive laws so that the response of the model to an input signal can approximate the response of the real system. The problem of identification of a model structure and estimation of its parameters can be formulated as a problem of learning a mapping between known input/output spaces. Almost all identification or approximation schemes can be mapped into a network. For example, the autoregressive moving average with exogenous input (ARMAX) model can be represented as a single-layer network with inputs comprising delayed system input/output data and error (Turner *et al.*, 1995). Since multilayer neural networks have good approximation capabilities, they provide a powerful tool for identification of unknown systems with nonlinearities. The neural network is composed of tapped delays of inputs and outputs, with a sufficient number of layers and neurons in each layer to be able to match the input/output behaviour of the corresponding nonlinear mapping of the plant. It implies that the nonlinear function of the plant is replaced by neural networks with fixed but unknown weight matrices, which is learnt using a suitable learning algorithm and available data set.

At the system identification stage, a neural network model of the plant to be controlled is developed. The identifier is composed of a multilayer neural network incorporated in parallel with a dynamical system, where the structure comes from the error function standard in the system identification and control literature. The structural information is actually contained in the neural network connectivity and weights. The system identification can be carried out in two ways:

- Forward plant identification model and
- Direct inverse identification model.

The first stage of plant identification in neural network control is to train a neural network to represent the forward dynamics of the plant. The error between the plant output y_p and the neural network model output y_m is used as the neural network training signal. The basic configuration for forward plant model identification is shown in Figure 5.5(a). The neural network plant model uses previous inputs and previous plant outputs (shown as delayed signals in the figure) to predict future values of the plant output. The network can be trained offline in batch mode, using data collected from the operation of the plant. Any of the backpropagation training algorithms discussed in Chapter 4 can be used for network training.

Perhaps one of the most widely applied neuro-model schemes is the direct inverse model approach. Once a neural network has been trained to learn the inverse of the plant, it can then be configured to control the plant directly. The basic configuration of the direct inverse model is shown in Figure 5.5(b). In the direct inverse model architecture, the network is trained offline using patterns obtained from the plant's open-loop (or closed-loop) characteristics.

Different models of multi-input multi-output (MIMO) nonlinear autoregressive moving average (NARMA) forms are common and cover a large range of systems in adaptive systems literature for the identification and control of linear systems. These models can be considered

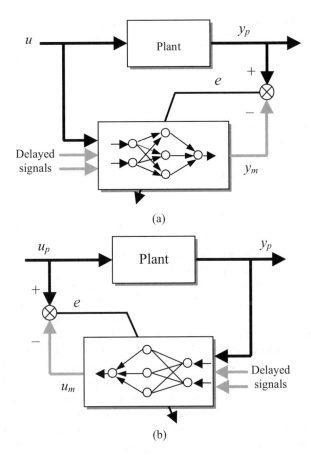

(a)

(b)

Figure 5.5 Plant identification using neural networks. (a) Forward plant identification; (b) Direct inverse model plant identification

as their generalization to nonlinear systems (Narendra and Parthasarathy, 1990; Nelles, 2001). For a detailed treatment of these models and their identification using neural networks, readers are referred to Sarangapani (2006).

5.3.2   Neural Networks for Control Design

Research on neural network-based control systems has received considerable attention over recent years. The widely used neural control structures are similar to those employed in adaptive control systems. A neural network is used to estimate the unknown nonlinear system. The control formulation is then designed using the estimated neural network. The estimation process uses the measured input/output from the system and is achieved through use of various types of neural network architectures, such as the multilayer perceptron (MLP), radial basis function (RBF), recurrent neural networks (RNN) and B-spline networks. The aim of this chapter is to introduce some of the typical neural network control structures. A second objective is to show how some of these control schemes can be implemented using the Neural Network Toolbox in MATLAB®. The schemes considered are:

- NN-based direct control;
- NN-based indirect control;
- Backpropagation through time control;
- NN-based direct inverse control;
- NN-based model predictive control;
- NN-based adaptive control;
- NARMA-L2 (feedback linearization) control.

5.3.2.1 NN-Based Direct (Specialized Learning) Control

In the direct (specialized learning) control architecture illustrated in Figure 5.6, the controller network is trained in an online way (goal-directed) to minimize some norm of error between the output and reference signal. The error is backpropagated through the plant at every sample to adjust the parameters of the network. This architecture was named specialized learning control by Psaltis *et al.* (1999). One advantage is that no identification is involved in this method. However, owing to the location of the plant, the Jacobian of the plant (i.e., $\frac{\partial y}{\partial u}$) is required (Saerens and Soquet, 1989), which is difficult to obtain if the plant dynamics is not known *a priori*. In order to avoid this, the elements of the Jacobian may be approximated by their signs, which are the orientations of the control parameters influencing the outputs of the plant. The training involves the response r as input signals to the NN controller. The error $e = r - y_p$ is backpropagated through the plant and used to adjust the NN controller. The objective of the training of the NN controller is to generate the required control signal u to drive the plant to the desired output such that the plant output y_p matches the reference signal r over the training epochs k, i.e., $\lim_{k \to \infty} \|r - y_p\| \leq \varepsilon$, where $\varepsilon \geq 0$. The disadvantage of the direct NN control is that the plant's initial stability is not guaranteed for this method of control.

Despite the disadvantages, many researchers have applied the direct adaptive control scheme to a variety of unknown nonlinear and non-affine systems with significant success and improved performance (Noriega and Wang, 1998; Park *et al.*, 2005).

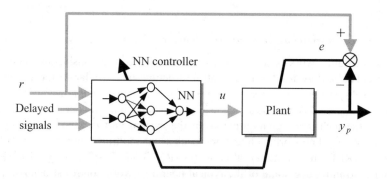

Figure 5.6 Direct (or specialized learning) control architecture

5.3.2.2 NN-Based Indirect Control

The main disadvantages in the direct control scheme were the plant's Jacobian and initial stability. When the plant inverse (or plant's Jacobian) is not well defined or the plant's initial stability is critical, an indirect control scheme is considerably more successful than direct control. In this architecture, a neuro-emulator of the plant is trained to represent the plant's response. The training of the neuro-emulator is shown in Figure 5.7(a), which is the same as forward plant identification. The difference between the plant output y_p and neuro-emulator response $\hat{y}$ is used to adjust the parameters of the neuro-emulator. Once the neuro-emulator is trained sufficiently, a neuro-controller is trained using the same reference signal r and the performance error $e_c = \hat{y} - r$. The performance error e_c is backpropagated through the neuro-emulator to adjust the parameters of the neuro-controller. Training of the neuro-controller is shown in Figure 5.7(b). This control scheme can be found under different names in the literature, such as feedforward inverse control (Narendra, 1995) and specialized inverse learning (Hunt *et al.*, 1992). The added advantage of indirect control is that the parameters of the neuro-emulator can

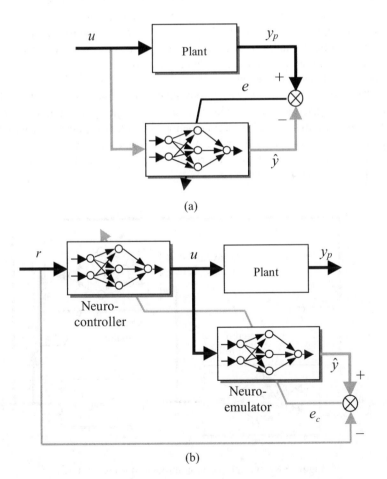

(a)

(b)

Figure 5.7 Indirect control architecture. (a) Training of neuro-emulator; (b) Training of neuro-controller

be readjusted online during operation of the controller if the neuro-emulator appears not to be accurate. A close variant of indirect control is the internal model control, in which an internal model is placed in parallel with the plant and a controller is used in series with the plant. The internal model is eventually a forward plant model as shown in Figure 5.5(a). Details of the internal model control architecture can be found in Sen *et al.* (1998).

Among the disadvantages of the indirect control scheme appears to be the robustness of the controller, as there is no feedback loop used in the control strategy. While training the neuro-controller using the same reference signal r and the performance error $e_c = \hat{y} - r$ (which is backpropagated through the neuro-emulator), poor convergence and unstable control may be a problem at the initial stage. Despite these disadvantages, indirect control is more successful than direct control and has found a wide range of applications (Nguyen and Widrow, 1990; Wu *et al.*, 1992; Khalid *et al.*, 1993; Yang and Linkens, 1994).

5.3.2.3 Backpropagation-Through-Time Control

There is another model of neural network control architecture using a backpropagation learning algorithm which was proposed by Jordan and Rumelhart (1990), Narendra and Parthasarathy (1990) and Nguyen and Widrow (1990). Werbos (1990b) classified this method as the backpropagation-through-time architecture. The architecture resembles many traditional adaptive control structures, called indirect adaptive control.

In this control scheme two neural networks are used to control the plant, as shown in Figure 5.8. The first neural network is a neuro-emulator. The emulator network can be trained offline using a generalized learning architecture or even online by injecting random inputs to learn the forward plant dynamics. The second neural network is a controller. This architecture allows training the neuro-controller online as the performance error $e = \|r - y_p\|$ can be backpropagated through the emulator at every sample. Some applications of this architecture can be found in Omatu *et al.* (1995).

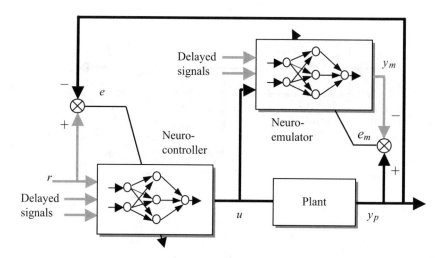

Figure 5.8   Backpropagation-through-time architecture

5.3.2.4   NN-Based Direct Inverse Control

The most widely applied neuro-control scheme is the direct inverse model neuro-control approach. As the term suggest, NN-based direct inverse control utilizes an inverse model of the controlled system, which is simply cascaded with the controlled system in order that the system results in an identity mapping the desired response (network input) and the output of the controlled system. Thus the network acts directly as a feedforward controller, and the output of the controlled system is equal to the desired output. The term 'direct inverse control' is adopted from Werbos (1990a). The modelling process shown in Figure 5.9(a) first constructs an inverse of the plant to estimate the inverse model output $\hat{u}$. The estimated output $\hat{u}$ is compared with the training signal u and the error $e = u - \hat{u}$ is used to train the inverse model. Once the inverse model is obtained, it is then cascaded with the plant as an open-loop controller. The parameters of the neuro-controller are adjusted directly. The direct inverse control architecture is shown in Figure 5.9(b). The reference signal r should cover a sufficiently large input/output space while building the inverse model.

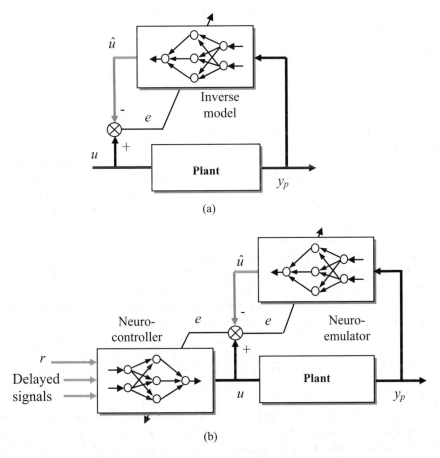

Figure 5.9   Direct inverse control architecture. (a) Inverse modelling; (b) Open-loop control

Direct inverse control is based on the assumption that there exists a one-to-one mapping from the input state to the output state and the plant must be open-loop stable. Problems are experienced with direct inverse control for several reasons summarized as follows (Sen *et al.*, 1999):

- lack of robustness, resulting from the fact that no direct feedback error is used in direct inverse control;
- inefficient learning, caused by improper operational range of data;
- actual operational data may be hard to define *a priori*.

To overcome the problem of robustness, some researchers use a conventional PI controller to incorporate the required feedback loop. This also helps reduce the sensitivity of the whole control system against inverse modelling (Sen *et al.*, 1999). Some researchers used an evaluation function comprised of output, input and reference signal (Cai, 1997). Khalid *et al.* (1995) investigated the performance of an inverse neural network controller for temperature regulation. They found the controller has encouraging advantages over a range of controllers such as fuzzy logic control, generalized predictive control and PI control.

5.3.2.5 Model Predictive Control

Predictive control (or model-based predictive control) was developed in the 1970s. The approach has proved to be stable for nonlinear systems (Mayne and Michalska, 1990). The model predictive controller (MPC) uses a neural network model to predict future plant responses to potential control signals. The first step in model predictive control is to determine the neural network plant model (using system identification). The neural network plant model is trained offline, in batch form, using any of the training algorithms (such as the backpropagation algorithm) discussed in Chapter 4. This is the same procedure for all the control architectures. At the current time instant, the NN model predicts $\tilde{y}$ over some time step into the future (or horizon) based on the future control signal $\tilde{u}$. An optimization algorithm then computes the control signal $\tilde{u}$ that optimizes the future plant performance. The controller, however, requires a significant amount of online computation, because an optimization algorithm is performed at each sample time to compute the optimal control input. The MPC method is based on the receding horizon technique (Soloway and Haley, 1996). The neural network model predicts the plant response over a specified time horizon. The predictions are used by a numerical optimization program to determine the control signal that minimizes the performance criterion defined in Equation (5.4) over the specified horizon:

$$J = \sum_{j=N_1}^{N_2} (y_r(t+j) - y_m(t+j))^2 + p \sum_{j=1}^{N_u} (\tilde{u}(t+j-1) - \tilde{u}(t+j-2))^2 \quad (5.4)$$

where the controller horizon N_1 is kept fixed at a value, the cost horizon N_2 is the number of time steps over which prediction errors are minimized and the control horizon N_u is the number of time steps over which the control increments are minimized. The variable $\tilde{u}$ is the tentative control signal, y_r is the desired response and y_m is the network model response. The control weighting factor p multiplies the sum of squared control increments in the performance

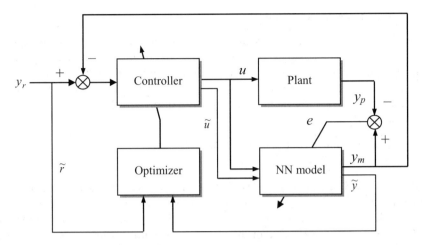

Figure 5.10   NN-based model predictive control

function, which means it determines the contribution that the sum of squares of the control increments has on the performance index. The generic block diagram in Figure 5.10 illustrates the process of model predictive control.

In principle, a straightforward approach would be to replace the controller and the optimizer by a single neural network (Hunt *et al.*, 1992). The NN can be trained indirectly using the approach shown in Figure 5.7. Figure 5.11 illustrates this version of the model predictive control process. The controller consists of the neural network plant model and the optimization block. The optimization block determines the values of $\tilde{u}$ that minimize J, and then the optimal u is input to the plant.

MPC has been widely accepted in industry as a powerful control strategy (Akesson and Toivonen, 2006; Yuzgec *et al.*, 2008). The potential problem of MPC is that the optimization may be computationally very demanding, especially for nonlinear systems.

The Neural Network Toolbox in MATLAB® provides an MPC controller block that can be used with Simulink® to implement any MPC control problem. An MPC controller has been implemented for the continuous stirred tank reactor (CSTR) (see Appendix E).

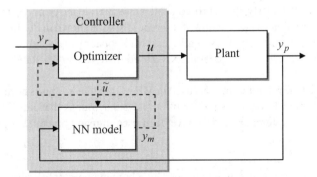

Figure 5.11   Block diagram of model predictive controller

5.3.2.6 NN-Based Adaptive Control

In conventional adaptive control, a regression matrix for the dynamic system needs to be computed that is computationally very expensive. On the contrary, NN-based adaptive control of nonlinear systems does not require *a priori* information about the dynamics of the system to be controlled. This makes the NN-based adaptive control of nonlinear systems a popular choice. In the same way as for traditional adaptive control, the NN-based adaptive control is classified into two types:

- Model reference adaptive control (MRAC) and
- Self-tuning control (STC).

The difference between MRAC and STC is that STC regulates the inner parameters of the controller directly according to the results of the forward and/or inverse model identification of the controlled system with the aim of satisfying a given performance index of the system. In MRAC, the desired performance of the closed-loop control system is described in terms of a stable reference model which is chosen based on some prior information about its input/output behaviour. The reference model gives the desired response to a command signal r. The objective is to make the plant output $y_p(t)$ match the reference model output $y_m(t)$ asymptotically, i.e., $\lim_{t \to \infty} \|y_m(t) - y_p(t)\| \le \varepsilon$, where $\varepsilon \ge 0$.

NN-based MRAC

The earliest approaches to adaptive control, introduced in the 1950s, were mainly model reference adaptive control (Landau, 1979). In the case of complex systems where the plant cannot be approximated by a linear time-invariant model or when its nonlinear parameters are unknown, an NN-based MRAC approach is the best option for control design (Haykin, 2009). The NN-based MRAC can be implemented in two ways – direct MRAC and indirect MRAC.

In the direct MRAC scheme, the parameters are adjusted in the same way as for the direct NN control shown in Figure 5.6. Direct MRAC attempts to keep the difference between the output of the controlled object and the output of the reference model $e_c(t) = \lim_{t \to \infty} \|y_m(t) - y_p(t)\|$ to a minimum. The structure of direct MRAC is shown in Figure 5.12(a).

In the indirect MRAC scheme, the method of indirect control shown in Figure 5.7 can be used. In this implementation a neuro-emulator first identifies the feedforward model of the controlled plant offline, then online learning and revision are done using the error defined by $e_i(t) = y_e(t) - y_p(t)$. The neuro-controller is adjusted using the error defined by $e_c(t) = y_p(t) - y_m(t)$. Clearly, the neuro-controller can provide the error $e_c(t)$ or the backpropagation path of its gradient. The structure of indirect MRAC is shown in Figure 5.12(b). Indirect MRAC was studied extensively by Narendra and Mukhopadhyay (Narendra, 1996; Narendra and Mukhopadhyay, 1997). In general, NN-based MRAC poses stability problems for nonlinear plants.

It is assumed that the reference models used in MRAC are linear. This is because determining general nonlinear models with desired input/output behaviour is difficult (Narendra, 1990). That is why the linear reference models have been used in many practical applications.

NN-based STC

Similar to tuning of parameters of a controller by a human operator, a neural network is used to tune the parameters of a conventional controller in the self-tuning control scheme (Astrom

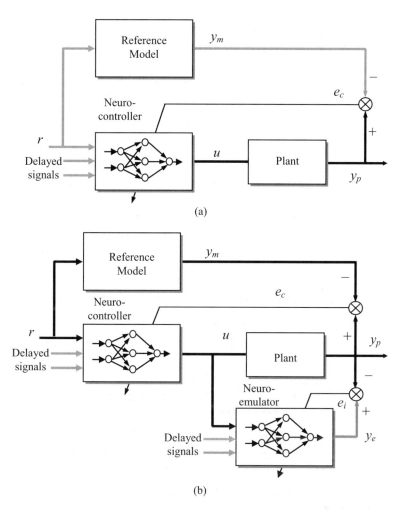

Figure 5.12 Model reference adaptive control architectures. (a) NN-based direct MRAC; (b) NN-based indirect MRAC

and Wittenmark, 1973). The neural network can be trained using the experiential data and then it is used in an online fashion. A generic NN-based self-tuning control scheme is shown in Figure 5.13.

There are two types of NN-based STC: direct STC and indirect STC. NN-based direct STC consists of a traditional controller and an NN model with high model accuracy. The NN model is developed using an offline identification procedure and then used in an online fashion. An example of NN-based direct STC for a PID controller is shown in Figure 5.14. The NN-based indirect STC consists of an NN controller and an NN model that can be tuned online. The structure of the NN-based indirect STC for a PID controller is shown in Figure 5.15. An example of an NN-based STC for a PID controller is derived in the following.

PID controllers have a long history of success in control engineering and applications in the industry. They have been proven to be robust, simple and stable for many real-world

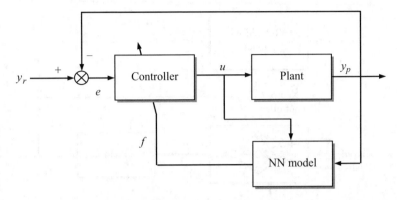

Figure 5.13   NN-based self-tuning control

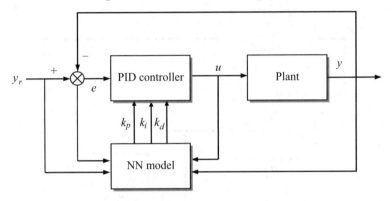

Figure 5.14   NN-based self-tuning PID control

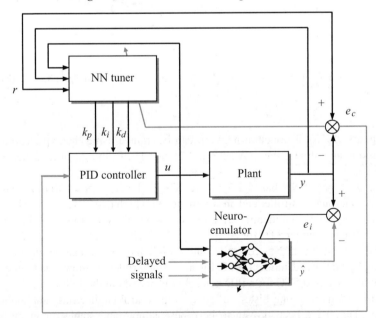

Figure 5.15   NN-based indirect STC for PID controller

applications. In Japan, almost 80% of the controllers used in the industry are PID controllers. In general, PID controllers provide a compromised performance from PD and PI controllers in terms of rise time, maximum overshoot and steady-state error. PD- and PI-type controllers have been discussed in Chapter 3. The equation of a generic PID controller is given as

$$u(t) = k_p e(t) + k_I \int_0^t e(\tau)d\tau + k_d \frac{de(t)}{dt} \tag{5.5}$$

where k_p is proportional, k_I is integral and k_d is derivative gain of the controller. In a discrete-time system, it is possible to approximate the control input as follows:

$$u(t) = u(t-1) + k_p(t)[e(t) - e(t-1)] + k_I(t)e(t)$$
$$+ k_d[e(t) - 2e(t-1) + e(t-2)] \tag{5.6}$$

The error is defined as

$$e(t) = r(t) - y(t) \tag{5.7}$$

where $r(t)$ is the desired plant output and $y(t)$ is the plant's actual output. In NN-based STC, the neural network adjusts the three gains k_p (proportional), k_I (integral) and k_d (derivative) of the PID controller, which are the three outputs of the neural network, to minimize the cost function as follows:

$$E = \frac{1}{2}e^2 \tag{5.8}$$

Using a three-layered neural network (i, j and k are the input, hidden and output layer, respectively), the learning rule for the PID gains is to be found. The outputs at the output layer are $O(1)$, $O(2)$ and $O(3)$ corresponding to k_p, k_I and k_d, respectively. Based on the gradient descent method, the weight update rules and the connection weights are defined as

$$\Delta w(t) = -\eta \frac{\partial E}{\partial w(t-1)} + \alpha \Delta w(t-1) + \beta \Delta w(t-2) \tag{5.9}$$

$$w(t) = w(t-1) + \Delta w(t) \tag{5.10}$$

where η is the learning rate, α is the momentum rate and β is the acceleration rate. The term $\frac{\partial E}{\partial w(t-1)}$ is calculated from the following equations:

$$\frac{\partial E}{\partial w(t-1)} = -[r(t) - y(t)]\frac{\partial y(t)}{\partial w(t-1)} = -e(t)\frac{\partial y(t)}{\partial w(t-1)} \tag{5.11}$$

Using the chain rule, we can define $\frac{\partial y(t)}{\partial w(t-1)}$ as

$$\frac{\partial y(t)}{\partial w(t-1)} = \frac{\partial y(t)}{\partial u(t-1)}\frac{\partial u(t-1)}{\partial w(t-1)} \tag{5.12}$$

$$\frac{\partial u(t-1)}{\partial w(t-1)} = \frac{\partial \mathbf{K}(t-1)}{\partial w(t-1)}\left[\frac{\partial u(t-1)}{\partial \mathbf{K}(t-1)}\right]^T \tag{5.13}$$

where $\mathbf{K}(t-1) = \left[k_p(t-1), k_I(t-1), k_d(t-1)\right]$ and $\frac{\partial y(t)}{\partial u(t-1)}$ is the Jacobian of the system. It is to be noted that $\mathbf{K}(t-1)$ is an output vector of the NN and the gradient with respect to $w(t-1)$ is given by

$$\frac{\partial \mathbf{K}(t-1)}{\partial w(t-1)} = \left[\frac{\partial k_p(t-1)}{\partial w(t-1)}, \frac{\partial k_I(t-1)}{\partial w(t-1)}, \frac{\partial k_d(t-1)}{\partial w(t-1)}\right] \tag{5.14}$$

Furthermore, we get from the control input

$$\frac{\partial u(t-1)}{\partial \mathbf{K}(t-1)} \equiv [u_1(t-1), u_2(t-1), u_3(t-1)] \tag{5.15}$$

where $u_1(t) = e(t) - e(t-1), u_2(t) = e(t), u_3(t) = e(t) - 2e(t-1) + e(t-2)$. Using Equations (5.11)–(5.13), the term $\frac{\partial E}{\partial w(t-1)}$ can finally be defined as follows:

$$\frac{\partial E}{\partial w(t-1)} = -e(t)\frac{\partial y(t)}{\partial u(t-1)}$$

$$\left[u_1(t-1)\frac{\partial k_P(t-1)}{\partial w(t-1)} + u_2(t-1)\frac{\partial k_I(t-1)}{\partial w(t-1)} + u_3(t-1)\frac{\partial k_d(t-1)}{\partial w(t-1)}\right] \tag{5.16}$$

The term $\frac{\partial y(t)}{\partial u(t-1)}$ is the system's Jacobian, which is to be approximated using suitable means. In indirect STC, the system's Jacobian is estimated using a neural network. Figure 5.15 shows the configuration of an indirect STC for a PID controller, where the NN tuner provides the three gains to the PID controller. The error $e_c = r - y$ is used to adjust the NN tuner. The neuro-emulator is used to estimate the system's Jacobian and the error $e_i = y - \hat{y}$ is used to adjust the emulator.

There have been many applications of NN-based STC for PI, PID and other conventional controllers reported in the literature (Omatu *et al.*, 1995; Wang and Chen, 1999; Potocnik and Grabec, 2000; Wang *et al.*, 2001).

5.3.2.7   NARMA-L2 (Feedback Linearization) Control

Analytical solution to complex and diverse nonlinear problems is difficult. A wide range of techniques do exist for this kind of problem, such as the graphic method, overtone linearized method, partial linearized method and feedback linearized method. The graphic method is suitable for one- or two-step nonlinear systems. The overtone linearized method is suitable for nonlinear systems such as continuous electrical forms, gear meshing, etc. The partial linearized method is used to perturb a nonlinear system to linearize about an operating point. Brockett (1978) first introduced the idea of transforming nonlinear systems into linear controllable systems. Since then there has been considerable interest in feedback linearization (Isidori, 1985; Hunt *et al.*, 1986). The feedback linearization is a systematic approach for nonlinear control system design. The central idea is to transform nonlinear system dynamics into linear dynamics by cancelling the nonlinearities so that known linear control techniques can be used to design controllers (Ge *et al.*, 1998).

NARMA-L2 is an approximate feedback linearization control method using neural networks proposed by Narendra and Mukhopadhyay (1997). The main drawback of this method is that the plant must either be in companion form, or be capable of approximation by a companion form model. It is referred to as feedback linearization control when the plant model has a companion form (Pukrittayakame *et al.*, 2002). The other model is referred to as NARMA-L2 control, when the plant model can be approximated by the same form.

NARMA-L2 control requires the least computation of all the architectures described in this section. The controller is simply a rearrangement of the neural network plant model. The plant model is trained offline in batch form. The only online computation is a forward pass through the neural network controller.

There are two steps involved in the NARMA-L2 control design: systems identification and control design.

Identification of the NARMA-L2 model

As with model predictive control, the first step in using feedback linearization (or NARMA-L2) control is the identification of the system to be controlled. A neural network can be trained to represent the forward dynamics of the system. In the identification process, the first step is to choose a model structure of the system in question. One standard model that is used to represent general discrete-time nonlinear systems is the nonlinear autoregressive moving average (NARMA) model described by the following equation:

$$y(k + 1) = \Phi \left[y(k), y(k - 1), \ldots, y(k - n + 1), u(k), u(k - 1), \ldots, u(k - n + 1) \right]$$

$$(5.17)$$

where $u(k)$ is the system input and $y(k)$ is the system output. In the identification phase, a neural network is trained for the approximation of the nonlinear function $\Phi(.)$. This is the same identification procedure used for the NN predictive controller. If the system output is to follow some reference trajectory, say $y(k + 1) = y_r(k + 1)$, the next step is to develop a nonlinear controller of the form

$$u(k) = \Gamma[y(k), y(k - 1), \ldots, y(k - n + 1), y_r(k + d), u(k - 1), \ldots, u(k - m + 1)]$$

$$(5.18)$$

The controller in Equation (5.18) has a drawback while creating the function $\Gamma(.)$. A neural network is to be trained using a dynamic backpropagation algorithm to minimize the mean squared error (Narendra and Parthasarathy, 1991). This can be quite slow. Narendra and Mukhopadhyay (1997) proposed a solution to this problem where a NARMA-L2 model is adapted to feedback linearization of the affine system described below:

$$y(k + 1) = f[y(k), y(k - 1), \ldots, y(k - n + 1), u(k - 1), \ldots, u(k - m + 1)]$$
$$+ g[y(k), y(k - 1), \ldots, y(k - n + 1), u(k - 1), \ldots, u(k - m + 1)] \cdot u(k)$$

$$(5.19)$$

This model is in companion form, where the next controller input $u(k)$ is not contained inside the nonlinearity. The advantage of this form is that the control input that causes the system output to follow the reference $y(k + 1) = y_r(k + 1)$ can be solved. The resulting controller using the NARMA-L2 model would have the form

$$u(k) = \frac{y_r(k + d) - f\,[y(k), y(k - 1), \ldots, y(k - n + 1), u(k - 1), \ldots, u(k - n + 1)]}{g[y(k), y(k - 1), \ldots, y(k - n + 1), u(k - 1), \ldots, u(k - n + 1)]}$$

(5.20)

Using Equation (5.20) directly can cause realization problems, because the control input $u(k)$ must be determined based on the output $y(k)$ at the same time. So, instead, the model in Equation (5.21) is used:

$$\begin{aligned}y(k + 1) &= f\,[y(k), y(k - 1), \ldots, y(k - n + 1), u(k), u(k - 1), \ldots, u(k - m + 1)] \\ &\quad + g[y(k), \ldots, y(k - n + 1), u(k), \ldots, u(k - m + 1)] \cdot u(k + 1)\end{aligned}$$

(5.21)

A generic NARMA-L2 controller based on Equation (5.21) is shown in Figure 5.16.

TDL in Figure 5.16 denotes a tapped delay line whose output vector has for its elements the delayed values of the input signals.

The NARMA-L2 approximate model is parameterized using two neural networks denoted $\hat{f}$ and $\hat{g}$, to be used to identify the system:

$$\begin{aligned}\hat{y}(k + 1) &= \hat{f}[y(k), y(k - 1), \ldots, y(k - n + 1), u(k - 1), \ldots, u(k - m + 1)] \\ &\quad + \hat{g}[y(k), y(k - 1), \ldots, y(k - n + 1), u(k - 1), \ldots, u(k - m + 1)] \cdot u(k)\end{aligned}$$

(5.22)

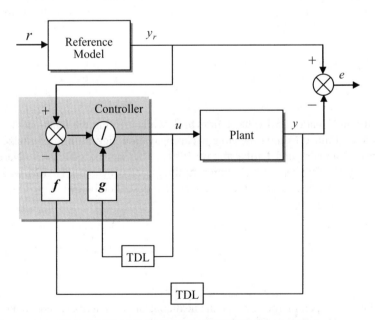

Figure 5.16   Block diagram of NARMA-L2 controller

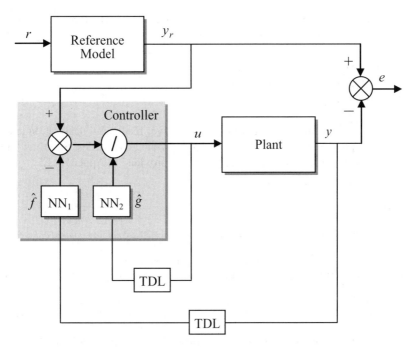

Figure 5.17   NARMA-L2 NN controller

NARMA-L2 neural controller

The NARMA-L2 controller design is simple, based on the NARMA-L2 plant model in Equation (5.22). Using the NARMA-L2 model, the controller is obtained as

$$u(k+1) = \frac{y_r(k+1) - \hat{f}\,[y(k), y(k-1), \ldots, y(k-n+1), u(k-1), \ldots, u(k-n+1)]}{\hat{g}[y(k), y(k-1), \ldots, y(k-n+1), u(k), u(k-1), \ldots, u(k-n+1)]}$$

$$(5.23)$$

This controller can be implemented with the previously identified NARMA-L2 plant model. The NARMA-L2 neural controller is shown in Figure 5.17.

5.4   MATLAB® Programs

Neural networks have been applied to various identification and dynamic control systems very successfully. The universal approximation capabilities of the multilayer perceptron make it a popular choice for modelling nonlinear systems and for implementing general-purpose nonlinear controllers. This section presents three examples of modelling, identification and adaptive identification of systems. The MATLAB® codes with associated descriptions are given in Appendix E.

The appendix then presents a brief description of the GUI of the Neural Network Toolbox. The section introduces three popular neural network architectures for prediction and control that have been implemented using Simulink® block sets. The three Simulink® block sets are:

- Model predictive control,
- NARMA-L2 (feedback linearization) control and
- Model reference control.

The three control architectures (i.e., the model predictive, feedback linearization and model reference control) are described with three examples. MATLAB® codes, Simulink® models, plots of simulation results and system descriptions are provided in Appendix E.

References

Akesson, B.M. and Toivonen, H.T. (2006) A neural network model predictive controller, *Journal of Process Control*, 16, 937–946.

Astrom, K.J. and Wittenmark, B. (1973) On self-tuning regulators, *Automatica*, 9, 185–199.

Brockett, R.W. (1978) Feedback invariants for nonlinear systems, *Proceedings of the IFAC World Congress*, Helsinki, Finland, pp. 1115–1120.

Cai, Z.-X. (1997) *Intelligent Control: Principles, Techniques and Applications*, World Scientific, Singapore.

Ge, S.S., Lee, T.H. and Harris, C.J. (1998) *Adaptive Neural Network Control of Robotic Manipulators*, World Scientific, Singapore.

Haykin, S. (1999) *Neural Networks – A Comprehensive Foundation*, Prentice-Hall, Upper Saddle River, NJ.

Haykin, S. (2009) *Neural Networks and Learning Machines*, 3rd edn, Prentice-Hall, Englewood Cliffs, NJ.

Hunt, K.J., Sbarbaro, D., Zbikowski, R. and Gawthrop, P.J. (1992) Neural networks for control systems: a survey, *Automatica*, 28, 1083–1112.

Hunt, L.R., Luksic, M. and Su, R. (1986) Exact linearization of input–output systems, *International Journal of Control*, 43(1), 247–255.

Isidori, A. (1985) *Nonlinear Control Systems*, Springer-Verlag, Berlin.

Jordan, M.I. and Rumelharrt, D.E. (1990) Forward models: supervised learning with distal teacher, *Cognitive Science*, 16, 313–355.

Kawato, M., Furukawa, K. and Suzuki, R. (1987) A hierarchical neural network model for control and learning of voluntary movements, *Biological Cybernetics*, 57, 169–185.

Kawato, M., Uno, Y., Isobe, M. and Suzuki, R. (1988) Hierarchical neural network model for voluntary movement with application to robotics, *IEEE Control Systems Magazine*, 8(2), 8–15.

Khalid, M., Omatu, S. and Yusof, R. (1993) MIMO furnace control with neural networks, *IEEE Transactions on Control Systems Technology*, 1, 238–245.

Khalid, M., Omatu, S. and Yusof, R. (1995) Temperature regulation with neural networks and alternative control schemes, *IEEE Transactions on Neural Networks*, 6(3), 572–582.

Kim, Y.H. and Lewis, F.L. (1998) Dynamic neural networks for closed-loop feedback control and estimation of uncertain nonlinear systems. In *Industrial and Manufacturing Systems, Vol. 4: Neural Network Systems Techniques and Applications*, C.T. Leondres (ed.), Academic Press, New York.

Landau, Y.D. (1979) *Adaptive Control – The Model Reference Approach*, Marcel Dekker, New York.

Lee, T.H., Hang, C.C., Lian, L.L. and Lim, B.C. (1992) An approach to the inverse non-linear control using neural networks, *Mechantronics*, 2(6), 595–611.

Ljung, L. (1999) *System Identification — Theory for the User*, 2nd edn, PTR Prentice-Hall, Upper Saddle River, NJ.

Mayne, D.Q. and Michalska, H. (1990) Receding horizon control of non-linear systems, *IEEE Transactions on Automatic Control*, 35, 814–824.

McClelland, W.S. and Rumelhart, D.E. (1986) *Parallel Distributed Processing*, MIT Press, Cambridge, MA and the PDP Research Group.

Miller, T.W., Sutton, R.S. and Werbos, P.J. (1990) *Neural Networks for Control*, MIT Press, Cambridge, MA.

Minsky, M. and Papert, S. (1959) *Perceptrons*, MIT Press, Cambridge, MA.

Narendra, K.S. (1990) Adaptive control using neural networks. In *Neural Networks for Control*, W.T. Miller, R.S. Sutton and P.J. Werbos (eds), The MIT Press, Cambridge, MA.

Narendra, K.S. (1995) Adaptive control: neural network applications. In *The Handbook of Brain Theory and Neural Networks*, M.A. Arbib (ed.), The MIT Press, Cambridge, MA, pp. 69–73.

Narendra, K.S. (1996) Neural networks for control: theory and practice, *Proceedings of the IEEE*, 84, 1385–1406.

Narendra, K.S. and Mukhopadhyay, S. (1997) Adaptive control using neural networks and approximate models, *IEEE Transactions on Neural Networks*, 8, 475–485.

Narendra, K.S. and Parthasarathy, K. (1990) Identification and control of dynamic systems using neural networks, *IEEE Transactions on Neural Networks*, 1(1), 4–27.

Narendra, K.S. and Parthasarathy, K. (1991) Gradient methods for the optimization of dynamical systems containing neural networks, *IEEE Transactions on Neural Networks*, 2(2), 252–262.

Nelles, O. (2001) *Nonlinear System Identification*, Springer-Verlag, Berlin.

Nguyen, D.H. and Widrow, B. (1990) Neural network for self-learning control systems, *IEEE Control System Magazine*, 10(3), 18–23.

Noriega, J.R. and Wang, H. (1998) A direct adaptive neural-network control for unknown nonlinear systems and its application, *IEEE Transactions on Neural Networks*, 9(1), 27–34.

Omatu, S., Khalid, M. and Yusof, R. (1995) *Neuro-control and Its Applications*, Springer-Verlag, London.

Park, J.-H., Huh, S.-H., Kim, S.-H., Seo, S.-J. and Park, G.-T. (2005) Direct adaptive controller for non-affine nonlinear systems using self-structuring neural networks, *IEEE Transactions on Neural Networks*, 16(2), 414–422.

Potocnik, P. and Grabec, I. (2000) Adaptive self-tuning neuro-control, *Mathematics and Computers in Simulation*, 51, 201–207.

Psaltis, D., Sideris, A. and Soquet, A. (1999) A multilayer neural network controller, *IEEE Control Systems Magazine*, 8(2), 17–21.

Pukrittayakame, A., De Jesus, O. and Hagan, M.T. (2002) Smoothing the control action for NARMA-L2 controllers, *Proceedings of the 2002 45th Midwest Symposium on Circuit and Systems (MWSCS-2002)*, Vol. 3, pp. 37–40.

Rosenblatt, F. (1958) The perceptron: a probabilistic model for information storage and organisation in the brain, *Psychology Review*, 65, 386–408.

Saerens, M. and Soquet, A. (1989) A neural controller, *Proceedings of the 1st IEE International Conference on Artificial Neural Networks*, London, pp. 211–215.

Sarangapani, J. (2006) *Neural Network Control of Nonlinear Discrete-Time Systems*, CRC Press, London.

Sen, P., Hearn, G.E. and Zhang, Y. (1998) Adaptive neural controller. In *Neural Network Systems Techniques and Applications, Vol. 4: Industrial and Manufacturing Systems*, C.T. Leondes (ed.), Academic Press, New York, pp. 273–344.

Soloway, D. and Haley, P.J. (1996) Neural generalized predictive control, *Proceedings of the 1996 IEEE International Symposium on Intelligent Control*, pp. 277–281.

Turner, P., Morris, J. and Montague, G. (1995) Applications of dynamic artificial neural networks in state estimation and nonlinear process control. In *Neural Network Application in Control*, G.W. Irwin, K. Warwick and K.J. Hunt (eds), IEE Control Engineering Series 53, IEE Press, London, pp. 141–159.

Ungar, L.H. (1996) A bio-reactor benchmark for adaptive network-based process control. In *Neural Networks for Control*, W.T. Miller, R.S. Sutton and P.J. Werbos (eds), The MIT Press, Cambridge, MA.

Wang, G.J. and Chen, T.C. (1999) A robust parameter self-tuning learning algorithm for multilayer feedforward neural network, *Neurocomputing*, 25, 167–189.

Wang, G.-J., Fong, C.-T. and Chang, K.J. (2001) Neural network based self-tuning PI controller for precision motion control of PMAC motors, *IEEE Transactions on Industrial Electronics*, 48(2), 408–415.

Werbos, P.J. (1974) Beyond regression: new tools for prediction and analysis in the behavioural sciences, Doctoral Dissertation, Applied Mathematics, Harvard University.

Werbos, P.J. (1990a) Neuro-control and related techniques. In *Handbook of Neural Computing Applications*, A. Maren, C. Harston and R. Pap (eds), Academic Press, San Diego, pp. 345–380.

Werbos, P.J. (1990b) Overview of designs and capabilities. In *Neural Networks for Control*, W.T. Miller III, R.S. Sutton and P.J. Werbos (eds), The MIT Press, Cambridge, MA, pp. 59–65.

Widrow, B. and Hoff, M.E. Jr. (1960) Adaptive switching circuits, *IRE Western Electric Show and Convention Record*, part 4 (August 23), pp. 96–104.

Wu, Q.H., Hogg, B.W. and Irwin, G.W. (1992) A neural network regulator for turbogenerators, *IEEE Transactions on Neural Networks*, 3, 95–100.

Yang, Y.Y. and Linkens, D.A. (1994) Adaptive neural network-based approach for the control of continuously stirred tank reactor, *IEE Proceedings D: Control Theory and Applications*, 141, 341–349.

Yuzgec, U., Becerikli, Y. and Turker, M. (2008) Dynamic neural network based model predictive control of an industrial baker's yeast process, *IEEE Transactions on Neural Networks*, 19(7), 1231–1242.

6

Evolutionary Computing

6.1 Introduction

We see a diversity of life on earth – millions of species each with its own unique behaviour patterns and characteristics or traits. All of these plants, animals, birds, fishes and other creatures have evolved, and continue evolving, over millions of years. They have adapted themselves to a constantly shifting and changing environment in order to survive. Those weaker and less fit members of species tend to die away, leaving the stronger and fitter to mate, create offspring and ensure the continuing survival of the species. Their lives are dictated by the laws of natural selection and Darwinian evolution – struggle for existence and survival of the fittest. Such an evolutionary process is shown in Figure 6.1.

And it is upon these ideas that evolutionary computing (EC) is based. Evolutionary computing is the emulation of the process of natural selection in a search procedure. In nature, organisms have certain characteristics that influence their ability to survive and reproduce. These characteristics are represented by encoding of information contained in the chromosomes of the organisms. New offspring chromosomes are created by means of mating and reproduction mechanisms. The end result will be offspring chromosomes that contain the best characteristics of each parent's chromosomes, which enable them to survive in an adverse environment. The process of natural selection ensures that more fit individuals have the opportunity to mate most of the time, leading to the expectation that the offspring will have similar or better fitness.

6.2 Evolutionary Computing

The population can be viewed simply as a collection of interacting creatures. As each generation of creatures comes and goes, the weaker ones tend to die away without producing children, while the stronger mate in the process of recombination to produce new and perhaps unique children with attributes from both parents to continue the evolutionary process. In nature, a diverse population within a species tends to allow the species to adapt to its environment

Computational Intelligence: Synergies of Fuzzy Logic, Neural Networks and Evolutionary Computing, First Edition.
Nazmul Siddique and Hojjat Adeli.
© 2013 John Wiley & Sons, Ltd. Published 2013 by John Wiley & Sons, Ltd.

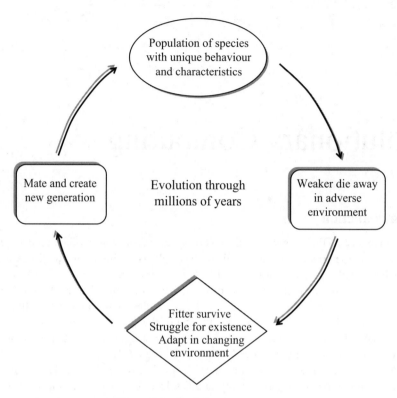

Figure 6.1   Evolution in the nature

with more ease. The process of evolution can be modelled algorithmically and simulated on a computer (Fogel, 1998). In the simplest form the model can be expressed as

$$\Pi(g+1) = \Psi[\Phi[\Pi(g)]] \tag{6.1}$$

The operations of random variation Φ and selection Ψ are applied on a population of $\Pi(g)$ at generation g to evolve to a new population of $\Pi(g+1)$ in the next generation. Useful variations have the highest chance of surviving in the struggle for existence, leading to a process of continual improvement (Darwin, 1859). The evolutionary algorithm cycle is shown in Figure 6.2. Successive application of the variations and selection drives the population towards particular optima in a search space. Evolutionary computation is the field that studies the properties of these algorithms and similar search procedures. Although the term 'evolutionary computation' was coined in the 1990s, the field has a long history. The origins of evolutionary computing can be traced back to the 1950s. A.S. Fraser was the first to conduct simulation of genetic systems using diploid organisms represented by binary strings (Fraser, 1957). The well-known statistician G.E.P. Box proposed an evolutionary approach to optimizing industrial production. The method, termed 'evolutionary operation', was first published in 1957 (Box, 1957). R.M. Friedberg was among the first to evolve computer programs, but reported the results as of limited success (Friedberg, 1958). But in another investigation, Friedberg et al. (1959) suggested that 'where we should go from here is not clear'. The field remained relatively

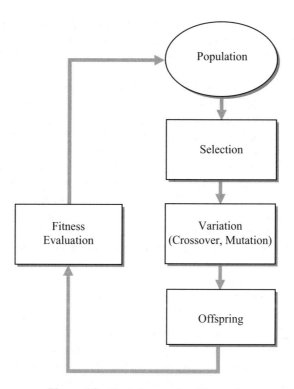

Figure 6.2   Evolutionary algorithm cycle

unknown to the broader scientific community for many years. The fundamental works of Fogel (1962), Holland (1962), Rechenberg (1965) and Schwefel (1968) had a great influence on the research community to accept evolutionary computation as a general concept for problem solving, especially for difficult optimization problems. Application of evolutionary algorithms was possible to solve difficult real-world optimization problems due to the fact that powerful computers were available in the 1980s. This eventually attracted a broader audience for evolutionary computing in the 1990s and in the following decades (Fogel *et al.*, 1996; Coella *et al.*, 2007; Deb, 2008).

6.3   Terminologies of Evolutionary Computing

6.3.1   *Chromosome Representation*

In nature, characteristics or traits of organisms are represented by long strings of information encoded in the chromosomes. The first design step in EC is commonly called chromosome representation, where each individual of a population represents a candidate solution to an optimization or search problem. The characteristics of an individual are represented by the chromosome, or genome. A chromosome can be thought of as a vector X consisting of m genes denoted by x:

$$X = \{x_1, x_2, x_3, \ldots, x_m\} \tag{6.2}$$

Each chromosome X represents a point in the m-dimensional search space, i.e., $X \in \Re^m$. A chromosome consists of a number of genes x_i, where the gene is the functional unit of inheritance. In terms of optimization, a gene represents one parameter of the optimization problem.

Objects forming possible solutions within the original problem context are referred to as phenotypes, while their encodings are called genotypes. A phenotype is the expressed behavioural traits of an individual in a specific environment. A genotype describes the genetic composition of an individual as inherited from its parents. In other words, it is a mechanism to store experiential evidence as gathered by parents. It is important to understand the difference between the phenotype space and the genotype space. The evolutionary search essentially takes place in the genotype space, whereas a set of good solutions is obtained from the phenotype space.

There are many synonymous terms for elements of individuals widely used in the EC literature. In EC, a genotype is defined as a string of genes (a biology-oriented terminology). These genes reside at various positions called a locus (plural loci) and have a value called an allele. In optimization or search problems, these place-holders are commonly called variables.

6.3.2 Encoding Schemes

The first task in EC is to find a mechanism to encode the genetic information of a population representing an entire search space of a problem domain into chromosomes. In fact, this is a mapping from the phenotype to the genotype space. Encoding schemes in any evolutionary algorithm (EA) should be such that the representation and the problem space are close together, i.e., a natural representation of the problem. This also allows the incorporation of knowledge about the problem domain into the EC system in the form of special genetic information and a set of operations. As yet, there has been no general method of choosing an encoding scheme; rather, it depends on the problem in question. Even two similar problems may require a completely different encoding scheme. But the literature indicates some rule of thumb that can be used to select an encoding scheme to be used for a particular EA. Any particular encoding should embody the fundamental building blocks that are important to the problem domain. It is very important to keep in mind – while encoding genetic information into chromosomes – that the encoding scheme should also be invertible, i.e., to each genotype there has to be at most one corresponding phenotype. The encoding scheme should also be amenable to a set of genetic operators (discussed in Section 6.4) that propagates these building blocks from generation to generation, i.e., from parents to children. A tractable mapping to phenotype should be ensured to allow fitness values to be calculated at a minimum computation cost. Most EAs represent their chromosomes as vectors of specific data types. In this section, some of the encoding schemes will be introduced that have been in use with some success in the field.

6.3.2.1 Binary Coding

The most commonly used chromosome representation in the EC is the binary coding scheme. For an n-dimensional search space, each individual consists of n variables with each variable in the parameter set encoded as a binary string and concatenated to form a chromosome. Some problems can be expressed very efficiently using binary coding, as follows:

$$X = \{(b_1, b_2, \ldots, b_l), (b_1, b_2, \ldots, b_l), \ldots\}, b_i \in \{0, 1\} \tag{6.3}$$

For example

| Chromosome A | 10110010 |
| Chromosome B | 11111110 |

In this example, chromosomes A and B have eight genes. The position or locus of the ith gene is simply the ith bit in the bit-string and the value or allele is given by the bit-string $A[i]$ or $B[i]$.

The problem existing in the binary coding lies in the fact that a long string always occupies the computer memory even though only a few bits are actually involved in the crossover and mutation operations. This is particularly the case when a lot of parameters need to be optimized and a higher precision is required for the final result. There are other pitfalls of using binary coding. The value of a bit may suppress the fitness contribution of other bits. This can also cause fitness insensitiveness to alleles. Some problems require higher-order genes than binary symbol sets or building blocks. Genetic operators can produce illegal solutions due to not being able to describe the search point. Also, empirical evidence suggests that a large Hamming distance in the representational mapping between adjacent values, as in the case of binary coding, can result in the search process being deceived or unable to efficiently locate the global minimum (Caruana and Schaffer, 1988). To overcome the inefficient occupation of memory and inefficient search process, there is increasing interest in alternative encoding strategies such as integer or real-valued encoding.

Example 6.1   The problem is to optimize a function $f(x_1, x_2, x_3)$ that takes real values between [0.0, 1.0] with each value represented by 8 digits. In binary coding the string {00000000} corresponds to the real value *0.0* and {11111111} corresponds to *1.0*. Now a chromosome represented by $\{x_1, x_2, x_3\}$ looks like

$$X = \{00000001 \quad 00101000 \quad 10110001\}$$
$$\downarrow \qquad\qquad \downarrow \qquad\qquad \downarrow$$
$$0.00390625 \qquad 0.15625 \quad 0.69140625$$
$$\downarrow$$
$$1/256 = 0.00390625$$

6.3.2.2   Gray Coding

While binary coding is frequently used, it has the disadvantage of Hamming Cliffs. A Hamming Cliff is formed when two numerically adjacent values have bit representations with a large Hamming distance. For example, consider the decimal numbers 7 and 8. The corresponding binary representations using 4-bits are $7 = 0111$ and $8 = 1000$, with a Hamming distance of 4. This causes a problem when a small change in variables should result in a small change in fitness. To overcome the problem of Hamming Cliffs, an alternative bit representation is to use Gray coding. Gray coding has the advantage over binary coding in that the Hamming distance between two successive numerical values is one. Binary numbers can easily be converted into Gray coding using the conversion

$$g = g_1 g_k$$
$$g_k = (b_{k-1} \oplus b_k) \tag{6.4}$$

where $b_k = b_1, b_2, \ldots, b_n$ and b_1 is the most significant bit in binary representation, $g_1 = b_1$ and $\oplus$ represents XOR operation.

Example 6.2 Represent the decimal numbers 1–8 in binary and Gray code showing Hamming Cliffs between two subsequent numbers. The difference in bit position is shown in italic in the box below.

Decimal	Binary	Hamming Cliff	Gray	Hamming Cliff
1	0001	–	0001	–
2	00*10*	2	00*1*1	1
3	001*1*	1	0010	1
4	0*100*	3	0*1*10	1
5	010*1*	1	011*1*	1
6	01*1*0	2	01*0*1	1
7	0111	1	0100	1
8	*1000*	4	*1*100	1

6.3.2.3 Real-valued Coding

The use of real-valued genes in EC is claimed to offer a number of advantages in numerical function optimization over binary encoding (Wright, 1991). The efficiency of EC is increased as there is no need to convert binary strings into real values before each function evaluation and hence there is no loss in precision caused by conversion. For a detailed description of real-valued encoding schemes, see Michalewicz (1992). When real values are used in chromosome representation, chromosomes are simply a string of real values

$$X = \{r, r, r\} \quad r \in \Re \tag{6.5}$$

For example

| Chromosome A | 456.1, 0.6879, 4.589 |
| Chromosome B | 456.34, 0.7968, 5.984 |

Example 6.3 The problem is to optimize a function $f(x_1, x_2, x_3)$ that takes real values between [0.0, 1.0] with each value represented by a real value rather than binary digits. Now the chromosome represented by $\{x_1, x_2, x_3\}$ looks like

$$X = \{0.00390625 \quad 0.15625 \quad 0.69140625\}$$

The advantage of real-valued coding over binary coding includes increased precision and the chromosome string becomes shorter. Also, real-valued coding gives greater freedom to use special crossover and mutation techniques.

6.3.2.4 Hybrid Coding

There are many heterogeneously structured problems that occur very often in the industry which have a large complex set of solutions. A simple homogeneous encoding scheme of chromosome representation – such as a binary string, encoded integers, permutation of symbols or expression trees – does not work out to a solution of such problems. Partitioning a problem into components is sometimes realistic in terms of implementation issues. A component is defined as a homogeneous collection of parameters or variable values of the same type or structure. A component can be, for example, a set of integers, floating points, trees, permutation strings, etc. The chromosome can thus be a combination of binary, real values and other expressions depending on the problem:

$$X = \{x_1, x_2, x_3\} = \{(b_1 b_2 \cdots b_l), (b_1 b_2 \cdots b_l), (r)\}, b_i \in [0, 1], r \in \Re \qquad (6.6)$$

An example of hybrid coding is shown below:

Chromosome A = {(100), (010), (8), (3)}
Chromosome B = {(001), (110), (C), (B)}

This type of coding poses extra constraints on the EA operators. Normal crossover can still be applied but the mutation operator has to be modified in that it reinitializes a gene depending on the coding of that gene.

6.3.2.5 Permutation Coding

Permutation problems require the optimal arrangement of a set of symbols in a list. The travelling salesperson problem (TSP) is such a problem, where a symbol represents a city and the arrangements of symbols in a list represent the order in which the person should visit each city for a circuit of all cities. A permutation encoding can be used in sequencing events represented by a list of distinct integers in the order they should occur. In permutation encoding, every chromosome is a string of numbers that represent a position in a sequence. Permutation coding is different from other EA chromosome representations in that the values or numbers do not occur more than once. An example of chromosomes with permutation encoding is shown below:

Chromosome A = {1 5 3 2 6 4 7 9 8}
Chromosome B = {a d e b f i c g h}

This representation, in fact, prohibits missing or duplicate allele values and facilitates a simple decoding mechanism. For some types of crossover and mutation, corrections must be made to leave the chromosome consistent (i.e., have a real sequence in it) for some problems. There are actually two classes of problems that can be represented by permutation coding. The first kind is the ordering problem in which events should occur in a fixed order, e.g., job shop scheduling. The second kind is the adjacency problem. A typical problem is the TSP. The problem is to find a complete tour of n given cities of minimal length. One complete tour is a permutation of n cities. It is obvious that new variation operators are required to conserve the permutation property. Building blocks in permutation-type problems are determined simultaneously by the qualitative nature of the fitness function, genetic operators and encoding.

6.3.2.6   Value Coding

Direct value encoding can be used in problems where some more complicated values such as real numbers are used. Use of binary encoding for this type of problem would be difficult. In the value encoding, every chromosome is a sequence of some values. Values can be anything connected to the problem, such as (real) numbers, charts or any objects. An example of chromosomes with value encoding is as follows:

> Chromosome A = {1.2324 5.3243 0.4556 2.3293 2.4545}
> Chromosome B = {ABDJEIFJDHDIERJFDLDFLFEGT}
> Chromosome C = {(back), (back), (right), (forward), (left)}
> Chromosome D = {NB, NB, ZO, ZO, PS, NS, NS, ZO}

Value encoding is a good choice for some special problems. However, for this encoding it is often necessary to develop some new genetic operators such as crossover and mutation specific to the problem.

6.3.2.7   Tree Coding

Tree encoding is used mainly for evolving programs or expressions, i.e., for genetic programming (discussed in section 6.6.4). In the tree encoding every chromosome is a tree of some objects, such as functions or commands in programming language. For example, an algebraic expression $\left(x + \frac{5}{y}\right)$ can be described by the tree encoding shown in chromosome A in Figure 6.3(a). Similarly, a computer command 'steps do until wall' is expressed using tree encoding shown in chromosome B in Figure 6.3(b). Tree encoding is useful for evolving programs or any other structures that can be encoded in trees. A programming language like LISP is often used for this purpose, since programs in LISP are represented directly in the form of a tree and can easily be parsed as a tree, so the crossover and mutation can be done relatively easily. The task is to find a function that would approximate given pairs of values.

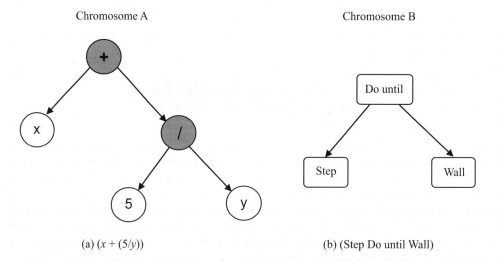

(a) $(x + (5/y))$ (b) (Step Do until Wall)

Figure 6.3   Tree coding of chromosome (a) algebraic expression; (b) computer command

| S|A|B|C|D | A|c|p|a|c | B|a|a|a|e | C|a|a|a|a | D|a|a|a|b |

Figure 6.4   Chromosome encoding a grammar

6.3.2.8   Grammar Coding

Grammatical coding was introduced by Kitano (1990) to train neural network architectures. A grammar is a set of rules that is applied to produce a set of structures (e.g., sentences in a natural language, programs in a computer language). A simple example is the following:

$$S \rightarrow aSb$$
$$S \rightarrow \epsilon$$

Here S is the start symbol and a non-terminal, a and b are terminals, and ϵ is the empty string terminal. $S \rightarrow \epsilon$ means that S can be replaced by the empty string. To construct a structure from this grammar, start with S and replace it with one of the allowed replacements given by the right-hand sides; take the resulting structure and continue until no non-terminals are left. For example:

$$S \rightarrow aSb \rightarrow a(aSb)b \rightarrow a(a(\epsilon)b)b \rightarrow aabb$$

Kitano applied this general type of grammar, called a 'graph-generation grammar', to represent the architecture of a neural network. A simple example of such a grammar-encoded chromosome is shown in Figure 6.4, where the right-hand side of each rule is a 2×2 matrix rather than a one-dimensional string. Capital letters are non-terminals and lowercase letters are terminals. Each terminal represents one of 16 possible 2×2 arrays consisting of 1s or 0s. For example, the terminals are defined as

$$a \rightarrow \begin{bmatrix} 0 & 0 \\ 0 & 0 \end{bmatrix}, b \rightarrow \begin{bmatrix} 0 & 0 \\ 0 & 1 \end{bmatrix}, c \rightarrow \begin{bmatrix} 1 & 0 \\ 0 & 1 \end{bmatrix}, \ldots, e \rightarrow \begin{bmatrix} 0 & 1 \\ 0 & 1 \end{bmatrix}, \ldots, p \rightarrow \begin{bmatrix} 1 & 1 \\ 1 & 1 \end{bmatrix}.$$

Thus, the mapping from genotype to phenotype follows as shown in Figure 6.5.

The matrix represents the directed graph shown in Figure 6.5, which is a feedforward neural network. A detailed description of grammar coding can be found in Kitano (1990, 1994). A useful application for the design of the architecture of neural networks with grammar-coding chromosome representation is discussed in Vonk et al. (1997).

6.3.3   Population

The role of a population in EC is to represent the search space in a specific coding scheme, discussed in Section 6.3.2, by a multi-set of genotypes. In a generation of population the individuals do not change; it is the population that changes or adapts to yield a solution. Once a coding scheme is chosen, a population is defined by a randomly generated set of chromosomes or individuals. The maximum number of individuals in a population is called the size of the population. The size of the population is an important parameter for any EA performance. Too small a population will limit the diversity (variability) to act on and too

$$S \Rightarrow \begin{bmatrix} A & B \\ C & D \end{bmatrix} \Rightarrow \begin{bmatrix} \begin{bmatrix} c & p \\ a & c \\ a & a \\ a & a \end{bmatrix} & \begin{bmatrix} a & a \\ a & e \\ a & a \\ a & b \end{bmatrix} \end{bmatrix} \Rightarrow$$

$$\begin{bmatrix} 1 & 0 & 1 & 1 & 0 & 0 & 0 & 0 \\ 0 & 1 & 1 & 1 & 0 & 0 & 0 & 0 \\ 0 & 0 & 1 & 0 & 0 & 0 & 0 & 1 \\ 0 & 0 & 0 & 1 & 0 & 0 & 0 & 1 \\ 0 & 0 & 0 & 0 & 0 & 0 & 0 & 0 \\ 0 & 0 & 0 & 0 & 0 & 0 & 0 & 0 \\ 0 & 0 & 0 & 0 & 0 & 0 & 0 & 0 \\ 0 & 0 & 0 & 0 & 0 & 0 & 0 & 1 \end{bmatrix} \Rightarrow$$

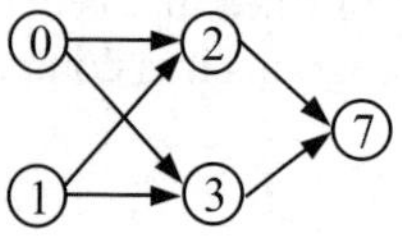

Figure 6.5   Mapping from genotype to phenotype

large a population is inefficient and slow due to the extra computation required. Typically, a population is composed of between 20 and 100 individuals (De Jong, 1975). A variant called micro-GA uses a very small population, approximately 10, with restrictive genetic operators in order to implement real-time execution (Karr, 1991). An example of a population for a search space is shown in Equation (6.7), with N individuals each of which consists of m genes:

$$P = \begin{bmatrix} {}^1X_1 & {}^1X_2 & \dots & {}^1X_m \\ {}^2X_1 & {}^2X_2 & \dots & {}^2X_m \\ \vdots & \vdots & \vdots & \vdots \\ {}^NX_1 & {}^NX_2 & \dots & {}^NX_m \end{bmatrix} \qquad (6.7)$$

Each chromosome X represents a point in the m-dimensional search space in $\Re^m$.

De Jong (1975) introduced the ideas of overlapping populations and crowding for some problem domains, for example, multimodal function optimization. In overlapping populations, new offspring replace similar solutions of the population. This maintains population diversity and therefore can help prevent premature convergence. This causes a crowding effect, which is introduced as a parameter called the crowding factor. The concept of crowding led to the ideas of 'niche' (Deb, 2008), 'speciation' (Eiben and Smith, 2003) and 'fitness sharing' (Cordon et al., 2001; Deb, 2008; Eiben and Smith, 2003) in EA. Biological species maintain a restrictive mating scheme, which allows individuals to mate within the same niche. This strategy helps formation of speciation. The strategy of forming speciation is generally termed 'niching' in EAs.

Initialization

The first population is seeded by randomly generated individuals. It is reasonable to discretize and place upper and lower bounds on the solution spaces for each of these parameters. An initial population of N chromosomes is generated using a random number generator that uniformly distributes numbers within the desired search space. An example of such a search space is shown in Figure 6.6. A variation is the extended random initialization procedure whereby a number of random initializations are tried for each individual and the one with the

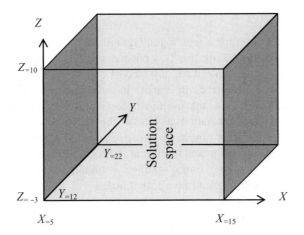

Figure 6.6   Search space in $\Re^3$

best performance is chosen for the initial population (Bralette, 1991). Other users have seeded the initial population with some individuals that are known *a priori*.

For example, let $g = f(x, y, z)$ be the function to be optimized. The parameters are x, y and z. The population P is initialized with a random number generator shown in Equation (6.8):

$$P = \begin{bmatrix} {}^1X & {}^1Y & {}^1Z \\ \vdots & \vdots & \vdots \\ {}^NX & {}^NY & {}^NZ \end{bmatrix} = \begin{bmatrix} 2.37 & 13.8 & -2.6 \\ \vdots & \vdots & \vdots \\ 11.5 & 20.2 & 9.7 \end{bmatrix} \tag{6.8}$$

6.3.4   Evaluation (or Fitness) Functions

A criterion is required to evaluate each individual's performance, which assigns a fitness value to each individual. Depending on the fitness value, it is decided whether the individual will go to the next generation to produce better populations of individuals or not. This eventually defines the basis for improvement over generations. The only thing that the fitness function must do is rank the individuals in some way by producing the fitness value. A common practice is to calculate the relative fitness of each individual, $F(x_i)$, from each individual's raw fitness, $f(x_i)$, relative to the fitness of the whole population, i.e.

$$F(x_i) = \frac{f(x_i)}{\sum\limits_{i=1}^{N} f(x_i)} \tag{6.9}$$

where N is the population size and x_i is the ith individual. This fitness value ensures that each individual has a probability of reproducing according to its relative fitness. The fitness function $f(x_i)$ must be defined by the user, which is very problem-dependent. Different problem-dependent fitness functions are discussed in Chapters 8 and 9. Evolutionary algorithms are in general optimization techniques, where an objective function is very often used to measure the optimization criterion. Therefore, the evaluation (or fitness) function is identical to the objective function.

6.3.5   Fitness Scaling

The evaluation of the relative fitness function in Equation (6.9) is simple and straightforward, but it poses three potential problems. Firstly, a positive relative fitness value calculated using Equation (6.9) may not be guaranteed for any kind of objective function. Secondly, proportionate selection can cause premature convergence in early generations due to the presence of an individual with high fitness value that eventually dominates a population. Thirdly, selection pressure decreases in the later generations as most of the individuals achieve the same fitness value. Therefore, readjustment of fitness values of solutions is essential to sustain a steady selective pressure on the population and to prevent premature convergence of the population to a sub-optimal solution. The use of fitness scaling is common among researchers. The calculated raw fitness $f(x_i)$ is transformed into a scaled fitness $\hat{f}(x_i)$, which is used for selection. There have been a variety of methods proposed by researchers (Cordon *et al.*, 2001). In linear scaling, the scaled fitness is defined as

$$\hat{f}(x_i) = c_0 f(x_i) + c_1 \tag{6.10}$$

where c_0 and c_1 are parameters, which can be static or dynamic, to be adjusted based on the raw fitness distribution over the current population. c_0 and c_1 are chosen such that the expected number of offspring for the best individual becomes $\alpha \approx 1.5 \dots 3.0$ and for the average individual becomes $\alpha \approx 1$. This determines the ratio

$$\frac{c_0}{c_1} = \frac{\alpha - 1}{f_{\max} - \bar{f}(x)} \tag{6.11}$$

where $f_{\max}$ is the best fitness and $\bar{f}(x)$ is the average fitness. The values of the parameters c_0 and c_1 can be determined from the relations in Equations (6.12) and (6.13) as long as the scaled fitness $\hat{f}(x_i)$ remains positive:

$$c_0 = \alpha - 1 \tag{6.12}$$

$$c_1 = f_{\max} - \bar{f}(x) \tag{6.13}$$

In sigma truncation scaling, the scaled fitness is defined as

$$\hat{f}(x_i) = \frac{f(x_i) - \bar{f}(x) - \sigma}{\sigma} \tag{6.14}$$

where σ is the standard deviation of the fitness values in the current population.

6.4   Genetic Operators

Each population of an EA produces a new generation of individuals, representing a set of new potential solutions to the optimization problem. The new generation is formed through the application of operators. There are three basic genetic operators found in every evolutionary computation. Although some algorithms may not employ the crossover operator, we will refer to them as evolutionary algorithms rather than genetic algorithms:

- Selection
- Crossover or recombination
- Mutation.

6.4.1   Selection Operators

The selection operator allows individuals (chromosomes) to be copied for possible inclusion in the next generation. The chance that a string will be copied is based on the individual's *fitness value*, calculated from a *fitness function*. For each generation, the selection operator chooses individuals that are placed into a *mating pool*, which is used as the basis for creating offspring for the next generation. There are many different types of selection operators. One can select the fittest and discard the worst, statistically selecting the rest of the mating pool from the remainder of the population. In general, selection is typically probabilistic, which offers better chances for individuals with high fitness to get selected into the mating pool. Low-fitness individuals are also often given a small chance. There are two important factors closely related to any EA: selective pressure and population diversity. A strong selective pressure essentially means focusing on best-fit individuals in the population. This in turn instigates a decrease in the population diversity, which may result in a premature convergence, whereas a weak selective pressure can make a search ineffective. It takes many generations to converge as the population increases. There are hundreds of variants of the selection scheme. None are right or wrong. In fact, some will perform better than others depending on the problem domain being explored. For the moment, we will look at the most commonly used selection methods in evolutionary algorithms.

6.4.1.1   Random Selection

Random selection is the simplest method, where individuals are selected randomly with no reference to fitness at all. Each individual, good or bad, has an equal chance or probability $\frac{1}{N}$ of being selected, where N is the size of the population. As a result, random selection has a low selective pressure causing a slow convergence to solution.

6.4.1.2   Proportional Selection

The chance of an individual being selected is proportional to the fitness value. A probability distribution proportional to fitness is created, and individuals are selected through sampling of the distribution

$$P(C_i) = \frac{f(C_i)}{\sum\limits_{i=1}^{N} f(C_i)} = F(C_i) \tag{6.15}$$

where $P(C_i)$ is the probability that an individual C_i will be selected, $f(C_i)$ is the fitness of an individual, N is the size of the population and $F(C_i)$ is the relative fitness of an

Table 6.1 Three individuals with corresponding fitness values

Individual (string)	Fitness value	Relative fitness	Percentage	No. selected
01001	5	$\dfrac{5}{26}$	19%	0
10000	12	$\dfrac{12}{26}$	46%	2
01110	9	$\dfrac{9}{26}$	35%	1

individual. That is, the probability of an individual being selected (e.g., to produce offspring) is directly proportional to the relative fitness value of that individual. This may cause an individual to dominate the production of offspring, thereby limiting diversity in the new population. Limiting the number of offspring produced by that single individual is of course possible.

In the *roulette wheel* method the relative fitness values are calculated (or fitness values are normalized by dividing each fitness value by the maximum fitness value). The probability distribution can then be thought of as a roulette wheel, where each slice has a width corresponding to the selection probability of an individual. Selection can then be visualized as the spinning of the wheel. To look abstractly at this method, consider the three individuals in Table 6.1. From this table, it is obvious that the string 10000 is the fittest, and should be selected for reproduction approximately 46% of the time. The string 01001 is the weakest, and should only be selected 19% of the time. The corresponding roulette wheel looks like Figure 6.7. The roulette wheel is spun three times, with the results indicating the string to be placed in the pool. It is obvious from this wheel that there is a good chance of the string 10000 being selected more than once. Multiple copies of the same string can exist in the mating pool. This is even desirable, since the stronger strings will begin to dominate, eradicating the weaker ones from the population. There are difficulties with this type of selection, as it can lead to premature convergence on a local optimum.

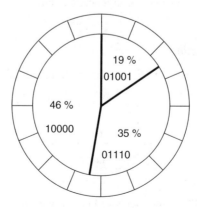

Figure 6.7 Roulette wheel selection

The following pseudocode can be used for the roulette wheel selection algorithm:

```
{
i ← 1                    ; set chromosome index to 1
val ← P(Cᵢ)              ; set val to P(Cᵢ)
ξ ← random(0,1)   ; choose a random value between (0,1)
while val < ξ
  {
  i++
  val = val + P(Cᵢ)
  }
return Cᵢ as selected individual
}
```

There are many techniques available in the literature for the roulette wheel selection method.

6.4.1.3 Tournament Selection

In tournament selection a group of k individuals is randomly selected from a population of N individuals. These k individuals then take part in a tournament and the performance of the selected individuals is evaluated. The individual with the best fitness is selected from the group. For crossover operation, two tournaments are held to select the two parents. The advantage of tournament selection is that the worst individuals of the population will not be selected and will therefore not contribute to the genetic construction of the next generation, and the best individual will not dominate in the reproduction process. As a result, the tournament selection has a low selective pressure which ensures an optimum solution. It is very important to remember that the size of k individuals is directly related to the selective pressure. It causes a very high selective pressure for $k = N$ and results in a low selective pressure for $k = 1$.

6.4.1.4 Rank-based Selection

Rank-based selection uses the rank ordering of the fitness values to determine the probability of selection and not the fitness values themselves. This means that the selection probability is independent of the actual fitness. Ranking therefore has the advantage that a highly fit individual will not dominate in the selection process as a function of the magnitude of its fitness.

One example of the rank-based selection is non-deterministic linear sampling, in which individuals are sorted by decreasing fitness value. The first individual is the fittest one. The selection of the individual C_i from N individuals using the selection operator is defined as

$$i = \text{random (random } (N))$$

$$C_i \leftarrow P(i) \tag{6.16}$$

The individual C_i is selected from the population P.

There are other nonlinear ranking techniques that have also been used, for example

$$P(C_i) = \frac{1 - e^{-r(C_i)}}{\mu} \tag{6.17}$$

$$P(C_i) = \rho \, (1 - \rho)^{N-1-r(C_i)} \tag{6.18}$$

Table 6.2   Mating pool

Strings in mating pool	Fitness (%)
10000	46
10000	46
01110	35

where $r(C_i)$ is the rank of the individual C_i, μ is a normalization constant and ρ is the probability of selecting the next individual. These nonlinear selection operators are biased towards the best individuals at the cost of possible premature convergence.

6.4.1.5   Elitism

Elitism is the selection of a set of individuals from the current generation to survive to the next generation. The number of individuals to survive to the next generation, without mutation, is referred to as the generation gap. If the generation gap is zero, the new generation will consist entirely of new individuals. For a positive generation gap, say k, k individuals will survive to the next generation. Elitism is generally used to prevent the loss of the fittest member in a generation. Therefore, a trace is kept of the current fittest individual and always copied to the next generation.

6.4.1.6   Mating Pool

After selecting the individuals, they are placed in a mating pool from which two parents are chosen randomly for crossover. Table 6.2 shows the mating pool after selecting the individuals according to proportional selection from Table 6.1. Multiple copies of string 10000 in the mating pool can occur due to its highest fitness value.

6.4.2   Crossover Operators

Once the mating pool is created, the next operator in the EA is the crossover operator. The term 'recombination' is also used in the more general case. Crossover in biological terms refers to mating of two parents to produce new chromosomes by blending of genetic information from the parent chromosomes. The analogy carries over to crossover in EAs, whereby new solutions are created from the information contained within two (or more) parent solutions. This is the primary mechanism of creating new solutions (i.e., chromosomes) with higher fitness values that survive to the next generation. Whether a chromosome will undergo a crossover operation or not is determined by the crossover probability p_c. The EA selects two individuals at random from the mating pool and applies a crossover operation according to p_c. Depending on the number of parents involved in the crossover operation, crossover can be divided into three classes.

- Asexual – when a single parent is involved in producing offspring.
- Sexual – when two parents are involved in producing offspring.
- Multi-parent – when more than two parents are involved in producing offspring. Bremermann *et al.* (1966) first proposed the multi-parent crossover for binary representation.

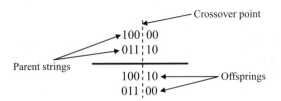

Figure 6.8   Crossover operation

The individuals selected as parents may be different or identical.

Crossover probability: The EA then calculates whether crossover should take place using a parameter called the *crossover probability* p_c. This is simply a probability value $p_c \in [0, 1]$ (calculated by flipping a weighted coin). The value of p_c is set by the user, and the suggested value ranges between 0.6 and 0.8 (although this value can be domain-dependent), which means that 60–80% of the new population will be formed by crossover. If EA decides not to perform crossover, the two selected strings are simply copied to the new population (they are *not* deleted from the mating pool). They may be used multiple times during crossover.

Crossover point: If crossover does take place, then a random splitting point is chosen in a string, the two strings are split and the split regions are mixed to create two (potentially) new strings. These child strings or offspring are then placed in the new population. The crossover point for a two-parent crossover is shown in Figure 6.8.

6.4.2.1   Single-point Crossover

A single point in the chromosome string is selected randomly. Holland (1975) suggested that segments of genes be swapped between parents to create offspring. The binary string from the beginning of the chromosome to the crossover point is copied from the first parent, and the rest is copied from the other parent. For example:

$$\textbf{P1} \rightarrow \textbf{11001}|011$$
$$+$$
$$\text{P2} \rightarrow 11011|\textbf{111}$$

- - - - - - - - - - - - - - - -

$$\textbf{O1} \rightarrow \textbf{11001}|\textbf{111}$$

The single-point crossover operation is illustrated in Figure 6.9:

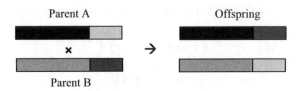

Figure 6.9   Single-point crossover

Parent A

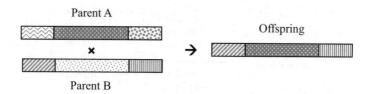

Offspring

Parent B

Figure 6.10   Two-point crossover

6.4.2.2   Two-point Crossover

Two crossover points are selected randomly, the binary string from the beginning of the chromosome to the first crossover point is copied from the first parent, the part from the first to the second crossover point is copied from the other parent and the rest is copied from the first parent again. For example:

$$P1 \rightarrow \mathbf{11}|0010|\mathbf{11}$$
$$\times$$
$$P2 \rightarrow 11|\mathbf{0111}|11$$
- - - - - - - - - - - - - - -
$$O1 \rightarrow 11|0111|11$$

The two-point crossover operation is illustrated in Figure 6.10.

6.4.2.3   *n*-point Crossover

One- or two-point crossover can be generalized to *n*-point crossover. *n* points are selected randomly to divide the chromosome into *n* segments of continuous genes. Offspring are created by combining alternative segments from the parents, for example three-point crossover is shown below:

$$P1 \rightarrow 11|0010|11|001$$
$$\times$$
$$P2 \rightarrow 11|0111|11|111$$
- - - - - - - - - - - - - - -
$$O1 \rightarrow 11|0111|11|111$$
$$O2 \rightarrow 11|0010|\mathbf{11}|001$$

6.4.2.4   Uniform Crossover

Syswerda (1989) introduced a new form of crossover called uniform crossover. Uniform crossover does not use any crossover point but instead creates offspring by swapping each corresponding allele of both parents. The swap is made with probability P_0. The bits of the strings are randomly copied from the first or from the second parent, as shown in the example below:

$$P1 \rightarrow 1\ \mathbf{1\ 0}\ 0\ 1\ 0\ \mathbf{1\ 1}$$
$$\times$$
$$P2 \rightarrow 1\ 1\ 0\ \mathbf{1\ 1\ 1}\ 0\ 1$$
- - - - - - - - - - - - - - -
$$O1 \rightarrow 1\ 1\ 0\ 1\ 1\ 1\ 1\ 1$$

The uniform crossover operation is illustrated pictorially in Figure 6.11:

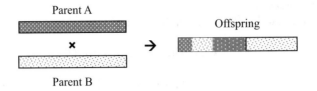

Parent A

Parent B

× →

Offspring

Figure 6.11   Uniform crossover

6.4.2.5   Arithmetic Crossover

Some arithmetic operation is performed to make a new offspring from two parents, for example:

$$P1 \rightarrow \textbf{11001011}$$
$$\oplus(XOR)$$
$$P2 \rightarrow 11011111$$
$$-----------$$
$$O1 \rightarrow 00010100$$

An XOR operation is performed on the parents for binary chromosomes.

Michalewicz (1992) described three types of arithmetic crossover: simple arithmetic, single arithmetic and whole arithmetic. In all three types, it works by taking the weighted sum of the two parental alleles x_i and y_i, i.e. the first offspring is $o_{i1} = \alpha.x_i + (1 - \alpha).y_i$ and the second offspring is $o_{i2} = (1 - \alpha).x_i + \alpha.y_i$ where $i = 1, \ldots, n$ are the alleles of the parental chromosomes. In simple arithmetic, choose a random point k in the chromosome, copy the first k alleles from both parents to both children and the rest of the alleles are obtained by arithmetic averaging of parents 1 and 2. In single arithmetic, choose a random allele k, copy all alleles from parents 1 and 2 to children 1 and 2 and take the arithmetic average of the kth allele from parents 1 and 2. In whole arithmetic, all alleles in children are calculated by arithmetic averaging of parents 1 and 2. The three crossover operators are illustrated below with three examples.

Simple arithmetic crossover
Simple arithmetic crossover is illustrated in the example below

$$P1 \rightarrow \textbf{0.2 0.0 0.4} \,|\, \textbf{0.5 0.1 0.9 0.7 0.5}$$
$$(avg)$$
$$P2 \rightarrow 01\ 0.2\ 0.9 \,|\, 0.3\ 0.5\ 0.2\ 0.8\ 0.3$$
$$------------------------------$$
$$O1 \rightarrow \textbf{0.2 0.0 0.4}|\ 0.4\ 0.3\ 0.55\ 0.75\ 0.4$$
$$O2 \rightarrow 0.1\ 0.2\ 0.9 \,|\, 0.4\ 0.3\ 0.55\ 0.75\ 0.4$$

Single arithmetic crossover
Single arithmetic crossover is illustrated in the example below

$$P1 \rightarrow \textbf{0.2 0.0 0.4} \,|\, \textbf{0.5} \,|\, \textbf{0.1 0.9 0.7 0.5}$$
$$(avg)$$
$$P2 \rightarrow 0.1\ 0.2\ 0.9 \,|\, 0.3 \,|\, 0.5\ 0.2\ 0.8\ 0.3$$
$$------------------------------$$
$$O1 \rightarrow \textbf{0.2 0.0 0.4} \,|\, \textbf{0.4} \,|\, \textbf{0.1 0.9 0.7 0.5}$$
$$O2 \rightarrow 0.1\ 0.2\ 0.9 \,|\, \textbf{0.4} \,|\, 0.5\ 0.2\ 0.8\ 0.3$$

Whole arithmetic crossover

Whole arithmetic crossover is illustrated in the example below

$$\text{P1} \rightarrow 0.2\ 0.0\ 0.4\ 0.5\ 0.1\ 0.8\ 0.7\ 0.5$$
$$\text{(avg)}$$
$$\text{P2} \rightarrow 0.1\ 0.2\ 0.8\ 0.3\ 0.5\ 0.2\ 0.7\ 0.3$$

- -

$$\text{O1} \rightarrow 0.15\ 0.1\ 0.6\ 0.4\ 0.3\ 0.5\ 0.7\ 0.4$$
$$\text{O2} \rightarrow 0.15\ 0.1\ 0.6\ 0.4\ 0.3\ 0.5\ 0.7\ 0.4$$

6.4.2.6   Linear Crossover

Linear crossover was introduced by Wright (1991). Two parents i and j are selected and three offspring are created by linearly combining the two parents. Three offspring represent three different search regions: one inside the two parents, one on the left side of parent $x_i(t)$ and one on the right side of parent $x_j(t)$, assuming parent $x_i(t)$ lies on the left of parent $x_j(t)$. The mechanism is shown in the example below:

$$\text{P1} \rightarrow x_i(t) = x_{i1},\ x_{i2}, x_{i3}, \ldots, x_{in}$$
$$\text{P2} \rightarrow x_j(t) = x_{j1},\ x_{j2}, x_{j3}, \ldots, x_{jn}$$

- -

$$\text{O1} \rightarrow o_1(t) = 0.5 * \lfloor x_i(t) + x_j(t) \rfloor$$
$$\text{O2} \rightarrow o_2(t) = \lfloor 1.5 * x_i(t) - 0.5 * x_j(t) \rfloor$$
$$\text{O3} \rightarrow o_3(t) = \lfloor -0.5 * x_i(t) + 1.5 * x_j(t) \rfloor$$

From the three offspring, the best two are accepted as offspring from this crossover operation.

6.4.2.7   Naïve Crossover

This crossover operator is similar to the single-point crossover operator used in binary chromosomes. The crossover point is chosen at the variable boundary of the selected parents and two offspring are created. The mechanism is shown in the example below:

$$\text{P1} \rightarrow x_i(t) = x_{i1},\ x_{i2}, x_{i3}, |x_{i4}, \ldots, x_{in}$$
$$\text{P2} \rightarrow x_j(t) = x_{j1},\ x_{j2}, x_{j3}, |x_{j4}, \ldots, x_{jn}$$

- -

$$\text{O1} \rightarrow o_1(t) = x_{i1},\ x_{i2}, x_{i3}, |x_{j4}, \ldots, x_{jn}$$
$$\text{O2} \rightarrow o_2(t) = x_{j1},\ x_{j2}, x_{j3}, |x_{i4}, \ldots, x_{in}$$

The operator does not have much search power, as the new search point lies on the variable boundaries.

6.4.2.8   Blend Crossover

Eshelman and Schaffer (1993) proposed the blend crossover for real-valued chromosomes. The operator blends two parents $x_i(t) = x_{i1},\ x_{i2}, x_{i3}, \ldots, x_{in}$ and $x_j(t) = x_{j1},\ x_{j2},$

$x_{j3}, \ldots, x_{jn}$ and generates offspring within the range $\lfloor x_i(t) - \alpha(x_j(t) - x_i(t)) \rfloor$ and $\lfloor x_j(t) + \alpha(x_j(t) - x_i(t)) \rfloor$. It then picks one offspring according to

$$x(t) = (1 - \gamma) x_i(t) + \gamma x_j(t) \tag{6.19}$$

where $\gamma = (1 + 2\alpha) u - \alpha$, $u \in [0, 1]$ and α is chosen arbitrarily. It is found that blend crossover performs best for $\alpha = 0.5$. The advantage of blend crossover is that it allows the search to focus when the population tends to converge in some small region.

6.4.2.9   Unfair Average Crossover

Nomura and Miyoshi (1996) proposed an unfair average crossover using two parents that produces two offspring. The crossover mechanism is illustrated in the example below:

$$P1 \rightarrow x_i(t) = x_{i1}, \; x_{i2}, x_{i3}, \ldots |x_{ik}, \ldots, x_{in}$$
$$P1 \rightarrow x_j(t) = x_{j1}, \; x_{j2}, x_{j3}, \ldots |x_{jk}, \ldots, x_{jn}$$

$$O1 \rightarrow x_i(t+1) = \begin{cases} (1+\alpha)x_i(t) - \alpha x_j(t) & i, j = 1, \ldots, k-1 \\ -\alpha x_i(t) + (1+\alpha)x_j(t) & i, j = k, \ldots, n \end{cases}$$

$$O1 \rightarrow x_i(t+1) = \begin{cases} (1-\alpha)x_i(t) + \alpha x_j(t) & i, j = 1, \ldots, k-1 \\ \alpha x_i(t) + (1-\alpha)x_j(t) & i, j = k, \ldots, n \end{cases}$$

The parameter $\alpha \in [0, 0.5]$ causes the operator to be biased towards one parent rather than preserving the mean of the parent solutions. k is the randomly chosen crossover point.

6.4.2.10   Crossover for Permutation Coding

Two genes may articulate meaningful information if they appear side by side (relative order). They may even articulate information if one gene precedes the other gene in the chromosome (absolute order) regardless of how many genes lie between. Syswerda (1989) inferred that the order as well as the position of genes in the permutation is meaningful. Therefore, permutation coding poses particular difficulties for crossover operation as a simple exchange of substrings between parents does not maintain the permutation properties of the chromosomes. The crossover operator has to comply with the semantic properties of the chromosome representation, that is, combining building blocks to form larger building blocks which share the phenotypical traits of the smaller building blocks. A number of specialized crossover operations have been devised to maintain the order in permutation coding, such as partially mapped crossover (PMX), edge crossover, order crossover, cycle crossover.

Partially mapped crossover (PMX) – PMX was first proposed by Goldberg and Lingle (1985) as a crossover operator for the travelling salesman problem. A variant of PMX is presented in Whitley (2000). Generalized PMX (GPMX) (Bierwirth et al., 1996) assembles one offspring from two parent chromosomes (donator and receiver). In this technique a substring is chosen from the donating chromosome. Then, all genes of the substring are deleted with respect to their index of occurrence in the receiving chromosome. The rest of the genes are copied onto the offspring. Figure 6.12(a) illustrates the mechanism of GPMX.

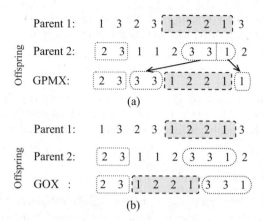

Figure 6.12   GPMX in comparison with GOX. (a) GPMX; (b) GOX

Order crossover (OX) – OX is based on the idea of preserving the relative order of positions present in parents. It begins in a similar way to PMX. The generalized OX (GOX) operator was first presented by Bierwirth (1995). GOX implants the substring into the receiver at the position where the first gene of the substring occurs (before deletion) in the receiver. Unlike GOX, GPMX implants the substring at the position where it occurs in the donator. Figure 6.12(b) illustrates the mechanism of GOX.

Edge crossover (EX) – EX is based on the idea of preserving edges that are present in one or more parent. There have been a number of revisions of EX over the years.

Cycle crossover (CX) – CX preserves information on the absolute position of elements in which they occur by dividing the elements into cycles, as shown in Figure 6.13. Offspring are created by selecting alternate cycles from each parent.

6.4.2.11   Crossover for Tree Coding

The crossover operation on tree coding is performed by randomly selecting subtrees of parents and swapping them. An example of crossover operation is shown in Figure 6.14. Two sub-trees, shaded in Figure 6.14(a), are chosen from the parents and swapped to create two offsprings shown in Figure 6.14(b). Asexual crossover is also possible in tree coding. Different variants of sexual and asexual crossover operation of tree coding are discussed in Section 6.6.4.

6.4.2.12   Fuzzy Connective-based and Recombination Operator

Some researchers have proposed fuzzified genetic operators such as fuzzy connective-based crossover (Herrera *et al.*, 1997; Herrera and Lozano, 2000, 2001) and soft genetic

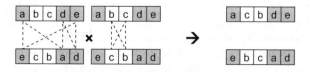

Figure 6.13   Cycle crossover

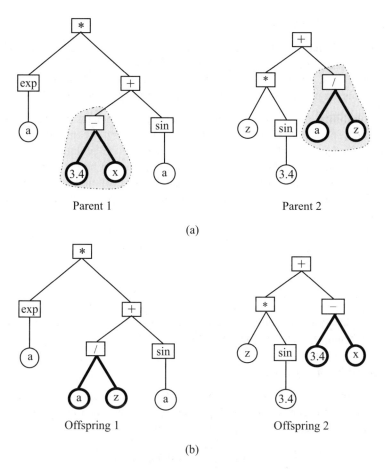

Parent 1   Parent 2

(a)

Offspring 1   Offspring 2

(b)

Figure 6.14   Crossover operation with two different parents. (a) Two parents in tree coding; (b) Two offspring created by sexual crossover operation

operators (Voigt *et al.*, 1995). It has been demonstrated by Herrera *et al.* (1997) that the fuzzy connective-based crossover operator is able to balance the exploitation and exploration and model the diversity of population. These operators are discussed in Section 8.4.2 of Chapter 8.

There have been some other crossover operators reported in the literature, such as simulated binary crossover developed by Deb and Agrawal (1995). This is a single-point crossover applied to two binary chromosomes and produces two offspring. Ono and Kobayshi (1997) suggested a unimodal normally distributed crossover with ellipsoidal probability distribution applied to three or more parents and producing two or more offspring. Tsutsui *et al.* (1999) proposed a simplex crossover where more than two parents take part in the operation. The centroid of the parents is calculated. Offspring are created along the line of the centroid and the parent with uniform distribution. Interested readers are directed to Beyer and Deb (2000) and Deb (2008) for an account of analysis on the different crossover operators.

6.4.3 Mutation Operators

Depending on the initial population chosen, there may not be enough variety of strings to ensure the EA sees the entire problem space. Or, the EA may find itself converging on strings that are not quite close enough to the optimum it seeks due to a bad initial population. Some of these problems are overcome by introducing a mutation operator into the EA.

Mutation probability: the *mutation probability*, $m_p \in [0, 1]$, dictates the frequency at which mutation occurs. The mutation probability should be kept very low (usually about 0.001) as a high mutation rate will destroy fit strings and degenerate the evolutionary algorithm into a random walk, with all the associated problems. Mutation can be performed either during selection or crossover (though crossover is more usual). The evolutionary algorithm checks to see if it should perform a mutation. If it should, it randomly changes the gene value to a new one. For example

Mutation

$1\ 0\ 0\ (0)\ 0 \rightarrow 1\ 0\ 0\ (1)\ 0$

There are a variety of mutation operators available in the literature. Some of them are discussed with examples in the following sections.

6.4.3.1   Mutation for Bit Coding

Inversion (for binary or Gray coding) – the most common mutation operator used for binary or Gray encoding considers each gene separately and allows each bit to flip with a small probability p_m, i.e., selected bits are inverted. For example

$1\ \mathbf{1}\ 0\ 0\ 1\ 0\ 0\ 1 \rightarrow 1\ \mathbf{0}\ 0\ 0\ 1\ 0\ 0\ 1$
Mutation

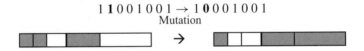

6.4.3.2   Mutation for Integer Coding

Random setting – a random cardinal value is chosen from a set of permissible intervals in each position with a small probability p_m. For example

$(2\ 1\ \mathbf{3}\ 0\ 5\ 1\ 7\ \mathbf{1}) \rightarrow (2\ 1\ \mathbf{5}\ 0\ 5\ 1\ 7\ \mathbf{2})$
Mutation

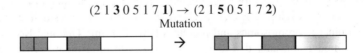

Creep mutation – a small value (positive or negative) is added to each gene with a small probability p_m. It is to be ensured that random values are sampled from a zero-mean symmetric distribution. Thus, creep mutation requires a number of parameters to control the distribution, which poses some extra constraints on the algorithm to find appropriate parameter settings.

6.4.3.3　Mutation for Real Coding

Since the allele values are continuous in a search space rather than discrete, it is common practice to change the allele value $x_i \in [L_i, U_i]$ to $x_i' \in [L_i, U_i]$ for each gene randomly within its domain specified by $[L_i, U_i]$. L_i and U_i are the lower and upper bounds, respectively. Two types of mutation can be distinguished from the literature.

Uniform mutation – values of $x_i' \in [L_i, U_i]$ are drawn random uniformly from within the respective bounds.

Non-uniform mutation – analogous to creep mutation in integer coding. It is designed to keep the change small by adding a small random value with zero-mean Gaussian distribution and user-defined variance. This can be defined as

$$x_i' = \lfloor x_i + \omega(n) \rfloor \in [L_i, U_i] \tag{6.20}$$

where $\omega(n)$ is a Gaussian random value with zero mean and user-defined variance. For example

$$(1.29\ 5.68\ \mathbf{2.86}\ \mathbf{4.11}\ 5.55) \rightarrow (1.29\ 5.68\ \mathbf{2.73}\ \mathbf{4.22}\ 5.55)$$

6.4.3.4　Mutation for Permutation Coding

Swap mutation or order changing – two genes are selected randomly and their positions exchanged. For example

$$(1\ \mathbf{2}\ 3\ 4\ 5\ 6\ \mathbf{8}\ 9\ 7) \rightarrow (1\ \mathbf{8}\ 3\ 4\ 5\ 6\ \mathbf{2}\ 9\ 7)$$

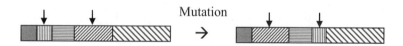

Insert mutation – two genes are chosen randomly and then moved next or close to each other by shuffling the others along. For example

$$(1\ \mathbf{2}\ 3\ 4\ 5\ 6\ \mathbf{8}\ 9\ 7) \rightarrow (1\ \mathbf{2}\ \mathbf{8}\ 3\ 4\ 5\ 6\ 9\ 7)$$

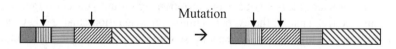

Scramble mutation – all or a subset of genes are chosen randomly and their positions scrambled. For example

$$(1\ \mathbf{2}\ 3\ 4\ 5\ 6\ \mathbf{8}\ 9\ 7) \rightarrow (1\ \mathbf{3}\ \mathbf{5}\ \mathbf{2}\ \mathbf{4}\ 6\ 8\ 9\ 7)$$

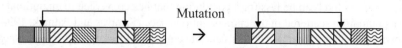

Inversion mutation – a subset of genes is selected randomly and the order of their positions reversed. For example

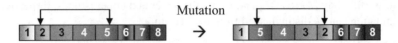

6.4.3.5 Mutation for Tree Coding

Changing operator, number – mutation in tree coding is the same as for other EAs. That is, the value of selected nodes is changed through some small random variation. For example

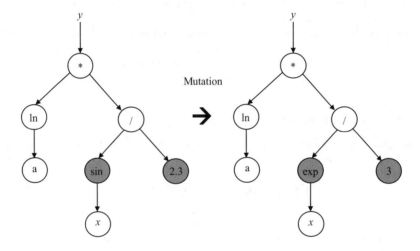

Further examples of different variants of mutations on tree coding are discussed in Section 6.6.4.

6.5 Performance Measures of EA

The standard performance measure of any optimization algorithm is the convergence property or rate of convergence. That is, the average number of generations that the EA requires to generate a solution with high fitness. In general, convergence means whether a solution can be found and how quickly. Because of the population-based stochastic nature of EA, the notion of convergence is different: an EA is said to be converged if the population converges to a uniform population consisting of P copies of a single individual which may or may not correspond to a global optimum. One possible way to measure the convergence of the EA is to observe the average fitness ($\bar{f}$) in relation to the maximum or best fitness (f_{best}) of the current population curves as a function of the number of samples or generations (Spears, 2000). The value of the term ($f_{best} - \bar{f}$) is likely to be less for a population that has converged to an optimal solution than for a population scattered in the entire solution space (Srinivas and Patnaik, 1994). It has been shown by Srinivas and Patnaik that EA converges to a local optimum with fitness value 0.5 with decreasing value of ($f_{best} - \bar{f}$), whereas a global optimum has a fitness value of 1.0.

6.6 Evolutionary Algorithms

The history of the field suggests that there are many different variants of evolutionary algorithms, but the common underlying idea behind all these algorithms is the same. Given is a population of individuals, the natural selection process causes a particular group of individuals or species to survive depending on the fitness. Typically, the variants are:

- Evolutionary programming (EP),
- Evolution strategies (ES),
- Genetic algorithms (GA),
- Genetic programming (GP),
- Differential evolution (DE),
- Cultural algorithm (CA).

In the last two decades EP, ES, GA, GP, DE and CA have been very popular with the scientific and research communities. These algorithms are discussed in subsequent sections.

6.6.1 Evolutionary Programming

Evolutionary programming, originally conceived by Lawrence J. Fogel in the 1960s, is a stochastic optimization strategy (Fogel, 1962). EP was then developed further by David Fogel in the 1990s (Fogel, 1991, 1995). EP differs substantially from GA and GP in that EP emphasizes the development of behavioural models and not genetic models. EP is derived from simulation of adaptive behaviour in evolution. That is, EP considers phenotypic evolution. The evolutionary process consists of finding a set of optimal behaviours from a space of observable behaviours. For this purpose, the fitness function measures the behaviour error of an individual with respect to the environment of that individual.

For EP, like GA, there is an underlying assumption that a fitness landscape can be characterized in terms of variables, and that there is an optimum solution (or multiple such optima) in terms of those variables. For example, if one were trying to find the shortest path in a travelling salesman problem, each solution would be a path. The length of the path could be expressed as a number, which would serve as the solution's fitness. The fitness landscape for this problem could be characterized as a hypersurface proportional to the path lengths in a space of possible paths. The goal would be to find the globally shortest path in that space, or more practically, to find very short tours very quickly.

The basic EP method involves four steps (repeated until a threshold for iteration is exceeded or an adequate solution is obtained):

1. Initialization of population. A population of individuals is created randomly, which uniformly covers the search space of the optimization problem. The number of individuals in a population is highly relevant to the speed of optimization, but no definite answers are available as to how many individuals are appropriate (other than >1) and how many individuals are just wasteful.
2. Mutation. Each individual is replicated into a new population. Each of these offspring are mutated according to a distribution of mutation types, ranging from minor to extreme, with a continuum of mutation types between. The severity of mutation is judged on the basis of the functional change imposed on the parents.

3. Evaluation. Each offspring is assessed by computing its fitness values $f(x_i)$ from the objective function by scaling them to positive values and sometimes by imposing some random alternation v_i. The fitness values actually quantify behavioural traits. Survival in EP is usually based on a relative fitness measure. Individuals that go into the next generation are selected based on relative fitness.

4. Selection. The purpose of the selection mechanism is to choose individuals from parents and offspring that survive to the next generation. Typically, a stochastic tournament is held to determine N individuals to be retained for the population of the next generation, although this is occasionally performed deterministically. There is no requirement that the population size be held constant, however, neither that only a single offspring be generated from each parent. Different selection strategies for EP are discussed later in this section.

A general EP algorithm is formulated as follows:

$g \leftarrow 0$; generation 0
$C(g) \leftarrow \{C(g)_n \mid n = 1,2, \ldots, N\}$; create and initialize population of size N
Do while (No convergence)
    {
    $f(x) \leftarrow$ evaluate $[C(g)]$; calculate fitness value
    $C'(g) \leftarrow$ mutate $[C(g)]$; generate offspring by mutation
    $C(g+1) \leftarrow$ select $[C(g) \cup C'(g)]$; create new population by selecting from old
                                                      ; population and offspring

    }
$g \leftarrow g+1$

EP differs from the other evolutionary algorithms in that no crossover operation is implemented. Only selection and mutation operators are applied to produce the new generation of population. Selection is based on competition and mutation is based on the amount of variation determined by a step size sampled from some probability distribution. There are a range of implementations of the two operators found in the literature. Some of them are discussed in the following section.

6.6.1.1   Mutation Operators

Mutation is applied to each of the individuals at a certain probability. The mutation operator to be used depends on the specific application. In general, the mutation is defined as

$$x_i'(t) \leftarrow x_{ij}'(t) = x_{ij}(t) + \Delta x_{ij}(t) \tag{6.21}$$

where $x_i'(t)$ is the offspring created from the ith individual $x_i(t)$ by mutating the jth gene of the ith individual $x_{ij}'(t)$. The step size $\Delta x_{ij}(t)$ is a random noise sampled from the probability distribution with a standard deviation of σ_i, defined as

$$\Delta x_{ij}(t) = \sigma_i * v \tag{6.22}$$

where v is a random variable with probability distribution $p(v) = \frac{1}{\sqrt{2\pi}} \exp(-\frac{v^2}{2})$ and σ_i is called the strategy parameter, which scales the contribution of the noise v. EP algorithms can fall into three categories, depending on the values of σ_i: dynamic, when σ_i varies over generations (Fogel *et al.*, 1991); non-adaptive, when σ_i remains constant; and self-adaptive, when σ_i is being learnt along with the optimization procedure (Fogel *et al.*, 1991).

6.6.1.2   Selection Operators

A selection operator is applied in EP to generate a new population from both the parent population and the offspring population, i.e., $\{C(g) \cup C'(g)\}$. That means both the parents and offspring compete to survive in a q-tournament ($q > 1$) selection. A group of individuals is randomly chosen from $\{C(g) \cup C'(g)\}$. The relative fitness of each individual of the group is calculated to measure the performance of the individual. This performance enables each individual to compete for survival to the next generation. There are a number of selection methods based on the relative fitness:

- Elitist – transfer the group of best individuals to the next generation. The remainder of the population is selected from the remainder of the parents and offspring.
- All individuals – parents and offspring have the same chance of being selected. Any of the selection operators discussed earlier can be used to create the new population.
- First cull – the worst parents and offspring are marked and then the remaining 'good' individuals are selected from the population.

Example 6.4: Finite-state machine   EP was originally developed to evolve finite-state machines (FSMs) for prediction and control. FSMs embody a class of abstract machines that represent the behaviour of sequential logic circuits. In other words, they are computer programs that describe a sequence of actions to be executed. Characterized by a set of inputs, outputs and internal states, a 'memory' of previous inputs is retained through the use of an FSM's internal states. FSMs, due to the outputs being dependent on the input and current state, can be defined by the ordered sextuple $M = \langle Q, I, O, \psi, \phi, q \rangle$, where Q is a finite set of internal states, I is a finite set of input symbols, O is a finite set of output symbols (alphabet of FSM), ψ is the transition (next state) function $\psi : Q \times I \rightarrow Q$, ϕ is the output function $\phi : Q \times I \rightarrow O$ and $q \in Q$ is the initial state. A three-state FSM is given in Figure 6.15 and the response of the FSM to a given input symbol is shown in Table 6.3, where $Q = \{1, 2, 3\}$ with $q = 3$ illustrated in the figure by a double circle, $I = \{0, 1\}$ and $O = \{a, b, c\}$.

A population of finite-state machines is exposed to the environment. As each input symbol is input to the machine, each output symbol is compared with the next input symbol. The worth of this prediction is measured with respect to the payoff function, which provides the fitness of the machine. Offspring machines are created by applying mutation operation to parent machines. For the FSM, there can be five possible mutation operations: (i) change initial state, (ii) add a state, (iii) delete a state, (iv) change a state transition, and (v) change an output symbol.

The states of the FSM can be represented using binary coding or value coding. In binary coding, each state of the FSM can be represented by a 6-bit string. The first 1-bit represents activation of the corresponding state {0 for not active, 1 for active}. The second 1-bit represents the input symbols {0, 1}. The third 2-bits represent the next state {01 for state 1, 10 for state 2, 11 for state 3}. The fourth 2-bits represent the output symbols {00 for a, 01 for b, 10 for c}.

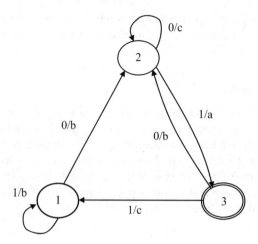

Figure 6.15   A three-state FSM

For the three states of the FSM in Figure 6.15, each individual will be 18 bits long. For example, an individual of the FSM will look like

Chromosome of FSM = {0, 1, 01, 10 | 0, 0, 10, 01| 1, 1, 11, 00}

For example, an initial population of five individuals using binary coding for the FSM will look like

$$P = \begin{pmatrix} 101110 \mid 110000 \mid 011001 \\ 111110 \mid 110110 \mid 101110 \\ 100011 \mid 110010 \mid 010111 \\ 101001 \mid 100011 \mid 001101 \\ 001111 \mid 110101 \mid 001111 \end{pmatrix}$$

Because of the length of 2-bits for the states and output symbols, extra caution must be taken to avoid the invalid state {00} and the invalid output symbol {11} during population generation and also mutation operation. This restriction needs additional effort in coding programs. An alternative to binary coding would be to choose value coding for chromosome representation. In value coding, each state of the FSM is represented by four values. The first value represents the present state value {1, 2, 3}, the second value represents the input symbol {0, 1}, the third value represents the next state {1, 2, 3}, the fourth value represents the output symbol {1 for a, 2 for b, 3 for c}. The values {a, b, c} can also be used instead of integer values. Each

Table 6.3   Response of FSM

Present state	3	2	3	1	1	2
Input	0	1	1	1	0	0
Next state	2	3	1	1	2	2
Output	b	a	c	b	b	c

individual for FSM will consist of 12 values. For example, an individual of the FSM in value coding will look like

Chromosome of FSM = {1, 0, 2, 2 | 2, 1, 3, 1| 3, 1, 1, 3}

For example, an initial population of five individuals using integer coding for the FSM will look like

$$
P = \begin{pmatrix}
3031 \mid 3122 \mid 1331 \\
3011 \mid 2013 \mid 2022 \\
3011 \mid 2121 \mid 1032 \\
1023 \mid 2022 \mid 2111 \\
3032 \mid 0111 \mid 1131
\end{pmatrix}
$$

Mutation: One of the five possible mutation operations mentioned earlier can be applied on a gene chosen randomly. The gene itself can be a present state, input value, next state or output value. For example:

Mutation {3031 | 3122 | 1331} → {3031 | 2122 | 1331}

For a population in binary coding, the mutation would be flipping of a bit. For example:

Mutation {101110 | 110000 | 011001} → {101111 | 110000 | 011001}

Fitness evaluation: The fitness value assigned to a given FSM behaviour, here evaluated through an input/output sequence perspective, is defined by the fitness function F in the form

$$
F = \sum_{i=1}^{N} w_i H_i \tag{6.23}
$$

where w_i is a weighting factor for fitness case i, N is the number of training sequences (TS) and H_i is the number of output hits due to the training sequence i (TS_i). Initially, the FSM must be in the reset (idle) state. In the sequence, for each input of the TS_i, its output is compared with the correct output. $H_i = 1$ in case of a match, otherwise $H_i = 0$. A partial matching can also be considered, where $H_i = 0.5$.

6.6.2   Evolution Strategies

Evolution strategies were developed as a method to solve parameter optimization problems by Rechenberg in the 1960s (Rechenberg, 1965) and further developed by Schwefel (1968, 1975, 1995). Evolution-strategic optimization is based on the hypothesis that during the biological evolution the laws of heredity have been developed for fastest phylogenetic adaptation. ESs imitate, in contrast to GAs, the effects of genetic procedures on the phenotype. The presumption for coding the variables in ES is the realization of a sufficiently strong causality, i.e., small changes of the cause must create small changes of the effect. The theory states that evolutionary progress takes place only within a very narrow band of the mutation step size.

The earliest ESs were based on a population consisting of one individual only. There was also only one genetic operator, mutation, used in the evolution process. However, the interesting idea was to represent an individual as a pair of real-valued vectors $v = (x, \sigma)$, where x represents a point in the search space and σ is a vector of standard deviations. Mutations are realized by replacing x with

$$x(t + 1) = x(t) + N(0, \sigma) \tag{6.24}$$

where $N(0, \sigma)$ is a vector of independent random Gaussian numbers with zero mean and standard deviation.

This is in accordance with the biological observation that smaller changes occur more often than larger ones. The offspring (mutated individual) is accepted as a new member of the population if and only if (iff) it has improved fitness and all constraints are satisfied. For example, if f is the objective function without constraints to be maximized, an offspring $(x(t + 1), \sigma)$ replaces its parent $(x(t), \sigma)$ iff $f(x(t + 1)) > f(x(t))$, otherwise, the offspring is eliminated and the population remains unchanged.

When implemented to solve real-valued function optimization problems, both typically operate on the real values themselves (rather than any coding of the real values as is often done in GAs). Multivariate zero-mean Gaussian mutations are applied to each parent in a population and a selection mechanism is applied to determine which solutions to remove (i.e., cull) from the population. The similarities extend to the use of self-adaptive methods for determining the appropriate mutations to use – methods in which each parent carries not only a potential solution to the problem at hand, but also information on how it will distribute new trials (offspring).

The following pseudo-code is an illustration of a general ES algorithm:

```
g ← 0                              ; generation 0
C_g ← {C_{g,n} | n = 1,2,...,N}    ; initialize population
F_EP ← evaluate (C_{g,n})          ; evaluate fitness of each individual

Do while no(convergence)
  {
   For l = 1 to λ (λ ≅ no. of offspring)
     {
      P_{g,n} (n2) ← select(C_{g,n})    ; select at random
      O_{g,λ} ← crossover (P_{g,μ})     ; crossover
      O_{g,λ} ← mutate (O_{g,μ})        ; mutate offspring
      F_EP ← evaluate (O_{g,λ})
     }
      C_{g,μ} ← select (C_{g,μ}, O_{g,λ})   ; select best μ individuals from parent
                                              and offspring
  }
  g ← g+1
```

There are a range of implementations of ES using selection, recombination and mutation operators found in the literature. In ES, the operator recombination is used rather than crossover. Some of them are discussed in the following section.

6.6.2.1 Selection Operators

ES typically uses deterministic selection in which the worst solutions are purged from the population based directly on their function evaluation. For each generation, λ offspring are generated from μ parents and mutated. After crossover and mutation the individuals for the next generation are selected. Two strategies have been developed:

- $(\mu + \lambda)$-ES. In this case the ES generates λ offspring from μ parents, with $1 \leq \lambda \leq \mu$. The next generation consists of the μ best individuals selected from μ parents and λ offspring. The $(\mu + \lambda)$-ES implements elitism to ensure that the fittest parents survive to the next generation.
- (μ, λ)-ES. In this case the next generation consists of the μ best individuals selected from λ offspring. (μ, λ)-ES requires that $1 \leq \mu < \lambda$. By doing this, the life of each individual is limited to one generation. This allows (μ, λ)-ES to perform better on problems with an optimum moving over time, or on problems where the objective function is noisy.

The notations (1+1)-ES, (1+λ)-ES, (1, λ)-ES, (m/μ, λ)-ES characterize evolution strategies with an increasing level of imitation of biological evolution. The letter m stands for the total number of parents, μ marks the number of parents, which will be recombined, and λ stands for the number of offspring.

6.6.2.2 Recombination Operators

There was no recombination operator in earlier ESs. In order to introduce recombination, Rechenberg proposed (1+1)-ES. Later on this was extended to $(\mu + 1)$-ES, which utilizes a crossover operator. Many forms of recombination have been implemented within ES. Again, the effectiveness of such operators depends on the problem at hand. The operators used in $(\mu+\lambda)$-ES and (μ, λ)-ES incorporate two-level learning: their control parameter σ is no longer constant, nor is it changed by some deterministic algorithm, but it is incorporated in the structure of the individual and undergoes the evolution process. To produce an offspring, the system acts in several stages.

- Select two individuals

$$(x^1, \sigma^1) = \left(\left(x_1^1, \ldots, x_n^1 \right), \left(\sigma_1^1, \ldots, \sigma_n^1 \right) \right)$$
$$(x^2, \sigma^2) = \left(\left(x_1^2, \ldots, x_n^2 \right), \left(\sigma_1^2, \ldots, \sigma_n^2 \right) \right)$$

- Apply a recombination (crossover) operator. There are two types of recombination: Discrete recombination – the new offspring is defined as

$$(x, \sigma) = \left(\left(x_1^{q_i}, \ldots, x_n^{q_i} \right), \left(\sigma_1^{q_i}, \ldots, \sigma_n^{q_i} \right) \right)$$

where $q_i = 1$ or $q_i = 2$, i.e., each component comes from either the first or the second pre-selected parent. Discrete recombination is shown in the example below:

1.2	0.3	0.0	1.7	0.8	1.2

Discrete recombination

0.8	0.5	1.0	1.1	0.2	1.2

1.2	0.5	1.0	1.1	0.8	1.2

Intermediate recombination – the vectors of two parents are averaged together, element by element, to form a new offspring where the new offspring is

$$(x, \sigma) = \left(\left((x_1^1 + x_1^2)/2, \ldots, (x_n^1 + x_n^2)/2 \right), \left((\sigma_1^1 + \sigma_1^2)/2, \ldots, (\sigma_n^1 + \sigma_n^2)/2 \right) \right)$$

Intermediate recombination is illustrated in the example below.

1.2	0.3	0.0	1.7	0.8	1.2

Intermediate recombination

0.8	0.5	1.0	1.1	0.2	1.2

1.0	0.4	0.5	1.4	0.5	1.2

6.6.2.3   Mutation Operators

Mutation is applied to the offspring (x, σ) obtained after recombination operation. The resulting new offspring is (x, σ), where $\sigma = \sigma.e^{N(0, \Delta\sigma)}$, $x = x + N(0, \sigma)$ and $\Delta\sigma$ is a parameter of the method.

The effects of these operators reflect the behavioural as opposed to the structural interpretation of the representation, since knowledge of the values of vector elements is used to derive new vector elements.

Example 6.5: Braitenberg vehicle   Moving around while avoiding obstacles is a key issue in autonomous agent research. Braitenberg has proposed a simple architecture for such tasks (Braitenberg, 1984). Figure 6.16 shows a control architecture inspired by a Braitenberg vehicle based on the idea that a sensor with high proximity activation accelerates the motor on the sensor's side. The activation of the left and right motors M_l and M_r is expressed by $M_l = \sum_{i=1}^{8} P_i w_i^l + w_0^l$ and $M_r = \sum_{i=1}^{8} P_i w_i^r + w_0^r$, respectively, where P_i denotes the activation of the proximity sensor i, w_i^l and w_i^r denote weights that connect the proximity sensors with the left and right motors, respectively. w_0^l and w_0^r represent the idle activation of the left and right motors, respectively, that are responsible for forward motion of the robot in the absence of an obstacle.

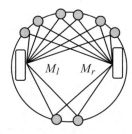

Figure 6.16   Control architecture for Braitenberg vehicle

The problem here is to determine the weights w_i^l, w_i^r, w_0^l, w_0^r, $i = 1, 2, \ldots, 8$ such that the robot is moving around while avoiding obstacles. Since no training patterns are available for training, the evolution strategy is an ideal candidate for this optimization problem.

In the evolution of a Braitenberg vehicle, the first important thing is to find an initial controller that exhibits meaningful behaviour. In order to exploit the morphology of the robot, the left and right parts of the controller weights are constrained to equal $w_i^l = w_i^r$ and $w_0^l = w_0^r$. This leads to a reduced search space of nine parameters. Thus, the chromosome of the Braitenberg vehicle will consist of nine weights and standard deviations σ_0 and σ_i as follows:

$$\{w_0, \sigma_0 \mid w_1, \sigma_1 \mid w_2, \sigma_2 \mid w_3, \sigma_3 \mid w_4, \sigma_4 \mid w_5, \sigma_5 \mid w_6, \sigma_6 \mid w_7, \sigma_7 \mid w_8, \sigma_8\}$$

All weights w_i, $i = 1, 2, \ldots, 8$ are initialized with very small negative random values, i.e., $w_i \in [-0.5, 0]$ and the idle activation weights w_0 are set to very small positive values, i.e., $w_0 \in [0, 0.1]$. An initial standard deviation is set to 0.5. The initial population of three individuals using real coding will thus look like

$$P = \begin{bmatrix} 0.07,0.5|, -0.21,0.5|, -0.23,0.5|, -0.45,0.5|, -0.32,0.5|,0.12,0.5|, -0.50,0.5|, -0.34,0.5|, -0.27,0.5 \\ 0.01,0.5|, -0.10,0.5|, -0.40,0.5|, -0.20,0.5|, -0.11,0.5|,0.00,0.5|, -0.12,0.5|, -0.41,0.5|, -0.50,0.5 \\ 0.09,0.5|, -0.12,0.5|, -0.30,0.5|, -0.23,0.5|, -0.21,0.5|,0.10,0.5|, -0.15,0.5|, -0.31,0.5|, -0.25,0.5 \end{bmatrix}$$

Recombination operation. There are two types of recombination operator used in ES: discrete and intermediate. In this example, intermediate recombination is applied, in which the vectors of two parents are averaged together, element by element, to form a new offspring where the new offspring is given by

Parent 1 0.07, 0.5|, −0.21, 0.5|, −0.23, 0.5|, −0.45, 0.5|, −0.32, 0.5|, 0.12, 0.5|, −0.50, 0.5|, −0.34, 0.5|, −0.27, 0.5

Parent 2 0.01, 0.5|, −0.10, 0.5|, −0.40, 0.5|, −0.20, 0.5|, −0.11, 0.5|, 0.00, 0.5|, −0.12, 0.5|, −0.41, 0.5|, −0.50, 0.5

- -

Offspring : 0.04, 0.5|, −0.15, 0.5|, −0.31, 0.5|, −0.32, 0.5|, −0.21, 0.5|, 0.06, 0.5|, −0.31, 0.5|, −0.37, 0.5|, −0.38, 0.5

Mutation. The mutation operator mentioned earlier can be applied on a gene chosen randomly. The resulting new offspring is (w, σ), where $\sigma = \sigma.e^{N(0,\Delta\sigma)}$ and $w = w + N(0, \sigma)$, for example, $w = -0.30 + N(0, 0.5) = -0.30 + 0.16 = -0.14$ and $\sigma = 0.5 * e^{N(0,0.5)} = 0.58$:

Mutation {0.09, 0.5|, −0.12, 0.5|, **−0.30, 0.50**|, −0.23, 0.5|, −0.21, 0.5|, 0.10, 0.5|, −0.15, 0.5|, −0.31, 0.5|, −0.25, 0.5}

- -

Offspring {0.09, 0.5|, −0.12, 0.5|, **−0.14, 0.58**|, −0.23, 0.5|, −0.21, 0.5|, 0.10, 0.5|, −0.15, 0.5|, −0.31, 0.5|, −0.25, 0.5}

In order to evolve a good Braitenberg vehicle, the fitness function has to incorporate motor speeds and distances to obstacles. The use of speed and distance in a fitness function may result in useless activity, such as spinning of the robot with high speed far from the obstacles. Therefore, a penalty term $\Delta v_t = |V_l - V_r|$ has been used in the fitness function to avoid spinning. Hence, the fitness measure for the evolution strategy at time t is defined as

$$f_t = V_t \left(1 - \sqrt{\Delta v_t}\right) \left(1 - I\hat{P}_t\right) \tag{6.25}$$

where $V_t = \frac{(V_l + V_r)}{2}$, $I\hat{P}_t = \max\{IP_i\}$, $i = 1, 2, \ldots, 8$. V_l and V_r are the left and right wheel speeds of the robot, respectively. The term $\left(1 - \sqrt{\Delta v_t}\right)$ ensures the two wheels rotate in the same direction, making it move forward. The fitness evaluation is carried out over a period of time, e.g., $t_{max} = 100$ times. The final fitness value over t_{max} will thus be

$$F = \sum_{t=1}^{t_{max}} V_t \left(1 - \sqrt{\Delta v_t}\right)\left(1 - I\hat{P}_t\right) \tag{6.26}$$

In this example, a (3+2)-ES has been applied to obtain a good set of controller weights for the Braitenberg vehicle. A (3+2)-ES generates two new offspring per generation and the next generation consists of the three best individuals selected from three parents and two offspring. Offspring 1 is created by intermediate recombination and offspring 2 is created by mutation. A typical run of the evolution of the Braitenberg vehicle can be performed by using the weights $w_i^l, w_i^r, w_0^l, w_0^r, i = 1, 2, \ldots, 8$ for 100 time steps and the final fitness F is computed. The offspring are evaluated and compared with parents $P_i, i = 1, 2, 3$. Out of the five individuals, the best three are selected again for the next generation population. The evolution process thus continues until the robot is able to move around avoiding obstacles.

6.6.3   Genetic Algorithms

What is known as a genetic algorithm today is the most widely applied and well-known evolutionary algorithm. This is attributed entirely to John Holland, whose extensive work in the field during the 1960s and 1970s made the GA a widely popular optimization methodology. John Holland is generally considered the father of GA. In GA, individuals are represented by means of strings (similar to the way genetic information is coded in organisms as chromosomes) (Holland, 1975). Each individual in the population represents a potential solution to the problem. Unlike other optimization techniques, GA does not require mathematical descriptions of the optimization problem, but instead relies on a cost function in order to assess the fitness of a particular solution to the problem in question (Goldberg, 1989). The GA then iteratively creates new populations from old by ranking the strings and interbreeding the fittest to create new strings, which are (hopefully) closer to the optimum solution of the problem in question. So, in each generation, the GA creates a set of strings from the bits and pieces of the previous strings, occasionally adding random new data to keep the population from stagnating. The end result is a search strategy that is tailored for vast, complex, multimodal search spaces. A genetic adaptive plan can then be defined as a quadruple

$$\Lambda = \{\Sigma, \Pi^N, \Phi, \Omega\} \tag{6.27}$$

where Σ is the coding format, Π^N is a population of size N, Φ is a fitness re-scaling algorithm and $\Omega = [\omega_1, \omega_2, \ldots, \omega_m]$ is the set of genetic operators.

The most common genetic operators in GA are selection, crossover and mutation. The genetic plan refers to the process through which successive populations are generated using evaluation, selection, mating and deletion. Let Ψ be a probability distribution over Π which is derived from the fitness of each trial, $\mu(A \in \Pi)$. A genetic plan can then be formally expressed as the mapping $\Lambda : (\Psi \times \Pi(g) \times \Omega) \to \Pi(g+1)$. The structure of the genetic algorithm can be stated as follows.

$t \leftarrow 0$
initialize $[P(t)]$
evaluate $[P(t)]$
do while (not termination-condition)
{
$P'_M(t) \leftarrow$ reproduce $[P_N(t)]$
    $P'_M(t) \leftarrow$ evaluate$[P'_M(t)]$
$Q \leftarrow$ select$[P'_M(t)]$
$P_N(t) \leftarrow$ replace$[Q]$
$P(t+1) \leftarrow [P_N(t)]$
$t \leftarrow t+1$
}
enddo

where $P_N(t)$ denotes a population of N individuals at generation t, $P'_M(t)$ denotes an offspring population of size M generated by means of reproduction, reproduction operators are such as crossover and mutation, and Q is an intermediate population in the mating pool.

6.6.3.1 Chromosome Representation

The classical chromosome representation scheme for a GA is a binary vector of fixed length. In the case of an n-dimensional search space, each individual consists of n variables with each variable encoded as a bit string. In the case of nominal-valued variables, each nominal value can be encoded as a D-dimensional bit vector, i.e., the variable can have 2^D nominal values. In the case of continuous-valued variables, each variable should be mapped to a D-dimensional bit vector, i.e., $\Phi : \Re \to \{0, 1\}^D$.

The range of continuous space needs to be restricted to a finite range $[\alpha, \beta]$. Using standard binary decoding, each continuous variable $C_{n,i}$ of chromosome C_n is encoded using a fixed-length bit string. For example, if $z \in [z_{min}, z_{max}]$ needs to be converted to a 30-bit representation, the following conversion can be used:

$$\left(2^{30} - 1\right) \frac{z - z_{min}}{z_{max} - z_{min}} \tag{6.28}$$

Binary coding is frequently used, but it has the disadvantage of Hamming Cliffs – formed when two numerically adjacent values have bit representations with a large Hamming distance. This causes a problem when a small change in variables should result in a small change in fitness. An alternative bit representation is to use Gray coding. Gray coding has the advantage over binary coding in that the Hamming distance between two successive numerical values is one. GAs have also been developed that use integer or real-valued representations. The advantage of real-valued coding over binary coding includes increased precision and the chromosome string becomes shorter. Also, real-valued coding gives a greater freedom to use special crossover and mutation techniques.

6.6.3.2   Selection Operators

The selection (reproduction) operator allows individual (strings) to be copied for possible inclusion in the next generation. The chance that a string will be copied is based on the individual's *fitness value*, calculated from a *fitness function*. The most commonly used selection methods are random, proportional, tournament, rank-based and elitism. These have been discussed in Section 6.4.1.

6.6.3.3   Crossover Operators

The aim of crossover is to produce offspring from two parents selected using a reproduction operator, which takes place at a certain probability called the *crossover probability* (C_p). The value of C_p is set by the user, and the suggested value ranges between 0.6 and 0.8, although this value can be domain-dependent. Several crossover operators are in wide use in GAs, such as single-point, uniform and multi-point crossover. These are discussed in Section 6.4.2.

6.6.3.4   Mutation Operators

There may not be enough variety of strings in the initial population generated randomly, which should be uniformly distributed within the entire problem space. The aim of mutation is to produce new genetic material in an existing individual and maintain genetic diversity at all generations of GA. The mutation probability should be kept very low (usually between 0.01 and 0.02). Widely used mutation operators are bit inversion (mainly used for binary or Gray coding) and adding a small number (usually used for real-value coding). Different mutation operators are discussed in Section 6.4.3.

Example 6.6: Simple function optimization problem   The problem is simply stated. Find the maximum value of the function in Equation (6.29). The GA to be used in this example is based almost exactly on the description given above. The population size will be 4, and strings of bits of length 5 will be used. A crossover probability of 0.6 is assumed and a mutation probability of 0.001. With such a low chance of mutation, it does not occur in the following example.

$$y = f(x) = -x^2 + 8x + 15 \qquad 0 \le x \le 25 \qquad\qquad (6.29)$$

In order to make things easy for us, we will assume that the maximum is between 0 and 25 (the actual maximum is at $x = 4$) and that the maximum is an integer value. See Figure 6.17.

Coding scheme: We can represent integer values in the range $[0 \ldots 31]$ with a 5-bit string, e.g.

String	Decoded value
00001	1
00101	5
10110	22

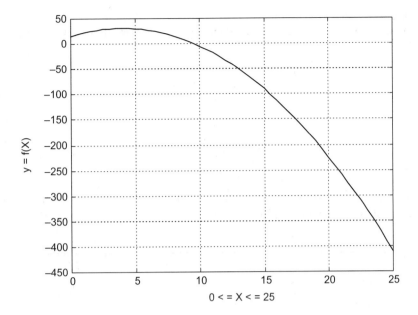

Figure 6.17 Values of $f(x)$

Fitness function: This will give the relative fitness values. The simplest method to employ here is to use the decoded x value to calculate the y coordinate and use the y coordinate as the fitness value:

$$f(x) = -x^2 + 8x + 15 \qquad (6.30)$$

Selection: The fitness value for string i, as a percentage, will be the y value at i divided by the sum of all the y values for every string. The fitness function for string i in the string population is as follows:

$$\text{Fitness value}_i = \frac{f_i}{\sum f} \qquad (6.31)$$

For example, say $y = x^2$ and we are trying to find the maximum value of the function between $[0 \ldots 31]$. Then the following strings would have the relative fitness indicated below:

String	x Value	$f(x)$	Relative fitness
00101	5	25	0.04
01101	13	169	0.25
10110	22	484	0.71

In reality, since the value of the function we want to minimize can take on negative values, the fitness function is slightly more complex than the one used above. However, in essence, the two remain equivalent.

Running the GA: First iteration

Firstly, we need to create a random population of strings. Say we start with the following:

String population
00010
00111
10110
01011

Now we perform selection. The fitness value of each string is calculated and the strings are selected the following number of times:

String	x Value	$f(x)$	Relative fitness value	No. of times selected
00010	2	27	0.35	1
00111	7	22	0.34	2
10110	22	−293	0.008	0
01011	11	−18	0.30	1

With these selections, the mating pool now looks like this:

String population
00111
00010
00111
01011

Finally, the crossover probabilities need to be calculated (two crossovers need to be performed to create a new population of two). The GA calculates that it should perform splitting twice on two sets of randomly selected genes. Crossover performs the following to create the new population:

Mating pool strings	New population
000 1 10	00011
001 1 11	00110
01 1 011	01010
00 1 010	00011

So, at the end of the first iteration, the new population looks like the following:

String population	x Value
00011	3
00110	6
01010	10
00011	3

Even after one generation of evolution, with no knowledge except for the relative fitness value, the GA has begun to quickly converge on the optimum value of 4. This is startling, considering the GA knows nothing about the problem space in which it searches. It is effectively blind. Yet, just by examining a measure of *goodness*, having a large number of points to examine simultaneously and having a large amount of randomization thrown in, the GA efficiently searches the problem space for possible answers.

6.6.4   Genetic Programming

Friedberg (Friedberg, 1958; Friedberg *et al.*, 1959) was among the first to evolve computer programs. The word 'evolution' was not used at that time, although the author intended to simulate evolution. Dunham and North pursued this line of research within IBM through the 1970s and 1980s up to the early 1990s (Dunham *et al.*, 1974). It was John R. Koza who applied the GA approach to perform an automatic derivation of equations, logical rules or program functions (Koza, 1992). He first used the term 'genetic programming' where, rather than representing the solution to the problem as a string of parameters as in a conventional GA, he used a tree encoding scheme or structure. The leaves of the tree, called *terminals*, represent input variables or numerical constants. Their values are passed to *nodes*, at the junctions of branches in the tree, which perform some arithmetic or program function before passing on the result further towards the root of the tree.

GP is much more powerful than genetic algorithms in that the output of the genetic algorithm is a quantity, while the output of the GP is another computer program. GP works best for several types of problem. The first type is where there is no ideal solution (for example, a program that drives a car). Furthermore, GP is useful in finding solutions where the variables are constantly changing.

There are five major steps in using GP for a particular problem. These need to be specified by the user.

- Set of terminals (e.g., the independent variables of the problem, zero-argument functions and random constants) for each branch of the to-be-evolved program.
- Set of functions for each branch of the to-be-evolved program.
- Fitness measure (for explicitly or implicitly measuring the fitness of individuals in the population).
- Selection of certain parameters for controlling the run.
- Selection of termination criterion and method for designating the result of the run.

GP typically starts with a population of randomly generated computer programs composed of the available programmatic ingredients within the hyperspace of valid programs. These programs are represented in the form of rooted trees. Langdon (1988) demonstrated in his excellent volume how a large collection of abstract data structures such as stacks, queues, lists, rings, etc. can be beneficial when evolving programs. GP iteratively transforms this population of computer programs into a new generation of the population by applying genetic operations such as crossover and mutation. These operations are applied to individual(s) selected from the population. The individuals are probabilistically selected to participate in the genetic operations based on their fitness. The iterative transformation of the population is executed inside the main generational loop of the run of GP. The structure of the GP can be stated as follows:

$$t \leftarrow 0 \qquad\qquad\qquad\qquad \text{; set generation}$$
$$[F(t)] \leftarrow \text{randomly initialize trees} \quad \text{; trees composed of functions and terminals}$$

do while (not termination-condition)
{
evaluate $[F(t)]$; using the problem's fitness measure
$Q \leftarrow$ select $[F(t)]$; with a probability based on fitness into the
                                      ; mating pool Q
$F_C'(t) \leftarrow$ crossover $[Q]$
$F_M'(t) \leftarrow$ mutate $[Q]$
$[f(t+1)] \leftarrow$ new population $\{F_C'(t), F_M'(t)\}$
$t \leftarrow t+1$
}

The single best program in the population produced during the run – the best-so-far individual – is harvested and designated as the result of the run. If the run is successful, the result may be a solution (or approximate solution) to the problem.

6.6.4.1   Fitness Evaluation

It is to be determined how good the individuals are at solving the given problem. As with GAs, the crossover and mutation operations are separate from the actual evaluation of the fitness, making the GP operators problem-independent. The fitness function is determined subjectively. For example, we could include the depth of the tree as a potential quality we wish to control, and therefore we could develop a fitness function which takes this into account.

The measurement of fitness is a rather nebulous subject. Since it is highly problem-dependent, we consider massaging the results to make fitness evaluation much easier, through a process known as *scaling*. Simply put, scaling standardizes the measurement of how fit a particular individual is with respect to the rest of the population.

6.6.4.2   Selection Operators

Based on the fitness value, selection operators is chosen for survival in one of two ways:

- Choose the individuals with the highest fitness for reproduction.
- Assign a probability that a particular individual will be selected for either mutation or crossover. This method of choice allows for more diversity. Some weak individuals may contain branches of code which are strong.

6.6.4.3 Crossover Operators

Koza (1995) considers crossover, along with reproduction, to be the two foremost genetic operations. It is mainly responsible for the genetic diversity in the population of programs. Similar to its performance under GA, crossover operates on two programs (or tress) and produces two child programs. Two random nodes are selected from within each program and then the resultant 'subtrees' are swapped, generating two new programs. These new programs become part of the next generation of programs to be evaluated. For the next couple of examples, parents are generated randomly using the terminal set specified by $\{a, b, x, y, z, 3.4\}$ with $\{a, b, x, y, z\} \in \Re$ and the function set given by $\{-, +, *, /, \sin, \exp, \ln\}$. A crossover operation with different parents is shown in Figure 6.18 and a crossover operation with identical parents is shown in Figure 6.19.

6.6.4.4 Mutation Operators

Several mutation operators have been developed for GP. The most frequently used operators are function node mutation, terminal node mutation, swap mutation, grow mutation, Gaussian mutation and trunk mutation.

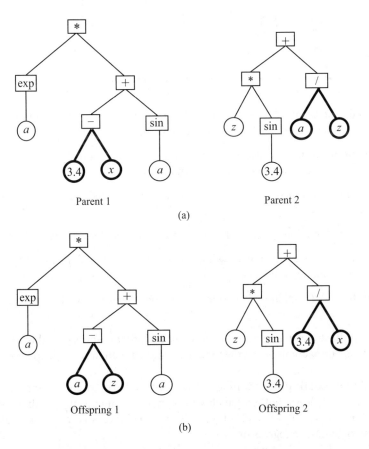

Figure 6.18 Crossover operation with two different parents

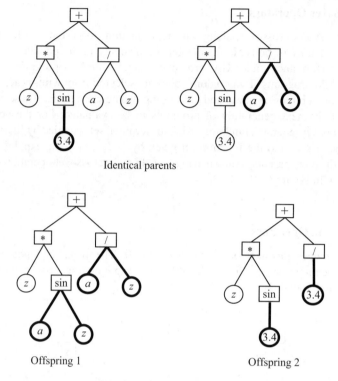

Figure 6.19   Asexual crossover operation with identical parents

Function node mutation: A non-terminal node or function node is randomly selected and replaced with a function selected randomly from the function set.

Terminal node mutation: A leaf node or terminal node is randomly selected and replaced with a new terminal node selected from the terminal set.

Swap mutation: Two function nodes are randomly selected and the arguments of the nodes are swapped.

Grow mutation: A randomly selected node is replaced by a randomly generated subtree.

Gaussian mutation: A terminal node with a constant value is randomly chosen and mutated by adding a Gaussian random value.

Trunk mutation: A function node is selected randomly and replaced by a random terminal node.

Figure 6.20 illustrates the concept of different mutation described in this section. Parent trees are generated randomly using the terminal and the function set defined above.

Example 6.7: Automatic programming   The problem is to find a computer program through evolution with a single variable x, which will output the value of the quadratic polynomial $y(x) = x^2 + x + 1$ with $x \in \Re$ and $-1 \le x \le +1$. This process is sometimes called system identification or symbolic regression.

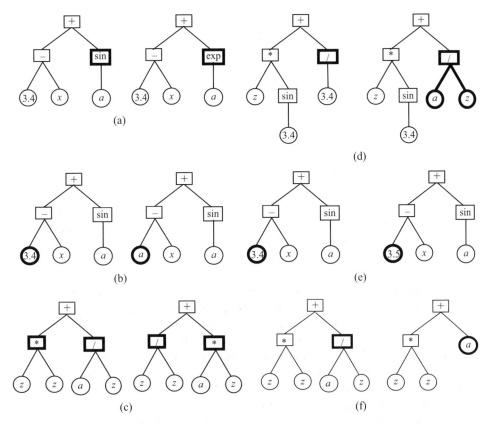

Figure 6.20 Widely used mutation in tree coding. (a) Function node mutation; (b) Terminal node mutation; (c) Swap mutation; (d) Grow mutation; (e) Gaussian mutation; (f) Trunk mutation

GP is the suitable EA to create the computer program for symbolic processing of the polynomial. That is, the goal is to create a computer program that matches certain numerical data. This is to find a mathematical function of one independent variable, therefore, the terminal set includes the independent variable x. Thus, the terminal set $T = \{X\}$.

The possible choice for the function set consists of the four ordinary arithmetic functions $\{-, +, *, /\}$. This choice is reasonable because mathematical expressions typically include these functions. Thus, the minimal function set is given by $F = \{-, +, *, /\}$. A common run-time problem of a program is division by 0. To avoid run-time error, the division function ($/$) is protected and the division function returns a value of 1 when division by 0 occurs and even 0 divided by 0 is also set to 1.

The purpose of the fitness measure is to specify what the human programmer wants. The high-level goal of this problem is to find a program whose output is equal to the values of the quadratic polynomial $x^2 + x + 1$. Therefore, the fitness assigned to a particular individual in the population for this problem must reflect how closely the output of an individual program comes to the target polynomial $x^2 + x + 1$. The fitness measure could be defined as the value of the integral (taken over values of the independent variable x between -1.0 and $+1.0$) of the absolute value of the differences (errors) between the value of the individual

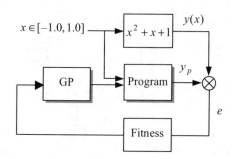

Figure 6.21   Genetic programming process for symbolic regression problem

mathematical expression and the target quadratic polynomial $x^2 + x + 1$. That is $f = \sum |e(x)|$ over $x = [-1.0, +1.0]$, where $e(x) = y(x) - y_p$. y_p is the program output. A smaller value of fitness (error) is better. A fitness (sum of absolute error) of zero would indicate a perfect match but if a program's fitness is less than 0.1, it will be an acceptable program in this case. The genetic programming process is shown in Figure 6.21.

For most problems of symbolic regression or system identification, it is not practical or possible to analytically compute the value of the integral of the absolute error. Thus, in practice, the integral is numerically approximated using dozens or hundreds of different values of the independent variable x in the range between -1.0 and $+1.0$. The population size in this small illustrative example will be just 4. In actual practice, the population size for a run of genetic programming consists of thousands of individuals. A population of four individuals is shown in Figure 6.22. The crossover operation will be performed on two individuals and the mutation operation will be performed on one individual.

The four individuals (a)–(d) in Figure 6.22 are evaluated and the outputs $y(x)$ and y_p are shown in Figure 6.23. The relative fitness (minimum is sought) of the individuals (a), (b), (c) and (d) in Figure 6.22 is 0.15, 0.22, 0.61 and 0.45, respectively.

The individuals (a) and (b) are fitter than the other two. Therefore, the individuals (a) and (b) are selected for crossover operation. The crossover operation is performed by swapping the selected subtrees, shown encircled in Figure 6.22. Two offspring are created and replace their parents, as shown in Figure 6.24(a,b). One individual is randomly chosen for mutation. This is shown in Figure 6.24(b). The new population after crossover and mutation is shown in Figure 6.25. The four individuals of the new population (Figure 6.25(a)–(d)) are evaluated

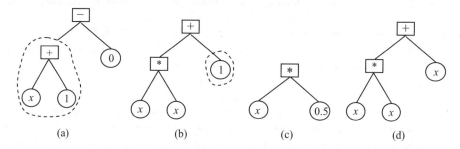

Figure 6.22   Initial population of four chromosomes at generation 0. (a) $x + 1$; (b) $x^2 + 1$; (c) $x*0.5$; (d) $x^2 + x$

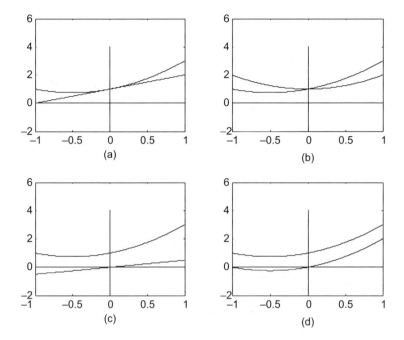

Figure 6.23 The fitness of the four individuals at generation 1. (a) $y = x + 1$, relative fitness $= 0.15$; (b) $y = x^2 + 1$, relative fitness $= 0.22$; (c) $y = x^*.5$, relative fitness $= 0.61$; (d) $y = x^2 + x$, relative fitness $= 0.45$

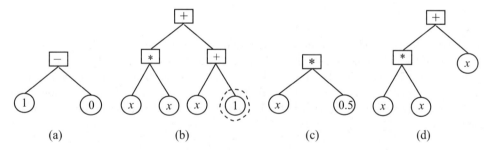

Figure 6.24 Population after crossover operation. (a) $1 - 0$; (b) $x^2 + x + 1$; (c) $x^*0.5$; (d) $x^2 + x$

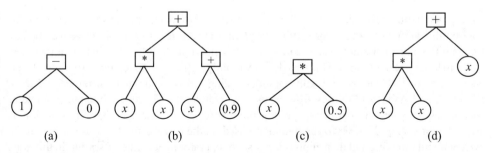

Figure 6.25 Population (after crossover and mutation operation) in generation 2. (a) $1 - 0$; (b) $x^2 + x + 0.9$; (c) $x^*0.5$; (d) $x^2 + x$

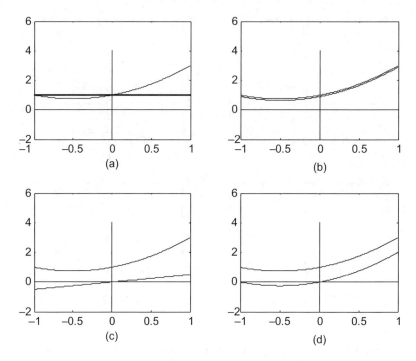

Figure 6.26 The fitness of the four individuals at generation 2. (a) $y = 1 - 0$, relative fitness $=$ 0.17; (b) $y = x^2 + x + 1$, relative fitness $= 0.03$; (c) $y = x^*.5$, relative fitness $= 0.61$; (d) $y = x^2 + x$, relative fitness $= 0.45$

and the outputs $y(x)$ and y_p are shown in Figure 6.26(a)–(d). The relative fitness (minimum is sought) of the individuals (a)–(d) is 0.17, 0.03, 0.61 and 0.45, respectively. As can be seen from Figures 6.25 and 6.26, the individual (b) is very close to the desired function and the fitness is less than 0.1. Therefore, GP is terminated at this stage. The example demonstrates the GP in a very simplified way but in the real simulation it may take several generations to converge.

6.6.5   Differential Evolution

DE is a population-based direct search algorithm which has mainly been used to solve continuous optimization problems (Storn, 1995, 1999; Storn and Price, 1997). DE was developed by Kenneth Price in an attempt to solve the Chebyshev polynomial fitting problem that had been posed to him by Rainer Storn. This was done by modifying genetic annealing, originally developed by Price (1994) to use a floating-point encoding scheme. The main difference between DE and other EAs is that DE uses differences of two randomly selected individuals (parameter vectors) as the source to perturb the vector population rather than a probability function as an evolution strategy. DE performs mutation based on the distribution of the solutions in the current population first and then applies a crossover operator to generate offspring. In this way, search directions and possible step sizes depend on the location of the individuals selected to calculate the mutation values.

The basic algorithm of DE is simple and straightforward and consists of the following four steps:

1. Initialization of parameter vectors.
2. Mutation with difference vectors.
3. Crossover operation: exponential (two-point modulo) or binomial (uniform).
4. Selection.

A practical optimization technique should satisfy three demands for any problem. First, the method should find the true global minimum. Second, the algorithm should ensure fast convergence. Third, the algorithm should have a minimum of control parameters. Considering these three demands, DE is a fast, simple technique, involves only three parameters and performs extremely well on a wide variety of test problems (Storn et al., 2005).

In the DE literature, a parent vector $x_i(t)$ from the current generation is called the target vector, a mutant vector obtained through the differential mutation operator is known as a donor vector $u_i(t)$ and finally an offspring formed by recombining the donor with the target vector is called a trial vector. The general structure of DE algorithm is described as follows:

$t \leftarrow 0$
initialize $[P(t)]$
do while (not termination-condition)
   {
   for each individual $x_i(t) \in P(t)$ do
     {
     evaluate fitness $f_i \leftarrow$ *evaluate* $[x_i(t)]$
     create donor vector $u_i(t) \leftarrow$ *mutate* $[x_i(t)]$
     create offspring $x_i'(t) \leftarrow$ *crossover* $[x_i(t)]$
     if $f\left[x_i'(t)\right] > f\left[x_i(t)\right]$; > meaning better than
      $P(t) \leftarrow P(t) + x_i'(t)$; add $x_i'(t)$ to $P(t)$
     else $P(t) \leftarrow P(t) + x_i(t)$; add $x_i(t)$ to $P(t)$
     }
   $t \leftarrow t+1$
   }

6.6.5.1 Chromosome Representation

The classical chromosome representation scheme for any EA is a vector of fixed length. In the case of an N-dimensional search space, each individual (also called parameter vector) in DE consists of N variables with each variable encoded as a real number. The ith individual (i.e., chromosome) of the population at generation t is an N-dimensional vector of a set of N parameters to be optimized. The ith chromosome of the DE is presented below:

$$x_i(t) = \left[x_{i,1}(t), x_{i,2}(t), \ldots, x_{i,N}(t)\right] \tag{6.32}$$

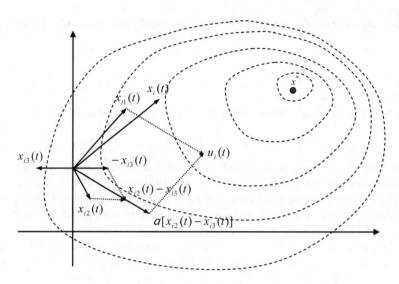

Figure 6.27   Mutation in differential evolution

6.6.5.2   Mutation Operators

Mutation in DE produces a donor vector by randomly selecting two individuals $x_{i,2}$ and $x_{i,3}$ from the population. The difference between the two is scaled by a scalar factor α and added to a target vector $x_{i,1}$ to form the donor vector $u_i(t)$. The process of the jth component of the mutation operation is described by

$$u_{i,j}(t) = x_{i,1}(t) + \alpha \lfloor x_{i,2}(t) - x_{i,3}(t) \rfloor \tag{6.33}$$

where $\alpha \in (0, \infty)$ is the scale factor that controls the amplification of the differential variation. The process of the mutation operation is shown in Figure 6.27.

6.6.5.3   Crossover Operators

At the heart of every direct search method is a strategy that generates variations of the parameter vectors. Once a variation is generated, a decision must be made whether or not to accept the newly derived offspring. The DE crossover is performed by a discrete recombination of the donor vector $u_i(t)$ and the parent vector $x_i(t)$ to produce offspring $x_i'(t)$. In general, crossover in DE is implemented as follows:

$$x_{ij}' = \begin{cases} u_{ij}(t) & \text{if } j \in C \\ x_{ij}(t) & \text{otherwise} \end{cases} \tag{6.34}$$

where $x_{ij}(t)$ represents the jth element of the vector $x_i(t)$ and C is the set of crossover points (element indices) selected in the variation. The crossover operation in DE is shown in Figure 6.28. The DE family uses two different schemes of crossover, namely, binomial and exponential.

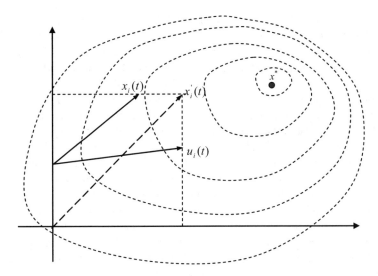

Figure 6.28   Crossover in differential evolution

Binomial crossover: The crossover points are selected randomly with a probability of p_c from the set of crossover points C. The probability p_c ensures a binomial distribution of the parameters inherited from the mutant.

Exponential crossover: A sequence of adjacent crossover points is selected, treating the list of potential crossover points as a circular array.

6.6.5.4   Selection Operators

The selection operator allows DE to keep the population size to a constant by deciding between the target and trial vector to be selected for possible inclusion in the next generation. If the trial vector has a better fitness value, it replaces its target vector in the next generation otherwise the parent is retained:

$$x_i(t+1) = \begin{cases} x_i'(t) & \text{if } f(x_i') > f(x_i) \\ x_i(t) & \text{otherwise} \end{cases} \tag{6.35}$$

The DE has been applied successfully to diverse domains of engineering and scientific problems. A study on the performance of DE and its comparison with other population-based optimization algorithms is reported in Vesterstrom and Thomson (2004), which shows that DE outperforms over several numerical benchmarks. Different variants of DE have been reported in the literature (Engelbrecht, 2007; Das and Suganthan, 2011).

6.6.6   *Cultural Algorithm*

Culture is the sum total of the learned behaviour of a population, which is generally considered to be the tradition of that population and transmitted from generation to generation. Some social researchers suggest that culture might be symbolically encoded and transmitted within and between populations, as an inheritance mechanism. Using this idea, Robert Reynolds developed a computational model (Reynolds, 1994, 1999a,b) called the cultural algorithm (CA). A cultural algorithm is a dual-inheritance mechanism where the population space represents the genetic traits and the belief space represents the cultural traits. These behavioural traits are passed from generation to generation using several socially motivated operators.

The concept is to preserve beliefs that are socially accepted and reject unacceptable beliefs in a population. A population of individuals is used in CA, similar to evolutionary algorithms discussed in earlier sections. Each of these individuals is, however, described in terms of a set of traits or behaviours. An evaluation function is required to evaluate the performance of each individual in solving a problem, analogously to the fitness function of EAs. A selection process is used to choose the parents to be evolved in the next generation. The evolution process is done with certain operators that tend to be domain-specific. The interactions between belief space and population are performed via a communication channel.

CA supports two modes of inheritance, one at the micro-evolutionary level in terms of traits and the other at the macro-evolutionary level in terms of beliefs. The two modes interact via a communications channel that enables the behaviour of individuals to change or the belief structure to be modified. The communication channel allows the belief structure to constrain the individuals' behaviour. Thus, CAs can be described in terms of three basic components, namely, the belief space, the population and the communications channel. The basic structure of CA algorithm is described by the following pseudo-code:

$t = 0;$
$P(t) \leftarrow$ initialize population
$BLF(t) \leftarrow$ initialize belief network
$CHL(t) \leftarrow$ initialize communication channel
evaluate $[P(t)]$
$t = 1$
repeat (until termination condition)
{
communicate $[P(t), BLF(t)]$
adjust $[BLF(t)]$
communicate $[BLF(t), P(t)]$
modulate fitness $[BLF(t), P(t)]$
$t = t + 1$
$P(t) \leftarrow$ select $[P(t - 1)]$
evolve $[P(t)]$
evaluate $[P(t)]$
}

CA has found many applications in different problem domains (Franklin and Bergerman, 2000; Reynolds et al., 2006). Different variants of CA have been reported in the literature (Reynolds and Peng, 2005; Engelbrecht, 2007).

6.7 MATLAB® Programs

MATLAB® provides the Genetic Algorithm and Direct Search Toolbox, which is a collection of functions to support direct search, genetic algorithm and simulated annealing for solving a variety of optimization problems. The toolbox provides two categories of tools: command line functions and GUI tools. The command line functions can also be used in an M-file to run the genetic algorithm many times with different option settings. In this section, only command line functions will be discussed and GA will be run from an M-file. All the toolbox functions are implemented using MATLAB® statements. The functions can also be extended using other functions and statements in combination with other toolboxes.

The toolbox provides three different types of approach to solving optimization problems, such as direct search, simulated annealing and genetic algorithms. The advantage of these methods is that they do not require any gradient information or higher derivatives of the objective function as opposed to traditional optimization methods. Direct search and simulated annealing compute a sequence of points and get closer and closer to the optimal point by testing them, whereas the genetic algorithm is a parallel search algorithm based on a population and natural selection. The evolutionary algorithms discussed in this chapter are all population-based optimization methods. Therefore, this section will specifically demonstrate a few examples of optimization problems using the genetic algorithm provided by the toolbox. MATLAB® codes for the examples and plots of results are provided in Appendix F.

References

Beyer, H.-G. and Deb, K. (2000) On the desired behaviour of self-adaptive evolutionary algorithms. In *Parallel Problem Solving from Nature* VI (PPSN-VI), pp. 59–68.

Bierwirth, C. (1995) A generalized permutation approach to job shop scheduling with genetic algorithms. *OR Spektrum*, 17, 87–92.

Bierwirth, C., Mattfield, D.C. and Kopfer, H. (1996) On permutation representation for scheduling problems, *Parallel Problem Solving from Nature*, 4, 310–318.

Box, G.E.P. (1957) Evolutionary operation: a method for increasing industrial productivity, *Applied Statistics*, 6(2), 81–101.

Braitenberg, V. (1984) *Vehicles: Experiments in Synthetic Psychology*, The MIT Press, Cambridge, MA.

Bralette, M.F. (1991) Initialisation, mutation, and selection methods in genetic algorithms for function optimisation. *Proceedings of ICGA* 4, pp. 100–107.

Bremermann, H., Rogson, M. and Salaff, S. (1966) Global properties of evolution processes. In *Natural Automata and Useful Simulations*, H. Pattee, E. Edlsack, L. Fein and A. Callahan (eds), Spartan Books, Washington, DC, pp. 3–41.

Caruana, R.A. and Schaffer, J.D. (1988) Representation and hidden bias: Gray vs. binary coding. *Proceedings of 6th International Conference on Machine Learning*, pp. 153–161.

Coella, C.A.C., Lamont, G.B. and Van Veldhuizen, D.A. (2007) *Evolutionary Algorithms for Solving Multi-objective Problems*, 2nd edn, Springer-Verlag, Berlin.

Cordon, O., Herrera, F., Hoffmann, F. and Magdalena, L. (2001) *Genetic Fuzzy Systems: Evolutionary Tuning and Learning of Fuzzy Knowledge Bases*, World Scientific, Singapore.

Darwin, C. (1859) *The Origin of Species by Means of Natural Selection or the Preservation of Favoured Races in the Struggle for Life*, Mentor Reprint 1958, New York.

Das, S. and Suganthan, P.N. (2011) Differential evolution: a survey of the state-of-the-art, *IEEE Transactions on Evolutionary Computation*, 15(1), 4–31.

De Jong, K.A. (1975) Analysis of the behaviour of a class of genetic adaptive systems, PhD Thesis, Department of Computer and Communications Sciences, University of Michigan.

Deb, K. (2008) *Multi-objective Optimisation using Evolutionary Algorithms*, 2nd edn, John Wiley & Sons, Chichester.

Deb, K. and Agrawal, S. (1995) Simulated binary crossover for continuous search space, *Complex Systems*, 9(2), 115–148.

Dunham, B., Lewitan, H. and North, J.H. (1974) Simultaneous solution of multiple problems by natural selection, *IBM Technical Disclosure Bulletin*, 17(7), 2191–2192.

Eiben, A.E. and Smith, J.E. (2003) *Introduction to Evolutionary Computing*, Springer-Verlag, Heidelberg.

Engelbrecht, A.P. (2007) *Computational Intelligence: An Introduction*, John Wiley & Sons, New York.

Eshelman, L.J. and Schaffer, J.D. (1993) Real-coded genetic algorithms and interval schemata. *Foundations of Genetic Algorithms* 2 (FOGA-2), pp. 187–202.

Fogel, D.B. (1991) *System Identification through Simulated Evolution: A Machine Learning Approach to Modelling*, Ginn Press, Needham Heights, MA.

Fogel, D.B. (1995) *Evolutionary Computation – Toward a New Philosophy of Machine Intelligence*, IEEE Press, New York.

Fogel, D.B. (1998) Evolutionary computation – the fossil record. In *An Introduction to Evolutionary Computation*, D.B. Fogel (ed.), IEEE Press, New York.

Fogel, D.B., Fogel, L.J. and Atmar, J.W. (1991) Meta-evolutionary programming. *Proceedings of 25th Conference on Signals, Systems and Computers*, Vol. 1, pp. 540–545.

Fogel, L.J., Owens, A.J. and Walsh, M.J. (1996) *Artificial Intelligence through Simulated Evolutionary*, John Wiley & Sons, Chichester.

Fogel, L.J. (1962) Autonomous automata, *Industrial Research*, 4, 14–19.

Franklin, B. and Bergerman, M. (2000) Cultural e-algorithms: concepts and experiments, *2000 IEEE Congress on Evolutionary Computation*, pp. 1245–1251.

Fraser, A.S. (1957) Simulation of genetic systems by automatic digital computers, I. Introduction, *Australian Journal of Biological Sciences*, 10, 484–491.

Friedberg, R.M. (1958) A learning machine: Part I, *IBM Journal of Research and Development*, 2(1), 2–13.

Friedberg, R.M., Dunham, B. and North, J.H. (1959) A learning machine: Part II, *IBM Journal of Research and Development*, 3, 282–287.

Goldberg, D.E. (1989) *Genetic Algorithms in Search, Optimization, and Machine Learning*, Addison Wesley, Boston, MA.

Goldberg, D.E. and Lingle, R. (1985) Alleles, loci and the travelling salesman problem. In *Proceedings of the 1st International Conference on Genetic Algorithms and Their Applications*, J.J. Grefenstette (ed.), Lawrence Erlbaum, Hillsdale, NJ.

Herrera, F. and Lozano, M. (2000) Gradual distributed real-coded genetic algorithms, *IEEE Transactions on Evolutionary Computation*, 4(1), 43–63.

Herrera, F. and Lozano, M. (2001) Adaptive genetic algorithms based on co-evolution with fuzzy behaviours, *IEEE Transactions on Evolutionary Computation*, 5(2), 149–165.

Herrera, F., Lozano, M. and Verdegay, J.L. (1997) Fuzzy connectives based crossover operators to model genetic algorithms population diversity, *Fuzzy Sets and Systems*, 92(1), 21–30.

Holland, J.H. (1962) Outline for a logical theory of adaptive systems, *Journal of ACM*, 3, 297–314.

Holland, J.H. (1975) *Adaptation in Natural and Artificial Systems*, University Michigan Press, Ann Arbor, MI.

Karr, C.L. (1991) Design of an adaptive fuzzy logic controller using a genetic algorithm. *Proceedings of the 4th International Conference on Genetic Algorithms*, Morgan Kaufmann, San Mateo, CA, pp. 450–457.

Kitano, H. (1990) Designing neural networks using genetic algorithms with graph generation system, *Complex Systems*, 4, 461–476.

Kitano, H. (1994) Neurogenetic learning: an integrated method of designing and training neural networks using genetic algorithms, *Physica D*, 75, 225–228.

Koza, J.R. (1992) *Genetic Programming: On the Programming of Computers by Means of Natural Selection*, The MIT Press, Cambridge, MA.

Langdon, W.B. (1998) *Genetic Programming and Data Structures: Genetic Programming + Data Structures = Automatic Programming*, Kluwer Academic, Dordrecht.

Michalewicz, Z. (1992) *Genetic Algorithms + Data Structures = Evolution Programs*, Springer-Verlag, Berlin.

Nomura, T. and Miyoshi, T. (1996) Numerical coding and unfair average crossover in GA for fuzzy rule extraction in dynamic environments. In *Fuzzy Logic, Neural Networks and Evolutionary Computation*, Y. Uchikawa and T. Furuhashi (eds), Lecture Notes in Computer Science, Vol. 1152, Springer-Verlag, Berlin, pp. 55–72.

Ono, I. and Kobayshi, S. (1997) A real-coded genetic algorithm for function optimisation using unimodal normal distribution crossover. *Proceedings of the 7th International Conference on Genetic Algorithms*, pp. 246–253.

Price, K. (1994) Genetic annealing, *Dr. Dobb's Journal*, October, pp. 127–132.

Rechenberg, I. (1965) *Cybernetic Solution Path of an Experimental Problem*, Royal Aircraft Establishment, Library Translation No. 1122, Farnborough, UK.

Reynolds, R.G. (1994) Introduction to cultural algorithms. In *Proceedings of the Third Annual Conference on Evolutionary Programming*, A.V. Sebald and L.J. Fogel (eds), World Scientific, Singapore, pp. 131–139.

Reynolds, R.G. (1999a) Cultural algorithms: theory and application. In *New Ideas in Optimisations*, D. Corne, M. Doriago and F. Glover (eds), McGraw-Hill, New York, pp. 367–378.

Reynolds, R.G. (1999b) *An Overview of Cultural Algorithms: Advances in Evolutionary Computation*, McGraw-Hill, New York.

Reynolds, R. and Peng, B. (2005) Cultural algorithms: computational model of how cultures learn to solve problems: an engineering example, *International Journal of Cybernetics and Systems*, 36, 753–771.

Reynolds, R.G., Peng, B. and Alomari, R.S. (2006) Cultural evolution of ensemble learning for problem solving. *2006 IEEE Congress on Evolutionary Computation*, July 16–21, Vancouver, Canada, pp. 1119–1126.

Schwefel, H.-P. (1968) *Projekt MHD-Strausstrhlrohr: Experimentelle Optimierung einer Zweiphasenduese*, Teil I, Technischer Bericht 11.034/68, 35, AEG Forschungsinstitute, Berlin.

Schwefel, H.-P. (1975) *Evolutionsstrategie und numerische Optimierung*, PhD Thesis, Department of Process Engineering, Technical University of Berlin.

Schwefel, H.-P. (1995) *Evolution and Optimum Seeking*, John Wiley & Sons, Chichester.

Spears, W.M. (2000) *Evolutionary Algorithms: The Role of Mutation and Recombination*, Springer-Verlag, Berlin.

Srinivas, M. and Patnaik, L.M. (1994) Adaptive probabilities of crossover and mutation in genetic algorithms, *IEEE Transactions on Systems, Man and Cybernetics*, 24(4), 656–667.

Storn, R. (1995) Constrained optimization, *Dr. Dobb's Journal*, May, pp. 119–123.

Storn, R. (1999) System design by constraint adaptation and differential evolution, *IEEE Transactions on Evolutionary Computation*, 3(1), 22–34.

Storn, R. and Price, K. (1997) Differential evolution – a simple and efficient heuristic for global optimisation over continuous space, *Journal of Global Optimisation*, 11(4), 431–459.

Storn, R., Price, K. and Lampinen, J. (2005) *Differential Evolution – A Practical Approach to Global Optimisation*, Springer-Verlag, Berlin.

Syswerda, G. (1989) Uniform crossover in genetic algorithms. ICGA89, pp. 2–9.

Tsutsui, S., Yamamura, M. and Higuchi, T. (1999) Multi-parent recombination with simplex crossover in real-coded genetic algorithms. *Proceedings of the Genetic and Evolutionary Computing Conference* (GECCO'99), pp. 657–664.

Vesterstrom, J. and Thomson, R. (2004) A comparative study of differential evolution, particle swarm optimisation, and evolutionary algorithms on numerical benchmark problems. *Proceedings of the 6th IEEE Congress on Evolutionary Computation*, IEEE Press, New York, pp. 1980–1987.

Voigt, H.M., Muehlenbein, H. and Cvetkovic, H. (1995) Fuzzy recombination for breeder genetic algorithms. *Proceedings of the Sixth International Conference on Genetic Algorithms* (ICGA'95), L. Eshelman (ed.), Morgan Kaufman, San Mateo, CA, pp. 104–111.

Vonk, E., Jain, L.C. and Johnson, R.P. (1997) *Automatic Generation of Neural Network Architecture using Evolutionary Computation*, World Scientific, Singapore.

Whitley, D. (2000) Permutations. In *Evolutionary Computation 1: Basic Algorithms and Operators*, T. Back, D.B. Fogel and Z. Michalewicz (eds), Institute of Physics Publishing, Bristol, pp. 274–284.

Wright, A. (1991) Genetic algorithms for real parameter optimisation. *Proceedings of the Foundations of Genetic Algorithms* 1 (FOGA-1), pp. 205–218.

7

Evolutionary Systems

7.1 Introduction

Optimum seeking is one of the central issues in science, engineering, industry, economy, business and even in everyday life. Every problem we solve, every product we design and produce and every single thing we do are the outcome of the best possible choice. A variety of tools and techniques have been developed and applied to manmade artificial systems for optimum seeking; meanwhile, optimum seeking in nature, biological and social systems takes place in a completely different way by means of natural evolution. In all optimum seeking in artificial or natural systems, there are goals or objectives to be satisfied and there are constraints to meet within which the optimum has to be found. Eventually, the optimum seeking can be formulated as an optimization problem. That is, it is reduced to finding the best solution measured by a performance index. The performance indices are functionals (often known as objective functions in many areas of computing and engineering) that vary from problem to problem. In general, a performance index can be given by

$$J(c) = \int_x Q(x, c)p(x)dx \tag{7.1}$$

where $Q(x, c)$ is the functional of the vector $c = (c_1, c_2, \ldots, c_N)$, which depends on the random sequence or process $x = (x_1, x_2, \ldots, x_N)$ with probability density function $p(x)$. The goal is to find the extremum of the functional $Q(x, c)$, i.e., the minimum or maximum depending on the problem. The expression in Equation (7.1) is generally known as the criterion of optimality. For ease of application for certain problems, the criterion can also be defined based on the averaging of $Q(x, c)$ with respect to time depending on x. If x is a random sequence, i.e., $x = \{x[n], \ n = 1, 2, \ldots, N\}$, then $J(c)$ is expressed as

$$J(c) = \lim_{N \to \infty} \frac{1}{N} \sum_{n=1}^{N} Q(x[n], c) \tag{7.2}$$

If x is a random process, i.e., $x = \{x[t], 0 \le t < \infty\}$, then $J(c)$ is expressed as

$$J(c) = \lim_{T \to \infty} \frac{1}{T} \int_0^T Q(x[t], c)dt \tag{7.3}$$

Computational Intelligence: Synergies of Fuzzy Logic, Neural Networks and Evolutionary Computing, First Edition.
Nazmul Siddique and Hojjat Adeli.
© 2013 John Wiley & Sons, Ltd. Published 2013 by John Wiley & Sons, Ltd.

The criterion of optimality, i.e., the term $Q(x, c)$, can have a different interpretation or physical meaning when implemented on a real system. For example, it is the deviation from the desired behaviour (or output) of a system in a control application. Thus, the solution to the optimality problem described by Equations (7.1)–(7.3) is now a problem of finding the vector $c = c^*$, also called the optimal vector, which satisfies $J(c)$. It is to be noted here that a process or system for which the optimality is sought can be deterministic or stochastic in nature.

It is now obvious that for any deterministic or stochastic system the criterion of optimality, i.e., the functional $J(c)$ in Equations (7.1)–(7.3), should be known explicitly with sufficient *a priori* information along with the constraints. If the functional $J(c)$ is differentiable, its extremum (i.e., maximum or minimum) can be obtained for the values of the parameter vector $c = (c_1, c_2, \ldots, c_N)$ when the partial derivatives $\partial J(c)/\partial c_v$, $v = 1, 2, \ldots, N$ are simultaneously equal to zero. That is

$$\nabla J(c) = \left(\frac{\partial J(c)}{\partial c_1}, \frac{\partial J(c)}{\partial c_2}, \ldots, \frac{\partial J(c)}{\partial c_N} \right) = 0 \tag{7.4}$$

The vectors $c = (c_1, c_2, \ldots, c_N)$ for which $\nabla J(c) = 0$ are called the stationary or singular vectors. The problem is that not all stationary vectors are optimal and they do not correspond to the desired solution, i.e., the desired extremum of the functional. Therefore, $\nabla J(c) = 0$ is only a necessary condition (Tsypkin, 1971). The sufficient conditions can be derived in the form of an inequality based on the determinant containing the partial derivatives of the second order of the functional with respect to $c = (c_1, c_2, \ldots, c_N)$. However, it is not worth doing even in cases where the computational effort is not huge. If there is only one extremum, the stationary vector corresponding to the maximum or minimum can be found from the physical conditions of the problem. The conditions of optimality define only local extrema. Finding the global extremum becomes extremely difficult when the number of such extrema is large.

There have been different methods for finding the unique optimal value of the vector c^*. Gradient-based optimization techniques use derivative information in determining the search direction. Among the gradient-based techniques, the steepest descent method and Newton's method are well known. Conjugate gradient, Gauss–Newton and Levenberg–Marquardt are well-known variants of these methods. There is no guarantee that a gradient-based descent algorithm will find the global optimum of a complex objective function within a finite time. All descent methods are deterministic, requiring the initial points to be selected randomly, which has a decisive effect on the final results. If the initial points are to be chosen randomly, then the approach must be stochastic in nature or derivative-free.

If the criterion of optimality $J(c)$ and its distribution are known, the approach for optimization is to be called ordinary. There exist many ordinary approaches and they are mainly analytic and algorithmic methods. These methods are suitable for simple problems of first and second order. Approximations are used for higher-order problems. Algorithmic methods seem not very promising for this kind of problem.

On the other hand, if the distribution is not known or not sufficient *a priori* information is available, then an adaptive approach is used for optimization. In an adaptive approach, current information is actively used to compensate the insufficient *a priori* information. When a process is unknown (i.e., when it is not certain whether the process is deterministic or stochastic), an adaptive approach is also applicable. The adaptive approach is mainly an algorithmic (or iterative) method where a unique optimal vector is sought in an iterative manner such that $\nabla J(c) = 0$. The algorithm is written as

$$c = c - \eta \nabla J(c) \tag{7.5}$$

where η is a scalar. The optimal vector $c = c^*$ can be found using the method of approximation derived iteratively from Equation (7.5):

$$c[n] = c[n-1] - \eta[n]\nabla J(c[n-1]) \tag{7.6}$$

where $n = 1, 2, \ldots, \infty$ and $\eta[n]$ is the step size of the iterative method. Starting with an initial vector $c = c[0]$, the iterative method will converge to an optimal vector c^*:

$$\lim_{n \to \infty} c[n] = c^* \tag{7.7}$$

The optimal vector c^* can be determined using the iterative procedure given by Equation (7.6). The problem arises when the gradient of the functional in Equation (7.5) cannot be computed in an explicit form. There exist such situations when the functional $J(c)$ is discontinuous, non-differentiable or dependence of the parameter vector c cannot be expressed explicitly (Tsypkin, 1971). If the gradient of the functional $\nabla J(c)$ in Equation (7.6) is not known but some realization of $\nabla Q(x, c)$ is known, for example estimation by measurements, then the algorithm of adaption can be written in the recursive form by substituting the gradient of the functional $\nabla J(c)$ with a sample of $\nabla Q(x, c)$:

$$c[n] = c[n-1] - \Gamma[n]\nabla Q(x[n], c[n-1]) \tag{7.8}$$

where $\Gamma[n]$ is a suitable step matrix and $\Gamma[n] = I\eta[n]$ where I is an identity matrix. To ensure convergence, certain constraints are imposed on $\Gamma[n]$. For the simplest adaptive algorithm, $\Gamma[n]$ is a diagonal matrix given by:

$$\Gamma[n] = I\eta[n] = \begin{bmatrix} \eta_1[n] & 0 & \cdots & 0 \\ 0 & \eta_2[n] & \cdots & 0 \\ \vdots & \vdots & \ddots & \vdots \\ 0 & 0 & \cdots & \eta_N[n] \end{bmatrix} \tag{7.9}$$

where $\eta[n]$ is a row vector. Owing to the diagonal property of the matrix $\Gamma[n]$, it is useful for the computation of the adaptive algorithm. In the case of the adaptive algorithm in Equation (7.8), $\nabla Q(x, c)$ is not equivalent to zero for an optimal vector $c = c^*$. Various algorithms have been proposed based on the selection of $\Gamma[n]$ or $\eta[n]$, such as Newton's algorithm or the steepest descent method.

But the problem is that, in real-world situations, the objective function and the constraints are often not analytically treatable or even not available in closed form (Baeck, 1996). In such situations, the algorithm in Equation (7.8) cannot be employed. The only possible solution of the optimization problem under such conditions is possibly the search methods. Obviously this suggests applying a stochastic method that is capable of searching a high-dimensional space. In the methods of search, a transition from $c[n-1]$ to $c[n]$ is based on a search step

$\eta\psi$, where ψ is a random vector uniformly distributed in N-dimensional ($c = [c_1, c_2, \ldots, c_N]$) space (Tsypkin, 1971). That is, the algorithm in Equation (7.8) can be expressed as

$$c[n] = c[n-1] - \eta\psi[n] \tag{7.10}$$

where $n = 1, 2, \ldots, M$ and M is the maximum number of iterations.

It is now quite clear that the gradient term in Equation (7.8) is replaced with a random step $\eta\psi$ in Equation (7.10). There are a variety of stochastic search methods that are applicable to the optimization of problems of this nature. These methods rely extensively on repeated evaluations of the objective function and use heuristic guidelines for estimating the next search direction. The guidelines used are very simply based on thermodynamics, such as simulated annealing (Kirkpatrick *et al.*, 1983), tabu-search (Glover, 1989), random search (Matyas, 1965) and the downhill simplex method (Nelder and Mead, 1965). The searching strategies used by simulated annealing, tabu-search, random search and downhill simplex search are local search techniques and use a generate-and-test search, manipulating one feasible solution based on physical characteristics. Without further explanation, it can be presumed that the selection of the initial values of c (i.e., $c[0]$) has a decisive effect on the final solution. In practice, knowing these initial values is nearly impossible. There is no known suitable heuristic approach, other than selecting them randomly.

A conclusion can be drawn based on the transition from Equation (7.5) (derivative-based approach) to Equation (7.10) (derivative-free approach) that even with these well-established analytical methods, the stochastic nature of all the approaches cannot be avoided. This assumption leads us to the choice of derivative-free stochastic optimization methods.

Evolutionary algorithms (Holland, 1975; Goldberg, 1989; Michalewicz, 1992; Fogel, 1995; Eiben and Smith, 2007; Engelbrecht, 2007; Deb, 2008) are based on natural evolution and biological analogy and work on a population of potential solution of the parameters (or structure) in parallel. EAs are particularly attractive due to their ability to explore an initially unknown search space and to exploit this information to guide the subsequent search over generations and identify useful subspaces in which the global minimum is located. The advantage of the implicit mechanism is that the search consists of a combination of high-performance building blocks discovered during past trials (De Jong, 1975). Very often the term 'adaptation' is used, which is a process of modifying the parameters or the structure of the system. The characteristic features of adaptation are accumulation and use of current information that compensates the insufficient *a priori* information for the purpose of optimization. Different variants of EAs, such as evolutionary programming, evolution strategies, genetic algorithms, genetic programming, differential evolution and cultural algorithms, have been discussed in Chapter 6. The purpose of this chapter is to explore EA-based approaches in different application and problem domains.

EAs have been applied to higher-dimensional and real-world complex problems ranging from simple optimization to multi-objective optimization and from simple evolution to symbiotic evolution of multiple species. In the last few decades, EAs have found wide-ranging application domains – specifically, multi-objective optimization, co-evolutionary systems with multiple populations and parallel evolutionary algorithms. The purpose of this chapter is to demonstrate a few examples in each domain. New ideas and concepts have enriched the EA paradigm, with new directions.

7.2   Multi-objective Optimization

Real-world problems are complex and the definition of optimality is not simple as they need to satisfy multiple competing objective functions at the same time. Moreover, some of these objectives may have conflicting relations with others, which in fact makes the optimization a difficult task. Problems requiring simultaneous optimization of more than one objective function are known as multi-objective optimization problems (MOOPs). They can be defined as problems consisting of multiple objectives, which are to be minimized or maximized while maintaining some constraints. Formally, they can be defined as:

$$\text{Minimize/maximize } f(x) \tag{7.11}$$

$$\text{Subject to } g_j(x) \geq 0, \, j = 1, 2, 3, \ldots, J \tag{7.12}$$

$$h_k(x) = 0, k = 1, 2, 3, \ldots, K \tag{7.13}$$

where $f(x) = \{f_1(x), f_2(x), \ldots, f_n(x)\}$ is a vector of objective functions, $x = \{x_1, x_2, \ldots, x_p\}$ is a vector of decision variables, n is the number of objectives and p is the number of decision variables. Here, the problem optimizes n objectives and satisfies J inequality and K equality constraints. This type of problem has no unique perfect solution. In traditional multi-objective optimization, it is very common to simply aggregate all the objectives together to form a single (scalar) fitness function. However, the obtained solution using a single scalar is sensitive to the weight vector used in the scaling process. This requires knowledge about the underlying problem which is not known *a priori* in most cases. Moreover, the objectives can interact or conflict with each other. Therefore, trade-offs are sought when dealing with such MOOPs, rather than a single solution. Most MOOPs do not provide a single solution; rather, they offer a set of solutions. Such solutions are the 'trade-offs' or good compromises among the objectives. In order to generate these trade-off solutions, an old notion of optimality called the 'Pareto-optimum set' (Ben-Tal, 1980) is normally adopted.

In multi-objective optimization, the definition of quality of solution is substantially more complex than for single-objective optimization problems. The main challenges in a multi-objective optimization environment are: converge as closely as possible to the Pareto-optimal front, and maintain as diverse a set of solutions as possible. The first task ensures that the obtained set of solutions is near optimal, while the second task ensures that a wide range of trade-off solutions is obtained.

Owing to the advantageous features of derivative-freeness and population-based approach to solutions of optimization problems, EAs are applied in MOOPs and the combination became known as a multi-objective evolutionary algorithm (MOEA). An MOEA will be considered good only if both the goals of convergence and diversity are satisfied simultaneously. The MOEA's population-based approach helps to preserve and emphasize the non-dominated diverse set of solutions in a population. The MOEA converges to a Pareto-optimal front with a good spread of solutions in some reasonable number of generations. Most MOEAs use the concept of domination to attain the set of Pareto-optimal solutions. In total absence of information for preferences of the objectives, solutions to multi-objective problems are compared using the notion of Pareto dominance (Corne *et al.*, 2000). For problems having more than one objective function, any two solutions $x^{(1)}$ and $x^{(2)}$ can have one of two possibilities: one dominates the other, or neither dominates the other. A particular solution $x^{(1)}$ with performance vector u is said to be dominant, or better than the solution $x^{(2)}$ with performance vector v, if both the following conditions hold: (i) the solution $x^{(1)}$ is no worse than $x^{(2)}$ in all objectives and (ii) the solution $x^{(1)}$ is strictly better than $x^{(2)}$ in at least one objective. This notion can be generalized in the following equation:

$$u \prec v \text{ iff } [\forall i \in \{1, 2, \ldots, n\}, u_i \leq v_i] \cap [\exists i \in \{1, 2, \ldots, n\} \,|\, u_i < v_i] \qquad (7.14)$$

where it holds that $u \prec v \Leftrightarrow x^{(1)} \prec x^{(2)}$. For a given finite set of solutions, we need to perform pairwise comparisons to find out which solutions dominate and which are dominated. From these comparisons, we can find a subset of the finite set of solutions such that any two solutions which do not dominate each other, and all the other solutions of the finite set, are dominated by one or more members of this subset. This subset is called the non-dominated set for the given set of solutions. A solution is said to be Pareto-optimal if it is not dominated by any other possible solution. This is described by:

$$x^{(1)} \in x_{PO} \text{ iff } \nexists x^{(2)} \in \Psi \,|\, x^{(2)} \prec x^{(1)} \qquad (7.15)$$

where x_{PO} is the set of Pareto optimal solutions and Ψ is the set of all feasible solutions. The Pareto front is the set of points in the criterion space that correspond to Pareto-optimal solutions.

In an MOEA, a randomly selected population is generated within a specific range. Each individual of the population is evaluated with the objective functions. Figure 7.1 shows many solutions trading off differently between the objectives for a two-objective minimization problem. Any two solutions from the feasible objective space can be compared. For a pair of solutions, it can be seen that one solution is better than the other in the first objective but worse in the second objective. The individuals that fall close to either axes or the origin of the two-dimensional objective space are better than those away from the axes or origin. In the objective

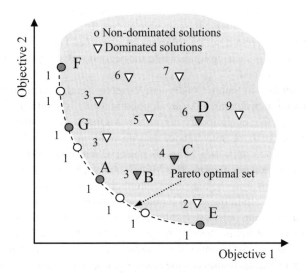

Figure 7.1 Dominated and non-dominated solutions with ranking

space, some individuals may be found (such as the individuals denoted by E, A, G and F in Figure 7.1) falling on an outer edge and close to the axes or origin and having one objective better than the other. For clarity, these individuals are joined by a dotted line in Figure 7.1. All the individuals lying on this curve form a set called the non-dominated solution set or Pareto-optimal set. The curve formed by joining these solutions is called the Pareto-optimal front.

Individuals A, E, F and G are called non-dominated because no other individuals provide better performance in the objective space. On the other hand, individuals falling away from edges, such as B, C and D, are called dominated solutions since many individuals provide better performance than these in terms of both objectives. The dominated and non-dominated solutions are shown in Figure 7.1. For example, individual A dominates individual B, similarly B dominates C and C dominates D in the objective space in terms of both objectives. In the process, each individual is ranked according to their degree of dominance. An individual's ranking equals the number of individuals better than it in terms of both objectives plus one. Individuals on the Pareto-optimal front (denoted by a small circle) are non-dominated and have a ranking of one. Individuals inside the Pareto-optimal front (denoted by a small triangle) have higher ranking than one. The numbers shown in Figure 7.1 correspond to their ranking. The main goal of an ideal multi-objective optimization is to find as many Pareto-optimal solutions as possible. Therefore, the objective of EAs would be to provide a diverse population of solutions.

Owing to the advantages of population-based EAs over the various difficulties in finding multiple Pareto-optimal solutions using classical multi-objective optimization techniques, currently researchers are exploiting EAs extensively for multi-objective optimization. A brief discussion on the difficulties of classical multi-objective optimization methods can be found in Deb (2008). The first application of EAs in multi-objective optimization was reported by Schaffer (1985). He proposed a vector-evaluated GA (also known as VEGA). Since then, research in MOEAs was dispersed up until the mid-1990s. Inspired by Goldberg's (1989) suggestion, different versions of MOEA were reported, such as the Niched Pareto genetic algorithm (NPGA) proposed by Horn *et al.* (1993, 1994). Fonseca and Fleming (1993) introduced a multi-objective GA (also known as MOGA) and the non-dominated sorting GA (NSGA) was developed by Srinivas and Deb (1994). Multi-objective evolutionary algorithms have proved to be very powerful tools for many complex problems and have become increasingly popular in a wide variety of application domains.

The basic principle of developing an EA-based algorithm is to use Pareto-based fitness to identify non-dominated individuals from the current population. Thus, an MOEA should guide the search towards a Pareto-optimal front and maintain diversity of known Pareto-optimal solutions. A generic algorithm of an MOEA would consist of the meta-level procedures shown in Figure 7.2.

The genetic diversity of a population can be lost due to the stochastic selection pressure. Fitness sharing based on the niching method can overcome this. The basic idea of fitness sharing is that all individuals within the same region (called a niche) share their fitness. Therefore, individuals in over-populated regions will experience a greater fitness decrease than isolated individuals. A new fitness function based on a ranking process has been suggested by Goldberg and Richardson (1987). A non-dominated sorting-based fitness-sharing technique was used in MOGA. Here, share counts are computed based on an individual's distance in the objective domain, but only between individuals with the same rank. Details of this method can be found in Fonseca and Fleming (1998a,b). The stochastic universal sampling method is used to select the best individuals. However, mating restrictions are employed in order to protect genetic drifts and premature convergence.

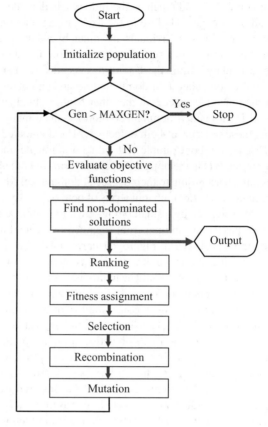

Figure 7.2   Flow diagram of a generic MOEA procedure

The general procedure for fitness assignment and sharing can be described as follows:

(i) Sort population according to ranking.
(ii) Assign fitness by interpolating from the best individual to the worst according to some function, in the form of a fitness function, such as linear or exponential. The lower the rank of an individual, the smaller the fitness of the individual.
(iii) Average the fitness assigned to individuals with the same rank, so that all of them are sampled at the same rate while keeping the global population fitness constant.

In the next few sections, we are going to discuss briefly the mainstream of MOEAs. In the discussion, the terms 'individual' and 'solution' will be used interchangeably.

7.2.1   Vector-Evaluated GA

VEGA is a straightforward extension of single-objective GA for multi-objective optimization problems, developed by Schaffer (1985). For an M-objective problem, the population is divided into M equal subpopulations randomly. Each subpopulation is evaluated against an objective function and assigned a fitness value. Selection is restricted only within subpopulations. Crossover and mutation operators are applied on the combined population.

7.2.2   Multi-objective GA

Fonseca and Fleming (1993) first proposed the non-dominated solutions of MOGAs. They also emphasized maintaining the diversity in non-dominated solutions. The fitness of the individuals (i.e., solutions) is computed and then each solution is checked for its domination in the population. A ranking procedure is carried out for all solutions in the population. An individual's ranking equals the number of individuals better than it in terms of all the objectives plus one. There must be at least one solution with rank 1 and the maximum rank of a solution cannot be more than the population size N. All ranks between 1 and N may not necessarily be assigned to solutions in a population. Non-dominated solutions are assigned a rank equal to 1. Fitness is assigned to a solution based on its rank. In order to maintain diversity among non-dominated solutions, Fonseca and Fleming (1993) used the niching technique for solutions of each rank. A pseudo-code for generic MOGA is given below:

```
gen ← 0
Initialize population
Evaluate objective values
Assign rank based on Pareto dominance
Compute niche count
Assign linearly scaled fitness
Assign shared fitness
Do while (gen < Max generation)
    {
    Select using Stochastic Universal Sampling
    Perform crossover
    Perform mutation
    Evaluate objective values
    Assign rank based on Pareto dominance
    Compute niche count
    Assign linearly scaled fitness
    Assign shared fitness
    gen ← gen + 1
    }
```

7.2.3   Niched Pareto GA

NPGA was proposed by Horn *et al.* (1993, 1994), where a tournament selection based on Pareto dominance is applied. In the tournament selection, two individuals are randomly chosen and compared against a subset of the population (typically 10%) for dominance. If one of them is dominated by the subset of population and the other is not, then the non-dominated individual is selected. If both the individuals are dominated or non-dominated by the subset of population, they are checked with the offspring population and the niche count is calculated. The individual with smaller niche count wins the tournament selection. The advantage of NPGA is that the fitness assignment to each individual is not needed.

7.2.4   Non-dominated Sorting GA

NSGA is a modification of the ranking procedure proposed by Srinivas and Deb (1994). After evaluating the population, the procedure is to sort the population according to non-dominance. The procedure divides the population into a number of mutually exclusive classes (i.e., non-dominated sets). All non-dominated individuals are classified into one class (or front). For example, a population for a two-objective minimization problem is classified into four fronts after non-dominated sorting, as shown in Figure 7.3. Obviously, the solutions in the first front are the best non-dominated set and the last set is the worst set. For example, front 1 is the best and front 4 is the worst in Figure 7.3. Therefore, the highest fitness is assigned to the best non-dominated front and the lowest fitness is assigned to the worst non-dominated front. The fitness assignment procedure starts from the first non-dominated set with the highest fitness equal to N (size of the population) and successively proceeds with lower fitness values to dominated sets. For example, the solutions of front 1 (in Figure 7.3) have a fitness value of 11 (population size). Assigning higher fitness values to better non-dominated solutions creates a selection pressure towards the Pareto-optimal front as these sets are closer to the Pareto-optimal front.

Maintaining diversity in the solution set (i.e., front) is important in MOEA and solutions should be well distributed within a front. NSGA preserves diversity among solutions of each non-dominated front using a sharing strategy. For example, the shared fitness of solution 4 (in Figure 7.3) in front 1 is 11 and the shared fitness of solutions 1, 2 and 3 would be 11/3 or 3.66. The fitness of solutions in the next front should start with a slightly smaller value than the minimum shared fitness value. Pseudo-code for a generic NSGA is given below:

gen ← 0
Initialize population
Evaluate objective values
Rank based on Pareto dominance in each class
Compute niche count

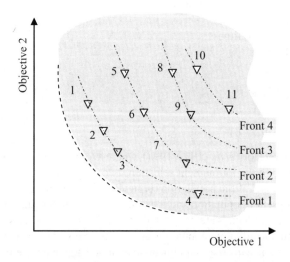

Figure 7.3   A population classified into four non-dominated classes

```
Compute shared fitness
Do while (gen < Max generation)
  {
  Selection using Stochastic Universal Sampling
  Perform crossover
  Perform mutation
  Evaluate objective values
  Rank based on Pareto dominance in each class
  Compute niche count
  Compute shared fitness
  gen ← gen + 1
  }
```

An improved version of NSGA, proposed by Deb *et al.* (2002), is known as NSGA-II. NSGA-II has been used as a foundation for many other multi-objective optimization algorithm design.

7.2.5   Strength Pareto Evolutionary Algorithm

In an attempt to combine the different features of VEGA, NPGA, NSGA and HLGA (not discussed above), Zitzler and Thiele (1999) proposed a new approach to find multiple Pareto-optimal solutions in parallel – called the strength Pareto evolutionary algorithm (SPEA). HLGA is a non-Pareto approach of aggregation by variable objective weighting methods proposed by Hajela and Lin (1992). HLGA will not be discussed further. SPEA maintains an external set for storing non-dominated solutions at each generation. A strength value is computed for each solution in this set. This strength value is similar to the ranking used in other methods. SPEA has the following distinct features:

(i) SPEA uses techniques such as storing the non-dominated solutions in an external set, Pareto dominance to assign fitness and pruning of the external set.
(ii) The fitness of an individual is determined from the solutions in the external non-dominated set.
(iii) All solutions stored in the external set take part in the selection process.
(iv) Pareto-based niching is used to preserve population diversity.

Pseudo-code for a generic SPEA is given below:

```
gen ← 0
Initialize population P
Create empty external set P'
Do while (gen < Max generation)
  {
  P' ← non-dominated solutions of P
  P' ← non redundant solutions of P'
  Prune P' (by means of clustering) if |P'| > N' (a given maximum)
  Evaluate individuals of P and P'
  Select from (P + P') until mating pool is full
```

Perform crossover
Perform mutation
gen ← gen+1
}

There is also a new version of SPEA called SPEA2 (Zitzler *et al.*, 2001). SPEA2 has three improved features compared with SPEA: SPEA2 uses a fine-grained fitness assignment strategy, nearest-neighbour density estimation technique for efficient search and improved storing to preserve boundary solutions.

NSGA-II and SPEA2 are the two benchmark methods widely used by the MOEA research community to compare the performance of new approaches and algorithms. There have been various other MOEAs reported in the literature over the last two decades. Knowles and Corne (2000) proposed a Pareto-archived evolution strategy (PAES) based on (1 + 1) evolution strategy. Van Veldhuizen and Lamont (2000) proposed a multi-objective messy genetic algorithm (MOMGA) based on messy GA. Corne *et al.* (2000) proposed a Pareto envelope-based selection algorithm (PESA) using a hypergrid division of phenotype space for maintaining selection diversity. The micro-genetic algorithm for multi-objective optimization, called micro-GA, was proposed by Coello Coello and Pulido (2005) using a small population with reinitialization process. A comprehensive review of all these evolutionary multi-objective optimization algorithms can be found in Coello Coello *et al.* (2007) and Deb (2008).

7.3   Co-evolution

Traditional evolutionary algorithms are not adequate for solving increasingly complex problems as they are highly simplified and abstract computational models of evolution that occur in nature. In these EAs, evolution is viewed as a process of adaptation of a population in a fixed environment. The environmental changes caused by populations are not taken into consideration. There are multiple species existing and interacting with each other in a natural environment. The selection, crossover and mutation mechanisms are far too simplified from natural ones. There are various feedback mechanisms between the individuals undergoing selection and crossover processes. Therefore, co-evolution is the complementary evolution of multiple species with interdependency or intertwined relationships with each other in an environment or eco-system (Watson and Pollack, 1999).

A complication of the Darwinian evolution is the fact that many species enter into close ecological relationships with other species. The species work together towards some common goals. However, in order to survive and produce offspring, they have to adapt themselves to the changing environment. The adaptive value of an organism is determined by the environmental niche. The characteristics of the niche are determined by the presence of a niche of the same species or other species (Eiben and Smith, 2003). The impact of the species on fitness depends on the inter-relationship between species. In this process, there are symbiotic relationships between species such as competition, exploitation and benefitting. The general term for this relationship is 'symbiosis', and it is a common phenomenon in co-evolution.

Before we discuss different forms of co-evolution, we will introduce different symbiotic relationships between species using graphical representations that have been used by researchers (Morrison, 1998). The nodes in Figure 7.4(a) represent individuals. The symbiotic relationship is denoted by a directed edge as shown in the figure. The relationship can be protagonist

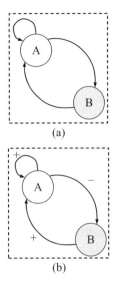

Figure 7.4 Symbiotic relationships between species. (a) Nodes represent species and edges represent relationship; (b) Protagonist (B $\overset{+}{\rightarrow}$ A), antagonist (A $\overset{-}{\rightarrow}$ B) and adaptism (A $\overset{+}{\rightarrow}$ A) relationships between species

(labelled +) or antagonist (labelled by –), as shown in Figure 7.4(b). All relationships between species and individuals can be represented and explained using the symbiosis graph.

While co-evolution occurs because of symbiotic relationships between species, there are differences between them. There are two types of co-evolutionary process identified:

- Cooperative
- Competitive

In cooperative co-evolution, different species cooperate with each other. This means species help each other to improve the fitness (survival strength). In competitive co-evolution, different species compete against each other to gain a fitness advantage at the expense of the other. The interplay of inverse fitness means a fitness gain for one species and a fitness loss for the other.

The relationships among species in cooperative co-evolution can have different forms (Watson and Pollack, 2001): commensalism, amensalism and mutualism.

In commensalism, one organism gets benefit with no significant detriment or benefit to the other organism. The symbiosis is between two organisms, called the host and the commensal. Figure 7.5(a) shows the commensalism, where A is the host and B is the commensal. The commensal derives benefit from the host (protagonist relationship denoted by B $\overset{+}{\rightarrow}$ A) without causing any harm to the host. The fitness of the host does not change as a result of the fitness change in the commensal. A change in the fitness of the host causes benefit or loss to the fitness of the commensal.

In amensalism, the symbiosis is between two individuals, called the host and the commensal. Figure 7.5(b) shows the amensalism where A is the host and B is the commensal. The amensal has no benefit from the host and is detrimentally affected by the host (antagonist relationship denoted by A $\overset{-}{\rightarrow}$ B), whereas the host is unharmed by the amensal. The fitness of the host does

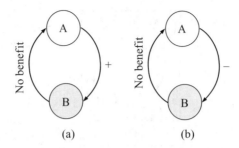

Figure 7.5 Symbiosis of commensalism and amensalism. (a) Commensalism; (b) Amensalism

not change as a result of the fitness change in the commensal. A change in the fitness of the host causes benefit or loss to the fitness of the commensal.

In mutualism, both or all species benefit from the relationship. This is known to be the first form of co-evolution. The species taking part in the mutualism are called symbiont. In mutualism both the species (symbionts) are benefitted (mutual relationship denoted $A \overset{+}{\leftrightarrow} B$). Figure 7.6 shows the symbiosis of mutualism. The fitness change in one symbiont will make a positive change in the fitness of the other.

The relationships among species in competitive co-evolution can have different forms: competition, predator–prey (parasitism) and adaptism.

In competition, different species fight each other for limited resources critical to fitness. In the competition one wins and takes all or the majority of the share of the resource. A positive change in the fitness of one will cause a negative change to the other. Figure 7.7 shows the symbiosis of the competing relationship between two species.

Predator–prey relationships are the best-known examples in co-evolution. In predator–prey, the predator derives benefit from the prey whereas the prey is detrimentally affected by the predator. That is, the change in fitness of the predator is always positive and that of the prey is always negative. The predator–prey symbiosis is shown in Figure 7.8. For example, the prey needs to develop a superior defending mechanism for survival (e.g., running faster, growing bigger shields, better camouflage technique, etc.). As a result, the prey has strong evolutionary pressure over the predator to develop better attacking strategies in future generations (e.g., stronger claws, better eye-sight, etc.). The success of one species is the failure of the other. The co-evolution results in a stepwise improvement of both species, with increased complexity of the co-evolutionary process.

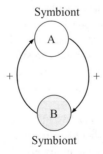

Symbiont

Symbiont

Figure 7.6 Symbiosis of mutualism ($A \overset{+}{\leftrightarrow} B$)

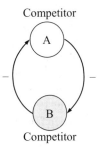

Figure 7.7 Symbiosis of competition (A $\overset{-}{\leftrightarrow}$ B)

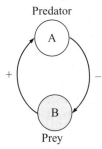

Figure 7.8 Symbiosis of predator–prey

In adaptism, species (or individuals) modify themselves or the environment without any assistance of other species that have an impact on their own, resulting in a change in fitness. The change in fitness can be positive or negative, as shown in Figure 7.9. The symbiotic relationships are denoted A $\overset{+}{\rightarrow}$ A or A $\overset{-}{\rightarrow}$ A. This is seen as co-evolution of another sort. These individuals are referred to as adaptive individuals, and have somehow been overlooked in the literature (Morrison, 1998).

7.3.1 Cooperative Co-evolution

Evolving full solutions causes the population to converge and fail to find solutions for high-dimensional large and complex problems. Maintaining population diversity in such high-dimensional problems is a challenge. One strategy would be to decompose the larger problem

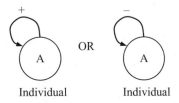

Figure 7.9 Symbiosis of adaptism

into tractable subproblems, each represented by a subpopulation (or species) of solutions. In cooperative co-evolution, a number of different species offer a way of interaction between solutions represented by each species and cooperate to come up with a solution to the larger problem. Thus, the evolution establishes solutions in diverse populations and is able to maintain diversity in prolonged evolution. Such cooperative co-evolution has been proposed by many researchers, for example, high-dimensional function optimization (Potter and De Jong, 1995a) and complex structures (Potter and De Jong, 1995b). These representations, adapted to the solutions currently in the population, have been found to speed up the search and provide even better solutions.

7.3.1.1 A Generic Algorithm for Cooperative Co-evolution

A mutual cooperative co-evolutionary algorithm is presented here. The fitness of an individual is calculated based on the ability to collaborate with individuals from other species. That is, individuals of different species collaborate with each other towards the solution of a common goal that defines their collective fitness. The problem is then how to distribute the fitness share among participating individuals from this collective effort. Potter and De Jong (1995a,b) proposed a general framework for evolving subpopulations independently and merging them to form a complete solution, which is evaluated to determine global fitness. The fitness share goes back to each subpopulation, reflecting their performance in the collaboration. A cooperative co-evolution should have the following characteristics:

(i) A species represents a subcomponent of a potential solution.
(ii) Complete solution is obtained by merging representative individuals of each of the species.
(iii) A fitness award (or credit assignment) is given to the species, proportional to their participation in the complete solutions.
(iv) Species (subpopulations) should themselves evolve when necessary.
(v) Standard EAs are used for evolution at species level.

A generic co-evolutionary algorithm is given here for cooperating subpopulations S_j, $j = 1, 2, \ldots, n$. Each subpopulation S_j is co-evolved in a round-robin fashion using standard EAs. The fitness of each individual of subpopulation S_j is computed by combining it with the current best subcomponent of the remaining subpopulations. This is the simplest form of fitness award.

```
t=0;
For j=1, ..., N
    S_j[t] ← Initialize
    For i=1, ..., n
        F_i(S_j[t]) ← evaluate(S_j[t])
    Endfor
Endfor
While not (termination condition)
    {
    For j=1, ..., N
        S_j[t] ← select(S_j[t − 1])
        S_j[t] ← recombine(S_j[t])
```

```
    For i=1, ..., n
        F_i(S_j[t]) ← evaluate(S_j[t])
    Endfor
  Endfor
  t=t+1
}
```

7.3.2   Competitive Co-evolution

Competitive co-evolution (CCE) works with the goal of producing an optimal species though competition. In general, CCE evolves two competing populations. Individuals in one population represent solutions to a problem, while individuals in the other population represent test cases.

7.3.2.1   Fitness Sampling and Relative Fitness Computation

The fitness of each individual is calculated independently. The relative fitness of individuals is calculated against a sample of individuals from the competing population. Two aspects are of importance for calculation of the relative fitness: which individuals are used from the competing population, and how these competing individuals are used in the fitness calculation. The relative fitness provides the number (score) of opponents that are beaten by an individual.
  The following sampling schemes are used to compute the fitness.

- All versus all – individual is tested against all individuals of the other population.
- Random – individuals are tested against a group of individuals selected from the other population.
- Tournament – uses relative fitness to select the best opponent individual.
- All versus best – individuals are tested against the best individual of the other population.
- Shared – sample is selected from the opponent individuals with maximum competitive shared fitness. The opponent is selected which beat the largest number of individuals from the competing population.

7.3.2.2   A Generic Competitive Co-evolution Algorithm

A generic co-evolutionary algorithm is given here for two competing populations A and B. Samples from each population, A_S and B_S, are chosen for calculation of the relative fitness $\bar{f}_{A_i}$ and $\bar{f}_{B_i}$. The individual populations are evolved for one generation (using standard EA) to produce offspring that undergo selection for the new generation:

```
  t ← 0;
  {A[t], B[t]} ← Initialize
  While not (termination condition)
    {
    For each A_i[t], i = 1, ..., N do
      Select B_S (a sample from B)
      Evaluate relative fitness f̄_A_i = f_A_i / f_B_S  with respect to B_S
```

$$\bar{f}_{A_i} = \frac{f_{A_i}}{f_{B_S}}$$

Endfor
For each $B_i[t]$, $i=1, \ldots, N$ do
  Select A_S (a sample from A)
  Evaluate relative fitness $\bar{f}_{B_i} = \frac{f_{B_i}}{f_{A_S}}$ with respect to A_S
Endfor
  Reproduce from A
  Reproduce from B
  $A[t]$=select($A[t]$)
  $B[t]$=select($B[t]$)
  $S=A[t] \cup B[t]$
  $t \leftarrow t+1$
  }Endwhile
Select the best individuals from solution population S

There have been many implementations of cooperative and competitive co-evolution reported in the literature so far. Some of these are under the generic term of symbiosis. Eguchi *et al.* (2006) reported a comprehensive study on multi-agent models based on symbiosis in ecosystems. The study was based on simulation of the symbiotic evolution of multi-agents and investigated the performance of the different relationships among the agents, such as mutualism, harm, predation and altruism. Goulermas and Liatsis (2003) reported a good application of adaptive symbiotic evolution for image–space matching and three-dimensional space analysis, where the super-problem was decomposed into a large set of small patches each represented by a separate species. All species evolved concurrently, maintaining a mutual relationship. Their study showed better performance compared to equivalent non-symbiotic optimizations. A good example of symbiotic evolution in designing a fuzzy controller was reported by Juang *et al.* (2000). The application showed that fuzzy system design and symbiotic evolution can complement each other and result in a robust controller within fewer trials and less computational time than classical GA.

7.4 Parallel Evolutionary Algorithm

Maintaining diversity in populations is very difficult and remains an open research issue in evolutionary algorithms. The most common methods are less aggressive selection strategy and high mutation rate. Poor selection strategies lead to slow convergence of the evolution. A high mutation rate only gives an artificial diversity through noise, and the search can turn into a random walk. Therefore, the mutation rate is kept low and introduces genetic material that may have been missing in the initial population or lost during crossover operations. In general, mutation is not a mechanism for creating population diversity. There have been some other methods developed by different researchers to enforce population diversity through adaptive mutation (Whitley *et al.*, 1994), crowding (De Jong, 1975), fitness sharing (Goldberg and Richardson, 1987), local mating (Collins and Jefferson, 1991) and implicit fitness sharing (Smith *et al.*, 1993; Horn *et al.*, 1994). But each of these techniques relies on external genetic functions and the diversity is achieved through very expensive operations.

Another inherent problem with the traditional evolutionary algorithm is that it cannot preserve different high-fitness individuals in a single population. When a suboptimal individual dominates the population, the selection strategy preserves such an individual by preventing genetic operations on it. As a result, premature convergence cannot be hindered for the evolutionary algorithm.

Using multiple and independent populations in parallel with occasional interchange (or migration) of individuals between populations would be an alternative approach to deal with high-dimensional problems. Each population can explore different parts of the search space. Diversity will be maintained via migration of high-fitness individuals from other populations and premature convergence can be delayed through diversity. Therefore, EAs are good candidates for parallelization, for effective searching of large-dimensional search problems. The basic idea behind parallel EAs is to take advantage of the divide-and-conquer approach. Parallel EAs are close to a biological metaphor, with diversity of structures and geographic locations incorporated into the population. In the literature, parallel EAs are mostly termed 'parallel GAs'. In the sequel, we will use 'parallel GA' and 'parallel EA' interchangeably.

There are many variations of parallel GAs based on the population structure and methods of recombination. In this section, we will discuss three broad categories of parallel GA:

- Global GA
- Migration GA
- Diffusion GA

7.4.1 Global GA

Global GA treats the entire population as single breeding and exploits parallelism in the implementations of the algorithm, i.e., applying genetic operators and evaluations of the objective functions for individuals. The master/slave architecture shown in Figure 7.10 demonstrates the parallelism implemented in global GA.

The master GA initializes and contains the entire population. Selection and fitness assignments are performed globally by the master GA, whereas the slave GAs perform recombination and mutation operations for producing offspring. Evaluation of objective functions for individuals is performed by individual slaves. Exploration of the inherent parallelism in a single population can achieve only a near-linear speedup for significantly complicated objective functions (Chipperfield and Fleming, 1994). Powerful computational resources may be employed

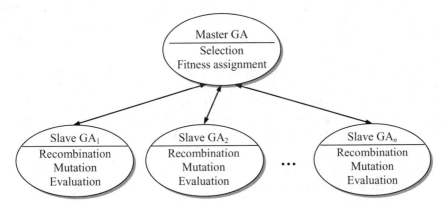

Figure 7.10 Architecture of the global GA

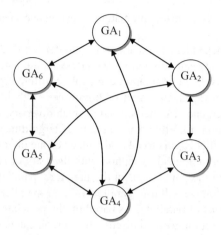

Figure 7.11   Architecture of the migration (island) GA

for a robust implementation of the master/slave GA, e.g., a number of parallel processors can be used in concurrent operation for genetic operator and objective function evaluation by the slave GAs. Such an implementation has been reported by Valdez *et al.* (2011), where a cluster of four computers was used to evolve a modular neural network. A variety of global GAs have been reported in the literature (Cantu-Paz, 1998, 2000, 2001a; Cantu-Paz and Goldberg, 2000).

7.4.2   *Migration (or Island) Model GA*

In natural evolution, the population is divided into subgroups (also called species) and individuals tend to mate and reproduce within their own species. There is always a possibility of mating occurring across the species. The migration (or island) model of the GA introduces the concept of dividing a large population into smaller semi-isolated (island) subpopulations or demes. Each subpopulation is a separate breeding unit and uses local selection and reproduction operations to evolve the subpopulation (species). Individuals of one subpopulation are allowed to migrate to other subpopulations. The migration pattern defines the genetic diversity of the global population, which is determined by the number of individuals migrating between subpopulations, the time interval of migration and the route of migration between population islands. A simple architecture of a migration GA is shown in Figure 7.11, with six population islands and migration paths.

The migration model is also known as a coarse-grained parallel GA. The main concept is to divide the population into a number of subpopulations and evolve each of the subpopulations using an independent GA. An additional procedure is used to handle the exchange of individuals (migration) between island populations at certain intervals using fixed migration routes. A simple pseudo-code is given below for the migration GA:

```
For i = 1 to n
  GA[i] ← initialize subpopulation
      While not (termination condition)
      {
      For i = 1 to n
```

```
{
Perform selection on GA[i]
Perform recombination on GA[i]
Perform evaluation on GA[i]
 For j =1 to n − 1 (not emigrating or immigrating to/from itself)
   {
   Send emigrants to GA[j]
   Receive immigrants from GA[j]
   }
 }
}
```

Sekaj and Perkacz (2007) have investigated the migration model with different forms of migration routes. Cantu-Paz and Goldberg (2000) reported a theoretical study and practical problems of migration GAs. A number of other studies on migration GAs have been reported in the literature (Cantu-Paz, 1998, 2000, 2001b; Yang *et al.*, 2004; Skolicki and De Jong, 2005). Migration model GAs are also known as distributed GAs, since they are implemented on distributed MIMD computers. They are also widely known as coarse-grained parallel GAs.

7.4.3 Diffusion GA

An alternative to the distributed population structure is the diffusion GA. In diffusion GAs, the population is partitioned into a large number of small subpopulations. Diffusion GAs are also known as fine-grained GAs because of the fine partitioning of the population and its treatment as a single continuous structure. There is no island for a subpopulation, rather a contiguous distribution of individuals. The population distribution of a diffusion GA is shown in Figure 7.12, where each individual $G_{j,k}$ is assigned a separate node on a toroidal

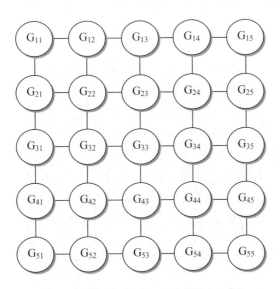

Figure 7.12 Architecture of diffusion GA

mesh of the parallel processing network. Each node uses individuals from the neighbouring nodes to apply genetic operations; for example, the individual node $G_{2,3}$ uses individuals from nodes $G_{1,3}$, $G_{2,2}$, $G_{3,3}$ and $G_{2,4}$. The selection is based on the fitness of the neighbouring individuals. A recombination operation is applied to produce a single offspring and replace the parent individual residing in the node. A simple pseudo-code is given below for the diffusion GA:

```
G[i,j] ← Initialize population
While not (termination condition)
{
For i = 1 to n
  For j = 1 to n
  {
  Perform evaluation on G[i,j]
    {
    Send individual to neighbours
    Receive individuals from neighbours
    }
  Perform selection
  Perform recombination
  }
}
```

In a few generations, the individuals start forming clusters of similar genetic material. In the fitness landscape of the population, these individuals will look like virtual islands. This scenario is shown in Figure 7.13. The darker shading in the figure represents higher fitness values of the individuals.

Robertson (1987) is known to have been the first to implement fine-grained GAs on a SIMD connection machine. One processor per individual was used for function evaluation, and global selection and recombination was implemented on a host machine. A significant

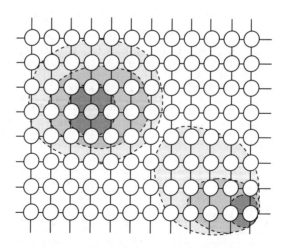

Figure 7.13 Virtual islands of high-fitness individuals

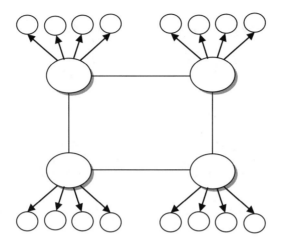

Figure 7.14   Hierarchical parallel GA

speedup was achieved, which inspired other researchers to implement diffusion model GAs. A notable implementation was reported by Muehlenbein (1989) and Gorges-Schleuter (1989) using an asynchronous parallel GA. The proposed architecture, called ASPARAGOS, is based on population genetics and implemented on a connected ring topology using transputers. One processor was assigned per individual. ASPARAGOS showed good performance for numerical optimization problems. A number of good implementations of fine-grained GAs were reported in the early 1990s (Baluja, 1993; Maruyama *et al.*, 1993). A number of other studies on diffusion GAs have been reported in the literature (Chipperfield and Fleming, 1994; Cantu-Paz, 1998, 2000, 2001a; Alba *et al.*, 2004).

7.4.4   Hybrid Parallel GA

The three basic forms of parallel GA, discussed in the above sections, have shown good performance when implemented properly. Analysis soon uncovered their limitations. Several

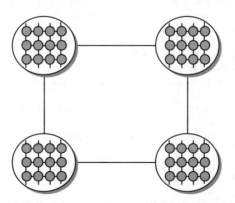

Figure 7.15   Coarse-grained and fine-grained combination of parallel GAs

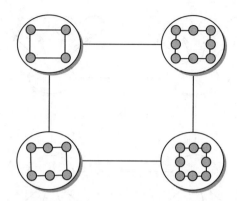

Figure 7.16   Coarse-grained and coarse-grained combination of parallel GAs

issues relating to communication topology or deme size, migration rate and frequency and search space segmentation remain unanswered. To overcome the limitations of the basic forms, researchers proposed various hybrid architectures by combining the three basic forms of architecture. Cantu-Paz and Goldberg (2000) proposed a set of simple rules for combining global GAs and migration GAs. They called this a hierarchical parallel GA, with master/slave demes. Figure 7.14 shows the hybrid architecture proposed by Cantu-Paz and Goldberg.

The architecture in Figure 7.14 has been implemented by Bianchini and Brown (1993) on transputers. Similarly, coarse-grained (migration model) and fine-grained (diffusion model) can be combined, as well as coarse-grained and coarse-grained. These two hybrid architectures are shown in Figures 7.15 and 7.16. Different combinations of parallel GAs have been investigated by Sekaj and Perkacz (2007). A good analysis on the hybrid architectures can also be found in Cantu-Paz and Goldberg (2000).

References

Alba, E., Luna, F. and Nebro, A.J. (2004) Advances in parallel heterogeneous genetic algorithms for continuous optimisation, *International Journal of Applied Mathematics and Computer Science*, 14, 317–333.

Baeck, T. (1996) *Evolutionary Algorithms in Theory and Practice*. Oxford University Press, New York.

Baluja, S. (1993) Structure and performance of fine-grain parallelism in genetic search. *Proceedings of the 5th International Conference on Genetic Algorithms*, pp. 155–162.

Ben-Tal, A. (1980) Characterization of Pareto and lexicographic optimal solutions. In *Multiple Criteria Decision Making: Theory and Application*, G. Fandel and T. Gal (eds), Lecture Notes in Economics and Mathematical Systems, Vol. 17, Springer-Verlag, Berlin, pp. 1–11.

Bianchini, R. and Brown, C. (1993) Parallel genetic algorithms on distributed-memory architectures, *Transputer Research and Applications*, 6, 67–82.

Cantu-Paz, E. (1998) A survey of parallel genetic algorithms, *Calculateurs Paralleles, Reseaux et Systems Repartis*, 10(2), 141–171.

Cantu-Paz, E. (2000) Markov chain models of parallel genetic algorithms, *IEEE Transactions on Evolutionary Computation*, 4(3), 216–226.

Cantu-Paz, E. (2001a) *Efficient and Accurate Parallel Genetic Algorithms*, Kluwer Academic, Dordrecht.

Cantu-Paz, E. (2001b) Migration policies, selection pressure, and parallel evolutionary algorithms, *Journal of Heuristics*, 7(4), 311–334.

Cantu-Paz, E. and Goldberg, D. (2000) Efficient parallel genetic algorithms: theory and practice, *Computer Methods in Applied Mechanics and Engineering*, 186(2–4), 221–238.

Chipperfield, A.J. and Fleming, P.J. (1994) *Parallel Genetic Algorithms: A Survey*, ACSE Research Report 518, Department of Automatic Control and Systems Engineering, University of Sheffield, UK.

Coello Coello, C.A. and Pulido, G.T. (2005) Multi-objective structural optimisation using a micro-genetic algorithm, *Structural and Multidisciplinary Optimisation*, 30(5), 388–403.

Coello Coello, C.A., Lamont, G.B. and Van Veldhuizen, D.A. (2007) *Evolutionary Algorithms for Solving Multi-Objective Problems*, 2nd edn, Springer-Verlag, Berlin.

Collins, R.J. and Jefferson, D.R. (1991) Selection in massively parallel genetic algorithms. *Proceedings of the 4th International Conference on Genetic Algorithms*, Morgan Kaufmann, San Mateo, CA, pp. 249–256.

Corne, D.W., Knowles, J.D. and Oates, M.J. (2000) The Pareto envelope-based selection algorithm for multi-objective optimisation. *Proceedings of the VI Conference on Parallel Problem Solving from Nature*, Paris, France, Lecture Notes in Computer Science, Vol. 1917, pp. 839–848.

De Jong, K.A. (1975) *Analysis of the behaviour of a class of genetic adaptive systems*, PhD Thesis, Department of Computer and Communications Sciences, University of Michigan, Ann Arbor, MI.

Deb, K. (2008) *Multi-objective Optimisation using Evolutionary Algorithms*, 2nd edn, John Wiley & Sons, Chichester.

Deb, K., Pratap, A., Agarwal, S. and Meyarivan, T. (2002) A fast and elitist multi-objective genetic algorithm: NSGA-II, *IEEE Transactions on Evolutionary Computation*, 6(2), 182–197.

Eguchi, T., Hirasawa, K., Hu, J. and Ota, N. (2006) A study of evolutionary multi-agent models based on symbiosis, *IEEE Transactions of Systems, Man and Cybernetics – Part B: Cybernetics*, 36(1), 179–193.

Eiben, A. and Smith, J. (2007) *Introduction to Evolutionary Computing*, 2nd edn, Springer-Verlag, Berlin.

Engelbrecht, A.P. (2007) *Computational Intelligence: An Introduction*, John Wiley & Sons, Chichester.

Fogel, D.B. (1995) *Evolutionary Computation – Toward a New Philosophy of Machine Intelligence*.

Fonseca, C.M. and Fleming, P.J. (1993) Genetic algorithms for multiobjective optimization: formulation, discussion and generalization. *Genetic Algorithms: Proceeding of the Fifth International Conference*, San Mateo, CA, pp. 416–423.

Fonseca, C.M. and Fleming, P.J. (1998a) Multi-objective optimization and multiple constraints handling with evolutionary algorithms – Part I: A unified formulation, *IEEE Transactions on Systems, Man and Cybernetics – Part A: Systems and Humans*, 28(1), 26–37.

Fonseca, C.M. and Fleming, P.J. (1998b) Multi-objective optimization and multiple constraints handling with evolutionary algorithms – Part II: Application example, *IEEE Transactions on Systems, Man and Cybernetics – Part A: Systems and Humans*, 28(1), 38–47.

Glover, F. (1989) Tabu Search – Part I, *ORSA Journal on Computing*, 1, 190–206.

Goldberg, D.E. (1989) *Genetic Algorithms in Search, Optimization, and Machine Learning*, Addison-Wesley, Reading, MA.

Goldberg, E. and Richardson, J. (1987) Genetic algorithms with sharing for multimodal function optimization. In *Proceedings of the Second International Conference on Genetic Algorithms*, J. Grefenstette (ed.), Lawrence Erlbaum Associates, Hillsdale, NJ, pp. 41–49.

Gorges-Schleuter, M. (1989) ASPARAGOS: An asynchronous parallel genetic optimisation strategy. *Proceedings of the 1st International Conference on Genetic Algorithms*, pp. 422–427.

Goulermas, J.Y. and Liatsis, P. (2003) A collective-based adaptive symbiotic model for surface reconstruction in area-based stereo, *IEEE Transactions on Evolutionary Computation*, 7(5), 482–502.

Hajela, P. and Lin, C.-Y. (1992) *Genetic Search Strategies in Multi-criterion Optimal Design*, Structural Optimization, Vol. 4, Springer-Verlag, New York, pp. 99–107.

Holland, J.H. (1975) *Adaptation in Natural and Artificial Systems*, University Michigan Press, Ann Arbor, MI.

Horn, J., Nafpliotis, N. and Goldberg, D. (1993) *Multi-objective Optimization using the Niched Pareto Genetic Algorithm*, Technical Report IlliGaL, Department General Engineering, University of Illinois at Urbana Champaign, Urbana, IL.

Horn, J., Nafpliotis, N. and Goldberg, D. (1994) A niched Pareto genetic algorithm for multi-objective optimization. *Proceedings of the First IEEE Conference on Evolutionary Computation*, pp. 82–87.

Juang, C.-F., Lin, J.-Y. and Lin, C.-T. (2000) Genetic reinforcement learning through symbiotic evolution for fuzzy controller design, *IEEE Transactions of Systems, Man and Cybernetics – Part B: Cybernetics*, 30(2), 290–302.

Kirkpatrick, S., Gelatt, C.D. and Vecchi, M.P. (1983) Optimization by simulated annealing, *Science*, 220, 671–680.

Knowles, J.D. and Corne, D.W. (2000) Approximating the non-dominated front using the Pareto archived evolution strategy, *Evolutionary Computation*, 8(2), 149–172.

Maruyama, T., Hirose, T. and Konagaya, A. (1993) A fine-grained parallel genetic algorithm for distributed parallel systems. *Proceedings of the 5th International Conference on Genetic Algorithms*, pp. 155–162.

Matyas, J. (1965) Random optimization, *Automation and Remote Control*, 26, 244–251.

Michalewicz, Z. (1992) *Genetic Algorithms + Data Structures = Evolution Programs*, 2nd edn, Springer-Verlag, Berlin.

Morrison, J. (1998) *Co-evolution and genetic algorithms*, PhD Thesis, School of Computer Science, Carleton University, Ottawa, Canada.

Muehlenbein, H. (1989) Parallel genetic algorithms, population genetics and combinatorial optimisation, parallelism, learning and evolution, In *Workshop on Evolutionary Models and Strategies*, J.D. Becker, I. Eisele and F.W. Mundemann (eds), Lecture Notes in Artificial Intelligence, Vol. 565, Springer-Verlag, Berlin, pp. 398–406.

Nelder, J. and Mead, R. (1965) The downhill simplex method, *Computer Journal*, 7, pp. 308–313.

Potter, M.A. and De Jong, K.A. (1995a) A cooperative co-evolutionary approach to function optimisation. *Proceedings of the 3rd Conference on Parallel Problem Solving from Nature*, Y. Davidor, H.-P. Schwefel and R. Manner (eds), pp. 249–257.

Potter, M.A. and De Jong, K.A. (1995b) Evolving complex structures via cooperative co-evolution. *Proceedings of the 4th Annual Conference on Evolutionary Programming*, The MIT Press, Cambridge, MA, pp. 307–317.

Robertson, G. (1987) Parallel implementation of genetic algorithms in classifier systems. In *Genetic Algorithms and Simulated Annealing*, L. Davis (ed.), Pitman, London, pp. 129–140.

Schaffer, J.D. (1985) Multiple objective optimizations with vector evaluated genetic algorithm. *Proceedings of International Conference on Genetic Algorithms and their Applications*, J. Grefenstett (ed.), Morgan-Kaufmann, New York, pp. 93–100.

Sekaj, I. and Perkacz, J. (2007) Some aspects of parallel genetic algorithms with population re-initialization. *IEEE Congress on Evolutionary Computation*, pp. 1333–1338.

Skolicki, Z. and De Jong, K. (2005) The influence of migration sizes and intervals on island models. *Proceedings of GECCO '05*, Washington, DC, June 25–29.

Smith, R.E., Forrest, S. and Perelson, A.S. (1993) Searching for diverse, cooperative populations with genetic algorithms, *Evolutionary Computation Journal*, 1(2), 127–149.

Srinivas, N. and Deb, K. (1994) Multi-objective function optimization using non-dominated sorting genetic algorithms, *Evolutionary Computation*, 2(3), 221–248.

Tsypkin, Y.Z. (1971) *Adaptation and Learning in Automatic Systems*, Z.J. Nikolic (transl.), Academic Press, New York.

Valdez, F., Melin, P. and Parra, H. (2011) Parallel genetic algorithms for optimisation of modular neural networks in pattern recognition. *Proceedings of the IEEE International Joint Conference on Neural Networks*, San Jose, CA, pp. 314–319.

Van Veldhuizen, D.A. and Lamont, G.B. (2000) Multi-objective optimisation with messy genetic algorithms. *Proceedings of the 2000 ACM Symposium on Applied Computing*, Como, Italy, pp. 470–476.

Watson, R.A. and Pollack, J.B. (1999) How symbiosis can guide evolution. *Proceedings of European Conference on A Life V*, D. Florean, J.-D. Nicoud and F. Mondada (eds), Springer-Verlag, Berlin, pp. 29–38.

Watson, R.A. and Pollack, J.B. (2001) Symbiotic composition and evolvability. *Proceedings of European Conference on ALife VI*, J. Kelemen and P. Sosik (eds), Springer-Verlag (LNAI), Berlin, pp. 480–490.

Whitley, D., Gordon, V.S. and Mathias, K. (1994) Lamarckian evolution, the Baldwin effect and function optimisation. In *Parallel Problem Solving from Nature – PPSN III*, Y. Davidor, H.-P. Schwefel and R. Manner (eds), Springer-Verlag, Berlin, pp. 6–15.

Yang, Y., Vincent, J. and Littlefair, G. (2004) *A Coarse-Grained Parallel Genetic Algorithm Employing Cluster Analysis for Multi-modal Numerical Optimisation*, Lecture Notes in Computer Science, Vol. 2936, Springer-Verlag, Berlin, pp. 229–240.

Zitzler, E. and Thiele, L. (1999) Multi-objective evolutionary algorithms: a comparative case study and the strength Pareto approach, *IEEE Transactions on Evolutionary Computation*, 3(4), 257–271.

Zitzler, E., Laumanns, M. and Thiele, L. (2001) SPEA2: Improving the strength Pareto evolutionary algorithm. *Proceedings of the Evolutionary Methods for Design, Optimisation, and Control with Applications to Industrial Problems*, EUROGEN 2001, Athens, Greece, pp. 95–100.

8

Evolutionary Fuzzy Systems

8.1 Introduction

Although fuzzy systems have been applied successfully to many complex industrial processes, they experience a deficiency in knowledge acquisition and rely to a great extent on empirical and heuristic knowledge, which, in many cases, cannot be elicited objectively. One of the most important considerations in designing fuzzy systems is the construction of the membership functions for each linguistic variable, as well as the rule base. In most existing applications, the fuzzy rules are generated by an expert in the area, especially for control problems with only a few inputs. The correct choice of MFs is by no means trivial but plays a crucial role in the success of an application. Previously, the generation of MFs had been a task mainly done either interactively, by trial and error, or by human experts. With an increasing number of inputs and linguistic variables, the possible number of rules for the system increases exponentially, which makes it difficult for the experts to define a complete set of rules and associated MFs for a reasonable performance of the system. There are many different methodologies available in the literature for systematic design of fuzzy systems (FS) and especially fuzzy logic controllers (FLC). However, three main methods have emerged: nonlinear systems analysis, neural fuzzy and direct optimization. The nonlinear systems analysis approach is beyond the scope of this book and not addressed further here. Interested readers are referred to Vidyasagar (2002). The neural fuzzy approach has been in widespread use and very popular among researchers. This approach will be discussed in Chapter 9. The direct optimization approach is mainly a direct search strategy for finding optimality either in the set of all design parameters or only a subset of parameters.

In the direct search approach, the design of a fuzzy system can be formulated as a search problem in a high-dimensional space where each point in the space represents a rule set, membership function, scaling functions and the corresponding system performance, that is, the performance of the system forms a hypersurface in the space according to given performance criteria. Thus, finding the optimal location of this hypersurface is a search problem, which is equivalent to developing the optimal fuzzy system design (Shi *et al.*, 1999). These characteristics make evolutionary algorithms a suitable method for searching the hypersurface rather than many other stochastic or derivative-free optimization methods such as simulated annealing (Kirkpatrick *et al.*, 1983), tabu-search (Glover, 1989), random search method (Matyas, 1965)

Computational Intelligence: Synergies of Fuzzy Logic, Neural Networks and Evolutionary Computing, First Edition.
Nazmul Siddique and Hojjat Adeli.
© 2013 John Wiley & Sons, Ltd. Published 2013 by John Wiley & Sons, Ltd.

and downhill simplex method (Nelder and Mead, 1965). In contrast to smart heuristics such as simulated annealing, tabu-search, random search and downhill simplex search are local search techniques which use a generate-and-test search manipulating one feasible solution based on a physical rather than a biological analogy. The EA works in parallel on a number of search points (potential solutions) and not on a unique solution, which means that the search method is rather global over the entire search space. An EA-based optimization technique has been applied to many control engineering problems with significant success in improving the system performance (Goldberg, 1989; Whidborne and Istepanian, 2001; Chou, 2006; Kwon and Sudhoff, 2006; Loop *et al.*, 2010).

Solving a particular optimization task using an EA requires the designer to decide on five components (also discussed in Chapter 6): (i) genetic (chromosome) representation of the solution space; (ii) creation of an initial population representative of the entire solution space; (iii) definition of a fitness function capable of describing the quality of solution; (iv) selection of an appropriate set of genetic operators; and (v) appropriate choice of EA parameter values such as population size, maximum number of generations and probabilities of genetic operators. Each of these components has parameters. There is no straightforward way to determine the EA parameter values. Empirical methods like trial-and-error are common practice in evolutionary computing, which is a time-consuming task. Therefore, a systematic approach to the choice of EA parameter values is very demanding. Considerable efforts have been made in developing heuristics for choosing these parameter values as well (Davis, 1989; Fogarty, 1989; Srinivas and Patnaik, 1994; Yun and Gen, 2003). The two issues discussed above lead to two synergistic combinations of fuzzy systems and evolutionary algorithms:

 (i) Evolutionary adaptive fuzzy systems,
(ii) Fuzzy adaptive evolutionary algorithms.

The objective of an EA adaptive fuzzy system is to adapt knowledge in fuzzy system design. The knowledge base of a fuzzy system is not a monolithic structure, rather it is composed of information about MFs, scaling parameters and the rule base. In an EA adaptive fuzzy system, the EA collaborates with a fuzzy system to tune, optimize or learn the parameters, membership functions and rule base of a fuzzy system. To assess the performance of the fuzzy system (or FLC), a plant is embedded within the loop. Owing to the computational effort and time required for an EA, it is obviously an offline approach. The EA adaptive fuzzy

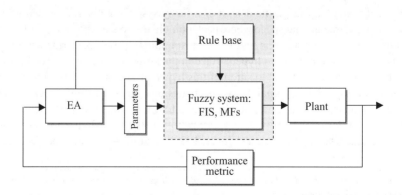

Figure 8.1   A generic EA adaptive fuzzy system

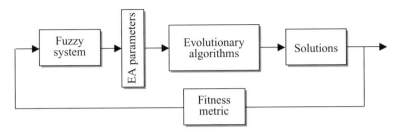

Figure 8.2   A generic fuzzy adaptive EA system

collaboration is shown in Figure 8.1. In fuzzy adaptive EA, the fuzzy system (or FLC) helps in adapting or controlling different parameters of the EA during the run, which has a decisive influence on the performance of the EA. The performance of the EA can be estimated from the quality of the solutions (e.g., best fitness, average fitness, convergence speed, etc.) and based on the estimation, the EA parameters are adjusted. This is obviously an online approach. The fuzzy EA combination is shown in Figure 8.2. This chapter will focus mainly on these two synergistic combinations.

8.2   Evolutionary Adaptive Fuzzy Systems

Large dimensionality, strong nonlinearity, non-differentiability, and noisy and time-varying objective functions are the associated factors involved in the optimization problem of a fuzzy system, which lead to difficult optimization tasks. In real-world situations the objective functions and constraints are often not analytically treatable or even available in a closed form (Baeck, 1996). The traditional approach in this class of problems is to develop a formal model that resembles the original functions closely enough in a real system, solvable by means of traditional mathematical methods. Most of these mathematical methods are gradient-based approaches. The unfortunate thing is that no gradient-based descent optimization algorithm is guaranteed to find the global optimum of a complex objective function within a finite period of time (Jang *et al.*, 1997). Moreover, selecting the initial points for the deterministic methods clearly has a decisive effect on the final result of the optimization algorithms being used. In practice, knowing such initial points is nearly impossible. Some of the issues are discussed in Chapter 7. If the initial points are to be randomly selected, then there is no strong argument for supporting the choice of employing a deterministic method. It would rather be better to employ a stochastic method that perturbs the final points when the method converges. Therefore, the optimization approach being used must be stochastic in nature. Furthermore, if the calculation of the gradient is time-consuming or difficult due to the complexity of the objective function, then the stochastic or derivative-free optimization methods mentioned earlier should be chosen. EAs are stochastic optimization methods based on the principle of Darwinian evolution theory. Different EAs are discussed in Chapter 6.

Efforts have been made to automate the construction and adjustment of MFs (Karr and Gentry, 1993), rule bases (Ishibuchi *et al.*, 1995; Chin and Qi, 1997) and scaling factors (Ahmed *et al.*, 2012) in various ways using evolutionary algorithms. In most cases, either the rule base is fixed and the parameters of the MFs are adjusted or the MFs are fixed and evolutionary algorithms optimize the rule base (Linkens and Nyongesa, 1995a,b). Some researchers have optimized the rule base, the MFs, scaling factors and system/controller

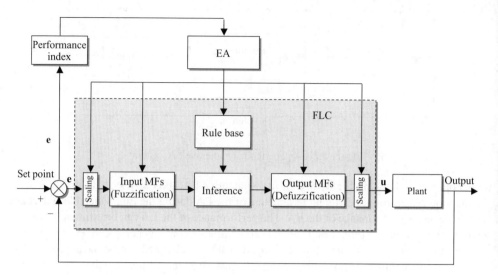

Figure 8.3   EA-based synthesis/optimization of FLC

parameters (Homaifar and McCormick, 1995), which seems somewhat redundant. A block diagram of the EA-based synthesis, optimization and tuning of a fuzzy system (e.g., a fuzzy controller) is shown in Figure 8.3.

The performance of the fuzzy systems is aggregated into a scalar performance index on which basis the EA selects the outperforming rule base, MFs or scaling parameters or their combinations. It is important in the design of the EA fuzzy system to decide which parts of the fuzzy system are subject to optimization (i.e., the different blocks in Figure 8.3). Different types of fuzzy system (e.g., Mamdani, Sugeno and Tsukamoto) are discussed in Chapters 2 and 3. EA fuzzy systems are discriminated along two main approaches (Hoffmann, 2000):

 (i) Evolutionary tuning of fuzzy system,
(ii) Evolutionary learning of fuzzy system.

Tuning deals with optimization of an existing fuzzy system with predefined MFs, rule base and scaling parameters, whereas learning deals with automated design (synthesis) of the fuzzy system, carrying out an elaborate search for a set of MFs, rule base and scaling functions which ensures an optimal performance of a fuzzy system.

8.2.1   Evolutionary Tuning of Fuzzy Systems

In general, the design of the MFs and rule base of a fuzzy system is carried out with the help of human experts with domain knowledge and experience. Very often, the design of the MFs and construction of the rule base using expert knowledge do not reflect the actual data distribution of the system, which is the main reason behind the poor performance. A readjustment or tuning of the MFs and rule base is essential for the improvement in performance and robustness of operation of the fuzzy system over the entire data range in changing operating conditions.

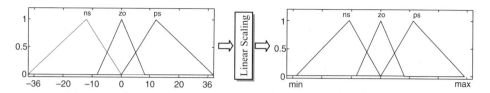

Figure 8.4 Normalization of MFs onto fixed range $E = [\min, \max]$

Therefore, a further enhancement of the performance is sought by tuning the scaling functions, input and output MFs, and rule base.

8.2.1.1 Tuning of Scaling Functions

Very often the designer uses normalized domains (universe of discourse) for input/output MFs for ease of design process and to reduce the design time, which again requires undergoing a scale transformation. The transformation maps the physical values of the process state variables into a normalized domain. The normalization of the input range of respective variables into a fixed relative distribution is shown in Figure 8.4. This is input normalization. Consequently, output denormalization maps the normalized values of the control output variables into their respective physical domains. Scaling functions transform the input and output MFs into the universe of discourse within which MFs are defined. The simple way to do this is to parameterize the scaling functions by a single scaling factor, or a lower and an upper limit in case of linear scaling and a contraction factor in case of nonlinear scaling. A linear scaling function of the form is given by:

$$f(x) = k_1 x + k_2, \ \forall x \in X, \ X = [X_{\min}, X_{\max}] \tag{8.1}$$

The scaling factor k_1 widens or reduces the operating range of the respective input or output variable and the corresponding gain. The parameter k_2 is an offset and shifts the operating range of the corresponding variable. Scaling allows it to define the MF over a normalized universe of discourse. Such linear scaling of MFs is shown in Figure 8.4. A disadvantage of the linear scaling function is that the MFs are distributed within a fixed upper and lower limit. The scaling does not affect the shape of the MFs. Nonlinear scaling provides a solution to this problem, whereby it modifies the relative distribution and shape of the MFs. Nonlinear scaling of the respective MFs into a fixed relative distribution with changed shape is shown in Figure 8.5. A nonlinear scaling function is of the form

$$f(x) = \text{sgn}(x) . |x|^{\alpha}, \text{ with } \alpha > 0, \ \forall x \in X, \ X = [X_{\min}, X_{\max}], \tag{8.2}$$

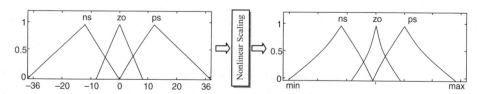

Figure 8.5 Nonlinear scaling to change shape of MFs

The parameter α increases (for $\alpha > 1$) or decreases (for $\alpha < 1$) the sensitivity in the region around the origin and has opposite effect at the boundaries of the operating range. Interested readers are directed to the book by Cordon *et al.* (2001) for a detailed description of the linear and nonlinear scaling functions.

From a control engineering point of view, the scaling factors describing the particular input normalization and output denormalization play an equivalent role, like that of the gain coefficients in a conventional controller. The gain coefficients are responsible for the performance defined in terms of rise time, settling time, steady-state error, overshoot and undershoot, stability, oscillation and deteriorated damping effects of the system.

There are basically two major approaches to the determination of the scaling factors: (i) heuristic and (ii) analytical. The heuristic approach is a trial-and-error method. Daugherity *et al.* (1992) gave a rule of thumb for tuning the scaling factors of a fuzzy system, particularly a fuzzy controller, where the performance criteria are defined in terms of desired value of rise time, overshoot and amplitude of oscillation and heuristic production rules are used for adjusting the scaling factors. The production rules can be explained as follows: if the system response is slower than the desired response, then the change of rise time should be positive. This means we need to increase the effect of error on the system. Therefore, we increase the proportional scaling factor. Similarly, if the overshoot or amplitude of oscillation is higher than the desired overshoot or amplitude of oscillation, then it needs to increase the effect of change of error, hence we increase the derivative scaling factor. The analytical approach aims to establish a relationship between the scaling factors and the closed-loop behaviour of the control process. In this case it is assumed that a mathematical model of the systems under control is available and the fuzzy model is considered as a nonlinear transfer element (Driankov *et al.*, 1993). Both the heuristic and analytical methods do not guarantee a global optimum value for the scaling factors. Therefore, an evolutionary tuning is desirable, which can ensure a global optimum and robustness of the solution of the fuzzy system (or the fuzzy controller).

Evolutionary tuning of the scaling function requires parameterization, which is done by means of two parameters discussed above: a scaling factor and an offset. Scaling corresponds to a linear transformation of the interval $[a, b]$ onto a normalized interval, e.g., $[-1, 0], [-1, +1]$, $[0, +1]$. A single scaling factor maps the interval $[-a, +a]$ to a symmetric normalized interval $[-1, +1]$. The scaling factor does not necessarily map the interval $[-a, +a]$ onto a normalized interval. It can be any arbitrary interval within the operating range of the inputs and outputs. Once this is achieved, the evolutionary algorithm needs the parameters to be encoded into a suitable chromosome representation within the upper and lower limits of the operating range. Partial coding of the individual parameters is concatenated to form the chromosome. For example, consider the two-input single-output fuzzy controller with scaling functions k_e, $k_{\Delta e}$ and k_u in Figure 8.6.

For linear scaling, the functions k_e, $k_{\Delta e}$ and k_u are parameterized according to Equation (8.1) as follows:

$$k_e = f(e) = k_1 e + k_2 \tag{8.3}$$

$$k_{\Delta e} = f(\Delta e) = k_3 \Delta e + k_4 \tag{8.4}$$

$$k_u = f(u) = k_5 u + k_6 \tag{8.5}$$

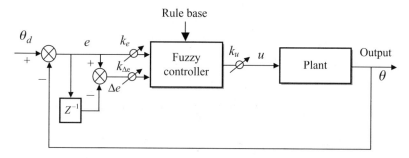

Figure 8.6 PD-like fuzzy controller with scaling functions

The performance of the fuzzy controller can be improved by optimizing the different scaling factors and offset values of k_i, $i = 1, 2, 3, \ldots, 6$. In general, most designers set the values of $\{k_2, k_4, k_6\}$ to zero and tune the scaling factors $\{k_1, k_3, k_5\}$ by trial and error, which is time-consuming. The other option is to optimize k_i, $i = 1, 2, 3, \ldots, 6$ by EA. The chromosome representation will look like that in Figure 8.7. k_i, $i = 1, 2, 3, \ldots, 6$ can be encoded as binary or real depending on the required precision for the system.

The objective can thus be expressed in terms of minimization of the system performance indices or in terms of desired value of rise time, overshoot and amplitude of oscillation, which are in common use. These include the integral of absolute error (IAE), integral of square error (ISE) and integral of time-weighted absolute error (ITAE). These criteria eventually include all three performance indices (rise time, overshoot and oscillation) implicitly in this definition. For example, in the case of ITAE, it is defined as

$$J(p) = \sum_{t=1}^{T} \Delta t \cdot |e(t)| \tag{8.6}$$

where $e(t)$ is the output error of the system. T is some reasonable number of time units by which the system can be assumed to have settled quite close to a set point. Δt takes care of ensuring a reasonable rise time and settling time for the FLC. Obviously the objective is to minimize $J(p)$ subject to the parameter set $p = \{k_1, k_2, \ldots, k_6\}$. Different objective functions suitable for any FLC or fuzzy system design are described in detail in Section 8.3.1.

Any standard genetic operators (such as crossover, mutation and selection) can be applied to the chromosome representation. Different crossover and mutation operators on binary and real-valued chromosomes are discussed in detail in Chapter 6.

$$\boxed{\{k_1, k_2, k_3, k_4, k_5, k_6\}}$$

Figure 8.7 Chromosome representation of scaling factors and offsets

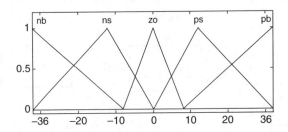

Figure 8.8   Mapping MFs onto domain $E = [-36, +36]$

8.2.1.2   Tuning of MFs

For computational efficiency and the need for performance analysis, a uniform representation of the MFs is required. Such uniform representation is achievable by employing MFs with uniform shape, parametric and functional definition. As discussed in Chapter 2, the most widely used and popular MFs are triangular, trapezoidal, sigmoidal, Gaussian, bell-shaped, etc. MFs can be classified into two main groups: piecewise linear and differentiable. Among piecewise linear MFs are the triangular and trapezoidal MFs while differentiable MFs are Gaussian, radial-basis, sigmoidal and bell-shaped. There is no exact method of selecting the shape of MFs. The choice of shape of MFs depends mainly on heuristic rules such as ease of parametric and functional description, computational cost and efficiency of manipulation of MFs. For example, triangular MFs are the most economic according to this heuristic rule. Once the shape is selected, one has to map the MFs onto the corresponding domain. For example, the input variable error of a fuzzy controller mapped onto the domain $E = [-36, +36]$ is shown in Figure 8.8.

This simple mapping of the MFs onto the domain for each linguistic term may not guarantee the optimal performance of the fuzzy system or an FLC. There are many reasons for this, for example, the shape of MFs, number of MFs, influence of overlapping (cross-point) of MFs, symmetry, width of MFs and distribution of MFs (Kovacic and Bogdan, 2006). Various combinations of shapes and distributions of MFs result in a varying performance of fuzzy systems. If the overlap between two neighbouring MFs is at a cross-point of 0 as shown in Figure 8.9, then a single rule is fired at a time (Driankov, 1993). In general, if the cross-point of two adjacent MFs is 0.5 as shown in Figure 8.10, then a fuzzy controller provides faster rise time, significantly less overshoot and less undershoot. In such cases, the shape of the MFs does not play a dominant role in the performance of a fuzzy system. It is found that the trapezoidal

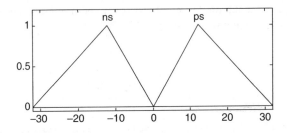

Figure 8.9   Overlapping of MFs at cross-point of 0

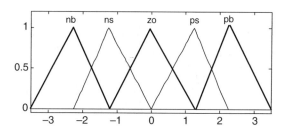

Figure 8.10   Overlapping of MFs at cross-point of 0.5

shape of MFs causes slower rise time (Driankov, 1993). Though these results are empirical in nature, in general an overlap of two adjacent MFs at a cross-point of 0.5 provides the best results, as has been reported by many researchers (Boverie *et al.*, 1991; Yager and Filev, 1994; Margaliot and Langholz, 2000; Kovacic and Bogdan, 2006).

The width of the MF is the sum of the left and right width of the MF. The left width is the length of the interval from the peak value to the point to the left where the membership value is zero and the right width is the length of the interval from the peak value to the point to the right where the membership value is zero. If the left width is equal to the right width then the MF is symmetric, otherwise it is asymmetric. The width and symmetry of MFs also play an influential and decisive role in the defuzzified value of the consequent MFs when using the centre-of-gravity or mean-of-maxima method of defuzzification. For example, the resulting defuzzified value will be close to the peak of the MF when a symmetrical MF is used in the case of a single-rule system. In contrast, the defuzzified value drifts away from the peak of the MF when an asymmetric MF is used (Driankov, 1993). The influence of the symmetric and asymmetric shape of MFs on defuzzification is shown in Figure 8.11.

To translate the membership functions to a representation useful as genetic material, they are parameterized with one to four coefficients and each of these coefficients constitutes a gene of the chromosome for evolutionary algorithms.

In fuzzy system design, one can frequently assume triangular MFs for which each MF can be specified by just a few parameters. In the case of a triangular MF, it is determined by three parameters: left position, peak and right position. Once the choice of MF is made, the input/output MFs of the variables are mapped onto the domain as shown previously in Figure 8.8. As discussed above, an overlapping (not more than 0.5) of the MFs is desired

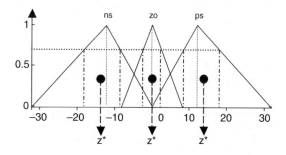

Figure 8.11   Influence of symmetric and asymmetric shape of MFs

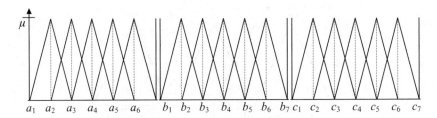

Figure 8.12   Parameterized input/output MFs

to ensure a good performance of the fuzzy system. Therefore, the left and peak position of the next MF is the same as the peak and right position of the previous MF. An example of parameterized triangular MFs of a two-input single-output fuzzy system with five MFs for each variable (e.g., the FLC in Figure 8.6) is shown in Figure 8.12. Seven parameters are needed to define five MFs for each input and output, that is, the five MFs each having 3-tuple parameters.

Input 1: $\{(a_1, a_2, a_3), (a_2, a_3, a_4), (a_3, a_4, a_5), (a_4, a_5, a_6), (a_5, a_6, a_7)\}$

Input 2: $\{(b_1, b_2, b_3), (b_2, b_3, b_4), (b_3, b_4, b_5), (b_4, b_5, b_6), (b_5, b_6, b_7)\}$

Output: $\{(c_1, c_2, c_3), (c_2, c_3, c_4), (c_3, c_4, c_5), (c_4, c_5, c_6), (c_5, c_6, c_7)\}$

There are 21 parameters in total for all inputs and outputs. A reduction in the number of parameters can be achieved by fixing the upper and lower limits of the domain for each input and output, as shown in Figure 8.13(a). Hence, the entire chromosome for MFs looks like in Figure 8.13(b).

$\{a_i, b_i, c_i\}$ with $i = 1, 2, 3, \ldots, 5$ can be encoded as binary or real depending on the precision required for the system. The problem associated with binary or integer coding is

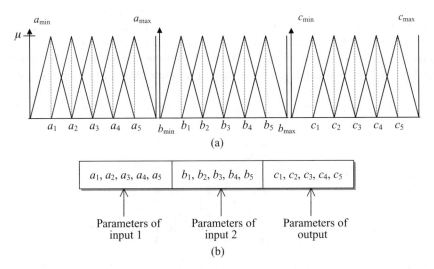

Figure 8.13   Chromosome representation for MFs. (a) Fixed upper and lower limit of MFs; (b) Chromosome representation of MFs

that it brings in inaccuracy when representing high-precision real-value MF parameters and even difficulties in mapping from genotypes to phenotypes. Moreover, a long bit string always occupies computer memory even though only a few bits are actually involved in the crossover and mutation operations. This is particularly the case when a large number of parameters need to be adjusted for the same problem and higher precision is required for the final result. To overcome the problem of inefficient use of computer memory, the real-valued chromosome representation of membership functions is preferred. Hatanaka *et al.* (2004) used GA to learn the MFs of a fuzzy system for nonlinear system identification. A real-valued encoding was used for the parameters of the trapezoidal MFs. For Gaussian, bell-shaped or sigmoidal MFs, it is the centre and width of the MFs that are used as the parameters of the MFs and coded into the chromosome as genetic material. The parametric Gaussian, bell-shaped and sigmoidal MFs are discussed in Chapter 2.

Application of genetic operators such as crossover, mutation and selection to real-valued or binary string chromosome representation is very straightforward and discussed in detail in Chapter 6.

The objective function can be defined in terms of minimization of the system performance indices, such as sum of squared error or mean squared error. A minimum of the sum of squared error or mean squared error does not guarantee other performance indices of the fuzzy controller. For example, the performance index can be defined in terms of a desired value of rise time, overshoot and amplitude of oscillation or steady-state error for a fuzzy controller. Widely used objective functions include the integral of absolute error (IAE), integral of square error (ISE) and integral of time-weighted absolute error (ITAE). Obviously the objective is to minimize $J(p)$ subject to the parameter set $p = \{\theta_1, \theta_2, \ldots, \theta_n\}$. Different objective functions suitable for the design of a fuzzy controller or fuzzy system are described in detail in Section 8.3.1.

8.2.1.3 Tuning of Rule Base

The fuzzy rule base is the backbone of any fuzzy system. The success and performance of a fuzzy system depends largely on the rule base. The rule base consists of if–then rules (also known as fuzzy implication, fuzzy conditional statement). An obvious question is how a set of rules can be derived. The designer depends mainly on expert knowledge or skilled operators available to provide the necessary knowledge. When no expert or skilled operators available, the rule base is constructed by operating the process directly. Moreover, the number of rules in a fuzzy system grows with the number of input/output variables and linguistic terms for each variable. As a result, it is difficult for a human expert to suggest a combination of input terms for an output variable as there could be certain combinations of input variables that do not appear in the dynamical system during operation. Rules constructed in this way are actually influenced by subjective decisions of the expert operator and result in incomplete, inconsistent, partly incorrect, redundant and sometimes useless rules, especially when the operating conditions are changed. This has great influence on the performance or optimal performance, for which it is required to refine and tune the rough rules. The refining or tuning of the fuzzy rule base can be formulated as a search problem in a high-dimensional space where each point in the space represents a rule set, MF and the corresponding system performance, that is, the performance of the system forms a hypersurface in the space according to given performance criteria. Owing to the difference in presentation of the rule base in Mamdani- and Sugeno-type fuzzy systems, EA-based tuning is applied in two different ways for Mamdani- and TSK-type (widely known

Table 8.1   Rule base for two-input one-output Mamdani system

	Input 2 (Y)				
Input 1 (X)	NB	NS	ZO	PS	PB
NB	PB	PB	PB	PS	ZO
NS	PB	PS	ZO	ZO	NS
ZO	PS	ZO	ZO	ZO	NS
PS	PS	ZO	ZO	NS	NB
PB	ZO	NS	NB	NB	NB

as Sugeno-type) fuzzy systems. It is mainly the difference in chromosome representation. The chromosome representation of the rule base for Mamdani-type fuzzy systems is carried out in the following way.

The linguistic variables are represented by integer values, for example 1 for NB, 2 for NS, 3 for ZO, 4 for PS and 5 for PB. Applying this code to the fuzzy rule base shown in Table 8.1, the encoded rule base shown in Table 8.2 is obtained. A chromosome is thus obtained from the decision table by going through row-wise and coding each output fuzzy MF as an integer in $\{1, 2, \ldots, n\}$, where n is the maximum number used to label the MFs defined for the output variable of the fuzzy system. In the case of $n = 5$, i.e., five linguistic variables (or MFs), the chromosome for the rule base of a Mamdani-type fuzzy system is shown in Figure 8.14.

The problem associated with binary coding of the rule base is that a long string always occupies the computer memory even though only a few bits are actually involved in the crossover and mutation operations. Another problem with binary coding is encountered in chromosome representation of the rule base when a mutation operation is applied. Mutation applied to a linguistic code of the rule base alters it to another valid linguistic code, which is restricted to a linguistic distance of two, i.e., up a level or down a level. This is illustrated

Table 8.2   Encoding of the rule base

	Input 2 (Y)				
Input 1 (X)	NB	NS	ZO	PS	PB
NB	5	5	5	4	3
NS	5	4	4	3	2
ZO	4	3	3	3	2
PS	4	3	2	2	1
PB	3	2	1	1	1

$$\{5\,5\,5\,4\,3\,|\,5\,4\,4\,3\,2\,|\,4\,3\,3\,3\,2\,|\,4\,3\,2\,2\,1\,|\,3\,2\,1\,1\,1\}$$

1st row   2nd row   3rd row   4th row   5th row

Figure 8.14   Chromosome representation of the rule base

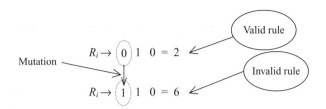

Figure 8.15   Problem in rule-base mutation using binary coding

in Figure 8.15. It requires three bits to represent integer values from 1 to 5 for five linguistic variables. Performing mutation on a single bit of the chromosome, it can change to a value 6, which is not a valid linguistic variable at all and thus will cause disruption to the fuzzy system. Such a jump of the value by mutation will be difficult to manage in binary coding. Integer coding is much easier to handle in any programming language. Therefore, an integer-valued coding is suggestive for chromosome representation of the rule base.

The chromosome representation of the rule base in a Sugeno-type fuzzy system will be different from that in a Mamdani-type fuzzy system as there are two different sets of parameters that are subject to tuning, i.e., parameters of the antecedent MFs and consequent parameters of the linear function. Assume the rule of the two-input single-output Sugeno-type fuzzy system

$$\text{IF } X \text{ is } A_i \text{ and } Y \text{ is } B_j \text{ THEN } Z_k = a_k X + b_k Y + c_k$$

with $i = 1, \ldots, N$, $j = 1, \ldots, N$, $k = 1, \ldots, M$ and $M \in N \times N$. The antecedent part of the Sugeno-type fuzzy system is similar to that of the Mamdani-type fuzzy system and it is the consequent linear function Z_k and its parameters $\{a_k, b_k, c_k\}$ that are different. There are two possible ways the rules can be tuned for a Sugeno-type fuzzy system: firstly, tuning of the linear function Z_k when the parameters $\{a_k, b_k, c_k\}$ are defined and secondly, tuning of the parameters of the linear function $\{a_k, b_k, c_k\}$. Tuning of the linear function Z_k would be the same as tuning Mamdani-type fuzzy rules where MFs of the consequent part of the Mamdani-type fuzzy system are replaced with Z_k. In other words, the linguistic variables (for example, in Table 8.1) are replaced with Z_k. For $k = 1, \ldots, 5$, the rule base will look like that shown in Table 8.3. Z_k $(k = 1, \ldots, 5)$ can be represented by integer values, for example 1 for Z_1, 2 for Z_2, 3 for Z_3, 4 for Z_4 and 5 for Z_5. Applying this code to the fuzzy rule base shown in Table 8.3, the encoded rule base shown in Table 8.4 is obtained. A chromosome

Table 8.3   Rule base for two-input single-output Sugeno system

Input 1 (X)	Input 2 (Y)				
	NB	NS	ZO	PS	PB
NB	Z_5	Z_5	Z_5	Z_4	Z_3
NS	Z_5	Z_4	Z_4	Z_3	Z_2
ZO	Z_4	Z_3	Z_3	Z_3	Z_2
PS	Z_4	Z_3	Z_2	Z_2	Z_1
PB	Z_3	Z_2	Z_1	Z_1	Z_1

Table 8.4 Encoding of the rule base

Input 1 (X)	Input 2 (Y)				
	NB	NS	ZO	PS	PB
NB	5	5	5	4	3
NS	5	4	4	3	2
ZO	4	3	3	3	2
PS	4	3	2	2	1
PB	3	2	1	1	1

$$\{5\,5\,5\,4\,3\,|\,5\,4\,4\,3\,2\,|\,4\,3\,3\,3\,2\,|\,4\,3\,2\,2\,1\,|\,3\,2\,1\,1\,1\}$$

1st row   2nd row   3rd row   4th row   5th row

Figure 8.16 Chromosome representation of the rule base of a Sugeno-type system

is thus obtained from the decision table by going through row-wise and coding each output function Z_k as an integer in $\{k = 1, \ldots, M\}$, where M is the maximum number used to label the linear functions defined for the output variable of the fuzzy system. In the case of $M = 5$, the chromosome for the rule base of a Sugeno-type fuzzy system is shown in Figure 8.16.

Since Z_k is a function of the parameters $\{a_k, b_k, c_k\}$, the second option of carrying out the tuning process would be to replace the Z_k with respective 3-tuple parameters in the rule base. Applying this, the rule base in Table 8.3 would look like that in Table 8.5. A chromosome is thus obtained from the decision table by going through the 3-tuple parameter sets and choosing the parameters of the five output functions Z_k, $\{k = 1, \ldots, M\}$, where $M = 5$ is the maximum number used to label the linear functions. In this case, the chromosome for the rule base of the Sugeno-type fuzzy system is shown in Figure 8.17, where $\{a_k, b_k, c_k\} \in \Re$ (set of real numbers).

An alternative would be to choose the parameters of the 25 output functions Z_k, $\{k = 1, \ldots, M\}$, with $M = 25$. In this case, the chromosome would consist of 25 times 3-tuple parameter sets shown in Figure 8.18, i.e., there will be 75 genes in the chromosome, which

Table 8.5 Rule base for two-input single-output Sugeno system

Input 1 (X)	Input 2 (Y)				
	NB	NS	ZO	PS	PB
NB	$\{a_5, b_5, c_5\}$	$\{a_5, b_5, c_5\}$	$\{a_4, b_4, c_4\}$	$\{a_4, b_4, c_4\}$	$\{a_3, b_3, c_3\}$
NS	$\{a_5, b_5, c_5\}$	$\{a_4, b_4, c_4\}$	$\{a_3, b_3, c_3\}$	$\{a_3, b_3, c_3\}$	$\{a_2, b_2, c_2\}$
ZO	$\{a_5, b_5, c_5\}$	$\{a_4, b_4, c_4\}$	$\{a_3, b_3, c_3\}$	$\{a_2, b_2, c_2\}$	$\{a_1, b_1, c_1\}$
PS	$\{a_4, b_4, c_4\}$	$\{a_3, b_3, c_3\}$	$\{a_3, b_3, c_3\}$	$\{a_2, b_2, c_2\}$	$\{a_1, b_1, c_1\}$
PB	$\{a_3, b_3, c_3\}$	$\{a_2, b_2, c_2\}$	$\{a_2, b_2, c_2\}$	$\{a_1, b_1, c_1\}$	$\{a_1, b_1, c_1\}$

$$\{a_1,b_1,c_1\}, \{a_2,b_2,c_2\}, \{a_3,b_3,c_3\}, \{a_4,b_4,c_4\}, \{a_5,b_5,c_5\}$$

Figure 8.17 Chromosome representation of the rule base of a Sugeno-type system

$$\{a_1,b_1,c_1\}, \{a_2,b_2,c_2\}, \{a_3,b_3,c_3\}, \{\cdots\} ,\ldots, \{\cdots\}, \{a_{25},b_{25},c_{25}\}$$

Figure 8.18 Chromosome representation of the rule base of a Sugeno-type system

would be an exhaustive search and make the tuning process somewhat redundant.

Though the rule base is the core of a fuzzy controller, the choice of the linguistic terms, MFs, number of MFs, their shape and distribution also greatly influence the behaviour of the fuzzy controller (Chang *et al.*, 1991; Kovacic and Bogdan, 2006; Qi and Chin, 1997). The optimal parameters of the MFs in the rule base will reflect the performance of the fuzzy controller. Therefore, the objective function can be defined in terms of minimization of the system performance indices. For example, it can be defined in terms of the desired value of the rise time, overshoot and amplitude of oscillation. An indirect measure of all the performance indices is the sum squared error for a fuzzy controller.

Most researchers apply standard genetic operators such as crossover, mutation and selection on the chromosome representation, shown in Figure 8.18, when a real-valued or binary encoding scheme is used.

Example 8.1 Consider a Sugeno-type fuzzy system with two inputs and a single output. The fuzzy system is described by the two input MFs shown in Figure 8.19, where $x_1 \cong$ error and $x_2 \cong$ change of error and the output functions defined by $z_1 = a_1x_1 + b_1x_2 + c_1$, $z_2 = a_2x_1 + b_2x_2 + c_2$, $z_3 = a_3x_1 + b_3x_2$ and $z_4 = a_4x_1 + b_4$, where $\{a_1, b_1, c_1\}$, $\{a_2, b_2, c_2\}$, $\{a_3, b_3\}$ and $\{a_4, b_4\}$ are the consequent parameters. An EA is to be applied for tuning the rule base of the fuzzy system assuming that the MFs $\{A_1, A_2, B_1, B_2\}$ do not need any further tuning. The behaviour of the system is described by the rule base shown in Table 8.6.

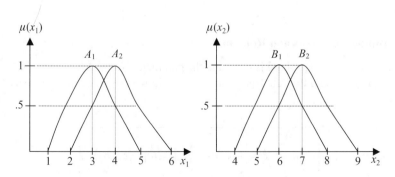

Figure 8.19 Membership functions for x_1 and x_2

Table 8.6   Rule base for a Sugeno-type FLC

	$x_2 \cong$ change of error	
$x_1 \cong$ error	B_1	B_2
A_1	z_1	z_3
A_2	z_2	z_4

Table 8.7   Rule base represented by the consequent parameters of the FLC

	$x_2 \cong$ change of error	
$x_1 \cong$ error	B_1	B_2
A_1	$\{a_1, b_1, c_1\}$	$\{a_3, b_3\}$
A_2	$\{a_2, b_2, c_2\}$	$\{a_4, b_4\}$

Since the fuzzy system is restricted to some predefined fuzzy if–then rules and fuzzy sets, the given task will be realized by assuming that the structure is fixed. Since the antecedent MFs do not require any further tuning, therefore, tuning the consequent part of the rule base will suffice for tuning the rule base (Siarry and Guely, 1998). Tuning the consequent part is equivalent to tuning the parameters of the four output functions z_k, $\{k = 1, \ldots, 4\}$, where the parameters of the output functions are $\{a_1, b_1, c_1\}$ for z_1, $\{a_2, b_2, c_2\}$ for z_2, $\{a_3, b_3\}$ for z_3 and $\{a_4, b_4\}$ for z_4. Applying these parameters to the rule base in Table 8.6, we obtain the rule base shown in Table 8.7.

The chromosome representation would look like Figure 8.20, where $\{a_k, b_k, c_k\} \in \Re$ (set of real numbers). Once the chromosome is defined, application of EP, ES and GA is straightforward. Any standard genetic operators (such as crossover, mutation and selection) can be applied to the chromosome when initialized with real-valued encoding. If the performance of the fuzzy controller is measured against the behaviour of a plant, then any of the fitness measures discussed in Section 8.3.1 can be used for this problem. For example, Kang *et al.* (2000) applied EP to determine an optimal rule set and parameters for fuzzy modelling and control. EP was used to simultaneously evolve the structure and the parameters of a fuzzy rule base for a given task with no predefined assumption about the rule-base structure and parameters.

8.2.1.4   Tuning Both MFs and Rule Base

In most cases, either the rule base is fixed and the parameters of the MFs are tuned or adjusted or the MFs are fixed and the rule base is tuned or optimized by EAs. Some researchers have

$$\{a_1, b_1, c_1\}, \ \{a_2, b_2, c_2\}, \ \{a_3, b_3\}, \ \{a_4, b_4\}$$

Figure 8.20   Chromosome representation of the rule base for a Sugeno-type fuzzy system

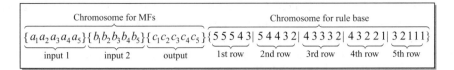

Figure 8.21 Chromosome representation of MFs and rule base

tuned the rule base, the MFs and the scaling factors simultaneously. In general, MFs are accommodated within a rule base during an optimization process. Therefore, optimizing both the MFs and the rule base is somehow redundant. In this case, a designer does not expect a drastic improvement in performance of the fuzzy system, rather a design simplification in a systematic manner.

The chromosome representation for MFs in Figure 8.13(b) and the rule base in Table 8.2 can be combined by concatenating the strings of the chromosomes for MFs and the rule base to make a simple chromosome representation. For example, such a representation is illustrated in Figure 8.21.

A homogeneous representation may be sought and binary bit strings can be used for both MFs and rule base. The homogeneous representation has the advantage that it has simple and existing genetic operators, which can be employed easily. Owing to the inaccuracy in MFs and difficulties in mapping rules, heterogeneous representation is preferred. For example, a real-valued coding scheme for MFs and binary or integer-valued coding scheme for the rule base can be used for the chromosome representation. It is more natural to represent a fuzzy system encoded in a chromosome in the way it appears in the implementation. Therefore, a heterogeneous coding scheme is proposed for the combined MFs and rule base as shown in Figure 8.21. Shi *et al.* (1999) used integers for rule base and real values for MFs. This increases complexity while applying genetic operators such as mutation. If the possible ranges of the parameters are not known in advance, conservatively large intervals for the parameter ranges are chosen (Setness and Roubos, 2000). But this does not guarantee the optimal solution lying within the chosen parameter ranges. Large parameter ranges also increase the number of bits per parameter, which makes the chromosome string too long – demanding computation time.

The crossover operation will be straightforward, but there are two different mutation operators for both parts of the chromosome string. The genes in the MF part of the chromosome will be replaced by a real value whereas the genes of the rule-base part of the chromosome will be changed either up a level or down a level from the integer value to avoid a possibly large deterioration in performance.

Any of the fitness functions defined in Section 8.3.1 can be used for evaluation of the fuzzy system. The objective should be to keep the number of rules to a minimum. A penalty term can also be included in the fitness function, as defined in Table 8.1. This will help reduce the number of rules by eliminating useless and redundant rules from the rule base.

8.2.2 Evolutionary Learning of Fuzzy Systems

One of the disadvantages of any fuzzy rule-based system is that it does not have any kind of knowledge acquisition or learning capability. Very often the acquired knowledge represents a partial, incomplete or incorrect description of the system. When acquired knowledge is not enough for the systems to be modelled or controlled, some kind of learning or adaptation is essential. Researchers have been striving to employ learning mechanisms that start from an empty or randomly generated knowledge and learn towards an optimal knowledge. Among these are neural networks simulating a fuzzy rule base and parameters and general-purpose optimization methods. Neural networks-based learning will be discussed in Chapter 10. The other possibility is learning by evolutionary methods. An influential paper by Hinton and Nowlan (1987) showed that learning can guide evolution and learning-evolution can work synergistically together. Evolutionary algorithms are general-purpose optimization algorithms that can be deployed as a learning tool for the fuzzy rule base and parameters of MFs. In this case, evolutionary learning is seen as an optimization or search problem requiring a simple scalar performance index. Evolutionary learning is suitable as it can incorporate *a priori* knowledge (Belarbi and Titel, 2000; Bonarini and Trianni, 2001). The *a priori* knowledge may be in the form of linguistic variables, MF parameters, fuzzy rules and number of rules. There can be two kinds of learning involved in this process: structure learning (i.e., rule base learning) and parameter learning (i.e., MF learning). Considering this, EA can be used for three levels of complexity of learning fuzzy rules:

- Michigan approach,
- Pittsburgh approach and
- Iterative rule learning approach.

8.2.2.1 Michigan Approach

In the Michigan approach, each chromosome in the population represents a single rule and a rule set is represented by the entire population. The foundation of the Michigan approach was laid by Holland (1975). There have been different implementations of the Michigan approach, mainly to learn an approximate fuzzy rule base and/or parameters of the MFs. The Michigan approach involves continuous (online) learning in a non-inductive problem represented by individual rules and MF parameters, learning through interaction with the environment and consequently adapting to it. An initial population of M fuzzy rules is generated randomly by specifying the possible combinations of antecedent MFs. A set of $K \subseteq M$ rules is selected as the rule base to apply on a plant to evaluate the performance of the rule base. Fitness is calculated for the rule base from the performance index. Here, EA plays the role of credit assignment by computing the fitness of the individual rules and/or MF parameters. The fitness of the K rules is shared among K individuals. New rule generation is accomplished by EA using genetic operators such as selection, crossover and mutation.

The Michigan approach is outlined as follows:

1. Generate a random initial population of M fuzzy {if . . . then} rules by specifying all possible combinations of antecedent MFs.
2. Select $K, K \subseteq M$, fuzzy {if . . . then} rules to the rule base of the fuzzy system as shown in Figure 8.24 and evaluate the performance of the rules, applying them to a plant.
3. Generate new individuals of fuzzy {if . . . then} rules by applying genetic operators such as selection, crossover and mutation.
4. Replace individuals with new individuals of the population.
5. Continue until a termination condition is satisfied.

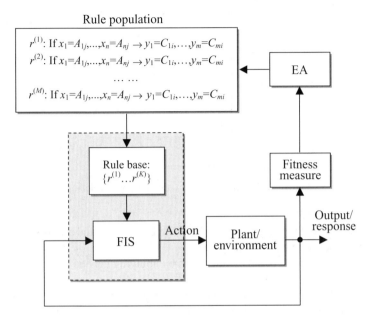

Figure 8.22 Working principle of Michigan approach

The Michigan approach is illustrated in Figure 8.22. The original Michigan approach was developed for classifier systems but this mechanism can be applied to any fuzzy system for learning its rule base using EA.

Each fuzzy {if... then} rule is coded as a string. For example, consider the three fuzzy rules below.

R_1: If X_1 is Small and X_2 is Medium and X_3 is Large Then Y is Low

R_2: If X_1 is Medium and X_2 is Medium and X_3 is Small Then Y is High

R_3: If X_1 is Small and X_2 is Large and X_3 is Medium Then Y is Low

The coded string of rules is shown in Figure 8.23, where Small=1, Medium=2, Large=3, Low=1 and High=2.

8.2.2.2 Pittsburgh Approach

In the Pittsburgh approach, each chromosome in the population represents the entire rule set, maintains a population of candidate rule sets and applies genetic operators (selection, crossover and mutation) to produce new generations of rule sets. Individuals compete among themselves through evolution and adapt to their environment. The Pittsburgh approach is outlined as follows:

1. Generate a random initial population of N fuzzy rule bases. The number of fuzzy {if... then} rules in the rule base is fixed.

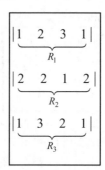

Figure 8.23   An example of a string-coded rule base

2. Select a rule base for the fuzzy system as shown in Figure 8.23 and evaluate the performance of each rule base, applying it to the plant.
3. Generate new individuals of fuzzy rule bases by applying genetic operators such as selection, crossover and mutation.
4. Replace individuals with new individuals of the population.
5. Continue until a termination condition is satisfied.

The Pittsburgh approach is illustrated in Figure 8.24 and the above mechanism can be applied to any fuzzy system for learning its rule base. Each fuzzy {if . . . then} rule is coded as a substring and a rule base is the concatenation of all substrings. For example, consider a fuzzy systems with three inputs $\{X_1, X_2, X_3\}$ and a single output Y with MFs (linguistic terms sets)

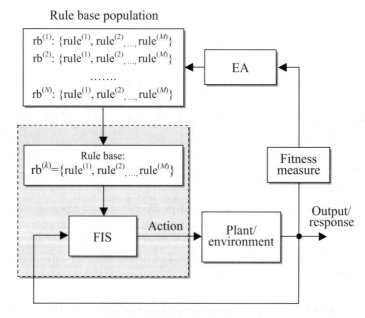

Figure 8.24   Pittsburgh approach

$$|1 \quad 2 \quad 3 \quad 1|2 \quad 2 \quad 1 \quad 2|1 \quad 3 \quad 2 \quad 1|\cdots|3 \quad 3 \quad 3 \quad 2|$$

$$\underbrace{\qquad}_{R_1} \quad \underbrace{\qquad}_{R_2} \quad \underbrace{\qquad}_{R_3} \quad \underbrace{\qquad}_{R_{27}}$$

Figure 8.25   Example of a string-coded rule base in the Pittsburgh approach

{Small, Medium, Large} and {Low, High}, respectively. There are 27 possible combinations of the three inputs $\{X_1, X_2, X_3\}$ with three MFs for each input. The entire rule base will consist of 27 rules as follows.

R_1: If X_1 is Small and X_2 is Medium and X_3 is Large Then Y is Low

R_2: If X_1 is Medium and X_2 is Medium and X_3 is Small Then Y is High

R_3: If X_1 is Small and X_2 is Large and X_3 is Medium Then Y is Low

. . .

. . .

. . .

R_{27}: If X_1 is Large and X_2 is Large and X_3 is Large Then Y is High

The coded string of the rule base is shown in Figure 8.25, where the input MFs are Small=1, Medium=2, Large=3 and the output MFs are Low=1 and High=2. The output MF is shown in the last digit of each rule.

8.2.2.3   Iterative Approach

In the iterative rule learning approach, known as the third approach to reduce the dimension of the search space by encoding individual rules like the chromosome in the Michigan approach, a new rule is adopted and added to the rule base in an iterative way during execution of the EA. The evolution takes cooperation between rules into account, as in the Pittsburgh approach. In this case, the EA needs to be run over several iterations to obtain a complete set of rules. The iterative approach is outlined as follows:

1. Generate a random initial population of M fuzzy {if . . . then} rules by specifying all possible combinations of antecedent MFs.
2. Select a single {if . . . then} rule randomly from the rule base of the fuzzy system as shown in Figure 8.26, which is empty initially (i.e., $rb = \emptyset$).
3. Evaluate the performance of the rule base by applying it to the plant.
4. Generate a new individual fuzzy {if . . . then} rule by applying genetic operators such as selection, crossover and mutation.
5. Add rule to the rule base or replace rule with a new rule if performance is not satisfactory.
6. Continue until a termination condition is satisfied.

The iterative approach is illustrated in Figure 8.26. The above mechanism can be applied to any fuzzy system for learning its rule base using EA.

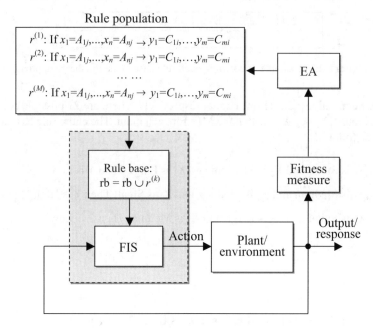

Figure 8.26   Iterative rule learning approach

In an iterative rule learning approach, each fuzzy {if... then} rule is coded as a string. For example, consider a fuzzy system with three inputs $\{X_1, X_2, X_3\}$ and a single output Y with MFs (linguistic terms sets) {Small, Medium, Large} and {Low, High}, respectively. There are 27 possible combinations of the three inputs $\{X_1, X_2, X_3\}$ with three MFs for each input. The entire rule base consists of 27 rules.

R_1: If X_1 is Small and X_2 is Medium and X_3 is Large Then Y is Low

R_2: If X_1 is Medium and X_2 is Medium and X_3 is Small Then Y is High

R_3: If X_1 is Small and X_2 is Large and X_3 is Medium Then Y is Low

...

...

...

R_{27}: If X_1 is Large and X_2 is Large and X_3 is Large Then Y is High

The coded string of rules is shown in Figure 8.27, where the input MFs are Small=1, Medium=2, Large=3 and the output MFs are Low=1 and High=2. The output MF is shown in the last digit of each rule.

The learning process in the iterative rule learning approach works in two stages. In the generation stage, it enforces the current best rule to be added incrementally to the intermediate rule base and rules that are covered by this rule are removed from the training set. In the

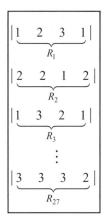

Figure 8.27   An example of a string of coded rules

post-processing stage, redundant or useless rules are eliminated in order to obtain a final rule base that demonstrates optimal performance.

8.3   Objective Functions and Evaluation

The performance of any EA-based fuzzy system depends mainly on the definition of an appropriate objective function or fitness measure. Although it is problem-dependent, finding a good general fitness measurement is quite important for evolving practical systems using EAs. Unlike traditional gradient-based methods, EAs can be used to evolve systems with any kind of fitness measurement function, including those that are non-differentiable and discontinuous. How to define the fitness measurement function for a system to be evolved is problem-dependent. The second issue is the evaluation of the fitness function. Obviously it is an offline procedure due to the nature of the problem and the computation involved.

8.3.1   Objective Functions

The procedure to evaluate the fuzzy system (i.e., MFs and rule base; also called the knowledge base of a fuzzy system) consists of submitting to a simulation model or real system, and returning an assessment value according to a given cost function J subject to minimization. In many cases J is determined as a summation over time of some instantaneous cost rate. As an example, a trial fuzzy system can be made to control the model of a process and then sum the errors over the response trajectory. The sum of errors is then directly related to the objective function (fitness) of the trial. The fitness of the trial is a measure of the overall worth of a solution, which takes into account the factors of an objective criterion; in this case, the performance of a fuzzy system implementable with the trial knowledge base. The objective is simply stated as the ability to follow a set point with minimal error. This objective can thus be expressed in terms of minimization of the system performance indices, which are in common use. These include the integral of absolute error (IAE), integral of square error (ISE) and integral of time-weighted absolute error (ITAE).

Assume a system with multiple inputs and outputs whose overall design effectiveness can be measured by just one output of the overall system, such as error. Finally, all MFs and the rule base can be expressed by some list of m (number of MFs and rules) parameters, $(p_1, p_2, \ldots, p_m) = p$, where each parameter takes only a finite set of values. In the case of IAE, it can be specified by the function

$$\min J(p) = \underset{\min}{E} = \sum_{k=1}^{n} |e(k)| \tag{8.7}$$

In the case of ISE, it is defined as

$$\min J(p) = \underset{\min}{E} = \sum_{k=1}^{n} e(k)^2 \tag{8.8}$$

In the case of ITAE, it is defined as

$$\min J(p) = \underset{\min}{E} = \sum_{k=1}^{n} \Delta t \cdot |e(k)| \tag{8.9}$$

where $e(k)$ is the output error of the system and n is some reasonable number of time units by which the system can be assumed to have settled quite close to a set point. Obviously the objective is to minimize $J(p)$ subject to p in the parameter space.

There are other objective functions such as the one used by Cho *et al.* (1997), defined as follows:

$$\min J(p) = \sum_{i=1}^{n} a_i C_i \tag{8.10}$$

where the a_i are the correction factors that adjust the dimensions or orders of the functions C_i. A proper selection of the correction factors is critical so that EAs provide effective solutions for multiple criteria problems.

Finding a good fitness measurement can make it easier for the EA to evolve to a useful system. Shi *et al.* (1999) defined the fitness function in terms of the difference between the maximum allowable error ($E_{\max}$) and the mean squared error or mean absolute error ($\bar{E}$):

$$\min J(p) = E_{\max} - \bar{E} \tag{8.11}$$

$$\bar{E} = \frac{1}{n} \sum_{k=1}^{n} e(k)^2 \quad \text{or} \quad \bar{E} = \frac{1}{n} \sum_{k=1}^{n} |e(k)| \tag{8.12}$$

The objective of any fuzzy controller is to drive the system's output to the desired set point in the shortest time possible and maintain the output at the desired value. Some researchers (Kang *et al.*, 2000) have defined the fitness function as

$$\min J(p) = \underset{\min}{E} = \sum_{t=1}^{T} \frac{|e(t).\Delta t|}{Y_d} \tag{8.13}$$

where t is the time index, T is the total time, $e(t)$ is the error, Δt is the sampling time and Y_d is the desired output or set point.

If the objective is to minimize the number of rules in the fuzzy system, then a penalty term can be used in the fitness function to force the solution to eliminate redundant or useless rules from the rule set. Maximizing F means minimizing the error E and the number of rules P, where E is defined by Equations (8.7)–(8.9):

$$\underset{\text{max}}{F} = \min J(p) = \frac{1}{E + P} \tag{8.14}$$

As mentioned earlier that the fitness function is problem-dependent. Therefore, the fitness functions defined in Equations (8.7)–(8.14) may not be suitable for problems such as fuzzy clustering or fuzzy pattern recognition. In general, the fitness function provides a measure for evaluating the performance of the fuzzy system with the selected set of fuzzy rule base, MFs or structure in the optimization or tuning process. Hatanaka *et al.* (2004) used a different fitness function for a system identification problem, arguing that the fitness function should be evaluated based on its phenotype performance. This should reflect the accuracy and quality of the global system model estimated by the mean squared error. The fitness function is then defined as

$$F_i = \log \left(\frac{mse_{\text{max}}}{mse_i} \right) \tag{8.15}$$

where F_i is the fitness of the ith individual and mse_i is the mean squared error of the model by the ith individual, defined by

$$mse_i = \frac{1}{N} \sum_{t=1}^{N} [y(t) - \hat{y}_i(t)]^2 \tag{8.16}$$

where $y(t)$ is the desired output and $\hat{y}_i(t)$ is the estimated output produced by the ith individual. mse_{max} is the maximum mean squared error of the population.

8.3.2   Evaluation

Optimization of a fuzzy system using EA can be distinguished by two different groups. The first group encompasses problems like modelling, identification, prediction and classification, where a desired behaviour of the system is sought. The performance index in this case is the error measure characterizing the difference between the desired and actual behaviour of the system. The second group encompasses problems like control, where the objective is to tune or learn the fuzzy system (described by MFs and the rule base) such that the plant shows the desired behaviour. The performance index in this case characterizes the behaviour of the closed-loop system. Evaluation or computation of the performance index is, therefore, critical depending on the problem at hand (closed-loop or open-loop) and can be performed either offline or online. The demand on computation time sometimes prohibits online application. Offline application of EAs is carried out through closed-loop simulation using a simplified model of the plant for fitness computation. The practical problem of implementation is how to evaluate each chromosome in the population. In this case, the fuzzy system is applied to the plant for each individual of the population. Its performance is evaluated using one of the objective functions (8.7)–(8.15) discussed in Section 8.3.1, depending on the problem at hand. Then, the obtained value is assigned to the individual's fitness. The time taken in the evaluation of genetic structures, especially in the case of a fuzzy system or fuzzy control, imposes restrictions on the size of the population and also the number of generations required to run the EA to a final solution. The optimal setting of population size and maximum generation is an issue related to implementation, which is discussed further in Section 8.4.

8.4   Fuzzy Adaptive Evolutionary Algorithms

In order to improve the performance of an EA, a fuzzy system is used to adapt the EA parameters and operators in fuzzy evolutionary algorithms. This leads to two variants of fuzzy EA:

- Adapt EA parameters using fuzzy logic,
- Adapt genetic operators of EA using fuzzy logic.

Standard GA in the 1980s was mainly based on binary chromosome representation, single-point crossover, bit-flip mutation and a roulette wheel selection mechanism. Parameter control was limited to crossover probability, mutation probability, size of tournament selection and population size. Researchers mostly found these parameter values by hand. De Jong recommended a set of parameter values based on extensive experiments on single-point crossover and bit mutation (De Jong, 1975). The parameter values recommended by De Jong are shown in Table 8.8. This can only provide reasonable performance in De Jong's test problems. On the other hand, Grefenstette proposed parameters for online and offline performance (Grefenstette, 1986). The best set of parameters obtained by Grefenstette is shown in Table 8.9.

The optimal and general set of parameters obtained by De Jong and Grefenstette cannot be generalized for all problem domains. It has also been found that the EA's behaviour is strongly determined by the balance between an exploiting and exploring relationship (EER). Poor parameter setting can cause the EER to be disproportionate, resulting in a lack of diversity in the population, which is the main cause of premature convergence. A secondary effect of disproportionate EER is a wasteful, time-consuming evolutionary procedure. This stresses the need for efficient techniques to find good parameter settings for a given problem. Mainly, we distinguish two major forms of parameter setting:

- Parameter tuning and
- Parameter control.

Table 8.8   Parameter values proposed by De Jong (1975)

Parameters	Values
Population size	50
Crossover probability	0.6
Mutation probability	0.001
Generation gap	100%
Scaling window	infinity
Selection strategy	elitist

Table 8.9 Parameter values proposed by Grefenstette (1986)

Parameters	Values online	Values offline
Population size	30	80
Crossover probability	0.95	0.45
Mutation probability	0.01	0.01
Generation gap	100%	90%
Scaling window	1	1
Selection strategy	elitist	non-elitist

In parameter tuning, good values of the parameters are sought using some heuristics (commonly by hand, i.e., trial and error) and the EA is run using the parameter values fixed during the run. Typically, one parameter is tuned at a time. Since parameters interact in a complex way, single-parameter tuning may result in a suboptimal solution. On the other hand, simultaneous tuning of multiple parameters leads to enormously time-consuming experimentations. The main technical pitfalls of parameter tuning can be summarized as follows:

- Parameters are all interdependent, requiring evaluation of all the different combinations of parameters, which is practically impossible.
- The process of parameter tuning is time-consuming.
- The selected parameter values are not necessarily optimal for a given problem.

The other option for parameter tuning would be to choose parameters by analogy, where parameter settings are chosen that have been proven successful in similar problems. However, similarity between problems does not necessarily always guarantee similarity between optimal parameter sets. Therefore, researchers are inclined to parameter control, whereby they can use some heuristic feedback mechanism from the current state of the search and modify the parameters accordingly. Eiben *et al.* (1999) classified the methods for modifying the parameters into three categories and proposed the taxonomy illustrated in Figure 8.28.

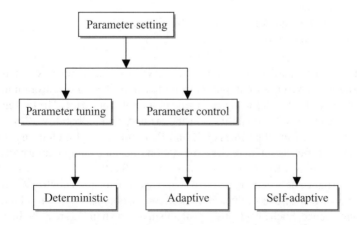

Figure 8.28 Taxonomy of parameter setting in EA

Adaptive techniques have been suggested to control the parameters of the EA during the run. The three widely accepted parameter control techniques are as follows.

- Deterministic parameter control: the parameter set is updated according to some deterministic rule, such as a time-varying schedule.
- Adaptive parameter control: the direction and/or magnitude of the parameter set are updated based on the feedback information from the search, such as a credit assignment.
- Self-adaptive parameter control: the parameter set is encoded into the chromosome representation and undergoes evolution along with the EA.

8.4.1　Fuzzy Logic-Based Control of EA Parameters

All three parameter control techniques can involve adapting any component of an EA such as chromosome representation, fitness function, variation operators (e.g., mutation and crossover) and their probabilities, selection operator, replacement operator and population size. There are many widely accepted control techniques but there has been a growing interest in combining FS and EA. The motivation is to control the EA parameters using EA performance measures or current control parameters as inputs to a fuzzy controller and compute new control parameters for the EA to be adaptive. The possible measures are diversity measures, best (maximum), average, worst (minimum) fitness, etc. Two diversity measures are widely used: genotypical diversity, which measures the average distance of the population from the best individual and phenotypical diversity, which measures the ratio between the best fitness and the average fitness. Current control parameters are also suggested as inputs to the FS. Possible outputs are control parameters or changes in them, such as crossover and mutation rate, population size and selective pressure. A mapping between the inputs and outputs should be established, then associated linguistic terms and a rule base describing the relationships between them should be defined. Finding a good rule base has been recognized as a difficult task, as reported in the literature. A brief review of adaptive EA using fuzzy control is reported in Schaffer and Morishima (1987), Xu and Vukovich (1993) and Herrera and Lozano (2001). Different combinations of FS-based EA components exist in the literature. Among them are two widely used approaches, which will be discussed in the following sections:

- FS to control crossover and/or mutation probability and
- FS to control population size.

The adaptive mechanism being sought must have two characteristics. Firstly, it should have the capacity to converge to an optimum on reaching the region of optimum solutions. Secondly, it must be able to explore a new region of the solution space. The two characteristics are dictated by the crossover probability (P_c) and the mutation probability (P_m), as well as the type of crossover. Increasing the values of P_c and P_m promotes exploration (meaning diversity of the population), which again increases exploitation (meaning computation cost). In general, large values for P_c (0.5–1.0) and small values for P_m (0.001–0.05) are chosen by trial and error to strike a balance between the two. An adaptive mechanism to control P_c and P_m can be devised in response to the fitness values of the solutions to obtain an optimal solution or prevent premature convergence. Such an adaptive mechanism is shown in Figure 8.29. In devising such mechanisms, it is important to identify the state of convergence of the EA according to which

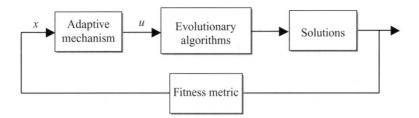

Figure 8.29 Adaptive mechanism for EA parameters

the values of P_c and P_m will be changed. The state of convergence can be observed by simply measuring the average fitness value $\bar{f}$ of the population in relation to the best (maximum) fitness value f_{best} of the population. The value of the term $(f_{best} - \bar{f})$ is likely to be less for a population that has converged to an optimal solution than for a population scattered in the entire solution space (Srinivas and Patnaik, 1994). It has been shown by Srinivas and Patnaik that the EA converges to a local optimum with a fitness value of 0.5 with decreasing value of $(f_{best} - \bar{f})$, whereas a global optimum has a fitness value of 1.0.

Srinivas and Patnaik (1994) proposed that when the value of $(f_{best} - \bar{f})$ decreases, P_c and P_m should be varied inversely with $(f_{best} - \bar{f})$. Thus, the relationship can be expressed as

$$P_c = \frac{k_c}{(f_{best} - \bar{f})} \tag{8.17}$$

$$P_m = \frac{k_m}{(f_{best} - \bar{f})} \tag{8.18}$$

where k_c and k_m are chosen arbitrarily and should be less than 1.0. The adaptive mechanism for EA shown in Figure 8.29 can be described by Equations (8.17) and (8.18). A variety of deterministic and adaptive mechanisms for controlling the parameters can be constructed using the relationships in Equations (8.17) and (8.18). Use of an automatic technique such as a fuzzy controller can reveal new high-performance EAs. The choice of fuzzy control is made not only because it is easy to design, but because new knowledge on the complex relationship between the different control parameters and their effects on performance is more understandable. The adaptive mechanism for the EA proposed here is a fuzzy system as illustrated in Figure 8.30.

The input x in Figure 8.30 can be any performance measure, like average fitness ($\bar{f}$), average fitness/best fitness ($\bar{f}/f_{best}$), worst fitness/average fitness ($f_{worst}/\bar{f}$), difference between best and average fitness ($f_{best} - \bar{f}$), current population size (N), or current control settings (s), i.e., $x = \{\bar{f}/f_{best}, f_{worst}/\bar{f}, f_{best} - \bar{f}, N, s\}$ and the output u can be any of the EA control parameters identified by Grefenstette, such as population size (N), crossover rate (P_c), mutation rate (P_m), generation gap (G), scaling window (W) or selection strategy (S) (Grefenstette, 1986), i.e., $u = \{N, P_c, P_m, G, W, S\}$. The performance of an EA can be expressed by the control parameters

$$f(EA) = g(u) = g(N, P_c, P_m, G, W, S) \tag{8.19}$$

The control parameters $u = \{N, P_c, P_m, G, W, S\}$ have a strong relationship with the performance measures and can best be described by

$$u(x) = g(\bar{f}/f_{best}, f_{worst}/\bar{f}, f_{best} - \bar{f}, N, s) \tag{8.20}$$

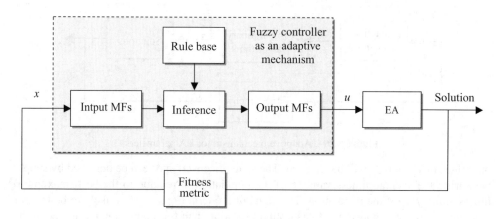

Figure 8.30   FL-based adaptive mechanism for EA parameters

The behaviour of the EA is strongly determined by the balance between exploitation and exploration. Poor parameter setting makes the relationship between exploitation and exploration disproportionate, causing a loss of diversity in the population (Last and Eyal, 2005). An immediate consequence of this is the premature convergence, leading to a non-optimal solution. One promising approach would be to control the crossover rate, mutation rate, population diversity and stopping condition of the EA using the relationships defined in Equations (8.19) and (8.20). A variety of fuzzy control techniques can be developed to control the population size with fixed crossover and mutation rates, to control the crossover and mutation rates with fixed population and to determine a stopping condition for the EA to be terminated. These combinations of FS and EA remain to be discussed.

De Jong designed two measures to quantify the performance of an EA: online performance to measure the ongoing performance of EA and offline performance to measure convergence. Online performance based on current control setting s is the running average of all fitness values up to a given time T, defined as

$$F_{on}(s) = \frac{1}{T} \sum_{t=1}^{T} f(t) \tag{8.21}$$

Offline performance based on current control setting s is the running average of the best fitness value up to a given time T, defined as

$$F_{off}(s) = \frac{1}{T} \sum_{t=1}^{T} f_{best}(t) \tag{8.22}$$

Researchers use both online and offline performance depending on the problem at hand.

8.4.1.1   Fuzzy System to Control Crossover and Mutation Probability

An EA expert formulates fuzzy control rules to adapt the EA parameters, and achieve a balance between exploitation and exploration through EA execution. Some researchers like to emulate natural evolution processes with a varying population size by introducing the concept of lifetime and age into the evolution process. Last and Eyal (2005) considered a lifetime extension of individuals as a state variable that controls the crossover probability. The proposed method takes into account the lifetime extension of the two parents and adapts the crossover probability according to a fuzzy rule base. Defining the rule base of the fuzzy system to eventually improve the performance of the EA is not straightforward. A qualitative measure of the state variables is taken to adapt the EA control parameters, such as crossover or mutation probability. The linguistic terms representing the range of each fuzzy variable, such as lifetime extension and crossover probability, are defined as

$$\text{Age} \in \{\text{Young, Middle\_aged, Old}\} \tag{8.23}$$

$$P_c \in \{\text{Low, Medium, High}\} \tag{8.24}$$

A lifetime can be allocated to each individual from initialization of the population, and each individual will grow older as it evolves with its generation. Tracking of the lifetime of each individual will make the EA computation-intensive and may not contain any useful genetic information, unless any fitness component is included in the lifetime calculation strategies. Last and Eyal (2005) proposed a bilinear allocation strategy. An individual has a maximum and minimum allowable lifetime. A lifetime is assigned to each individual i according to its fitness, as follows:

$$\text{lifetime}(i) = \begin{cases} \text{lifetime}_{\min} + \eta \cdot \frac{f(i) - f_{worst}}{\bar{f} - f_{worst}} & \text{if } \bar{f} \geq f(i) \\ \frac{1}{2}(\text{lifetime}_{\min} + \text{lifetime}_{\max}) + \eta \cdot \frac{f(i) - \bar{f}}{f_{best} - \bar{f}} & \text{if } \bar{f} < f(i) \end{cases} \tag{8.25}$$

where $\eta = \frac{1}{2}(\text{lifetime}_{\max} - \text{lifetime}_{\min})$, $\text{lifetime}_{\max}$ and $\text{lifetime}_{\min}$ are the maximum and minimum allowable lifetimes of individuals, respectively.

A rule base can be developed by trial and error or by an expert. Table 8.10 shows a rule base for a two-parent lifetime extension and crossover probability. Eyal (2003) has developed such a rule base through extensive experiments. The MFs for the linguistic terms are defined in Figure 8.31. The rule base will provide satisfactory results for similar problem domains only when the MFs are defined close to the MFs in Last and Eyal (2005).

Using the lifetime extension of the parents as inputs and crossover probability as output, a fuzzy controller can be designed as shown in Figure 8.32. Similarly, the mutation probability can also be controlled.

Table 8.10 FL rule base for crossover probability

Parent 2 Age	Parent 1		
	Young	Middle-aged	Old
Young	Low	Medium	Low
Middle-aged	Medium	High	Medium
Old	Low	Medium	Low

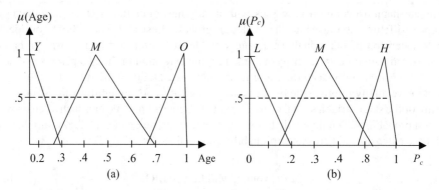

Figure 8.31   MFs for lifetime extension and crossover probability. (a) MFs for lifetime extension; (b) MFs for crossover probability. Here the symbols used are: Y for Young, M for Middle-aged, O for Old, L for Low, M for Medium and H for High

Some researchers use the current generation run by an EA and the actual population size for a two-input single-output fuzzy controller, which can adapt the crossover or mutation rates. The linguistic terms representing the generation, population size and crossover and mutation probability are defined in Equations (8.26)–(8.28). MFs for the linguistic terms are not shown for this fuzzy controller. The two inputs are generation and population size and the output is crossover or mutation probability. The rule base for the two-input single-output fuzzy controller is shown in Table 8.11. The two-input single-output fuzzy controller for this case is not shown here but will be similar to the fuzzy controller shown in Figure 8.32, with inputs of generation and population size and crossover or mutation probability as the single output.

$$\text{Generation} \in \{\text{Short, Medium, Large}\} \tag{8.26}$$

$$\text{Population\_size} \in \{\text{Small, Medium, Large}\} \tag{8.27}$$

$$P_m \text{ or } P_c \in \{\text{Low, Medium, High}\} \tag{8.28}$$

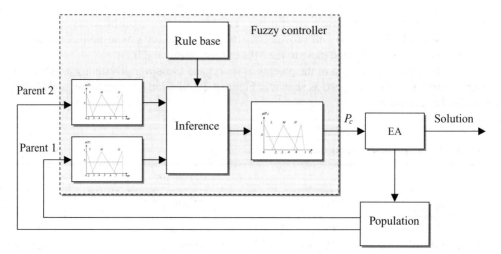

Figure 8.32   Fuzzy control of crossover or mutation probability

Table 8.11   Fuzzy controller's rule base for crossover/mutation probability

	Population size		
Generation	Small	Medium	Large
Short	Medium	Low	Low
Medium	High	High	Medium
Large	Very High	Very High	High

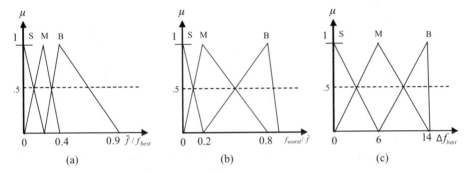

Figure 8.33   Input MFs for $\bar{f}/f_{best}$, $f_{worst}/\bar{f}$ and Δf_{best}. (a) MFs for average/best fitness ($\bar{f}/f_{best}$); (b) MFs for worst/average fitness ($f_{worst}/\bar{f}$); (c) MFs for change in best fitness (Δf_{best})

Lee and Takagi (1993) used input variables such as average fitness/best fitness ($\bar{f}/f_{best}$), worst fitness/average fitness ($f_{worst}/\bar{f}$), change in best fitness (Δf_{best}) and output variables such as change in crossover rate (ΔP_c), change in mutation rate (ΔP_m) to design a three-input single-output fuzzy controller to control the crossover or mutation rate. The linguistic terms representing the input and output variables are defined in Equations (8.29) and (8.30), i.e., mostly three MFs are used for each input and output variables. For example, the input MFs are shown in Figure 8.33 and the output MFs are shown in Figure 8.34. The universes of

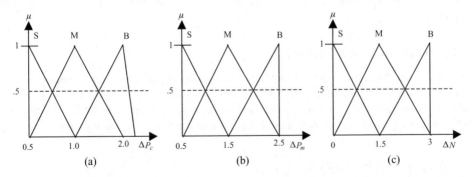

Figure 8.34   Output MFs for ΔP_c, ΔP_m and ΔN. (a) MFs for change of crossover rate (ΔP_c); (b) MFs for change of mutation rate (ΔP_m); (c) MFs for change of population size (ΔN)

Table 8.12   Rule base for change in mutation rate

Inputs			
$\bar{f}/f_{best}$	$f_{worst}/\bar{f}$	Δf_{best}	Output ΔP_m
S	S	S	S
S	S	B	M
S	M	M	M
S	M	B	B
S	B	S	M
M	S	S	M
M	S	M	B
M	S	B	M
M	M	S	M
M	B	S	B
M	B	M	M
M	B	B	M
B	S	S	S
B	S	B	B
B	M	B	B
B	B	B	S

discourse of the input/output variables shown in Figures 8.33 and 8.34 may vary depending on the optimization problem at hand.

$$\{\bar{f}/f_{best}, f_{worst}/\bar{f}, \Delta f_{best}\} \in \{\text{Small, Medium, Big}\} \qquad (8.29)$$

$$\{\Delta P_c, \Delta P_m\} \in \{\text{Small, Medium, Big}\} \qquad (8.30)$$

The rule base for the three-input $\{\bar{f}/f_{best}, f_{worst}/\bar{f}, \Delta f_{best}\}$ single-output $\{\Delta P_m\}$ fuzzy controller is shown in Table 8.12. A generic rule would read

IF $\bar{f}/f_{best}$ is S and $f_{worst}/\bar{f}$ is S and Δf_{best} is B THEN ΔP_m is M

The fuzzy controller is shown in Figure 8.35. The output ΔP_m is to be considered in Table 8.12 for controlling the mutation rate. In a similar way, a fuzzy controller for crossover change can be developed.

8.4.1.2   Fuzzy Control of Population

De Jong and Goldberg have been researching the effect of population size on EA performance; for example, as the current population size grows, the sensitivity to mutation rate decreases and the best mutation rate to use also decreases (De Jong and Spears, 1990; Goldberg *et al.*, 1992). A typical fuzzy rule to control the EA parameters relating to population size N may be as follows:

IF $(\bar{f}/f_{best})$ is Big THEN increase (N)

IF $(f_{worst}/\bar{f})$ is Small THEN decrease (N)

IF (P_m) is Small AND (N) is Small THEN increase (N)

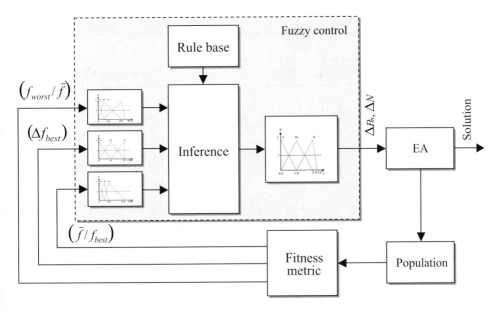

Figure 8.35   Fuzzy control of mutation rate/population size

Here, an increase or decrease in N means a change of the population by adding or subtracting a small value ΔN to or from N. Using the suggestions of De Jong and Goldberg, researchers designed fuzzy controllers to control the population size, crossover rate or mutation rate. The three input variables used are $(\bar{f}/f_{best})$, $(f_{worst}/\bar{f})$ and Δf_{best} and the output variable is ΔN (Lee and Takagi, 1993). The linguistic terms representing the output variable are defined in Equation (8.31), i.e., three MFs are used for the output variable. For example, the input MFs are shown in Figure 8.33(a–c) and the output MFs are shown in Figure 8.34(c).

$$\{\Delta N\} \in \{\text{Small, Medium, Big}\} \tag{8.31}$$

The universes of discourse of the input/output variables shown in Figures 8.33 and 8.34 may vary depending on the application, as the definition of the fitness function is very subjective to problems. For a three-input single-output fuzzy controller, the input and output variables are $\{\bar{f}/f_{best},\ f_{worst}/\bar{f}, \Delta f_{best}\}$ and $\{\Delta N\}$, respectively. A rule base for the fuzzy controller to control the population size would look like that in Table 8.13. A generic rule would read:

IF $\bar{f}/f_{best}$ is B and $f_{worst}/\bar{f}$ is M and Δf_{best} is S THEN ΔN is B

The fuzzy controller for the population change is the same as shown in Figure 8.35, with the output variable ΔN.

The current EA parameters $\{P_c,\ P_m,\ N\}$ are multiplied by the change in EA parameters $\{\Delta P_c,\ \Delta P_m,\ \Delta N\}$ obtained from the fuzzy controller. The three-input single-output fuzzy control for the population size is shown in Figure 8.35. This is the same as that used for control of the crossover or mutation probability, except the output is ΔN in this case. As the number of inputs and outputs of a fuzzy system increases, the rule base increases exponentially. The three-input single-output fuzzy controller with three MFs for each variable requires 81 rules to

Table 8.13   Rule base for change in population

	Inputs		
$\bar{f}/f_{best}$	$f_{worst}/\bar{f}$	Δf_{best}	Output ΔN
S	S	B	B
S	M	S	B
S	M	M	S
S	M	B	B
S	B	S	S
S	B	B	M
M	S	S	S
M	S	M	M
M	S	B	S
M	M	S	S
M	M	M	M
M	M	B	B
B	S	S	M
B	S	B	S
B	M	S	B
B	M	M	B
B	M	B	B

process, which is certainly very time-consuming. There are different strategies for reduction of the rule base. Some techniques to handle the dimensionality problem have been discussed in Chapter 3.

Poluzzi *et al.* (1997) proposed a fuzzy knowledge base (fuzzy government) to control the EA parameters dynamically and to detect the emergence of a solution so that any undesired behaviour of the evolutionary process can be avoided. They also developed a fuzzy estimate of the fitness of individuals based on the recorded outcomes of competitions taking into account three quantities: number of competitions taken part in by an individual (denoted c), number of times an individual wins (denoted w) and number of times an individual succeeds (denoted s). The membership function of the fitness is then defined as

$$\mu_f(x) = N(a, b)\, x^a\, (1 - x)^b \tag{8.32}$$

where $N(a, b) = \frac{(a+b)^{a+b}}{a^a b^b}$ is a normalization factor with $a = w + s$ and $b = c - s$. The fuzzy fitness helps avoid useless competition among individuals and the fitness of individuals in the population is aggregated to provide population statistics to be used by the fuzzy controller. The statistics that can be used as inputs to the fuzzy controller are: genotypic diversity (D_Γ), phenotypic diversity (D_Φ), maximum fitness (f_{max}), average fitness ($\bar{f}$), minimum fitness (f_{min}), fitness range ($f_{max} - f_{min}$), tie rate (P_{tie}), success rate (P_{succ}), actual mutation rate (P_{m_act}) and time out rate ($P_{t/o}$). The outputs of the fuzzy controller are: mutation rate (P_m), crossover rate (P_c), selective strategy (S) and window of success (W). There are other measures such as emergence and premature convergence. The statistics and parameters can be used for any EA and applications. All the statistics and parameters are normalized and six MFs are

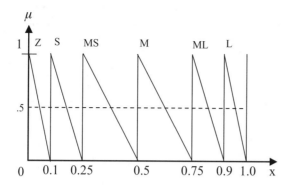

Figure 8.36 Six MFs defined for all inputs $x \in [-1, 1]$

defined within the normalized range $[-1, 1]$. The MFs are defined using the threshold function $\mu(x)$ for all $x \in [-1, 1]$:

$$\mu(x) = \begin{cases} 1 & x \leq \alpha \\ (\beta - x)/(\beta - \alpha) & \alpha < x < \beta \\ 0 & x \geq \beta \end{cases} \qquad (8.33)$$

The six MFs {Z, S, MS, M, ML, L} can be defined by choosing suitable values for $\{\alpha, \beta\} \in [0, 1]$ in Equation (8.33). The MFs {Z, S, MS, M, ML, L} are shown in Figure 8.36 and the corresponding values of $\{\alpha, \beta\}$ are shown in Table 8.14.

Linguistic hedges such as *Very* and *More or less* can also be used to intensify or dilute the MFs defined in Figure 8.36, respectively. The objective of the fuzzy system (controller) is to keep the EA in dynamical equilibrium. Therefore, the mapping between the population statistics defined earlier and the EA parameters defined by the fuzzy system will inversely relate the mutation rate to the population diversity (i.e., the higher the mutation rate, the less diverse the population). A generic set of rules would read

IF $NOT(P_{tie})$ is MS and P_{tie} is M THEN P_m is 0.01

IF $NOT(P_{succ})$ is S and P_{succ} is MS THEN W is 0.01

IF $Very(NOT(P_{t/o}))$ is MS THEN *Emergence* is 0.5

IF $NOT(P_{t/o})$ is $Very(M)$ THEN *Emergence* is 1.0

Table 8.14 MFs with corresponding values of $\{\alpha, \beta\}$

MF	α	β
Z	0.0	0.1
S	0.1	0.25
MS	0.25	0.5
M	0.5	0.75
ML	0.75	0.9
L	0.9	1.0

Table 8.15 A generic rule base for fuzzy control of population

Inputs						Outputs		
If $(P_{tie}, \bar{P}_{tie}, P_{succ}, \bar{P}_{succ}, \bar{P}_{t/o}, (\bar{P}_{t/o})^2)$						Then (P_m, W, E)		
P_{tie}	$\bar{P}_{tie}$	P_{succ}	$\bar{P}_{succ}$	$\bar{P}_{t/o}$	$(\bar{P}_{t/o})^2$	P_m	W	E
Z						0.0		
MS	Z					0.004		
M	MS					0.01		
	M					0.1		
		S					0.0	
		MS	S				0.01	
			MS				0.02	
				MS				0.5
					M^2			1.0
				MS			0.02	

A rule base is developed for the fuzzy system as shown in Table 8.15. $(\bar{P}_{t/o})^2$ is read as *Very*($NOT(P_{t/o})$), M^2 is read as *Very*(M) and *Emergence* is represented by E in the figure.

Some researchers considered a two-input single-output fuzzy controller to control the values of two different variables, such as a parameter establishing the frequency of application of crossover operators and a parameter establishing the selective pressure performed by the selection operator (Herrera and Lozano, 1996). The rule base is generally derived by a human expert following the behaviour of encouraging exploration when the population diversity is low and encouraging exploitation when the population diversity is high. There have been lots of researches on controlling the EA parameters reported in the literature in the last two decades, under the umbrella of adaptive systems. Interested readers are directed to read the two classic papers by De Jong (1980) and Grefenstette (1986), where the problems are discussed of designing an adaptive system and optimizing a complex system using a genetic approach by controlling the parameters of EA.

8.4.2 Fuzzy Logic-Based Genetic Operators of EA

One of the major problems of an EA is the premature convergence caused by the loss of critical alleles in the chromosomes due to selection, crossover and mutation and their probabilities. In order to solve the problem, some kind of mechanism is needed in the genetic process. There are several strategies, such as modified selection and crossover operators and adaptation of control parameters. Adaptation of control parameters has been addressed in Section 8.4.1. Genetic operators such as crossover, mutation and selection are mainly numeric operations. Some researchers have proposed fuzzified genetic operators such as fuzzy connective-based crossover (Herrera *et al.*, 1997; Herrera and Lozano, 2000, 2001) and soft genetic operators (Voigt *et al.*, 1995). This is a modified version of arithmetic crossover. Three types of arithmetic crossover (e.g., simple, single and whole) were discussed in Section 6.4.2 of Chapter 6. All three types can be modified with max/min arithmetic crossover. A generic example of arithmetic crossover will make clear how to apply the operator on two parents. Let us consider two chromosomes $C_i(t) = \{c_{i,1}, \ldots, c_{i,k}, \ldots, c_{i,M}\}$ and $C_j(t) = \{c_{j,1}, \ldots, c_{j,k}, \ldots, c_{j,M}\}$ representing the ith and jth individuals of the population. Averaging and max/min arithmetic crossover can be applied to produce four offspring. Applying the averaging operation will generate two offspring, where $0 \leq \lambda \leq 1$ and $k = 1, 2, \ldots, M$:

$$\text{Parent 1: } C_i(t) = \{c_{i,1}, \ldots, c_{i,k}, \ldots, c_{i,M}\}$$

$$\text{Parent 2: } C_j(t) = \{c_{j,1}, \ldots, c_{j,k}, \ldots, c_{j,M}\}$$

- -

$$\text{Offspring 1: } C_{1,k}(t+1) = (1 - \lambda)\,c_{i,k} + \lambda c_{j,k}$$

$$\text{Offspring 2: } C_{2,k}(t+1) = (1 - \lambda)\,c_{j,k} + \lambda c_{i,k}$$

Applying the max/min operation will generated two offspring as follows:

$$\text{Parent 1: } C_i(t) = \{c_{i,1}, \ldots, c_{i,k}, \ldots, c_{i,M}\}$$

$$\text{Parent 2: } C_j(t) = \{c_{j,1}, \ldots, c_{j,k}, \ldots, c_{j,M}\}$$

- -

$$\text{Offspring 3: } C_{3,k}(t+1) = \max\{c_{i,k}, c_{j,k}\}$$

$$\text{Offspring 4: } C_{4,k}(t+1) = \min\{c_{i,k}, c_{j,k}\}$$

For more details on these operations, interested readers are directed to Herrera *et al.* (1997). It has been demonstrated by Herrera *et al.* that the fuzzy connective-based crossover operator is able to balance the exploitation/exploration and model the diversity of population, and can avoid premature convergence.

References

Ahmad, S., Siddique, N.H. and Tokhi, M.O. (2012) Evolutionary tuning of modular fuzzy controller, *International Journal of Computational Intelligence and Applications*, 11(2), 1–23.

Baeck, T. (1996) *Evolutionary Algorithms in Theory and Practice*, Oxford University Press, New York.

Belarbi, K. and Titel, F. (2000) Genetic algorithm for design of a class of fuzzy controllers: an alternative approach, *IEEE Transactions on Fuzzy Systems*, 8(4), 398–405.

Bonarini, A. and Trianni, V. (2001) Learning fuzzy classifier systems for multi-agent coordination, *Information Sciences*, 136, 215–239.

Boverie, S., Demaya, B. and Titly, A. (1991) Fuzzy logic control compared with other automatic control approaches. *Proceedings of 30th IEEE-CDC Conference on Decision and Control*, Brighton, pp. 1212–1216.

Chang, T.C., Hasegawa, K. and Ibbs, C.W. (1991) The effects of membership function on fuzzy reasoning, *Fuzzy Sets and Systems*, 41, 169–186.

Chin, T.C. and Qi, X.M. (1997) Genetic algorithms for learning the rule base of fuzzy logic controller, *Fuzzy Sets and Systems*, 97, 1–7.

Cho, H.-J., Cho, K.-B. and Wang, B.-H. (1997) Fuzzy–PID hybrid control: automatic rule generation using genetic algorithms, *Fuzzy Sets and Systems*, 92, 305–316.

Chou, C.-H. (2006) Genetic algorithm-based optimal fuzzy controller design in the linguistic space, *IEEE Transactions on Fuzzy Systems*, 14(3), 372–385.

Cordon, O., Herrera, F., Hoffmann, F. and Magdalena, L. (2001) *Genetic Fuzzy Systems: Evolutionary Tuning and Learning of Fuzzy Knowledge Bases*, World Scientific, Singapore.

Daugherity, W.C., Rathakrishnan, B. and Yen, J. (1992) Performance evaluation of a self-tuning fuzzy controller. *Proceedings of the First IEEE International Conference on Fuzzy Systems*, San Diego, pp. 389–397.

Davis, L. (1989) Adapting operator probabilities in genetic algorithms. *Proceedings of Third International Conference on Genetic Algorithms*, pp. 61–69.

De Jong, K.A. (1975) *Analysis of the behaviour of a class genetic adaptive system*. PhD Thesis, Department of Computer and Communications Sciences, University of Michigan, Ann Arbor, MI.

De Jong, K.A. (1980) Adaptive system design: a genetic approach, adaptive systems, *IEEE Transactions on Systems, Man and Cybernetics*, 10(9), 566–574.

De Jong, K.A. and Spears, W.M. (1990) An analysis of interacting roles of population size and crossover in genetic algorithms. *Proceedings of the 1st Workshop on Parallel Problem Solving in Nature* (PPSN '90), Dortmund, Germany, pp. 38–47.

Driankov, D., Hellendoorn, H. and Reinfrank, M. (1993) *An Introduction to Fuzzy Control*, Springer-Verlag, Berlin.

Eiben, A.E., Hinterding, R. and Michalewicz, Z. (1999) Parameter control in evolutionary algorithms, *IEEE Transactions on Evolutionary Computation*, 3(2), 124–141.

Eyal, S. (2003) *A fuzzy-based age extension of genetic algorithm*. Master's Thesis, Department of Information Systems Engineering, Ben-Gurion University of the Negev, Israel.

Fogarty, T.C. (1989) Varying the probability of mutation in genetic algorithms. *Proceedings of Third International Conference on Genetic Algorithms*, pp. 104–109.

Glover, F. (1989) Tabu Search – Part I, *ORSA Journal on Computing*, 1, 190–206.

Goldberg, D.E. (1989) *Genetic Algorithms in Search, Optimization and Machine Learning*, Addison-Wesley, Reading, MA.

Goldberg, D.E., Deb, K. and Clark, J.H. (1992) Genetic algorithms, noise, and sizing of populations, *Complex Systems*, 6(4), 333–362.

Grefenstette, J.J. (1986) Optimisation of control parameters for genetic algorithms, *IEEE Transactions on Systems, Man and Cybernetics*, 16(1), 122–128.

Hatanaka, T., Kawaguchi, Y. and Uosaki, K. (2004) Nonlinear system identification based on evolutionary fuzzy modelling, *IEEE Congress on Evolutionary Computing*, 1, 646–651.

Herrera, F. and Lozano, M. (1996) Adaptation of genetic algorithm parameters based on fuzzy logic controllers. In *Genetic Algorithms and Soft Computing*, F. Herrera and J.L. Verdegay (eds), Studies in Fuzziness and Soft Computing, Vol. 8, Physica-Verlag, Berlin, pp. 95–125.

Herrera, F. and Lozano, M. (2000) Gradual distributed real-coded genetic algorithms, *IEEE Transactions on Evolutionary Computation*, 4(1), 43–63.

Herrera, F. and Lozano, M. (2001) Adaptive genetic algorithms based on co-evolution with fuzzy behaviours, *IEEE Transactions on Evolutionary Computation*, 5(2), 149–165.

Herrera, F., Lozano, M. and Verdegay, J.L. (1997) Fuzzy connectives based crossover operators to model genetic algorithms population diversity, *Fuzzy Sets and Systems*, 92(1), 21–30.

Hinton, G.E. and Nowlan, S.J. (1987) How learning can guide evolution, *Complex Systems*, pp. 495–502.

Hoffmann, F. (2000) Evolutionary algorithms for fuzzy control system design, *Proceedings of the IEEE*, 89(9), 1318–1333.

Holland, J.H. (1975) *Adaptation in Natural and Artificial Systems*, University of Michigan Press, Ann Arbor, MI.

Homaifar, A. and McCormick, E. (1995) Simultaneous design of membership functions and rule sets for fuzzy controllers using genetic algorithms, *IEEE Transactions on Fuzzy Systems*, 3(2), 129–139.

Ishibuchi, H., Nozaki, K., Yamamoto, N. and Tanaka, H. (1995) Selecting fuzzy if–then rules for classification problems using genetic algorithms, *IEEE Transactions on Fuzzy Systems*, 3, 260–270.

Jang, J.-S.R., Sun, C.-T. and Mizutani, E. (1997) *Neuro-Fuzzy and Soft Computing*, Prentice-Hall, Englewood Cliffs, NJ, pp. 335–363.

Kang, S.-J., Woo, C.-H., Hwang, H.-S. and Woo, K.B. (2000) Evolutionary design of fuzzy rule base for nonlinear system modelling and control, *IEEE Transactions on Fuzzy Systems*, 8(1), 37–45.

Karr, C.L. and Gentry, E.J. (1993) Fuzzy control of pH using genetic algorithms, *IEEE Transactions on Fuzzy Systems*, 1(1), 46–53.

Kirkpatrick, S., Gelatt, C.D. and Vecchi, M.P. (1983) Optimization by simulated annealing, *Science*, 220, 671–680.

Kovacic, Z. and Bogdan, S. (2006) *Fuzzy Controller Design: Theory and Application*, CRC Press, Boca Raton, FL.

Kwon, C. and Sudhoff, S.D. (2006) Genetic algorithm-based induction machine characterization procedure with application to maximum torque per amp control, *IEEE Transactions on Energy Conversion*, 21(2), 405–415.

Last, M. and Eyal, S. (2005) A fuzzy-based lifetime extension of genetic algorithm, *Fuzzy Sets and Systems*, 149, 131–147.

Lee, M.A. and Takagi, H. (1993) Dynamic control of genetic algorithms using fuzzy logic techniques. *Proceedings of the 5th International Conference on Genetic Algorithms* (ICGA'93), Urbana-Champaign, IL, pp. 76–83.

Linkens, D.A. and Nyongesa, H.O. (1995a) Genetic algorithms for fuzzy control, Part 1: Offline system development and application, *IEE Proceedings of Control Theory and Application*, 142(3), 161–176.

Linkens, D.A. and Nyongesa, H.O. (1995b) Genetic algorithms for fuzzy control, Part 2: Online system development and application, *IEE Proceedings of Control Theory and Application*, 142(3), 177–185.

Loop, B.P., Sudhoff, S.D., Żak, S.H. and Zivi, E.L. (2010) Estimating regions of asymptotic stability of power electronics systems using genetic algorithms, *IEEE Transactions on Control Systems Technology*, 18(5), 1011–1022.

Margaliot, M. and Langholz, G. (2000) *New Approaches to Fuzzy Modelling and Control*, World Scientific, Singapore.

Matyas, J. (1965) Random optimization, *Automation and Remote Control*, 26, 244–251.

Nelder, J. and Mead, R. (1965) The downhill simplex method, *Computer Journal*, 7, 308–313.

Poluzzi, R., Rizzotto, G.G. and Tettamanzi, A.G.B. (1997) An evolutionary algorithm for fuzzy controller synthesis and optimization based on SGS-Thomson's WARP fuzzy processor. In *Genetic Algorithms and Fuzzy Logic Systems: Soft Computing Perspectives*, E. Sanchez, T. Shibata and L.A. Zadeh (eds), World Scientific, Singapore, pp. 71–89.

Qi, X.M. and Chin, T.C. (1997) Genetic algorithms based fuzzy controller for higher order systems, *Fuzzy Sets and Systems*, 91, 279–284.

Schaffer, J.D. and Morishma, A. (1987) An adaptive crossover mechanism for genetic algorithms. *Proceedings of Second International Conference on Genetic Algorithms*, pp. 36–40.

Setness, M. and Roubos, H. (2000) GA-fuzzy modelling and classification: complexity and performance, *IEEE Transactions on Fuzzy Systems*, 8(5), 509–522.

Shi, Y., Eberhart, R. and Chen, Y. (1999) Implementation of evolutionary fuzzy systems, *IEEE Transactions on Fuzzy Systems*, 7, 109–119.

Siarry, P. and Guely, F. (1998) A genetic algorithm for optimizing Takagi–Sugeno fuzzy rule bases, *Fuzzy Sets and Systems*, 99, 437–471.

Srinivas, M. and Patnaik, L.M. (1994) Adaptive probabilities of crossover and mutation in genetic algorithms, *IEEE Transactions on Systems, Man and Cybernetics*, 24(4), 656–667.

Vidyasagar, M. (2002) *Nonlinear Systems Analysis*, 2nd edn, Prentice-Hall, Englewood Cliffs, NJ.

Voigt, H.M., Muehlenbein, H. and Cvetkovic, H. (1995) Fuzzy recombination for breeder genetic algorithms. *Proceedings of the Sixth International Conference on Genetic Algorithms (ICGA'95)*, L. Eshelman (ed.), Morgan Kaufman, New York, pp. 104–111.

Whidborne, J.F. and Istepanian, R.S.H. (2001) Genetic algorithm approach to designing finite-precision controller structures, *IEE Proceedings – Control Theory and Applications*, 48(5), 377–382.

Yager, R.R. and Filev, D.P. (1994) *Essential of Fuzzy Modelling and Control*, John Wiley & Sons, Chichester.

Yun, Y. and Gen, M. (2003) Performance analysis of adaptive genetic algorithms with fuzzy logic and heuristics, *Fuzzy Optimisation and Decision Making*, 2, 161–175.

9

Evolutionary Neural Networks

9.1 Introduction

Layered feedforward neural networks have become very popular, for several reasons: they have been found in practice to generalize well and there are well-known training algorithms such as Widrow–Hoff, backpropagation, Hebbean, winner-takes-all, Kohonen self-organizing map which can often find a good set of weights. Despite using minimal training sets, the learning time very often increases exponentially and they often cannot be constructed (Muehlenbein, 1990). When global minima are hidden among the local minima, the backpropagation (BP) algorithm can end up bouncing between local minima without much overall improvement, which leads to very slow training. BP is a method requiring the computation of the gradient of error with respect to weights, which again needs differentiability. As a result, BP cannot handle discontinuous optimality criteria or discontinuous node transfer functions. BP's speed and robustness are sensitive to parameters such as learning rate, momentum and acceleration constant, and the best parameters to use seem to vary from problem to problem (Badi and Homik, 1995). A method called momentum decreases BP's sensitivity to small details in the error surface. This helps the network avoid getting stuck in shallow minima which would prevent the network from finding a lower-error solution (Vogt et al., 1988).

The automatic design of artificial neural networks has two basic sides: parametric learning and structural learning. In structural learning, both the architecture and parametric information must be learned through the process of training. Basically, we can consider three models of structural learning: constructive algorithms, destructive algorithms and evolutionary computation. Constructive algorithms (Gallant, 1993; Honavar and Uhr, 1993; Parekh et al., 2000) start with a small network (usually a single neuron). This network is trained until it is unable to continue learning, then new components are added to the network. This process is repeated until a satisfactory solution is found. These methods are usually trapped in local minima (Angeline et al., 1994) and tend to produce big networks. Destructive methods, also known as pruning algorithms (Reed, 1993), start with a big network that is able to learn but usually ends in over-fitting and try to remove the connections and nodes that are not useful. A major problem with pruning methods is the assignment of credit to structural components of the network in order to decide whether a connection or node must be removed. Both methods, constructive

Computational Intelligence: Synergies of Fuzzy Logic, Neural Networks and Evolutionary Computing, First Edition.
Nazmul Siddique and Hojjat Adeli.
© 2013 John Wiley & Sons, Ltd. Published 2013 by John Wiley & Sons, Ltd.

and destructive, limit the number of available architectures, which introduce constraints in the search space of possible structures that may not be suitable for the problem. Although these methods have been proved useful in simulated data (Thodberg, 1991; Depenau and Moller, 1994), their application to real problems has been rather unsuccessful (Hirose *et al.*, 1991; Hassibi and Stork, 1993; Kamimura and Nakanishi, 1994).

Several researchers have begun to research robust methods for overcoming these kinds of problems. One such method may be the application of EAs. Evolutionary neural network systems mainly means the design and training of neural networks by evolutionary algorithms. The interest in combinations of neural networks and evolutionary search procedures has grown rapidly in recent years. There are several arguments in favour of applying EAs to NN optimization (weights and/or topology), as EAs have the potential to produce a global search of the parameter space and thereby avoid local minima. Also, it is advantageous to apply EAs to problems where gradient information is difficult or costly to obtain. This implies that EAs can potentially be applied to reinforcement learning problems with sparse feedback for training NNs with non-differentiable neurons. The only obvious disadvantage with EAs is their slow time scale.

In EAs it is not only the algorithm, representation and operators used for the problem but also the strategy parameter values and operator probabilities to be chosen which influence the performance (meaning a good set of solutions) and the convergence (meaning a good set of solutions in good time, i.e. finding the solution efficiently). The process of finding the appropriate parameter values and operator probabilities is a time-consuming task. Researchers have put in considerable amounts of effort and experimented with various problems from a particular domain and attempted to tune the strategic parameters. It is obvious that an EA is intrinsically a dynamic and adaptive process. Moreover, there are also drawbacks of the traditional approach in the sense that strategic parameters are static and an inappropriate choice of parameters may lead to suboptimal performance. Tuning of the parameters will cost a significant amount of time and the optimal parameter value may also vary during the evolution process. A detailed discussion on EA parameters and their settings is provided in Section 8.4 of Chapter 8. Therefore, an adaptive mechanism for the strategy parameters is more desirable as the deterministic parameter setting varies from problem to problem. Different fuzzy logic-based approaches to adaptive mechanisms for EA parameter settings have been discussed in Section 8.4.1 of Chapter 8. The main difficulties of the fuzzy logic-based approaches are the construction of the MFs and the rule-based learning, which mainly require expert knowledge. In a similar way, NNs can also be applied to develop adaptive mechanisms for EAs. The advantage of NNs over fuzzy logic-based approaches is that an NN is capable of learning from experiential data. Adaptive learning of EA parameters has been reported and discussed by many researchers in various conferences and journals in a dispersed way. It needs to be addressed in a more structured way.

Various schemes for combining EAs and NNs have been proposed and tested by many researchers in the last two decades (Yao and Liu, 1998; Cantú-Paz and Kamath, 2005; Jung and Reggia, 2006) but the literature is scattered among a variety of journals, proceedings and technical reports. Mainly, three types of combination have been reported hitherto in the literature:

- Supportive combination,
- Collaborative combination and
- Amalgamated combination.

In supportive combination, EAs and NNs are used sequentially where one is the primary problem solver and the other is secondary. In collaborative combination, they are used simultaneously where both EAs and NNs solve the problem together. In amalgamated combination, the EA search mechanism is represented in an NN paradigm. The supportive, collaborative and amalgamated combinations are discussed in the following sections.

9.2 Supportive Combinations

An EA uses a population representative of the entire solution space of an optimization problem at hand. The EA then converges depending on the population, and different strategic parameters of the EA subject to a predefined fitness metric. In contrast, an NN uses an experiential data set representative of the input/output space. The learning then converges depending on the data set, learning parameters and architecture of the NN subject to a predefined performance metric. The performances of both EAs and NNs can be improved and the convergence accelerated if an appropriate population of data sets and strategic parameters or learning parameters can be found. A supportive combination between the EA and NN can help in finding these parameters. The supportive combination typically involves the use of one of these technologies to prepare data for use by the other. In other words, one technology plays the primary role and the other plays a supporting role to solve the problem. The supportive mechanism can be one of two ways:

- Neural networks to assist evolutionary algorithm (NN-EA) and
- Evolutionary algorithm to assist neural networks (EA-NN).

9.2.1 NN-EA Supportive Combination

In the case of NN-EA, the concept is that there seems to be some natural grouping within the problems. There are certain sets of heuristics that make better starting points for some groups than for others. The NN's job is to learn this grouping and suggest starting points for any evolutionary algorithms. In this case, neural networks are mostly used as pattern associators matching the descriptions of the incoming problem with a good parameter set. These neural networks are trained using standard BP algorithms. The diagram shown in Figure 9.1 explains the supportive combination of NN and EA, where the NN produces the initial population from the raw data for an EA. Kadaba *et al.* (1991) used an NN to produce an initial population for genetic algorithms (GA), where the GA plays the role of finding a set of good parameters for a heuristic procedure and finding a good set of selection-heuristics for a vehicle routing problem.

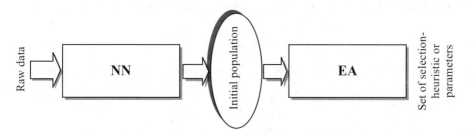

Figure 9.1 NN for grouping and suggesting initial population for EA

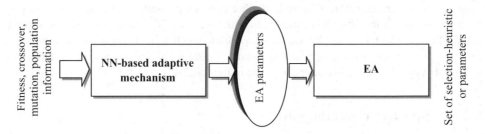

Figure 9.2   NN-based adapting parameter control of EA

The performance of an EA depends mainly on the choice of strategy parameter values. Finding a good set of strategy parameters in an EA is a time-consuming trial-and-error process. There are also technical disadvantages associated with traditional methods, such as that a user's choice in setting the parameters can be a source of errors and/or suboptimal performance, and optimal parameter values may vary during evolution. Therefore, an automated modification of the parameter values is sought during execution of an EA by using some heuristic rule or feedback information from the current state of the EA. An NN can assist in learning the parameter values of the EA. Such a supportive combination is shown in Figure 9.2.

Grefenstette (1986) identified the EA strategy (or control) parameters such as population size (N), crossover rate (P_c), mutation rate (P_m), generation gap (G), scaling window (W) and selection strategy (S), i.e., the possible set of control variables can be any of the $\{N, P_c, P_m, G, W, S\}$. Different performance indices can be computed from the fitness values of the individuals of the current population such as average fitness ($\bar{f}$), average fitness/best fitness ($\bar{f}/f_{best}$), worst fitness/average fitness ($f_{worst}/\bar{f}$), difference between best and average fitness ($f_{best} - \bar{f}$) or the current population size (N). That is, the possible set of controller inputs can be any of the $\{\bar{f}, \bar{f}/f_{best}, f_{worst}/\bar{f}, f_{best} - \bar{f}, N\}$. De Jong and Goldberg have been researching the effect of population size on EA performance. They found that as the current population size grows, the sensitivity to mutation rate decreases and the best mutation rate to use also decreases (De Jong and Spears, 1990; Goldberg *et al.*, 1992). The relationship between performance measures and strategy (or control) parameters has been discussed in Chapter 8, and led to the design of a fuzzy controller. The main difficulty in developing a multi-input multi-output (MIMO) fuzzy logic controller is the huge rule base. To overcome the problem of processing a huge rule base, a MIMO control mechanism using an NN can be used to control the EA parameters as shown in Figure 9.3. The NN's job here is to learn the strategy parameters of the EA using the error function derived from the difference between the desired or expected solution and the current solution. The standard BP with error feedback can be applied to train the NN (Kadaba and Nygard, 1990; Kadaba *et al.*, 1991).

9.2.2   EA-NN Supportive Combination

Most researchers have found it more natural to use EA to support NN. In the case of EA-NN, the supportive mechanism can be divided into three categories according to which stage they are used in the process:

- EA to select input features or to transform the feature space used by the NN classifier;
- EA to select the learning rules or parameters that control learning in the NN;
- EA to analyse an NN.

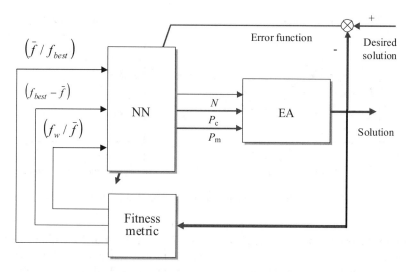

$$\left(\bar{f} / f_{best}\right)$$

$$\left(f_{best} - \bar{f}\right)$$

$$\left(f_w / \bar{f}\right)$$

Figure 9.3   NN-based adaptive control of population size, crossover and mutation rate

9.2.2.1   Feature Space Selection

An important issue in NN-based pattern classification and recognition problems is the feature selection. Given certain features or measurements of an object, one would like to determine which of these features is best suited for classifying or identifying the object from a given set of patterns or objects. One can naively attempt to examine each of the features to establish a preference ordering for the features. But if there are P features, then there are 2^P feature sets to be examined, which will be a time-consuming, tedious task. Moreover, it may be the case that the single best feature S^* is not present in the P features, i.e., $S^* \notin P$. Generally, the feature set is assumed to be a representative sample of the entire input space and we select a subspace $R^n \subseteq R^P$ in which features in R^n can be assigned to one of the N classes with minimum error. The minimum error is the expected number of misclassifications over some test set in R^n (Brill *et al.*, 1990). The transformation of the features may also be possible in R^n with minimum error. EAs are used in such cases to guide the search for optimal combination and transformation of the input features to NNs to satisfy the criteria of minimum inputs, faster training and accurate recall. EA has been used in preparing data for NN in two ways:

- Transforming the feature space and
- Selecting a subset of restricted features.

In the first approach, transforming the feature space has mainly been applied to nearest-neighbour-type algorithms. A very common example of transformation would be the temperature x^f measured in Fahrenheit scale being transformed into Celsius scale x^c using the formula $x^c = (x^f - 32)/1.8$. A further example may help in perceiving the meaning of transformation of data: let $x_i = [x_{ik}]$, $i \in P$, $k \in K$ be a feature vector, where P is the set of features and K is the set of variables. The Euclidean space R^n (where $n = |K|$) consists of all n-dimensional vectors of the form $x = [x_1, x_2, \ldots, x_n]$. The feature vector $x = [x_1, x_2, \ldots, x_n]$ can be transformed into $z = [z_1, z_2, \ldots, z_n]$ by applying a formula such as $z_{ik} = (x_{ik} - a_k)/b_k$, where a_k

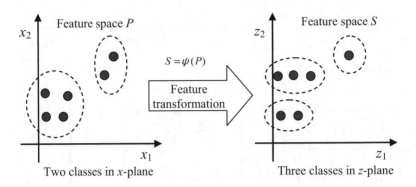

Figure 9.4 Two different classifications based on feature transformation

is the shift of the origin and b_k is the change of scale factor. By transformation is meant the shifting, rotation and scaling of the data to be aligned in such a manner that intraclass differences are diminished and interclass differences are magnified (Kelly and Davis, 1991; Mirkin, 1996; Yao, 1999). Figure 9.4 shows an example of two different classifications obtained by applying feature transformation on the same set of features. Interested readers are referred to Mirkin (1996) for a variety of transformation techniques used in mathematical classification and clustering.

As can be seen from Figure 9.4, inter-data distances and data clusters vary depending on which space x or z is considered. This reflects the contradiction between the features, their geometrical representation and comparability. The problem of finding a mechanism to aggregate the incomparable variables and their distance is a major problem in classification and clustering. By letting the EA choose the rotation and scaling parameters, the data is aligned for appropriate classification and clustering. An NN is used for the classification, where an EA provides the appropriate transformation. Such a supportive combination of EA-NN is shown in Figure 9.5, where the NN uses a subset of features for classification transformed by the EA. Here, S is the set of transformed features, i.e., $S = \psi(P)$, where $\psi(.)$ is a transformation function and P is the set of all features. F is the fitness measure for the EA.

Verma and Zhang (2007) transformed all the feature values to be positive and then normalized these values to the range [0, 1] prior to digital mammogram classification for breast

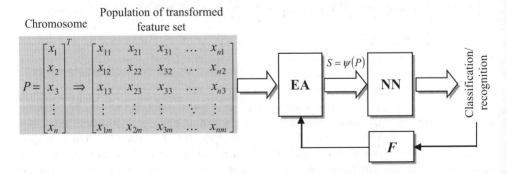

Figure 9.5 EA-based feature transformation for NN

cancer. Ramasubramanian and Kannan (2006) transformed alphabetic attributes of the data set into appropriate numeric values and then normalized the numeric values to the range [0.05, 0.95] in their application.

Very often the feature set to be used for training the NN contains redundant, useless and irrelevant data, which leads to higher computation costs without improving the performance of the NN and can even sometimes degrade the performance. A practical example of such a problem is the task of selecting a subset of clinical tests. Each test involves a financial cost, diagnostic value and associated risks. Other examples of such applications are large-scale data mining using NNs. In the second approach, therefore, a set of input features is selected as a restricted feature set that will improve the performance of a neural network classifier as well as reduce the computational requirements. Elimination of the irrelevant and redundant features, in other words selecting a subset of features, i.e., $S \subseteq P$, will help reduce the size of the NN, reduce the training time and can also improve the accuracy of the NN. Some applications require important variables to be distinguished from other variables. When no *a priori* knowledge is available to guide the selection, the EA can help in selecting inputs for neural networks for such applications.

Chromosome representation for the EA is straightforward. A binary-coded chromosome is used for the EA. This is also called the wrapper approach (Kohavi and John, 1997). The traditional binary bit strings are used as individuals to form a population for the EA search. Each position in a bit string is associated with an input variable, i.e., a feature which may or may not be selected as an input to the neural network depending upon the value of 1 or 0 in that position. Therefore, the number of bits in a string equals the number of variables (features) listed for the EA search, and the number of 1's in a string equals the number of variables selected by the EA as the inputs to the neural networks. That is, $x_i = 1$, if $x_i \in P$ and $x_i = 0$, if $x_i \notin P$ with $i = 1, 2, \ldots, n$. Siedlecki and Sklansky (1988, 1989) were the first to introduce the pioneering work on feature selection formulation using EAs. Since then, there have been many researches reported in the literature (Brill *et al.*, 1990; Vafaie and De Jong, 1993; Brotherton and Simpson, 1995; Yang and Honavar, 1998). For a population size of m, the supportive combination for feature selection by EA is shown in Figure 9.6. The individuals are evaluated by training the NN using the feature subset $S \subseteq P$ with a predefined fixed architecture for NN. The resulting accuracy is used as the fitness value of the individual. The main drawback of this approach is the high computation time required to train each network

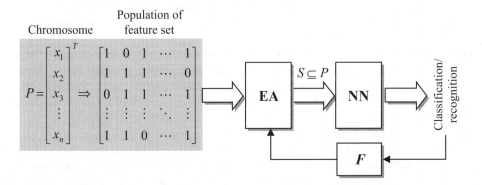

Figure 9.6   EA-base feature selection for NN

classifier using the features specified by the chromosome. Cantú-Paz and Kamath (2005) performed an empirical evaluation of different combinations of EA and NN for classification problems of 15 public-domain and artificial data sets. They used a binary-coded GA with one bit for each feature to represent a subset of features that are being used to train an NN. A population of $\lfloor 3\sqrt{l} \rfloor$ individuals was initialized uniformly at random with an enforcement of a minimum population size of 20. Here, l is the chromosome length. A uniform crossover with probability 1.0 and mutation with probability $1/l$ were used. The fitness function used was the generalization ability of the NN. The results indicate that EA feature selection proved significantly accurate and significantly reduced the feature set.

For a given number of training cycles, the NN will have different performances (measured by the recall error of the network) for different sets of input variables. The NN performance and the number of variables selected as network inputs can be used to evaluate the fitness function whose value will be used to guide the selection process in the EA. A general procedure can be outlined as below:

1. Create an initial population of individuals, transformed features or bit strings of features, each of which can be interpreted as a group of selected inputs to the NN.
2. Set up and train the NN with the inputs selected in step 1.
3. Transfer the training error and the number of inputs selected to the fitness function to evaluate the fitness for each NN.
4. Perform the EA operation to create a new population.
5. Repeat steps 2–4 until convergence or max generation is reached.

The fitness function F for both the feature transformation and feature selection (in Figures 9.5 and 9.6, respectively) mechanism can be defined by combining two criteria: one is the accuracy of the classification realized by the NN and the other is the cost of classification (Yang and Honavar, 1998). The accuracy is calculated from the percentage of patterns that are correctly classified by NN. Several measures of classification cost are suggested, e.g., the cost of performing necessary tests in medical diagnosis applications. For simplicity, the fitness function can be defined as

$$F(S) = \text{accuracy}(S) - \frac{\text{cost}(S)}{\text{accuracy}(S) + 1} + \text{cost}_{\max} \tag{9.1}$$

Here, accuracy(S) is the percentage of patterns that are correctly classified by the NN, cost(S) is the computational or test cost involved for the patterns and cost$_{\max}$ is the maximum cost that may be involved in performing the pattern classification correctly. The goal of any fitness function should be to guide the search for fewer features, faster training and higher accuracy. Both fewer features and smaller training errors produce higher fitness function values. Therefore, some researchers have defined the fitness function as the function of three independent variables: the number of selected features (i.e., inputs), the training error and the generation indices.

In an EA with small population, there are few superstrings with much higher fitness values, which will lead the superstrings to take over a significant proportion of the population in a single generation and cause premature convergence. One way to prevent such premature convergence is to maintain the population diversity. To keep the diversity of the population, the fitness function should be scaled down in the early stages to reduce the fitness difference among individuals and be scaled up in the later stage. Including the generation indices in the fitness function helps in scaling the fitness appropriately and guiding the search over generations. There are three commonly used scaling procedures: linear scaling, sigma truncation and power law scaling (Guo and Uhrig, 1992). The fitness function used by Guo and Uhrig (1992) in their study is defined as follows:

$$F = \left(1.0 - e^{-(r-1)^{0.15(g+1)}}\right).e^{-0.01err^{-0.01err^{0.7(g+1)^{1/3}}}} \tag{9.2}$$

where $r \doteq \dfrac{\text{number of features}}{\text{number of selected features}}$, err is the NN training error and g is the generation index. Many researchers have proposed an evaluation function based on the hypothesis that the relevant feature is highly correlated with the response variables and less correlated with the other features in the feature subset (Ozdemir et al., 2001).

9.2.2.2   Finding Parameters and Learning Rule

The most widely used and well-known learning algorithms for NN are backpropagation, Widrow–Hoff, Hebbian and Kohonen self-organizing map. Among them, BP is known to implement a gradient descent method which has the drawbacks of being slow for large problems and being susceptible to becoming stuck in local minima or in a plateau, can stop at early convergence or can take too long time to converge due to heuristic selection of control parameters. The control parameters are learning rate $\eta, 0 < \eta \leq \Re$, momentum $\alpha, 0 \leq \alpha \leq \Re$ and acceleration $\beta, 0 \leq \beta \leq \Re$ as defined in Equation (9.3). $\Re$ is a positive real value. Several researchers have investigated different modifications to speed up the convergence of the BP algorithm by applying dynamic adaptation of the learning rate such as adding a momentum and acceleration term in the weight update rule as described by Equation (9.3):

$$\Delta w_i(t) = -\eta \frac{\partial E}{\partial w_i} + \alpha \Delta w_i(t-1) + \beta \Delta w_i(t-2) \tag{9.3}$$

Dynamic adaptation of the learning rate can be done by line search in the gradient direction, dynamically adjusting the learning rate commonly for all weights or separately for each weight (Caudell and Dolan, 1989; Kamarthi and Pittner, 1999). Therefore, the weight update $\Delta w_i(t)$ depends mainly on the learning parameters $\{\eta, \alpha, \beta\}$ and the local gradient $\frac{\partial E}{\partial w}$. The local gradient depends mainly on the first derivative term of the activation function $\varphi(.)$, i.e., $\varphi'(.)$. The factor $\varphi'(.)$ involved in the computation of the local gradient $\delta_j(t)$ (see derivation of BP algorithm in Chapter 4) depends solely on the activation function $\varphi(.)$ associated with hidden-layer neurons. This means that the set of activation functions $\varphi(.)$ plays a vital role in the weight updating in the BP learning rule. BP's speed and robustness are sensitive to several of its control parameters such as $\{\eta, \alpha, \beta\}$ and the activation function $\varphi(.)$. A detailed description of control parameters and different activation functions is given in Chapter 4. There is no straightforward way to choose these parameters and the best parameters to use seem to vary from problem to problem. Mostly, these control parameters are determined by trial and error, which depends on the type of architecture being used. Different variants of the Hebbian leaning rule have been proposed to deal with different architectures. However, determining an optimal set of control parameters and learning rule becomes difficult when the type of architecture is not known *a priori*. Therefore, the NN needs to adjust the control parameters and learning rule adaptively according to the problem and its architecture rather

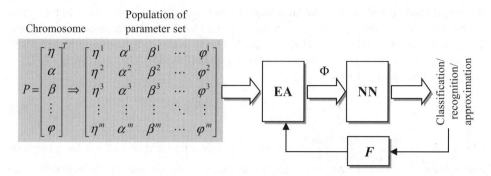

Figure 9.7   EA for parameter learning and learning rule

than having a fixed design. Several researchers used EA to learn the control parameters and learning rule of the NN (Harp *et al.*, 1989; Kim *et al.*, 1996; Taheri and Mohebbi, 2008). Figure 9.7 illustrates a supportive combination of EA and NN, where EA provides the NN parameters and the learning rule. In the diagram, Φ represents the set of NN parameters and learning rule, that is $\Phi = \{\eta, \alpha, \beta, \varphi_h, \varphi_o\}$. φ_h and φ_o are the activation functions of the hidden and output layers of the NN, respectively. It is assumed that the activation functions are the same for the same layer and may vary from layer to layer.

To evaluate the effectiveness of the set of parameters, the NN is trained for a fixed number of iterations. During the training process, the NN will achieve different performance levels (measured by the recall error of the network) for different sets of learning and control parameters. A typical procedure of the evolution of the learning rules can be outlined as below:

1. Create an initial population of individuals consisting of learning and control parameters, each of which can be interpreted as a group of selected parameters to the NN.
2. Set up and train the NN for a fixed number of iterations.
3. Transfer the training error and the set of parameters to evaluate the fitness for each NN.
4. Perform EA operations to create a new population.
5. Repeat steps 2–4 until convergence or max generation is reached.

It is assumed that the architecture of the NN is predefined and fixed in the above procedure (Belew *et al.*, 1991; Kim *et al.*, 1996). If varied NN architectures are to be considered for a near optimal learning rule in the evolution, the fitness should be the average fitness of the set of architectures (Jacob, 1988; Harp *et al.*, 1989). Some researchers applied an EA to extract rules of reinforcement learning and used these rules to train the NN (Zitar and Hassoun, 1995). As the performance of any NN also depends on the architecture, optimization of the architecture is an important issue in the combination of EA and NN. The evolution of the architecture is addressed in Section 9.3.2 in this chapter.

Another key issue is how to encode the dynamic behaviour of a learning rule into static chromosomes. A universal representation scheme would be impractical due to the prohibitively long computation time required to search the large learning rule space. Two assumptions that have been made often on learning rules are: (i) weight updating is dependent on local information such as activation of input and output nodes and current connection weights; (ii) learning rule is the same for all connections in the NN. Based on the two assumptions, a learning rule can be assumed to be a linear function of the local variables and their products (Yao, 1995, 1999) and defined as follows:

$$\Delta w(t) = \sum_{k=1}^{n} \sum_{i_1,i_2,\ldots,i_k=1}^{n} \left(\theta_{i_1,i_2,\ldots,i_k} \prod_{j=1}^{k} x_{i_j}(t) \right) \tag{9.4}$$

where t is time, Δw is the weight change, $x_1, x_2, \ldots, x_n$ are local variables and the θ_i's are real-valued coefficients that determine the learning rules and undergo evolution. The number of terms in Equation (9.4) involved in the evolution is large, which makes the evolution impractical. Therefore, a small subset of terms is to be determined and a real-valued chromosome representation is to be used for the coefficients to be evolved. Using a linear combination of four local variables and their six pairwise products with no third- or fourth-order terms, ten coefficients and a scale parameter encoded in a binary string, an EA discovered the well-known delta learning rule and some of its variants within 1000 generations of the evolution (Chalmers, 1990; Fontanari and Meir, 1991). Fontanari and Meir used four local variables and seven terms, which included one first-order, three second-order and three third-order terms in their EA.

9.2.2.3   Explaining and Analysing Neural Networks

One inherent drawback of the solutions offered by NN to dynamical systems is their limited explanation capability. The solutions are hard to trace back or explain and are often due to random factors. Instead of using an EA to construct a better NN in terms of performance and architecture, a few researchers have used an EA to help explain or analyse the NN. In order to explore the 'decision surface' of an NN, the EA can be used to discover input patterns that result in maximum or nearly maximum activation values for given output neurons. The input patterns are represented in the chromosome by a set of real values between 0.0 and 1.0. The EA is to discover three different types of vectors: (i) maximum activation vectors, meaning output node is activated; (ii) minimum activation vectors, meaning output node is off; and (iii) decision vectors, meaning output node is at the decision threshold. Multiple runs of any EA with different random seeds can be used to find a set of vectors of each type (Eberhart, 1992).

The NN model of a dynamical system processes the initial condition information over time while moving through a sequence of states. An attractor is a state towards which the system evolves over time from an initial condition. The set of conditions for an attractor is called the basin of attraction. The attracting sets may be represented algebraically as an n-dimensional vector. In physical systems the n dimensions may be, for example, two or three positional coordinates for each of one or more physical entities. An attractor can be a point, a finite set of points, a curve, a manifold or even a complicated set with a fractal structure known as a strange attractor. Some researchers used EAs to analyse the basins of attraction of a correlation associate memory model of an NN. The recall process of the attractor basin shows threshold and monotonous transition phenomena, which can be represented by a polynomial function of degree 2. The polynomial is a characteristic measurement of the attractor basin. Suzuki and Kakzu (1991) used GA to determine the optimal coefficients of the polynomial function, i.e., for the characteristic measurement.

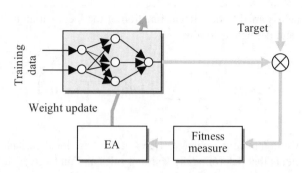

Figure 9.8   Learning weights of NN

9.3   Collaborative Combinations

In collaborative combinations, EAs and NNs work to solve a problem together. Among the collaborative approaches, there are two main groupings.

- Nguyen and Widrow demonstrated that weights and biases generated with certain constraints result in a faster learning speed for an NN (Nguyen and Widrow, 1990). Also, known weight training procedures for NNs are biased towards the data and parameter sets used for training. Very often these training procedures do not guarantee a global optimal weight set for the NN. There have been attempts to use evolutionary search to find appropriate connection weights in fixed architectures. Such a general mechanism of training NN weights using EA is shown in Figure 9.8. The issues relating to weight training are discussed in Section 9.3.1.
- Finding an appropriate architecture (topology) of an NN for a given problem is a time-consuming trial-and-error task. Alternatively, EAs have been used to find the optimal network architecture and are then trained and evaluated using known learning procedures (BP, Widrow–Hoff, Hebbian, SOM, etc.). Such a general collaborative combination to determine the network topology using an EA is shown in Figure 9.9. The issues relating to network topology are discussed in Section 9.3.2.

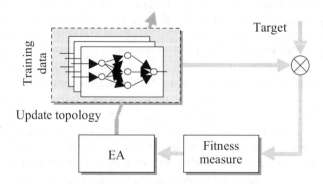

Figure 9.9   Learning architecture

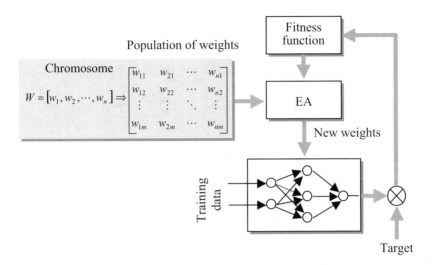

Figure 9.10   Weight training of an NN using EA

9.3.1   EA for NN Connection Weight Training

Supervised or unsupervised learning in an NN has mostly been formulated as a weight training process in which efforts are made to find an optimal set of connection weights according to some optimality criteria. To overcome the shortcomings in gradient descent, Widrow–Hoff or Hebbian learning, a global search procedure like EA can be used effectively in the training process as an evolution of connection weights towards an optimal set defined by a fitness function (Fogel *et al.*, 1990; Schaffer *et al.*, 1992; Zhao and Higuchi, 1996; Nikolaev and Iba, 2001, 2003). The other advantage is that an EA can handle large, complex, multimodal and non-differentiable functions.

The EA approach to weight training in an NN consists of three phases: chromosome representation of connection weights, definition of a fitness function and definition of genetic operators in conjunction with the representation scheme. Different representations and applicable genetic operators can lead to different training performance. A typical weight training process for the NN using an EA is shown in Figure 9.10.

9.3.1.1   Chromosome Representation

A major issue that is vital in the evolutionary training approach is to decide on the representation scheme of the connection weights. Different representations can lead the EA to quite different training performance in terms of training time and accuracy. The most convenient representation of connection weights is with binary strings. In such a representation scheme each connection weight is represented by some binary bits of certain length.

The most convenient and straightforward chromosome representation of connection weights and biases is in string form. In such a representation scheme, each connection weight and bias is represented by some value $\{w, b\} \in \Re$ where $\Re$ can be a real number or a binary number. An example of such a string representation scheme for a feedforward NN with five neurons is shown in Figure 9.11.

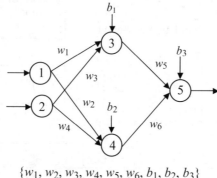

$$\{w_1, w_2, w_3, w_4, w_5, w_6, b_1, b_2, b_3\}$$

Figure 9.11   Chromosome represented in string form

9.3.1.2   Binary Representation

The advantage of binary representation is its simplicity and ease of applying genetic operators such as crossover (single-point or uniform) and mutation (bit inversion). In binary representation, each connection weight or bias $\{w, b\} \in \Re$ is represented by a number of bits of length l such that $2^l \approx \Re$. An NN is then encoded by concatenating all the connection weights and biases. A rule of thumb is to order the neurons of the NN in some way, as shown in Figure 9.11. Separating inputs to the same hidden node far apart in the binary representation would increase the difficulty of constructing useful feature detectors as the hidden nodes in an NN act as a feature extractor. The binary encoding of connection weights need not be uniform, as adopted by many researchers. It can also be Gray, exponential or a more sophisticated encoding. A limitation of binary representation is the precision of discretized connection weights. If too few bits are used to represent weights, training may take an extremely long time or even fail. On the other hand, if too many bits are used, the chromosome string for a large NN becomes very long, which will prolong the evolution dramatically and make the evolution impractical. It is still an open issue how to optimize the number of bits for each connection weight, range encoded and encoding scheme used. A dynamic encoding scheme can be adopted to alleviate those problems. An example of binary coding of the chromosome for the NN shown in Figure 9.11 can be as follows:

String representation: $\{w_1, w_2, w_3, w_4, w_5, w_6, b_1, b_2, b_3\}$

Binary coding: $\{0100 \mid 1010 \mid 0011 \mid 1010 \mid 0010 \mid 0001 \mid 1101 \mid 1001 \mid 1011\}$

A detailed discussion on different genetic operators applicable to binary coding is presented in Chapter 6.

9.3.1.3   Real-Value Representation

To overcome these shortcomings of the binary representation scheme, real numbers were proposed $\{w, b\} \in \Re$, i.e., $\Re$ is a real number per connection weight or bias. The chromosome is then represented by concatenating these numbers as a string. For example, a real number representation of the chromosome for the NN shown in Figure 9.11 is given by

String representation: $\{w_1, w_2, w_3, w_4, w_5, w_6, b_1, b_2, b_3\}$

Real coding: $\{1.91 \mid 2.55 \mid 1.9 \mid 3.12 \mid 0.88 \mid 0.91 \mid 1.1 \mid 1.9 \mid 0.98\}$

The advantages of real coding are many-fold, such as shorter string length with increased precision. Various kinds of crossover and adaptive crossover are applicable here. The standard mutation operation in binary strings cannot be applied directly in the real representation scheme. In such circumstances, an important task is to carefully design a set of genetic operators suitable for the real encoding scheme. For example, mutation in real number chromosome representation can be as follows:

$$w_i(t) = w_i(t-1) \pm \text{random}(0, 1) \tag{9.5}$$

$$b_i(t) = b_i(t-1) \pm \text{random}(0, 1) \tag{9.6}$$

Montana and Davis (1989) defined a large number of domain-specific genetic operators incorporating many heuristics about training NNs. A detailed discussion on different genetic operators applicable to real coding can be found in Chapter 6.

9.3.1.4   Matrix Representation of Chromosome

Another way of representing a chromosome for a feedforward NN is that an NN can be thought of as a weighted digraph $G = \{E, V\}$ with no closed paths and described by an upper or lower diagonal adjacency matrix with real-valued elements, where E is the set of all edges of the graph and V is the set of vertices (neurons in NN) in the digraph. The nodes in the NN should be in a fixed order according to layers. An adjacency matrix is an $N \times N$ array in which elements

$$n_{ij} = 0 \quad \text{if } \langle i, j \rangle \notin E \quad \forall i \leq j \tag{9.7}$$

$$n_{ij} \neq 0 \quad \text{if } \langle i, j \rangle \in E \quad \forall i \leq j \tag{9.8}$$

where $i, j = 1, 2, \ldots, N$ and $\langle i, j \rangle$ is an ordered pair and represents an edge or link between neurons i and j, and N is the total number of neurons in the network. The biases of the network are represented by the diagonal elements of the matrix expressed as

$$n_{i,j} \neq 0 \quad \forall i = j \tag{9.9}$$

Thus, an adjacency matrix of a digraph can contain all information about the connectivity, weights and biases of a network. For example, the adjacency matrix for the three-layered feedforward NN with bias, shown in Figure 9.12, is illustrated in Figure 9.13.

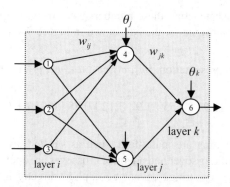

Figure 9.12   Digraph of a three-layered NN

Nodes 1, 2 and 3 are the input neurons and they are connected to hidden layer neurons 4 and 5. These connection weights are represented by $\{w_{14}, w_{15}, w_{24}, w_{25}, w_{34}, w_{35}\}$. The hidden-layer neurons are connected to the only output neuron 6 and the connection weights are represented by $\{w_{46}, w_{56}\}$. The hidden-layer and output-layer neurons have biases. The biases are represented by $\{w_{44}, w_{55}, w_{66}\}$. An example of real-valued chromosome representation in matrix form is shown in Figure 9.14.

A layered feedforward network is one such that a path from input node to output node will have the same path length. Thus, an n-layered neural network has path length n. The added advantage of the matrix representation is that it can be used for a fully recurrent network as well. In this case the matrix will be a full matrix in that the weights and biases are the elements as defined below:

$$n_{ij} \neq 0 \quad \text{if } \langle i, j \rangle \in E \quad \forall i \neq j \text{ (for weights)} \tag{9.10}$$

$$n_{i,j} \neq 0 \quad \forall i = j \text{ (for bias)} \tag{9.11}$$

The matrix or two-dimensional representation of the genes has the advantage of having the positional information of the connectivity of the network. For example, the incoming connections or outgoing connections of a neuron are kept in the rows or in the columns. Theoretical analysis suggests that the performance of an EA is better for a chromosome representation

	Input			Hidden		Output
node	1	2	3	4	5	6
1	0	0	0	w_{14}	w_{15}	0
2	0	0	0	w_{24}	w_{25}	0
3	0	0	0	w_{34}	w_{35}	0
4	0	0	0	w_{44}	0	w_{46}
5	0	0	0	0	w_{55}	w_{56}
6	0	0	0	0	0	w_{66}

Figure 9.13   Matrix representation of chromosome

	Input			Hidden		Output
node	1	2	3	4	5	6
1	0	0	0	0.9	0.6	0
2	0	0	0	1.1	0.4	0
3	0	0	0	1.5	1.9	0
4	0	0	0	0.3	0	0.1
5	0	0	0	0	0.2	1.2
6	0	0	0	0	0	1.0

Figure 9.14   Real-valued chromosome represented in matrix form

where the functionally similar genes are kept close together. The population can be thought of as a matrix of layers, as shown in Figure 9.15. The diagonal elements $\{w_{11}, w_{22}, \ldots, w_{nn}\}$ represent the bias of the NN. Any standard genetic operators can be applied to the chromosome representation scheme shown in Figure 9.15. For example, crossover could be implemented by swapping rows (incoming connections), columns (outgoing connections) or swapping a submatrix (functional groups of neurons), which will prevent disruption of the closeness of similar genes within the chromosome.

Siddique and Tokhi (2001) developed different types of crossover and mutation operations for matrix representation of chromosomes, as discussed in the following sections.

9.3.1.5 Weight Crossover

From programming point of view, handling matrices is simple. Rows and columns can be swapped or modified easily. Two types of crossover are shown in the example here, firstly row-wise crossover and secondly column-wise crossover. In row-wise crossover, an offspring is generated by choosing alternate rows (or swapping rows) from parent chromosome matrices. This is shown in Figure 9.16. It is similarly done column-wise, as shown in Figure 9.17.

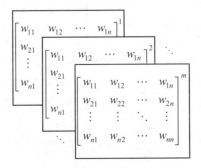

Figure 9.15   Population of weight matrices

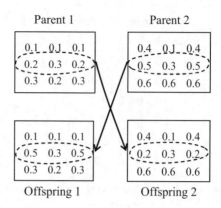

Figure 9.16   Row-wise crossover operation in matrix representation

9.3.1.6   Weight Mutation

A weight value from the set of weights and biases is selected with a certain probability and its value is modified by a random value. The mutation for weights and biases is shown in Equations (9.12) and (9.13), respectively:

$$w_{ij}(g) = w_{ij}(g) \pm \text{random}(0, 1) \tag{9.12}$$

$$w_{ii}(g) = w_{ii}(g) \pm \text{random}(0, 1) \tag{9.13}$$

Here, g is the generation index.

9.3.1.7   Bias Swap

Siddique and Tokhi (2001) applied bias swap to matrix chromosomes. Bias is an adjustable scalar parameter added to the summation of the neuron contributing to shift the output of the

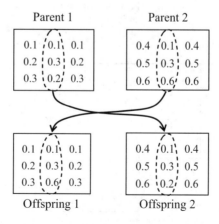

Figure 9.17   Column-wise crossover operation in matrix representation

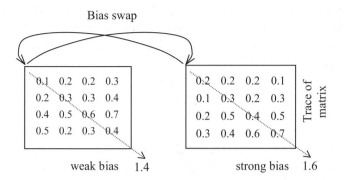

Figure 9.18   Bias mutation operator in matrix representation

neuron left or right depending on the sign of the bias. Adaptation of the bias can be carried out to verify the influence of bias on the output of the network. Weak biases are replaced by strong biases as determined by the trace of the adjacency matrix defined below:

$$T(g) = \sum_{i=1}^{n} w_{ii}(g) \tag{9.14}$$

Strong biases have a greater trace value than weak biases. That is, if $T_1(g) < T_2(g)$ then the biases in parent 1 are weaker. The entire weak biases (diagonal elements of the chromosome matrix) are replaced with entire strong biases. The bias swap is shown in Figure 9.18. Similarly, bias crossover can be implemented by swapping the diagonal elements between two matrices.

9.3.1.8   Fitness Function for Weight Evolution

The fitness can be defined as the minimum of the sum squared error (SSE) or mean square error (MSE) of the network over a set of training data after training the network for a fixed number of iterations:

$$f(NN) = \sum_{P} e^2 \tag{9.15}$$

$$f(NN) = \frac{1}{P} \sum_{P} e^2 \tag{9.16}$$

Here the network is predefined and fixed, denoted NN and defined by $NN = \langle A, W, \Phi \rangle$, where A is the architecture, W is the set of weights, Φ is the set of activation functions and P is the number of patterns used for training.

Some researchers used a fitness function based on the sample counter-changing method (Gao, 2003). As to network individual training of each generation, the whole set of training sample is not used, i.e., a part of the training sample set (say about 80% of the sample) is randomly chosen to train the individual of each generation. So, the training sample set used (denoted x^a in the equation) for the neural network of each generation is changed, and then the fitness of the individual whose generalization capacity is poor will be smaller while the fitness of the individual whose generalization capacity is strong will become large. Consequently, the performance of the whole NN model is improved through selection. The error function of the neural network is expressed as follows:

$$E = \frac{1}{2} \sum_{a=1}^{N} \sum_{k=1}^{M} \left[y_k \left(w_k; x^a \right) - t_k^a \right]^2 \tag{9.17}$$

where y_k is the network output and t_k^a is the target output for the sample set x^a. The individual fitness of the neural network is expressed by the following transformation of the error function of the neural network:

$$f(NN) = \frac{1}{1+E} \tag{9.18}$$

Some researchers have defined the fitness as the number of correctly labelled instances returned by NN among inputs (Tong and Mintram, 2010). This fitness function may be better suited for feature selection than weight training.

9.3.2 EA for NN Architectures

It is well known that an NN's architecture (or topology) plays a significant role in the NN's information-processing abilities. The architecture of an NN includes the topological structure or connectivity and the transfer function of each node of the NN. Unfortunately, there is no systematic way to design an optimal architecture for a particular task and they are mostly designed by experienced experts through tedious trial-and-error processes. For example, given a learning task, an NN with few connections (inputs to hidden layer), hidden neurons and linear node transfer (activation) functions may not be able to show the prediction or approximation capability due to limited information-processing ability, while an NN with large number of connections (inputs to hidden layer), hidden neurons and nonlinear node transfer functions may demonstrate a significant performance improvement.

Figure 9.19 shows two different RBF network architectures (discussed in detail in Chapter 4) with one input and three inputs (one input and two delayed inputs) for two-step prediction

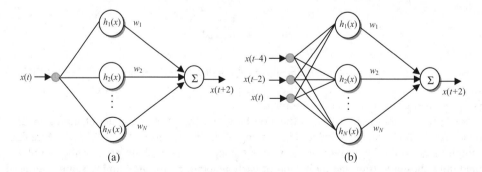

(a)   (b)

Figure 9.19   Two different RBF networks. (a) 1-1 I/O RBF; (b) 3-1 I/O RBF

of MacKey–Glass time series. The RBF network in Figure 9.19(b) has three times more connections from input to hidden layer than that of the network in Figure 9.19(a). The respective prediction errors of the network architectures are shown in Figure 9.20. Similarly, varying the number of neurons in the hidden layer will also cause a change in the performance (Billings and Zheng, 1994).

Therefore, the optimal NN architecture can be viewed as a complex search problem in the design space according to some optimality criterion such as NN performance, minimal number of neurons, fast learning, simplicity of connectivity, etc. and thus forms a complex surface that has to satisfy the optimality criteria. The surface has the following characteristics according to Miller *et al.* (1989) and Yao (1999):

- The surface is infinitely large since the number of possible neurons and connections is unbounded.
- The surface is non-differentiable since changes in the number of neurons or connections are discrete and can have a discontinuous effect on the performance.
- The surface is complex and noisy since the mapping from the NN's architecture to performance after training is indirect, strongly epistatic and dependent on initial conditions.
- The surface is deceptive, since NNs with similar architecture may have dramatically different information-processing abilities and performances.
- The surface is multimodal, since NNs with quite different architectures can have very similar performance and capabilities.

Finding an optimal architecture is equivalent to finding the optimal point on the complex surface. As the EA is a parallel and stochastic search technique, these characteristics make the EA a better candidate for searching the surface (Yao, 1993b). Theoretical analyses suggest that the EA can quickly locate high-performance regions on the surface. The key issue in applying the EA is to decide how much information about the architecture should be encoded into the chromosome representation for the EA to be successful. There are several ways to encode the NN architecture into chromosomes suitable for different variants of EA. These encoding schemes are divided into the following approaches (Yao, 1993a,b, 1999; Vonk *et al.*, 1997; Igel and Stagge, 2002; Igel and Kreutz, 2003; Teoh *et al.*, 2006):

- Direct encoding
- Indirect encoding
  - o Parametric encoding
  - o Grammar encoding
  - o Tree encoding
- Fractal encoding of connectivity.

9.3.2.1   Direct Encoding

In direct encoding, all information about the architecture can be represented as a binary string that directly represents the NN architecture with one-to-one correspondence between the genes and the connectivity of the NN, i.e., the entire network structure is encoded into the chromosome. This kind of representation is called a direct encoding scheme. In a direct encoding scheme, a network can be represented by an $N \times N$-dimensional connectivity matrix

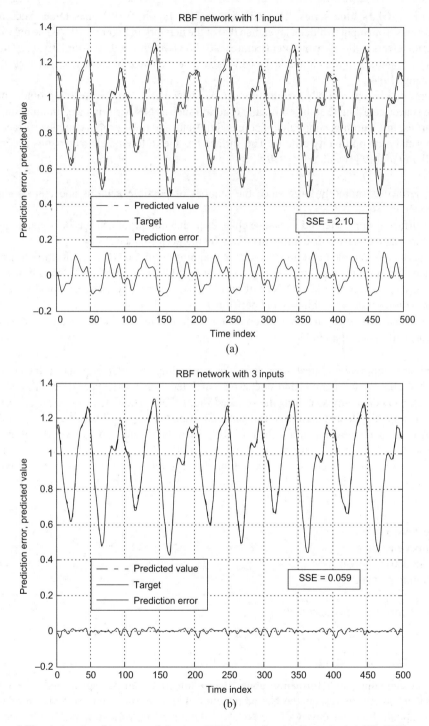

Figure 9.20 Approximation error of the two RBF networks. (a) Error in 1-1 I/O RBF networks; (b) Error in 3-1 I/O RBF networks

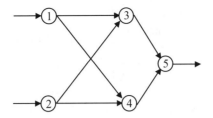

Figure 9.21 Connectivity of a feedforward NN

$C = [c_{ij}]$, $i = 1, 2, \ldots, N$ and $j = 1, 2, \ldots, N$ that constraints connections between the N neurons of the network, where $c_{ij} = 1$ indicates the presence of a connection from node i to node j and $c_{ij} = 0$ indicates that there is no connection between nodes i and j. Figure 9.21 shows the connectivity of a feedforward network of five neurons. The connection matrix (also called the adjacency matrix) is then converted to a bit string genotype of length $N \times N$ by concatenating the successive rows, as shown in Figure 9.22. A common problem is that the chromosome length becomes very large with increasing network size, which results in high computation cost and slow performance. This is a common problem with any large NN.

It can be seen that the connectivity matrix in Figure 9.22 for a feedforward NN is a lower-diagonal matrix. It will be an upper-diagonal matrix if the order of the nodes is changed. Using the lower- or upper-diagonal matrix, the length of the chromosome can be reduced significantly for feedforward NN architectures. It is obvious that this coding scheme can handle both feedforward and recurrent NN architectures.

The direct encoding also suffers from competing conventions, meaning that the same architecture can be represented by many connectivity matrices. The bit-string representation also causes functionally similar genes to go far apart from each other, i.e., outgoing connections. This is caused by the mapping from a two-dimensional representation of structural information into a one-dimensional chromosome. But the incoming connectivity information remains close together, which may help improve the performance of the EA according to the schema theorem. In order to improve the performance, the schema theorem (Holland, 1975) suggests keeping functionally similar genes together and not dispersing them by genetic operators. Individuals are translated into networks and evaluated using standard training procedures. A typical evolution process for the NN architecture using a direct encoding scheme will look like that in Figure 9.23.

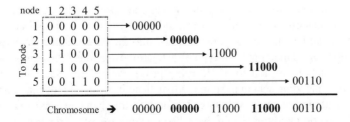

Figure 9.22 Chromosome representation of the connectivity matrix

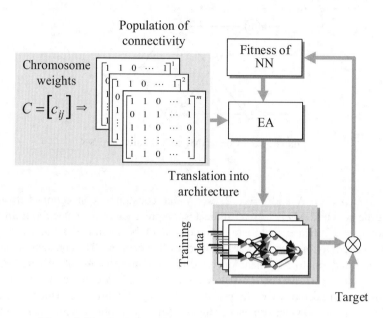

Figure 9.23   Evolution of NN architecture using direct encoding

One major problem with the evolution of an NN architecture is the permutation problem, which represents functionally equivalent NN architectures in different genotype representations. The permutation problem is quite vulnerable to the crossover operation and hence some researchers have adopted only the mutation operation in the evolution of the architecture. Though the direct encoding scheme is suitable for deterministic handling of small NN architecture, it does not scale well since a large architecture require large matrices to represent the architecture, which makes the evolution slower. To reduce the size of the chromosome, domain knowledge is used to restrain the search space.

The flexibility of direct encoding of the architecture has no limitation, such as differentiable or continuous, on how the fitness is defined. The training error pertaining to the architecture, such as training time and error, is often used in the fitness definition. The complexity of the architecture, such as number of neurons and connections, is also used in the fitness definition.

9.3.2.2   Indirect Encoding

Long bit strings of chromosomes in direct encoding are problematic to handle and demand higher computational effort. An alternative would be to use the most important features of an architecture that can be encoded or parameterized into the chromosome, such as the number of nodes, number of connections and type of activation functions (Dasgupta and McGregor, 1992a,b). Other details are left to the learning process to decide. Such a representation is called an indirect encoding scheme. An immediate benefit of an indirect encoding scheme is a compact representation of a relatively large NN architecture, which is easy to handle and easy to apply with genetic operators. There have been many researches reported in the literature on how to represent the architecture of an NN, so that EA operations can be applied without much computational effort. These approaches can be classified as parametric representation, deterministic developmental rules representation, fractal subsets representation and hidden-node representation, which are discussed below.

Parametric encoding
In parametric encoding, the chromosomes do not contain the detail of the network topology. Rather, they contain abstract information about the network topology such as number of hidden layers, number of neurons per hidden layer and type of activation functions and associated learning parameters. The number of inputs and outputs of an NN is almost always predefined and it is the number of hidden layers, number of neurons in the hidden layer and associated activation functions that determine the architecture of the NN. Certainly, such an encoding scheme has some restrictions on the network topology, e.g., feedforward architecture, full connectivity between consecutive layers or no connection from input layer to output layer. Harp *et al.* (1990) encoded network parameters in one long string of bits. The bit string is composed of one or more segments. Each segment represents an area (possibly a layer) and its efferent connectivity (or projections). Each segment is an area specification substring and consists of two parts. The first part is of fixed length and contains layer-specific information, the number of nodes in the layer, organization of the nodes and the learning parameters associated with the nodes. The second part consists of one or more projection specification fields, where each field describes the efferent connectivity between areas, connection density, target area address, organization of connections and learning parameters associated with connection weights.

An important aspect of the parametric representation by Harp *et al.* is the exploration of learning parameters along with the connectivity parameters. The searching of the learning rules for an architecture obtained by an EA is another computationally intensive task. Though the parametric representation reduces the chromosome length, the search space is limited to only a subset of the whole feasible architecture space (Yao, 1999). The parametric method requires more *a priori* information about the architecture of the NN, so that the construction of the NN architecture can be carried out from the optimized parametric description.

Grammar (or developmental rule) encoding
Though the parametric representation can reduce the length of bit string in coding of the NN architecture, it suffers from scalability as the number of nodes increases in a real-world problem. The grammar encoding representation is a shift from the direct optimization of the NN architecture to optimization of the developmental rule. Grammar encoding was introduced by Kitano (1990, 1994) to train the NN architecture. A grammar is a set of rules that is applied to produce a set of structures, e.g., sentences in a natural language, programs in a computer language. A simple example is the following grammar:

$$S \rightarrow aSb \qquad\qquad (9.19)$$

$$S \rightarrow \epsilon$$

Here, S is the start symbol and a non-terminal, a and b are terminals and ϵ is the empty string terminal. $S \rightarrow \epsilon$ means that S can be replaced by the empty string. To construct a structure from this grammar, start with S and replace it by one of the allowed replacements given by the

Figure 9.24 Chromosome using grammar encoding

right-hand sides; take the resulting structure and continue until no non-terminals are left. For example:

$$S \rightarrow aSb \rightarrow a(aSb)b \rightarrow a(a(\; \in \;)b)b \rightarrow aabb \tag{9.20}$$

Kitano applied this general type of grammar, called a 'graph-generation grammar', to represent the architecture of a neural network. A simple example of such a grammar-encoded chromosome is shown in Figure 9.24, where the right-hand side of each rule is a 2×2 matrix rather than a one-dimensional string. Uppercase letters are non-terminals and lowercase letters are terminals. Each terminal represents one of 16 possible 2×2 arrays consisting of 1's or 0's, where a 0 means no connectivity and a 1 means connectivity. The developmental rule is a compact genotypical representation and capable of preserving promising building blocks. The definitions of the non-terminals and terminals are given in Figure 9.25.

The developmental rules are repetitively applied to initial non-terminal elements until they contain terminal elements, i.e., until the connectivity pattern of an NN architecture is completely specified. A typical procedure of constructing a connectivity matrix from the production rules is shown in Figure 9.25. The final matrix represents the connectivity of an NN with eight neurons, as shown in Figure 9.26.

$$A \rightarrow \begin{pmatrix} c & p \\ a & c \end{pmatrix} \quad B \rightarrow \begin{pmatrix} a & a \\ a & e \end{pmatrix} \quad C \rightarrow \begin{pmatrix} a & a \\ a & a \end{pmatrix} \quad D \rightarrow \begin{pmatrix} a & a \\ a & b \end{pmatrix}$$

$$a \rightarrow \begin{bmatrix} 0 & 0 \\ 0 & 0 \end{bmatrix}, b \rightarrow \begin{bmatrix} 0 & 0 \\ 0 & 1 \end{bmatrix}, c \rightarrow \begin{bmatrix} 1 & 0 \\ 0 & 1 \end{bmatrix}, \ldots, e \rightarrow \begin{bmatrix} 0 & 1 \\ 0 & 1 \end{bmatrix}, \ldots, p \rightarrow \begin{bmatrix} 1 & 1 \\ 1 & 1 \end{bmatrix}$$

Figure 9.25 Developmental rule in grammar coding

$$S \Rightarrow \begin{bmatrix} A & B \\ C & D \end{bmatrix} \Rightarrow \begin{bmatrix} \begin{bmatrix} c & p \\ a & c \end{bmatrix} & \begin{bmatrix} a & a \\ a & e \end{bmatrix} \\ \begin{bmatrix} a & a \\ a & a \end{bmatrix} & \begin{bmatrix} a & a \\ a & b \end{bmatrix} \end{bmatrix} \Rightarrow \begin{pmatrix} \begin{bmatrix} 1 & 0 & 1 & 1 \\ 0 & 1 & 1 & 1 \\ 0 & 0 & 1 & 0 \\ 0 & 0 & 0 & 1 \\ 0 & 0 & 0 & 0 \\ 0 & 0 & 0 & 0 \\ 0 & 0 & 0 & 0 \\ 0 & 0 & 0 & 0 \end{bmatrix} & \begin{bmatrix} 0 & 0 & 0 & 0 \\ 0 & 0 & 0 & 0 \\ 0 & 0 & 0 & 1 \\ 0 & 0 & 0 & 1 \\ 0 & 0 & 0 & 0 \\ 0 & 0 & 0 & 0 \\ 0 & 0 & 0 & 0 \\ 0 & 0 & 0 & 1 \end{bmatrix} \end{pmatrix}$$

Figure 9.26 Genotype representation of the connection matrix

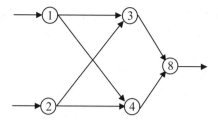

Figure 9.27   Phenotype representation of an NN architecture

The phenotype representation of the NN architecture can be constructed from the description provided by the genotype representation of the connection matrix in Figure 9.26. The neurons 5, 6 and 7 have no connections with the other neurons, so they are discarded from the NN architecture. A detailed description of the developmental rules and construction of the NN architecture using production rules can be found in Vonk *et al.* (1997) and Yao (1999). Thus the mapping (translation) from a genotype (as shown in Figure 9.26) to a phenotype representation of the NN architecture is shown in Figure 9.27.

It is clearly demonstrated in Figures 9.26 and 9.27 that the NN architecture can be constructed from the set of rules shown in Figure 9.25. An optimized or suitable architecture for a given application can be obtained by evolving a set of developmental rules. The whole rule set can be encoded as an individual (the so-called Pittsburgh approach) or each rule can be encoded as an individual (the so-called Michigan approach). For details of the Pittsburgh and Michigan approaches, see Chapter 8. The rule set in Figure 9.25 can have non-terminals ranging from A to Z and from a to p and they are involved in the evolution process. The rule set with terminal elements $\{0, 1\}$ is not involved in the evolution. A population is generated randomly from the chromosome representation of the rule set with non-terminal elements (shown in Figure 9.28). The evolution process using the developmental rules is shown in Figure 9.28. The matrix grammar representation scheme relies on three different spaces (shown in grey boxes in Figure 9.28): the representation space (chromosomes), the intermediate space consisting of connectivity matrices and the evaluation space (network structures). The mapping from the representation space to the intermediate space and the mapping from the intermediate space to the network structure space suffer from competing conventions. The matrix grammar encoding scheme is unable to generate any arbitrary feedforward network structure. The connectivity matrix must contain some kind of regularity, so that the evaluation space is only part of the complete problem space. Good results have been reported in the literature, but the method does not allow recursive rules. The method is also not very good at evolving detailed connectivity of the architecture.

Tree encoding (or GP-coding)

Tree encoding using GP offers another approach to encode an NN architecture into a chromosome (Sanger, 1991). A feedforward neural network is represented by a connected tree structure consisting of a function set F and a terminal set T. The function and terminal sets of the GP are discussed in Chapter 6. The advantage of tree encoding is that the topology and the connection weights can be defined within the structure. The terminal set T is made up of data inputs D to the NN and random real values R to be used as connection weights. The terminal set is defined as

$$T = \{D, R\} \tag{9.21}$$

Population of
production rules

Fitness of
NN

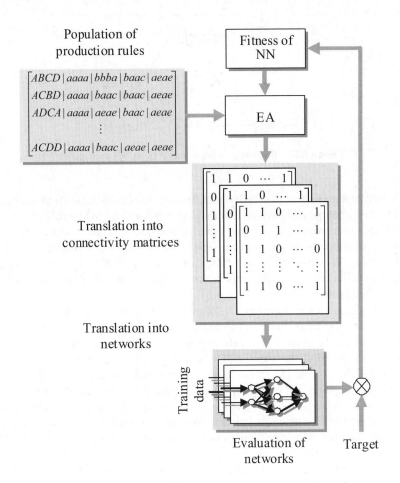

Translation into
connectivity matrices

Translation into
networks

EA

Training
data

Evaluation of
networks

Target

Figure 9.28   Evolution of NN architectures using production rules

For example, the terminal set T for a two-input NN is $T = \{D_0, D_1, R\}$. The function set F is made up of a processing function P representing a neuron (processing unit) and a weight function W. P performs the weighted sum of its inputs and forwards it to an activation function (also known as a node transfer function). For example, the function set F for an NN is $F = \{P, W\}$. The values of the weights are represented by random constant values R and their values are only modified by crossover or mutation operators.

Example 9.1   An example of a chromosome representation using tree encoding for a two-input single-output NN is given in Figure 9.29.

$$(P(W(P(W, -0.656, D_1)(W, 1.59, D_0)), 1.015)(W, 1.453(P(W, 1.703, D_1)(W, -0.828, D_0))))$$

Figure 9.29   Chromosome representation using GP

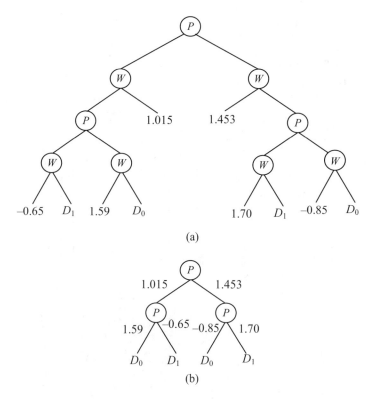

(a)

(b)

Figure 9.30   Tree-coding representation of NN architecture. (a) Tree-coding (genotype) representation of two-input single-output NN; (b) Condensed tree coding of NN

The tree representation of the chromosome in Figure 9.29 is shown in Figure 9.30(a). The tree in Figure 9.30(a) is further reduced in Figure 9.30(b) showing the inputs, nodes, output and connection weights. The phenotype representation of the NN architecture is then defined by translating the genotype representation in Figure 9.30. The final two-input single-output NN architecture is shown in Figure 9.31.

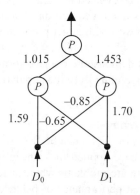

Figure 9.31   Phenotypic representation of NN architecture

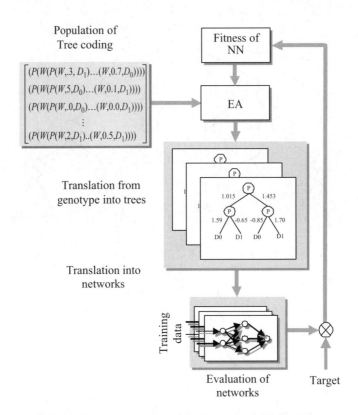

Figure 9.32   Evolution of NN architecture using tree coding

The evolution process using the tree coding is shown in Figure 9.32. A one-bit adder problem was to be solved using an NN. Koza and Rice (1991) applied tree coding to evolve the NN architecture. Good results have been reported in the literature (Koza and Rice, 1991; Vonk *et al.*, 1995, 1997). Severe restrictions are to be imposed on the network topology, i.e., only tree-structured networks can be evolved using this approach. As a result, the number of arguments of a function is always fixed, i.e., neurons can only have two inputs. This means that tree coding does not scale up well with larger problems. The advantage of tree coding is that the topology and weights are evolved simultaneously. The risk of this advantage is that an NN with optimized topology can demonstrate poor performance (poor fitness value), which will result in elimination of the individual from the population.

Fractal representation for architecture

There is strong evidence that parts of the human body (e.g., the lung) are fractally structured (West, 1988). Some researchers have applied the concept of such fractal structures to NN architecture. It is argued that there are problems where specially tailored network structures are essential to achieve the desired performance or behaviour. Merrill and Port (1991) presented a method to derive such network structures. Fractal representation of NN connectivity is biologically more plausible than the rule-encoding representation. The representation combines three properties that are desirable in a configuration search: plasticity, stability and biological plausibility. Each node represents three real parameters, namely an edge code, an input coefficient and an output coefficient, to specify each node in a connectivity pattern. The edge code is represented by a single real number in $[-1, 1]$ and all edge decisions are based upon that number. The input and output coefficients are chosen from the interval $[2, \infty)$. If two nodes are in successive layers of the network, there is a potential edge between them. The nodes then compete to form an edge. The process depends on the behaviour of a deterministic dynamic system, which is sufficiently unstable in the interval of interest that a small change in the control parameters of the system can change the outcome of the competition. The competition is a partition of the interval $[-1, 1]^2$ into two basins with fractal boundary. The dynamic system is then defined as

$$x_i(t + 1) \leftarrow (c_i(1 - |x_i(t)|)) - 1 \tag{9.22}$$

ec_1 and ec_2 denote the input and output codes, c_1 and c_2 denote the input and output coefficients of two nodes, respectively and $x_1 = (ec_1 + ec_2)/2$ and $x_2 = (ec_1 - ec_2)/2$ are two parameters of the dynamical system. If the pair of codes due to any initial condition $\{x_1, x_2\}$ corresponding to the two nodes is in one partition, then the edge is created, otherwise the edge is not created. In order to reduce the research parameters, the network is divided into cliques and the edge coefficients are distributed linearly through the clique. The optimal coefficients of the cliques are searched. Each network consists of entirely standard semi-linear nodes; each node computes the weighted inputs added with a bias and uses a sigmoidal activation function to transfer the node output. An evolutionary search can be employed over the space of all possible edge configurations of a network to learn some pattern. It is unlikely that the fractal representation scheme will have better scalability than the rule-encoding method.

9.3.2.3   Fitness Function for Architecture Evolution

There have been many fitness functions used for evolving NN architectures. The simplest fitness function can be the sum squared error (SSE) or mean squared error (MSE) defined in Equations (9.15) and (9.16) over a validation set containing P patterns after training the network for a fixed number of iterations. Prechelt (1994) suggested a different fitness function to make the error measure less dependent on the size of the validation set and the number of outputs neurons and proposed adopting the percentage of the mean squared error defined as follows:

$$f(\text{NN}) = E = 100 . \frac{o_{\max} - o_{\min}}{P.M} \sum_{i=1}^{P} \sum_{k=1}^{M} [y_k - t_k]^2 \tag{9.23}$$

Here, $o_{\max}$ and $o_{\min}$ are the maximum and minimum values of the outputs, P is the number of patterns, M is the number of outputs, y_k and t_k are the actual and desired outputs of the k th node of the network.

The fitness function in Equation (9.23) does not take into account the size of the architecture or the number of connectivities. Considering the relative connectivity of the network architecture, the following fitness function can be used for evaluation of the NN architecture generated by the EA:

$$f(\text{NN}) = \frac{E(\text{NN})}{N_{out}} + \alpha.\frac{C(\text{NN})}{C_{\max}} + \beta.P(\text{NN}) \tag{9.24}$$

where $E(\text{NN})$ is the cumulative squared error of the NN on the training set, N_{out} is the number of outputs, $C(\text{NN})$ is the number of connectivities in NN, $C_{\max}$ is the maximum possible connectivity in NN and $P(\text{NN})$ is the number of connectivities to be pruned for a specific NN (e.g., feedforward or recurrent). α and β are two weighting parameters for the connectivity and pruning terms, which can be chosen arbitrarily depending on the problem at hand.

9.3.3 EA for NN Node Transfer Functions

In general, the transfer function (or activation function) of a neuron in the architecture of an NN is assumed to be fixed and chosen arbitrarily by an expert. The same transfer function is used for all neurons of the same layer and very often different transfer functions are used for different layers. The choice of transfer function is by no means a trivial task, as the transfer function is an important part of the architecture and has significant impact on the performance of the NN (Stork *et al.*, 1990; DasGupta and Schniter, 1992; Tong and Mintram, 2010). Most neural network applications for supervised learning use sigmoidal or radial basis functions as the gradient information for these transfer functions is easy to obtain. The difference in transfer functions for nodes could be large (e.g., ranging from a hard-limiting threshold function to a Gaussian function) and could be small (e.g., just a change in the slope parameter of the sigmoidal function). However, applications may require more complex kinds of neuron or transfer function, such as product neurons. There are many applications reported in the literature that have tried non-logistic-function neurons, for example, threshold neurons (Bornholdt and Graudenz, 1991; Collins and Jefferson, 1991; Koza and Rice, 1991), linear neurons (Bergmann and Kerszberg, 1987), Grossberg field neurons (Lehr and Weaver, 1987) and biologically motivated neurons (Dress and Knisley, 1987). Computation of the gradient information for these neurons would be far more costly. EA-based training of these neural networks with non-logistic-function neurons is a viable alternative. Stork *et al.* (1990) were the first to apply EA to the selection of a node transfer function in a neural network. The transfer function used in this application was more complex than the usual sigmoidal function and was specified in the genotype representation of the chromosome. Liu and Yao (1996) applied EP for the selection of sigmoidal or Gaussian nodes in a neural network, where the EP allowed growth or shrinking of the neural network by adding or deleting a sigmoidal or Gaussian node. Experimentation on a set of benchmark problems demonstrated good performance.

Another issue with the sigmoidal node transfer function is its shape, which is assumed fixed throughout the network. However, parameters such as the optimal shape of the sigmoidal function are determined by trial and error or heuristically in most cases. There have been few studies on the optimal shape of the sigmoidal function. Yamada and Yabuta (1992) proposed an auto-tuning method for the sigmoidal function shape in order to apply it to a servo-control system. Their method is based on the steepest descent method and confirmed the characteristics

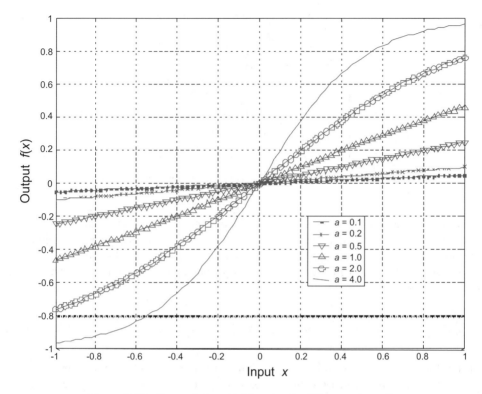

Figure 9.33   Shape of sigmoidal function for different values of a

and practicality of the method with simulation results. The usual tan-sigmoidal function $f(x)$ is defined as

$$f(x) = \frac{1 - e^{-ax}}{1 + e^{-ax}} \qquad (9.25)$$

where x is the network output and a defines the shape of the activation function. The shape of the sigmoidal function for different values of shape parameter a is shown in Figure 9.33. The activation function is defined in Equation (9.25) and the parameter a defines the shape of the sigmoidal function. The use of different shapes of sigmoidal function can lead to different weights and biases during learning with the backpropagation algorithm, which is experienced in many applications (Yamada and Yabuta, 1992). That is, the shape of sigmoidal functions should be fixed during execution of the backpropagation algorithm. This type of activation function is characterized by its gain (slope) and seriously affects the control characteristics. If this gain tuning is used in control applications, the network output may become unstable in certain cases. When the usual sigmoidal function is used only in the hidden layer, sigmoidal function shape tuning is the same as weight tuning. A mathematical proof is given in Yamada and Yabuta (1992). Therefore, sigmoidal function shape tuning in a neural network can contribute significantly in improving performance of the neural system or neuro-controller.

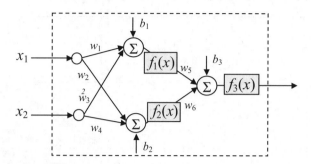

Figure 9.34   Single-neuron network with nonlinear activation function

Consider a neural network with three neurons, two inputs and a single output as shown in Figure 9.34. The two activation functions $f_1(x)$ and $f_2(x)$ at the hidden layer are tan-sigmoidal functions with shaping parameters a_1 and a_2, respectively and the activation function $f_3(x)$ at the output layer is a linear function with scaling parameter a_3 (we can also call it a shaping parameter):

$$f_1(x) = \frac{1 - e^{-a_1 x}}{1 + e^{-a_1 x}}, \quad f_2(x) = \frac{1 - e^{-a_2 x}}{1 + e^{-a_2 x}}, \quad f_3(x) = a_3 x \tag{9.26}$$

The transfer functions of the three neurons in the network can be different due to the shaping parameters $\{a_1, a_2, a_3\}$, which will impose restrictions on using the backpropagation algorithm for training of the network. Moreover, training of the weights $\{w_1, w_2, \ldots, w_6\}$ and biases $\{b_1, b_2, b_3\}$ of the network along with the sigmoidal function shape parameters $\{a_1, a_2, a_3\}$ using the backpropagation learning algorithm would be computationally intensive and cumbersome. Evolutionary learning can automatically decide the optimal shape of the sigmoidal function as well as optimize the weights and biases.

The decision on how to encode the transfer functions in the chromosome representation depends on the *a priori* information and computational time allowed for training the network. In general, nodes in the same layer tend to have the same type of transfer function with possible differences in the parameter set, e.g., the parameters $\{a_1, a_2\}$ of $\{f_1(x), f_2(x)\}$ in layer two of Figure 9.34. Nodes in different layers can have different transfer functions, e.g., $\{f_1(x), f_2(x)\}$ in layer two are tansigmoidal functions and $f_3(x)$ in layer three is linear. The training is possible in two ways: firstly, using EA for optimizing transfer functions and weight training (weights and bias) using the backpropagation algorithm and secondly, using EA for both the function and weight training. Use of EA and backpropagation would be slow and computationally exhaustive. Furthermore, it may not guarantee optimal network performance due to the nonlinear relationship between the two parameter spaces. Evolution of both transfer functions and weights at the same time would be advantageous since they constitute a complete architecture. Encoding weights and parameters of the transfer functions into the same chromosome would make it easier to find the optimal performance of the network by exploring the two sets of parameter space. The chromosome representation is very straightforward, as shown in Figure 9.35.

The different techniques of evolution discussed earlier in this chapter can be applied for the weights, biases and function shape parameters. One simple example is shown in Figure 9.36.

$$\{w_1, w_2, ..., w_6\}, \{b_1, b_2, b_3\}, \{a_1, a_2, a_3\}$$

Figure 9.35   Chromosome for weights and function shape parameters

The fitness function for the evolution in Figure 9.36 could be any of the fitness functions used for weight learning, such as SSE or MSE defined in Equations (9.15)–(9.18).

The connectivity of the network is assumed to be fixed during learning of the weights and parameters in the evolution shown in Figure 9.36. Some researchers suggested evolving the topology, connection weights and parameters at the same time (Hwang *et al.*, 1997).

9.3.4   EA for NN Weight, Architecture and Transfer Function Training

The major problem with designing an NN architecture using indirect coding, such as the parameterized, grammar and tree encoding schemes (discussed in Section 9.3.2), is that the weight training has to be done after a near-optimal architecture is found. This causes a noisy fitness evaluation due to the fact that the fitness is measured against the architecture with a full set of weights whereas the indirect encoding only provides the architecture without any weight information. For example, consider the encoding of the architecture shown in Figure 9.37(a). The architecture (genotype) encoded into the string is translated into a network (phenotype) shown in Figure 9.37(b). There exists a one-to-many mapping from genotype to phenotype. This contributes noise in the fitness evaluation. For example, to compute the fitness of the architecture, random weights are generated as shown in Figure 9.37(c) and used in the evaluation of the architecture against a set pattern, which is the fitness value of the architecture. Therefore, the fitness obtained in this way is called noisy. The noisy fitness comes from two sources: random initialization of weights for the training results in different fitness values and different training algorithms produce different fitness values for the same set of weights. In order to reduce the noise, the network (i.e., architecture) has to be trained several times with randomly generated weights and an average fitness is to be computed for the network's

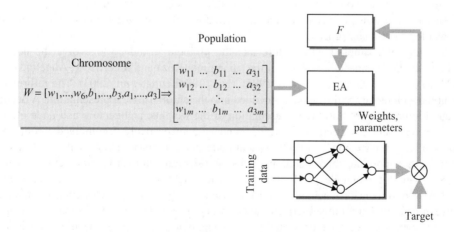

Figure 9.36   Evolution of weights and function shape parameters

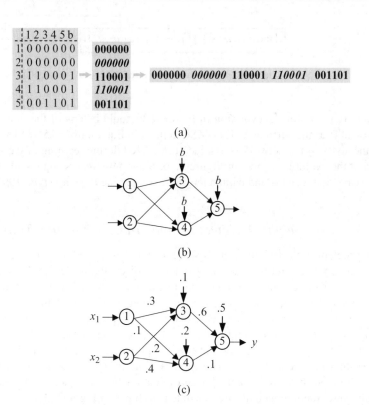

Figure 9.37 Mapping from genotype to phenotype for fitness evaluation. (a) NN in genotype (representation) space; (b) NN in phenotype (evaluation) space; (c) Random weights assigned for fitness evaluation

mean fitness. As a result, the overhead computation for fitness evaluation is high and hence not feasible for designing a large network architecture (Yao and Liu, 1997). It is evident that evolution of the network architecture alone without weight information is computationally expensive and results in inaccurate fitness, which does not guarantee an optimal network after the whole evolutionary process.

Therefore, the simplest way to alleviate the problem would be to evolve the network architecture and connection weights at the same time (Koza and Rice, 1991; Angeline *et al.*, 1994; Maniezzo, 1994; Liu and Yao, 1995; Yao, 1999; Leung *et al.*, 2003). The information about the architecture and weights is encoded into the chromosome representation. A detailed discussion on the chromosome representation of the weights, architecture and node transfer function is presented in Sections 9.3.1, 9.3.2 and 9.3.3, respectively. A combined chromosome representation for weights, connectivity and node transfer function can easily be developed from the three previous representations. Let's consider a feedforward neural network with N neurons with different node transfer functions. The weights of the network can be represented by an $N \times N$-dimensional weight matrix $W = (w_{ij})_{N \times N}$ that constrains the connection weight between neurons i and j in subsequent layers, where $i \neq j$. The biases of the nodes form an $N \times 1$-dimensional vector represented by the weights w_{ii} (i.e., $i = j$, meaning the diagonal elements of the matrix W). The architecture (topology) of the network can be represented by an $N \times N$-dimensional connection matrix $C = (c_{ij})_{N \times N}$ that constrains connections between

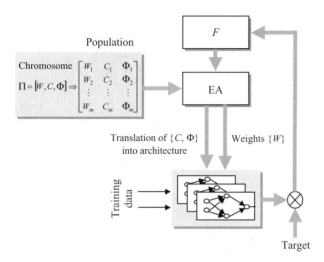

Figure 9.38 Evolution of architecture, weights and function parameters

the N neurons, where $c_{ij} = 1$ indicates the presence of a connection from node i to node j and $c_{ij} = 0$ indicates no connection. The node transfer functions with their parameters can be represented by an $N \times p$-dimensional matrix $\Phi = (\phi_{ij})_{N \times p}$, where p is the number of parameters of the node transfer function.

Thus, the chromosome for the weight, architecture and node transfer function $\Pi = [W, C, \Phi]_{N \times (2N+p)}$ is an $N \times (2N + p)$-dimensional matrix. It has been shown in Sections 9.3.1, 9.3.2 and 9.3.3 that handling the matrix as a chromosome is simple. Converting the matrix into a long string of values would be cumbersome without any benefit to computation, precision or evolution. The process of simultaneous evolution of weights, architecture and transfer function is shown in Figure 9.38 with a population size of m. Genetic operators such as crossover and mutation can be applied with convenience using the matrix representation of the chromosome (Siddique and Tokhi, 2001). Some examples are given in Section 9.3.1. Hwang *et al.* (1997) evolved weights, network architecture and node transfer functions simultaneously. Angeline *et al.* (1994) and Yao and Liu (1997) applied EP to learn the architecture and weights of the network with only a mutation operator. The crossover operator appears not to be useful and renders problems like competing conventions, also called the permutation problem. The permutation or competing convention problem is discussed in Section 9.5.

The fitness function for the evolution of weights, architecture and node transfer function should ensure minimal SSE or MSE and minimal topology. Therefore, any combination can be used of the fitness functions defined in Equations (9.15)–(9.18) for weight learning and the fitness function defined in Equations (9.23)–(9.24) for architecture learning.

9.4 Amalgamated Combination

In supportive and collaborative combinations of EA and NN, the characteristics of both the EA and NN are distinguishable from each other. These combinations are investigated in Sections 9.2 and 9.3. However, the vast majority of these combinations do not represent a functional integration of both technologies that are found in neuro-fuzzy systems (discussed in Chapter 10). In amalgamated combination, the EA's search mechanism is represented in

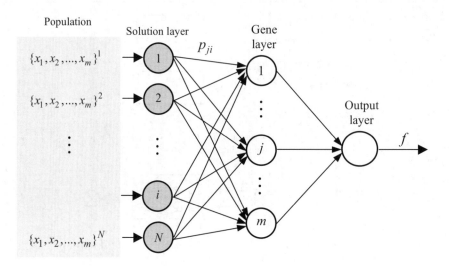

Figure 9.39   Schematic of an amalgamated EA-NN combination

a structure similar to the architecture of a feedforward neural network proposed by Koeppen *et al.* (1997a,1997b). The proposed architecture is called the neural evolutionary strategy system (NESSY). The hybrid architecture (we use the term 'amalgamated combination' of EA and NN with a view to reflecting the purpose of this chapter) consists of three layers of neuron-like processing units. The first layer is the solution layer. Each neuron in this layer represents a solution vector or individual of the population. The second layer is the gene layer. Each neuron represents a gene of the chromosome. That means, if there are n variables to be optimized, then there are n neurons in this layer. All the neurons in the second layer together present a complete solution to the optimization problem. There is one neuron in the third layer, which provides the value of the objective function for single-objective optimization problems. If it is a multi-objective optimization problem, then there will be multiple neurons in this layer representing each objective with one neuron.

An iteration of the NN training is equivalent to one generation of the EA cycle. The genetic operations are performed as the chromosomes are passed through to the gene layer from the input layer. The working principles are described below and illustrated in Figure 9.39.

Every neuron in the solution layer contains a chromosome indicated by $\{x_1, x_2, \ldots, x_m\}^i$, $i = 1, 2, \ldots, N$, where N is the population size and m is the number of genes in the chromosome. All the neurons in the solution layer comprise the population of the EA. Each neuron i in the solution layer is assigned a fitness f_i calculated from the objective function value x^i.

Every neuron in the gene layer represents a gene of the chromosome x_j, $j = 1, 2, \ldots, m$. The number of neurons in the gene layer corresponds to the number of genes in the chromosome, i.e., the size of the gene layer corresponds to the size of the chromosome as indicated by m in the figure. The generation layer is the counterpart of the selection operation in conventional EA.

The solution layer and the gene layer are fully connected and the connections are represented by the weights p_{ji}, which are considered as probabilities in this hybrid architecture. Every neuron in the gene layer randomly selects a neuron of the solution layer based on the probabilities p_{ji}. All weights are initialized randomly in [0, 1]. For example, the probability

p_{32} indicates that neuron 3 in the gene layer selects neuron 2 of the solution layer (i.e., chromosome 2) with probability p_{32}. The important thing to note here is that each neuron j in the gene layer chooses exactly one neuron i in the solution layer with probability proportional to p_{ji}. After the association is made, each neuron in the gene layer is assigned a fitness g_j.

The single neuron in the third layer represents the lower bound of the fitness of the chromosome. The lower bound is represented by the output neuron state O. The neurons in the gene layer are all connected to the output layer with weight unity (i.e., no special weights in the connectivity here). The fitness values of all solutions produced by the gene layer are summed up and compared with the lower bound. The relative error of every generation-layer neuron is backpropagated and the weights p_{ji} are updated according to the learning rule. The weight update is carried out according to

$$p_{ji}(t) = p_{ji}(t-1) - \eta \frac{f_i - g_j}{O} \qquad (9.27)$$

where η is the learning rate and O represents the state of the output neuron used to normalize the difference in fitness value. It can be the best fitness of the solution neurons at generation t. As can be seen from Equation (9.27), if $f_i > g_j$ then p_{ji} is increased, otherwise p_{ji} is decreased. During learning (i.e., evolution), high weights are assigned to good solutions. This improves the probability of repetitive choice of good solutions in the gene layer.

Since the solution layer contains the population, two genetic operators are applied to the solution-layer neurons. These are the transduction operator and the mutation operator. Each solution-layer neuron is modified by comparing its fitness with that of an arbitrary neuron in the gene layer. If the fitness of the gene-layer neuron is better, the corresponding gene in the solution-layer neuron is replaced. This operation is called the transduction operation. The notion came from bacterial genetics, which is equivalent to the crossover operation in traditional EA. The mutation operation is necessary when the transduction operation is applied. Mutation is performed by adding a random number to a gene with zero-mean Gaussian distribution. There are two parameters that control the mutation operator: the mutation probability p_m and the standard deviation of the Gaussian distribution σ. There are four parameters in the NESSY algorithm: structural parameter (i.e., size of solution layer), learning rate η, mutation probability p_m and standard deviation of Gaussian distribution σ.

9.5   Competing Conventions

EA operates on strings called chromosomes, or often called genotypes. Before evaluating a given genotype, it is first mapped onto a solution to the task space called a phenotype. For example, the width, height and depth of a box are represented by three parameters. The mapping from genotype to phenotype representation can then be viewed as a construction of a box from three measurements $\{A, B, C\}$ meaning width, height and depth, respectively, as shown in Figure 9.40.

If this mapping is many-to-one, then the different genotypes map onto the same or equivalent phenotypes, i.e., permuting the three values will not change the shape or the volume of the box. That is, the same evaluation space is obtained by very different chromosomes from the representation space. Standard crossover between two such chromosomes having the same convention will be unlikely to result in useful offspring; in this case, it will result in the same box as illustrated in Figure 9.41.

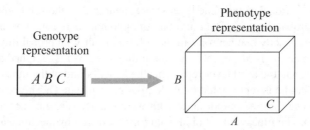

Figure 9.40   Mapping from genotype to phenotype

A similar problem occurs when applying an EA to a neural network structure – called competing conventions, also known as the permutation problem. This problem is of particular interest when designing network architecture (or topology) and has been reported by a number of researchers (Whitley *et al.*, 1990; Belew *et al.*, 1991). The topology of NNs can be encoded into a genotype using a direct or indirect encoding scheme. In direct encoding, the connectivity in the NN is represented explicitly in the genotype, e.g., in the adjacency matrix. Thus, the effect of the genetic operator, especially crossover, on the genotypes is more or less very direct on the phenotype space. In indirect encoding, connectivity in the NN is represented in the genotype using parametric information or a rule of construction process for the NN. Therefore, the effect of the genetic operator on the structure (i.e., the phenotype space) is not obvious. The problem of indirect encoding is that functionally equivalent NNs can be represented by numerous genotypes where the hidden neurons are ordered differently. Genetic operators

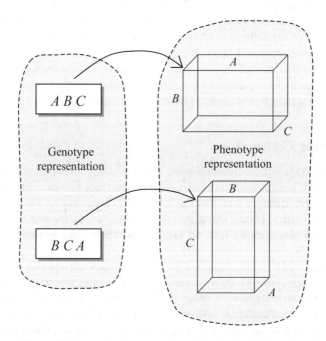

Figure 9.41   Mapping from genotype to the same phenotype

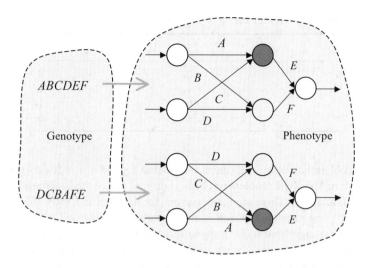

Figure 9.42 Mapping from genotype to phenotype of NN

generally produce NNs by simply permuting the hidden neurons. Up to now there has been neither a proven best representation/operator setting nor a constructive way to generate a good setting for a given problem except some general design heuristics (Igel and Stagge, 2002). The known genetic operation results in different genotype representations of NNs but when the representations are mapped onto phenotypes, they form the same network. For example, consider the two NN architectures in Figure 9.42. The only difference in the phenotypes is the switching of the two hidden nodes and such permuting of the hidden nodes of a feedforward network does not alter the function and will exhibit the same fitness. Applying a crossover operation to functionally equivalent network structures will not produce any better offspring. The number of competing conventions grows exponentially with increasing number of hidden neurons in a network as each permutation represents a different ordering of hidden neurons and so represents a different convention. Some researchers, therefore, avoid the crossover operation and use only mutations in the evolution of architectures (Palmes *et al.*, 2005).

Example 9.2   An XOR gate is to be realized using a feedforward neural network. The configuration of the network is shown in Figure 9.43. The truth table for the two-input XOR is shown in Table 9.1, which is to be used as training pattern. The weights of the NN circuit are to be trained using an EA.

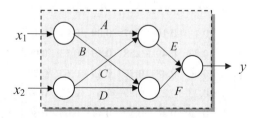

Figure 9.43   NN circuit for an XOR gate

Table 9.1   Truth table for an XOR gate

x_1	x_2	y_d
0	0	0
0	1	1
1	0	1
1	1	0

The XOR problem has historically been considered a good test of a network model and learning algorithm. The XOR problem has been chosen as one of the benchmark problems for many EA-NN simulations. There are many reasons for this choice. Firstly, the XOR problem is one of the simplest problems, which is not linearly separable and complex enough for the backpropagation algorithm to be trapped in local minima without reaching the global optimum. Secondly, there are significant numbers of researches reporting analytical work on XOR which claim that the XOR problem exhibits local minima, a view that is widely accepted in the neural network literature (Dayhoff, 1990; Gori and Tesi, 1992; Cetin et al., 1993). A simple hand-calculated example of GA has been contrived to demonstrate the methodology step by step for weight training a feedforward NN for an XOR gate.

The connectivity of the NN is shown using symbols for weights, $\{A, B, C, D, E, F\} \in R$. There is no bias used in this network. A real-valued chromosome representation is chosen for the weights of the NN and can be represented as follows:

$$\text{Chromosome} = \{A, B, C, D, E, F\} \in R \tag{9.28}$$

A population of five individuals of real values is generated randomly, as shown in Figure 9.44. A small population is chosen for this hand-calculation of the GA. Each row represents an individual or chromosome c_i. The columns indicate the weights of the connectivity between nodes, as shown in Figure 9.44. The fitness function considered here is the mean of sum squared error, which is subject to being minimized, as defined below:

$$f(c_i) = \frac{1}{N} \sum_{k=1}^{N} e(k)^2 \tag{9.29}$$

$$
\begin{array}{cccccc}
A & B & C & D & E & F
\end{array}
$$

$$
\prod = \begin{bmatrix}
0.1 & 0.19 & 0.27 & 0.09 & 0.91 & 0.17 \\
0.23 & 0.49 & 0.58 & 0.10 & 0.73 & 0.06 \\
0.56 & 0.48 & 0.36 & 0.39 & 0.63 & 0.99 \\
0.23 & 0.23 & 0.29 & 0.67 & 0.52 & 0.37 \\
0.76 & 0.34 & 0.19 & 0.47 & 0.03 & 0.80
\end{bmatrix}
$$

Figure 9.44   Population of NN weights

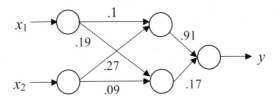

Figure 9.45   NN with weights

where $e(k) = y_d(k) - y(k)$ and $y(k)$ is the network output for each pattern. In this case the optimal value is a minimum. It is common to have the GA process provide the maximum value of the fitness function. It is reasonable to have the fitness function be the reciprocal of $f(c_i)$, i.e., $f_i = 1/f(c_i)$.

The NN looks like the network shown in Figure 9.45 using the first individual of the weights from the population shown in Figure 9.44.

The four input patterns of $\{x_1, x_2\}$ are propagated through the NN and the output values are calculated. A general equation is used for this calculation of the output y as follows:

$$y_i(k) = \psi\left(\phi[x_1(k) * A_i + x_2(k) * C_i] * E_i + \phi[x_1(k) * B_i + x_2(k) * D_i] * F_i\right) \quad (9.30)$$

where $i = 1, \ldots, 5$ represents the individuals in the population and $k = 1, \ldots, 4$ represents the input patterns of the truth table. $\phi(.)$ and $\psi(.)$ are the linear activation functions for the hidden layer and the output layer, respectively. Since linear functions are used, they are not shown in the following calculations. The first pattern $\{0\ 0\}$ is propagated through the NN and the output $y(1)$, error $e(1)$ and squared error $e(1)^2$ are calculated as follows:

$$y(1) = [\{x_1(0.1) + x_2(0.27)\}0.91] + [\{x_1(0.19) + x_2(0.09)\}0.17]$$
$$y(1) = [\{0(0.1) + 0(0.27)\}0.91] + [\{0(0.19) + 0(0.09)\}0.17] = 0$$
$$e(1) = y_d(1) - y(1) = 0 - 0 = 0$$
$$e(1)^2 = 0$$

Similarly, the rest of the three input patterns are propagated through the NN and the outputs, errors and mean of sum of squared errors are calculated, as shown in Table 9.2. The fitness of the first individual is calculated using Equation (9.30) and shown in Table 9.2.

Table 9.2   Computation of the fitness of individual 1

k	Input pattern	$y(k) = (x_1 * A + x_2 * C)E + (x_1 * B + x_2 * D)F$	$e(k) = y_d(k) - y(k)$	$e(k)^2$
1	00–0	0	0–0	0
2	10–1	0.26	1–0.26	0.547
3	10–1	0.123	1–0.123	0.769
4	11–0	0.383	0–0.383	0.146

$$msse = \text{mean of sum squared error } \tfrac{1}{4} \sum_{k=1}^{4} e(k)^2 = \mathbf{0.3656}$$

Table 9.3   Relative fitness of the individuals

$msse$	$f_i = 1/msse$	$f_i/\sum f_i$	Times selected	Mating pool
0.3656	2.7352	0.2044	$c_1 \times 1$	$\{.10\ .19\ .27\ .09\ .91\ .17\}$
0.3407	2.9354	0.2194	$c_2 \times 1$	$\{.23\ .49\ .58\ .10\ .73\ .06\}$
0.5639	1.7733	0.1325	$c_3 \times 0$	$-$
0.3395	2.9452	0.2201	$c_4 \times 1$	$\{.23\ .23\ .29\ .67\ .52\ .37\}$
0.3343	$\mathbf{2.9912}_{\text{best}}$	0.2235	$c_5 \times 2$	$\{.76\ .34\ .19\ .47\ .03\ .80\}$
				$\{.76\ .34\ .19\ .47\ .03\ .80\}$

The fitness value of each individual and its relative fitness are calculated. This is tabulated in Table 9.3. In this population, the best fitness is $f_{best} = 2.9912$ and the average fitness is $\bar{f} = 2.6761$. The reproduction process consists of copying individuals proportional to their fitness values. This means that individuals with higher fitness values (in this case the reciprocal of the mean sum of squared error, denoted $f_i = 1/msse$) have higher chances of being selected for the crossover operation. The selection probability for an individual may be defined as

$$p_i = \frac{f_i}{\sum f_i} \tag{9.31}$$

The mating pool of the next generation is selected according to the probability p_i (shown in the third column of Table 9.3). The selection of the individual c_i in the mating pool is shown in the fourth column of Table 9.3 according to the probability p_i. Random crossover mates (shown in bold) are chosen for each individual in the mating pool. A single-point crossover is performed on the pair of individuals and 10 offspring are created. Two genes of two individuals are randomly chosen for mutation. The mating pool, randomly selected crossover mate, crossover operation and mutation operator are shown in Table 9.4. The mutated value of the gene is shown encircled.

After crossover and mutation, 10 individuals are created which form the population of the second generation. The mean sum of squared error, fitness, relative fitness (i.e., selection

Table 9.4   Mating pool, crossover mate, crossover and mutation operation

Mating pool	Random crossover mate	Offspring
$\{.10\ .19\ .27\ .09\ .91\ .17\}$	$\{.10\ .19\ .27\ \mathbf{I}.09\ .91\ .17\}$ $\{\mathbf{.23\ .23\ .29}\mathbf{I}.67\ .52\ .37\}$	$\{.10\ \textcircled{.29}\ .27\mathbf{.67\ .52\ .37}\}$ $\{\mathbf{.23\ .23\ .29}\ .09\ .91\ .17\}$
$\{.23\ .49\ .58\ .10\ .73\ .06\}$	$\{.23\ .49\mathbf{I}.58\ .10\ .73\ .06\}$ $\{\mathbf{.76\ .34}\mathbf{I}.19\ .47\ .03\ .80\}$	$\{.23\ .49\ \mathbf{.19\ .47\ .03\ .80}\}$ $\{\mathbf{.76\ .34}\ .58\ .10\ .73\ .06\}$
$\{.23\ .23\ .29\ .67\ .52\ .37\}$	$\{.23\ .23\ .29\ .67\mathbf{I}52\ .37\}$ $\{\mathbf{.10\ .19\ .27\ .09}\mathbf{I}.91\ .17\}$	$\{.23\ .23\ .29\ .67\ \mathbf{.91\ .17}\}$ $\{\mathbf{.10\ .19\ .27\ .09}\ 52\ .37\}$
$\{.76\ .34\ .19\ .47\ .03\ .80\}$	$\{.76\ \mathbf{I}.34\ .19\ .47\ .03\ .80\}$ $\{\mathbf{.23}\ \mathbf{I}.49\ .58\ .10\ .73\ .06\}$	$\{.76\ \mathbf{.49\ .58\ .10\ .73\ .06}\}$ $\{\mathbf{.23}\ .34\ .19\ .47\ \textcircled{.33}\ .80\}$
$\{.76\ .34\ .19\ .47\ .03\ .80\}$	$\{.23\ .23\ .29\mathbf{I}.67\ .52\ .37\}$ $\{\mathbf{.10\ .19\ .27}\mathbf{I}.09\ .91\ .17\}$	$\{.23\ .23\ .29\ \mathbf{.09\ .91\ .17}\}$ $\{\mathbf{.10\ .19\ .27}.67\ .52\ .37\}$

Table 9.5 Population of second generation and its fitness

New population	$msse$	$f_i = 1/msse$	$f_i / \sum f_i$	Times selected
{.10 .29 .27 **.67 .52 .37**}	.3452	2.8968	0.1022	$c_1 \times 0$
{**.23 .23 .29** .09 .91 .17}	.3407	2.9351	0.1035	$c_2 \times 0$
{.23 .49 **.19 .47 .03 .80**}	.3382	2.9565	0.1043	$c_3 \times 2$
{**.76 .34** .58 .10 .73 .06}	.3788	2.6398	0.0931	$c_4 \times 0$
{.23 .23 .29 .67 **.91 .17**}	.3360	2.9758$_{best}$	0.1050	$c_5 \times 2$
{**.10 .19 .27** .09 .52 .37}	.3852	2.5961	0.0916	$c_6 \times 0$
{.76 **.49 .58 .10 .73 .06**}	.3815	2.6215	0.0925	$c_7 \times 0$
{.23 .34 .19 .47 .33 .80}	.3398	2.9433	0.1038	$c_8 \times 1$
{.23 .23 .29 **.09 .91 .17**}	.3407	2.9351	0.1035	$c_9 \times 0$
{**.10 .19 .27**.67 .52 .37}	.3513	2.8465	0.1004	$c_{10} \times 0$

probability p_i) for each individual of the second generation are calculated, as shown in Table 9.5. The mating pool of the next generation is selected according to the probability. The fifth column of Table 9.5 shows the number of copies chosen for each individual. In this population, the best fitness is $f_{best} = 2.9785$ and the average fitness is $\bar{f} = 2.8347$. As can be seen, the best fitness has decreased in the second generation but the average fitness of the population in the second generation has increased significantly. Different performance indices such as $\bar{f}$, $\bar{f}/f_{best}$, $f_{worst}/\bar{f}$, $f_{best} - \bar{f}$ and N are discussed in Section 9.2.1. It is indicative that $f_{best} - \bar{f}$ has improved with an increase in the population size N, which is a measure of the convergence of the GA. The evolution continues until the maximum generation when the value of $f_{best} - \bar{f}$ reaches a steady value and the NN is able to behave like an XOR gate; that is, it can generate the truth table given in Table 9.1 fairly closely.

References

Angeline, P.J., Saunders, G.M. and Pollack, J.B. (1994) An evolutionary algorithm that constructs recurrent neural networks, *IEEE Transactions on Neural Networks*, 5(1), 54–65.

Badi, P.F. and Homik, K. (1995) Learning in linear neural networks: a survey, *IEEE Transactions on Neural Networks*, 6(4), 837–858.

Belew, R.K., McInerney, J. and Schraudolph, N. (1991) Evolving networks: using genetic algorithm with connectionist learning. *Proceedings of Second Conference on Artificial Life*, New York, pp. 511–547.

Bergmann, A. and Kerszberg, M. (1987) Breeding intelligent automata. *IEEE International Conference on Neural Networks*, San Diego, CA, Vol. II, pp. 63–70.

Billings, S.A. and Zheng, G.L. (1994) *Radial Basis Function Network Configuration using Genetic Algorithms.* Research Report No. 521, Department of Automatic Control and Systems Engineering, University of Sheffield.

Bornholdt, S. and Graudenz, D. (1991) *General Asymmetric Neural Networks and Structure Design by Genetic Algorithms*, Deutsches Electronen Symchrotron (DESSY'91), Hamburg, Germany.

Brill, F.Z., Brown, D.E. and Martin, W.N. (1990) *Genetic Algorithms for Selection for Counterpropagation Networks*, University of Virginia, Institute of Parallel Computing, Charlottesville, VA, Technical Report IPC-TR-90-004.

Brotherton, T.W. and Simpson, P.K. (1995) Dynamic feature set training of neural nets for classification. In *Evolutionary Programming IV*, J.R. McDonnell, R.G. Reynolds and D.B. Fogel (eds), MIT Press, Cambridge, MA, pp. 83–94.

Cantú-Paz, E. and Kamath, C. (2005) An empirical comparison of combinations of evolutionary algorithms and neural networks for classification problems, *IEEE Transactions on Systems, Man and Cybernetics, Part B: Cybernetics*, 35(5), 915–927.

Caudell, T.P. and Dolan, C.P. (1989) Parametric connectivity: training of constrained networks using genetic algorithms. *ICGA'89*, pp. 370–374.

Cetin, B.C., Burdick, J.W. and Barhen, J. (1993) Global descent replaces gradient descent to avoid local minima problem in learning with artificial neural networks. *Proceedings of IEEE International Conference on Neural Networks*, Piscataway, NJ, Vol. 2, pp. 836–842.

Chalmers, D.J. (1990) The evolution of learning: an experiment in genetic connectionism. *Proceedings of 1990 Connectionist Models Summer School*, D.S. Touretzky, J.L. Elman and G.E. Hilton (eds), San Mateo, CA, pp. 81–90.

Collins, R. and Jefferson, D. (1991) An artificial neural network representation for artificial organisms. In *Parallel Problem Solving from Nature*, H.-P. Schwefel and R. Maenner (eds), Springer-Verlag, Berlin.

DasGupta, D. and McGregor, D.R. (1992a) Designing neural networks using the structured genetic algorithm. *Proceedings of the International Conference on Artificial Neural Networks (ICANN)*, I. Aleksander and J. Taylor (eds), Elsevier Science, Brighton, pp. 263–268.

DasGupta, D. and McGregor, D.R. (1992b) Designing application-specific neural networks using the structured genetic algorithm. *International Workshop on Combinations of Genetic Algorithms and Neural Networks*, Baltimore, MD, pp. 87–96.

DasGupta, B. and Schnitger, G. (1992) Efficient Approximation with Neural Networks: A Comparison of Gate Functions. *Technical Report, Department of Computer Science*, Pennsylvania State University.

Dayhoff, J.E. (1990) The exclusive-OR: a classic problem. In *Neural Network Architectures: An Introduction*, Van Nostrand Reinhold, New York, pp. 76–79.

De Jong, K.A. and Spears, W.M. (1990) An analysis of interacting roles of population size and crossover in genetic algorithms. *Proceedings of 1st Workshop on Parallel Problem Solving in Nature (PPSN'90)*, Dortmund, Germany, pp. 38–47.

Depenau, J. and Moller, M. (1994) Aspects of generalization and pruning. *Proceedings of World Congress on Neural Networks*, Vol. III, pp. 504–509.

Dress, W.B. and Knisley, J.R. (1987) A Darwinian approach to artificial neural systems. *IEEE Conference on Systems, Man and Cybernetics*, New York, pp. 572–577.

Eberhart, R.C. (1992) The role of genetic algorithms in neural network query-based learning and explanation facilities. *Proceedings of the IEEE International Workshop on Combinations of Genetic Algorithms and Neural Networks (COGANN-92)*, Baltimore, MD, pp. 169–183.

Fogel, D.B., Fogel, L.J. and Porto, V.W. (1990) Evolving neural networks, *Biological Cybernetics*, 63(6), 487–493.

Fontanari, J.F. and Meir, R. (1991) Evolving a learning algorithm for the binary perceptron, *Network*, 2(4), 353–359.

Gallant, S. (1993) *Neural-Network Learning and Expert Systems*, MIT Press, Cambridge, MA.

Gao, W. (2003) Study on new evolutionary neural network. *Proceedings of the Second International Conference on Machine Learning and Cybernetics,* Wan, pp. 1287–1292.

Goldberg, D.E., Deb, K. and Clark, J.H. (1992) Genetic algorithms, noise, and sizing of populations, *Complex Systems*, 6(4), 333–362.

Gori, M. and Tesi, A. (1992) On the problem of local minima in backpropagation, *IEEE Transactions on Pattern Analysis and Machine Intelligence*, 14, 76–85.

Grefenstette, J.J. (1986) Optimisation of control parameters for genetic algorithms, *IEEE Transactions on Systems, Man and Cybernetics*, 16(1), 122–128.

Guo, Z. and Uhrig, R.E. (1992) Using genetic algorithms to select inputs for neural networks. *Proceedings of the IEEE International Workshop on Combinations of Genetic Algorithms and Neural Networks* (COGANN-92), Baltimore, MD, pp. 223–234.

Harp, S.A., Samad, T. and Guha, A. (1989) Towards the genetic synthesis of neural networks. *Proceedings of the 3rd International Conference on Genetic Algorithms (ICGA'89)*, Morgan Kaufmann, San Mateo, CA, pp. 360–369.

Harp, S.A., Samad, T. and Guha, A. (1990) Designing application specific neural networks using genetic algorithms. In *Advances in Neural Information Processing Systems 2*, D.S. Touretzky (ed.), Morgan Kaufmann, San Mateo, CA, pp. 447–454.

Hassibi, B. and Stork, D. (1993) Second order derivatives for network pruning: optimal brain surgeon, *Advances in Neural Information Systems*, 5, pp. 164–172.

Hirose, Y., Yamashita, K. and Hijiya, S. (1991) Backpropagation algorithm which varies the number of hidden units, *Neural Networks*, 4, 61–66.

Holland, J.H. (1975) *Adaptation in Natural and Artificial Systems*, University of Michigan Press, Ann Arbor, MI.

Honavar, V. and Uhr, V.L. (1993) Generative learning structures for generalized connectionist networks, *Information Science*, 70(1/2), 75–108.

Hwang, M.W., Choi, J.Y. and Park, J. (1997) Evolutionary projection neural networks. *Proceedings of the 1997 IEEE International Conference on Evolutionary Computation* (ICEC'97), pp. 667–671.

Igel, C. and Kreutz, M. (2003) Operator adaptation in evolutionary computation and its application to structure optimization of neural networks, *Neurocomputing*, 55(1/2), 347–361.

Igel, C. and Stagge, P. (2002) Effects of phenotypic redundancy in structure optimization, *IEEE Transactions on Evolutionary Computation*, 6(1), 74–85.

Jacob, R.A. (1988) Increased rates of convergence through learning rate adaptation, *Neural Networks*, 1(3), 295–307.

Jung, J.-Y. and Reggia, J.A. (2006) Evolutionary design of neural network architectures using a descriptive encoding language, *IEEE Transactions on Evolutionary Computation*, 10(6), 676–688.

Kadaba, N. and Nygard, K.E. (1990) Improving the performance of genetic algorithms in automated discovery of parameters. In *Proceedings of the Seventh International Conference on Machine Learning*, B.W. Porter and R.J. Mooney (eds), Morgan Kaufmann, San Mateo, CA, pp. 140–148.

Kadaba, N., Nygard, K.E. and Juell, P.L. (1991) Integration of adaptive machine learning and knowledge-based systems for routing and scheduling applications, *Expert Systems with Applications*, 2(1), 15–27.

Kamarthi, S.V. and Pittner, S. (1999) Accelerating neural network training using weight extrapolations. *Neural Networks*, 12, 1285–1299.

Kamimura, R. and Nakanishi, S. (1994) Weight-decay as a process of redundancy reduction. *Proceedings of World Congress on Neural Networks*, III, pp. 486–489.

Kelly, J.D. and Davis, L. (1991) Hybridizing the genetic algorithm and the K nearest neighbors classification algorithm. In *Fourth International Conference on Genetic Algorithms*, R.K. Belew and L.B. Booker (eds), Morgan Kaufmann, San Mateo, CA, pp. 377–383.

Kim, H.B., Jung, S.H., Kim, T.G. and Park, K.H. (1996) Fast learning method for backpropagation neural network by evolutionary adaptation of learning rates, *Neurocomputing*, 11(1), 101–106.

Kitano, H. (1990) Designing neural networks using genetic algorithms with graph generation system, *Complex Systems*, 4, 461–476.

Kitano, H. (1994) Neurogenetic learning: an integrated method of designing and training neural networks using genetic algorithms, *Physica D*, 75, 225–238.

Koeppen, M., Teunis, M. and Nicholay, B. (1997a) A neural network that uses evolutionary learning. *Proceedings of the 1997 IEEE International Conference on Evolutionary Computation* (ICEC'97), Indianapolis, MN, pp. 635–639.

Koeppen, M., Teunis, M. and Nicholay, B. (1997b) NESSY – an evolutionary learning neural network. *Proceedings of the 2nd International ICSC Symposium on Soft Computing (SOCO'97)*, Nimes, France, pp. 243–248.

Kohavi, R. and John, G. (1997) Wrappers for feature subset selection, *Artificial Intelligence*, 97(1/2), 273–324.

Koza, J.R. and Rice, J.P. (1991) Genetic generation of both the weights and architecture for a neural network. *IEEE International Joint Conference on Neural Networks*, Seattle, WA, Vol. II, pp. 397–404.

Lehr, S. and Weaver, J. (1987) A developmental approach to neural network design. In *IEEE International Conference on Neural Networks*, M. Caudill and C. Butler (eds), San Diego, CA, Vol. II, pp. 97–104.

Leung, F., Lam, H., Ling, S. and Pam, P. (2003) Tuning of the structure and parameters of a neural network using an improved genetic algorithm, *IEEE Transactions on Neural Networks*, 14(1), 79–88.

Liu, Y. and Yao, X. (1995) A population-based learning algorithm which learns both architecture and weights of neural networks. *Proceedings of ICYCS'95 Workshop on Soft Computing*, pp. 29–38.

Liu, Y. and Yao, X. (1996) Evolutionary design of artificial neural networks with different nodes. *Proceedings of the 1996 IEEE International Conference on Evolutionary Computation (ICEC'96)*, Nagoya, Japan, pp. 670–675.

Martienzzo, V. (1994) Genetic evolution of the topology and weight distribution of neural networks, *IEEE Transactions on Neural Networks*, 5(1), 39–53.

Merrill, J.W.L. and Port, R.F. (1991) Fractally configured neural networks, *Neural Networks*, 4, 53–60.

Miller, G.F., Todd, P.M. and Hedge, S.U. (1989) Designing neural networks using genetic algorithms. *Proceedings of the 3rd International Conference on Genetic Algorithms and their Applications*, J.D. Schaffer (ed.), Morgan Kaufmann, San Mateo, CA, pp. 379–384.

Mirkin, B. (1996) *Mathematical Classification and Clustering*, Kluwer Academic, Dordrecht.

Montana, D.J. and Davis, L. (1989) Training feedforward neural network using genetic algorithms. *Proceedings of 11th International Joint Conference on Artificial Intelligence*, Morgan Kaufmann, San Mateo, CA, pp. 762–767.

Muehlenbein, H. (1990) Limitations of multi-layer perceptron networks – steps towards genetic neural networks, *Parallel Computing*, 14, 249–260.

Nguyen, D. and Widrow, B. (1990) Improving the learning speed of 2-layer neural networks by choosing initial values of adaptive weights. *International Joint Conference on Neural Networks*, 3, 21–26.

Nikolaev, N. and Iba, H. (2001) Regularization approaches to inductive genetic programming, *IEEE Transactions on Evolutionary Computation*, 5, 359–375.

Nikolaev, N. and Iba, H. (2003) Learning polynomial feedforward neural networks by genetic programming and backpropagation, *IEEE Transactions on Neural Networks*, 14(2), 337–350.

Ozdemir, M., Embrechts, M.J., Arciniegas, F., Breneman, C.M., Lockwood, L. and Bennett, K.P. (2001) Feature selection for in-silico drug design using genetic algorithms and neural networks. *IEEE Mountain Workshop on Soft Computing in Industrial Applications*, Blacksburg, VA, pp. 53–57.

Palmes, P.P., Hayasaka, T. and Usui, S. (2005) Mutation-based genetic neural network, *IEEE Transactions on Neural Networks*, 16(3), 587–600.

Parekh, R., Yang, J. and Honavar, V. (2000) Constructive neural-network learning algorithms for pattern classification, *IEEE Transactions on Neural Networks*, 11, 436–450.

Prechelt, L. (1994) *Proben 1 – A Set of Neural Network Benchmark Problems and Benchmarking Rules*, Fakultaet fuer Informatik, University of Karlsruhe, Germany, Technical Report 21/94.

Ramasubramanian, P. and Kannan, A. (2006) A genetic algorithm based neural network short-term forecasting framework for database intrusion prediction system, *Soft Computing*, 10, 699–714.

Reed, R. (1993) Pruning algorithms – a survey, *IEEE Transactions on Neural Networks*, 4, 740–747.

Sanger, T.D. (1991) A tree-structured adaptive network for function approximation in high-dimensional spaces, *IEEE Transactions on Neural Networks*, 2, 285–293.

Schaffer, J.D., Whiteley, D. and Eshelman, L.J. (1992) Combinations of genetic algorithms and neural networks: a survey of the state of the art. *International Workshop on Combinations of Genetic Algorithms and Neural Networks*, Baltimore, MD, pp. 1–37.

Siddique, N.H. and Tokhi, M.O. (2001) Training neural networks: backpropagation vs genetic algorithms. *IEEE International Joint Conference on Neural Networks*, Washington, DC, pp. 2673–2678.

Siedlecki, W. and Sklansky, J. (1988) On automatic feature selection, *International Journal of Pattern Recognition and Artificial Intelligence*, 2(2), 197–220.

Siedlecki, W. and Sklansky, J. (1989) A note on genetic algorithms for large-scale feature selection, *Pattern Recognition Letters*, 10(5), 335–347.

Stork, D.G., Walker, S., Burns, M. and Jackson, B. (1990) Pre-adaptation in neural circuits. *Proceedings of the International Joint Conference on Neural Networks*, Washington, DC, Vol. I, pp. 202–205.

Suzuki, K. and Kakazu, Y. (1991) An approach to the analysis of the basins of associative memory model using genetic algorithms. In *Fourth International Conference on Genetic Algorithms*, R.K. Belew and L.B. Booker (eds), Morgan Kaufman, San Mateo, CA, pp. 539–546.

Taheri, M. and Mohebbi, A. (2008) Design of artificial neural networks using genetic algorithm to predict collection efficiency in Venturi scrubbers, *Journal of Hazardous Materials*, 157(1), 122–129.

Teoh, E.-J., Tan, K.C. and Xiang, C. (2006) Estimating the number of hidden neurons in a feedforward network using the singular value decomposition, *IEEE Transactions on Neural Networks*, 17(6), 1623–1629.

Thodberg, H.H. (1991) Improving generalization of neural networks through pruning, *International Journal of Neural Systems*, 1(4), 317–326.

Tong, D.L. and Mintram, R. (2010) Genetic algorithm-neural network (GANN): a study of neural network activation functions and depth of genetic algorithm search to feature selection, *International Journal of Machine Learning and Cybernetics*, 1, 75–87.

Vafaie, H. and De Jong, K. (1993) Robust feature selection algorithms. *Proceedings of the 1993 IEEE International Conference on Tools with AI*, Boston, MA, pp. 356–363.

Verma, B. and Zhang, P. (2007) A novel neural-genetic algorithm to find the most significant combination of features in digital mammograms, *Applied Soft Computing*, 7(2), 612–625.

Vogt, T.P., Mangis, J.K., Rigler, A.K., Zink, W.T. and Alkon, D.L. (1988) Accelerating the convergence of the backpropagation method, *Biological Cybernetics*, 59, 257–263.

Vonk, E., Jain, L.C., Veelenturf, L.P.J. and Johnson, R.P. (1995) Automatic generation of neural network architecture using evolutionary computation. *Electronic Technology Direction to the Year 2000*, IEEE Computer Society Press, pp. 142–147.

Vonk, E., Jain, L.C. and Johnson, R.P. (1997) *Automatic Creation of Neural Network Architecture using Evolutionary Computation*, World Scientific, Singapore.

West, B. (1988) The fractal structure of human lung. *Proceedings of the Conference on Dynamic Patterns in Complex Systems*, S. Kelso (ed.), Lawrence Erlbaum, New York.

Whitley, D., Starkweather, T. and Bogart, C. (1990) Genetic algorithms and neural networks: optimizing connections and connectivity, *Parallel Computing*, 14, 347–361.

Yamada, T. and Yabuta, T. (1992) Neural network controller using auto-tuning method for nonlinear functions, *IEEE Transactions on Neural Networks*, 3(4), 595–601.

Yang, J. and Honavar, V. (1998) Feature subset selection using a genetic algorithm, *IEEE Intelligent Systems*, 13(2), 44–49.

Yao, X. (1993a) A review of evolutionary artificial neural networks, *International Journal of Intelligent Systems*, 8(4), 539–567.

Yao, X. (1993b) Evolutionary artificial neural networks, *International Journal of Neural Systems*, 4(3), 203–222.

Yao, X. (1995) Evolutionary artificial neural networks. In *Encyclopaedia of Computer Science and Technology*, Vol. 33, A. Kent and J.G. Williams (eds), Marcel Dekker, New York, pp. 137–170.

Yao, X. (1999) Evolving artificial neural networks, *Proceedings of the IEEE*, 87(9), 1423–1447.

Yao, X. and Liu, Y. (1997) A new evolutionary system for evolving artificial neural networks, *IEEE Transactions on Neural Networks*, 8(3), 694–713.

Yao, X. and Liu, Y. (1998) Making use of population information in evolutionary artificial neural networks, *IEEE Transactions on Systems, Man and Cybernetics, Part B: Cybernetics*, 28(3), 417–425.

Zhao, Q. and Higuchi, T. (1996) Evolutionary learning of nearest-neighbour MLP, *IEEE Transactions on Neural Networks*, 7(3), 762–767.

Zitar, R.A. and Hassoun, M.H. (1995) Neuro-controllers trained with rules extracted by genetic assisted reinforcement learning system, *IEEE Transactions on Neural Networks*, 6, 859–878.

10

Neural Fuzzy Systems

10.1 Introduction

In general, there are two main but apparently separate methodological developments relevant to computational intelligence: fuzzy logic systems and neural networks. Fuzzy logic systems try to emulate human-like reasoning using linguistic expression, whereas neural networks try to emulate the human brain-like learning and storing information on a purely experiential basis. Both the methodologies have been successfully applied in many complex and industrial processes though they experience a deficiency in knowledge acquisition.

The most important considerations in designing fuzzy systems are the construction of the membership functions (MF) and constructing the rule-base and have been a tiring process. The choice of MFs also plays a decisive role in the success of an application. But there is no automated way of constructing the MFs. They are mainly done by trial and error, or by human experts. As is well recognized, rule acquisition has been and continues to be regarded as a bottleneck for implementation of any kind of rule-based system. In most existing applications, fuzzy rules are generated by an expert in the area, especially for systems with only few inputs. With an increasing number of inputs, outputs and linguistic variables, the possible number of rules for the system increases exponentially, which makes it difficult for experts to define a complete set of rules and associated MFs for reasonable system performance. In Chapter 8, the construction of MFs, generation of rule base and tuning of scaling parameters using evolutionary algorithms were investigated. Evolutionary algorithms are the suitable choice where no *a priori* information about the MFs and the rule base is available. There have been many successful applications of evolutionary fuzzy systems reported in the literature, and Chapter 8 presents an overview of these techniques and their applications. As is well known, evolutionary algorithms are a slow process and the performance of an evolutionary algorithm depends inherently on the size of the population and the number of generations required for a solution to be robust for specific problems. Some designers may not like this. Then the problem is, if there is no expert knowledge available for constructing the MFs and the rule base, they must be constructed from environmental data, which may or may not be available. A second issue in fuzzy systems is processing of the rule base consisting of $R = n_1 \times n_2 \times \cdots \times n_N$ rules with n_i, $i = 1, 2, \ldots, N$, the number of MFs (primary fuzzy sets) for each of the N inputs. The processing of such a huge rule base is time-consuming. Consequently, computing

Computational Intelligence: Synergies of Fuzzy Logic, Neural Networks and Evolutionary Computing, First Edition.
Nazmul Siddique and Hojjat Adeli.
© 2013 John Wiley & Sons, Ltd. Published 2013 by John Wiley & Sons, Ltd.

the outputs using known defuzzification procedures (such as the centre of gravity method) also takes up significant time, which in some applications can degrade the system response (Siddique, 2002). The problem of such defuzzification methods has been eliminated by the use of Sugeno-type or Tsukamoto-type fuzzy systems, where each consequent MF is replaced by a polynomial or monotone function with two to three parameters (even more for higher-order nonlinear systems) (Tsukamoto, 1979; Takagi and Sugeno, 1985). This imposes a further set of consequence parameters to be estimated. The objective at this stage of the development of fuzzy systems is to facilitate the construction of MFs and the rule base, and minimize the processing time for generating outputs by a simple rule-processing and defuzzification procedure. That is, an automated way of designing and tuning fuzzy systems is deemed necessary. Therefore, the main interest of this chapter lies in constructing, learning or training fuzzy MFs, rule base and tuning parameters using a suitable neural network technology and required information such as experiential data, partial numerical data, partial linguistic data and/or description.

In contrast, most successful applications of neural networks are used for pattern recognition, signal processing, modelling and control of complex systems due to their flexible structures and available learning algorithms. The learning procedure is not knowledge-based, rather data-driven. If a set of experiential data is available, the neural system can be trained with sufficient accuracy, which builds a mapping between inputs and outputs. The input/output mapping is encoded into a layered neural structure which we call a neural network (NN). A variety of NN architectures and learning algorithms are available, suitable for a variety of offline or online applications. Different architectures and learning algorithms are discussed in detail in Chapter 4. Different architectures and learning procedures require different experiential data. For example, a supervised learning algorithm for a feedforward NN requires pairs of data sets consisting of inputs and the desired outputs. Unsupervised training algorithms require only input data. An NN generally requires a long training time depending on the architecture, but recall involves only a single pass. NNs such as the probabilistic NN (PNN), radial-basis function NN (RBFNN) and generalized regression NN (GRNN) (see Chapter 4 for details) require less training time (generally they require a single pass), which is better suited for online applications. New methodologies are to be found for the correct combinations of both technologies for new applications.

There is no straightforward rule for the choice of network size. The number of neurons in the input and output layer is determined by the problem and data samples. The hidden layers and number of neurons in each layer mainly decide the size of the NN. Some NN architectures have a predetermined number of layers, such as the Hopfield, ART-1, Kohonen SOM and LVQ NN. Backpropagation NNs have three layers. In general, the exact size of the hidden layer in a three-layer NN is not a critical parameter and the training time does not vary much for similar-sized hidden layers. Increasing the size of the hidden layer sometimes provides feature detectors. Too few neurons in the hidden layer fail to map features correctly or too many neurons hinder the generalization. Nonlinearity of a system can be represented by a higher number of hidden layers that also increases the processing powers of NNs. Large NNs usually require a large training sample, longer training time and higher computation time. It is of immense importance that the training sample should be representative of the entire input/output space, otherwise the NN does not generalize well. NNs can work with incomplete data sets, but missing data can cause a problem. Redundant data sets or useless data points will cause unnecessary training time and computation costs. It is important how data are represented or translated to an NN. It is very often useful to convert the observed or measured data into another form to be meaningful to NNs. For example, temperature can be represented by actual measured values or can be translated into linguistic terms such as 'very cold', 'cold', 'mild', 'warm' and 'hot'. NNs are very sensitive to magnitudes of values, for instance, when the difference between the maximum and minimum value is too high. It is very often useful and necessary to normalize or scale the continuous data into a suitable range when the natural range of the data is different from the NN's operating range or reference range.

Therefore, research efforts have been exerted to combine the features of both technologies. Firstly, we need to analyse the features of fuzzy systems and neural networks to find out what features allow them to be supportive of or collaborative with each other. It is clear from the above discussion that there are features in both systems that can be combined towards a clever system. Fuzzy systems and neural networks are both mathematical model-free systems and contain their own advantages and pitfalls. The main objective of neural fuzzy systems would be combining them into a system to maximize the desirable properties and reduce the disadvantages of both systems. However, subjective phenomena (such as perceptions and reasoning) are often regarded beyond the domain of conventional neural network theory. Fuzzy systems possess great power in representing linguistic and structured knowledge by fuzzy sets and fuzzy reasoning for modelling uncertainties associated with human cognition, thinking and perception. The limitation of a fuzzy system is that it usually relies on domain experts to acquire knowledge. On the other hand, neural networks are powerful in representing nonlinear mappings into its structure. Knowledge is generally acquired by training using a set of experiential data. Therefore, the integration of fuzzy systems and neural networks is believed to be capable of modelling systems without much *a priori* information and associated with uncertainties that can learn from experiential data. Paradigms based upon this integration are believed to have considerable potential in control systems, adaptive systems and autonomous systems. The advantages and comparative features of both technologies are highlighted in Table 10.1.

10.2 Combination of Neural and Fuzzy Systems

A neuro-fuzzy system finds the parameters of a fuzzy system by means of learning methods obtained from neural networks. The most important reason for combining neural networks with fuzzy systems is their learning capability. Such a combination should be able to learn linguistic rules and/or membership functions or tune existing ones. Learning in this case means

- Creating a rule base,
- Adjusting MFs from scratch and
- Determining other system parameters.

Table 10.1 Comparison of neural and fuzzy systems

Neural networks (NN)	Fuzzy systems (FS)
• No mathematical model required.	• No mathematical model required.
• Acquire knowledge usually from samples and knowledge is encoded into the network structure.	• Acquire knowledge from domain experts and knowledge is represented by rule base.
• Supervised and unsupervised learning algorithms available.	• No learning algorithms available but simple implementation possible.
• Rules cannot be extracted.	• Rules must be available.
• Capable of learning from experiential data.	• Capable of working without much *a priori* information.

The combination can be a fuzzy-neural system to support a neural network. A fuzzy-neural system is a neural network that uses a fuzzy approach to enhance learning capabilities and improve performance. The term 'fuzzy-neural system' is not common in the literature. A fuzzy-neural system controls inputs to the NN, different learning parameters and architectural parameters of a neural network to minimize the training effort and computation cost (Hu and Hertz, 1994; Ishibuchi *et al.*, 1993, 1995). Controlling in this case means

- Controlling learning speed,
- Scaling input data and
- Adjusting other architectural parameters.

In general, three kinds of combination between neural networks and fuzzy systems are distinguished in a survey of the literature:

- Cooperative neuro-fuzzy systems,
- Concurrent neuro-fuzzy systems and
- Hybrid neuro-fuzzy systems.

A cooperative system is considered as a pre-processor, wherein an NN learning mechanism determines the FS membership functions or fuzzy rules from the training data and then the NN recedes into the background, leaving the FS to work independently. In a concurrent system, the NN and FS assist each other in determining the required parameters continuously. In a hybrid system, the FS is represented in a special NN-like architecture so that a learning algorithm can be applied to the fuzzy system. The three synergisms of NN and FS are discussed in detail in the following sections.

10.3   Cooperative Neuro-Fuzzy Systems

The cooperative combination lies in the determination of certain parameters of a fuzzy system (mentioned above) by a neural network and vice versa, where both the neural network and the fuzzy system work independently of each other and provide necessary support. In the late 1980s both the neural and fuzzy technologies were well established and research started in combining the NN and FS to improve the overall system performance and design time of various system designs and consumer products. There have been many consumer products available in the market since the 1990s which use both NN and FS in a variety of cooperative combinations. Although the two systems can be combined in many possible ways, the following combinations have found applications in different consumer products since the 1990s and are in widespread use (Takagi, 1992, 1995, 1997):

- NN and FS as development tool,
- NN and FS as correcting mechanism,
- NN and FS in cascade combination.

The three categories will be discussed in two broad groups: cooperative FS-NN systems and cooperative NN-FS systems. In FS-NN systems, the FS provides prerequisite information for an NN to control (or solve) a plant (or problem). In NN-FS systems, the NN provides prerequisite information or learns parameters, MFs or rules for an FS. Cooperation in both directions at the same time may seem attractive, but computationally may not be an advantage or a viable approach.

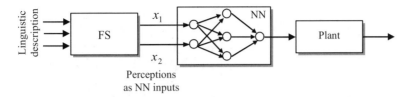

Figure 10.1   Cascade combination of cooperative FS-NN system

10.3.1   *Cooperative FS-NN Systems*

In this cooperation, a fuzzy system translates linguistic statements or descriptions into suitable perceptions or estimates system parameters in the form of input data to be used by an NN.

10.3.1.1   Cascade Combination

The FS-NN combination has been implemented in many applications of consumer products by Hitachi, Sanyo and Toshiba. A generic cascade combination of FS-NN is shown in Figure 10.1. Figure 10.2 shows a cascade combination of FS and NN where the FS estimates system parameters required by the NN. The NN provides the control output for the plant using the parameters supplied by the FS and sensor information. An example application is an electric fan developed by Sanyo, which detects the location of its remote controller with three infrared sensors $\{s_1, s_2, s_3\}$ and changes the direction of the fan to a user's location. An FS estimates the distance between the remote controller and the fan. The estimated distance and the ratio of sensor information are used by the NN to determine the angle between the fan and the remote controller assuming the user is in close proximity to the remote. A two-stage estimation has been applied because the three infrared sensor values change on varying the distance of the remote controller, even though the angle remains the same.

10.3.1.2   Developing Tool Type

In general, the learning rate of an NN using backpropagation or other techniques is fixed during the training process. The convergence of such training algorithms can be accelerated through learning rate adaptation. An FS can be used to dynamically adapt the learning rate using the error information. Some researchers proposed heuristic rules for adaptation of the learning parameters (Jacobs, 1988; Haykin, 1999). They suggested that every adjustable network parameter of the cost function should have its own learning rate parameter to be

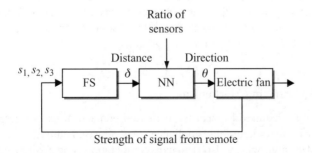

Figure 10.2   Cascade combination – FS collaborating with NN

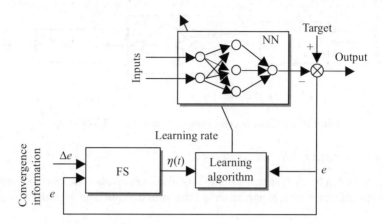

Figure 10.3 Fuzzy system to control the learning rate in an NN

varied from iteration to iteration. When the derivative of the cost function with respect to weight has the same sign for consecutive iterations, the learning rate parameter should be increased and it should be decreased when the sign alters for several consecutive iterations. Halgamuge *et al.* (1994) reported a fast training approach by adapting the learning rate of an NN training algorithm using an FS. A simple developing tool-type combination between FS and NN is shown in Figure 10.3, where the FS provides the appropriate learning parameters using the convergence information based on e and Δe or other derivative information for the NN to accelerate the training.

10.3.2   Cooperative NN-FS Systems

In NN-FS cooperation, an NN is used as supportive technology to determine or estimate different parameters of an FS such as MFs, rule base, scaling factors and rule weighting from available experiential or sensor data. It is also important to make sure that the available data are sufficient for extracting the desired parameters and training of the NN. The process of determination of the parameters can be offline or online during the operation of the FS. Besides this cooperation, an NN can be used as a pre-processor or post-processor to an FS where the structure of the FS is fixed and predetermined. The role of the NN here is to improve the performance of the combined system. Additionally, there are other important issues to be considered, such as whether the cooperation is for an existing FS to be modified or if the FS has to be designed completely. There have been various techniques, learning algorithms and heuristic approaches reported in the literature over the last two decades (Takagi and Hayashi, 1991; Yager, 1994; Yea *et al.*, 1994; Takagi, 1995). Some useful and important techniques using NNs will be discussed in the following sections.

10.3.2.1   NN as Correcting Mechanism

More sensor data are required to meet the increased demand of users for smoother control, precision and sensitivity of household appliances. An increased number of inputs complicates the fuzzy control design, which also demands computation time. To reduce the processing

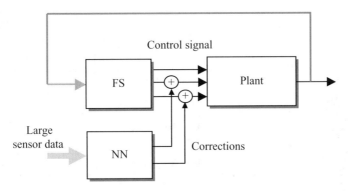

Figure 10.4   Correcting output mechanism

time by the fuzzy controller, fewer inputs are used by the FS and the larger number of sensor inputs is dedicated to the NN. This saves substantial processing time for the FS. By using the NN, the larger sensor data are handled and necessary corrections to the fuzzy system are made. The model in Figure 10.4 shows the NN-FS combination to correct the output of an FS using an NN to increase the precision. This kind of combination has been implemented by many companies for consumer products – such as a washer/drier machine by Hitachi and a microwave oven by Sanyo.

10.3.2.2   NN for Determining MFs (Developing Tool Type)

One of the features of a fuzzy system is that it separates the inference engine part from the fuzziness of the linguistic terms used. The inference engine consists of *if... then* rules and the fuzziness is represented by the linguistic terms defined using suitable MFs. Therefore, it is easy to tackle knowledge which is expressed as rules using qualitative (or linguistic) terms. But the construction of the MFs (the meaning of the terms) remained a difficult task for the designer. Poor performance of an FS is mainly caused by improperly defined MFs. A widely accepted approach is the trial-and-error method, which is mostly a time-consuming method. Therefore, the problem of constructing MFs has been a central issue in FS design with a number of subjective, statistical and neural approaches being proposed. If experiential data are available for the system, the NN clustering approach can be used to extract parameter values of the MFs. In general, an NN in an NN-FS cooperative system determines the number of rules by clustering the data for designing the fuzzy system. Using this clustered data, a neural network decides on a multi-dimensional, nonlinear MF, and this network is then used as a generator of the MF. One-dimensional MFs can be constructed based on the parameters, such as cluster centres and distance metric from multi-dimensional data clusters. One useful contribution of this approach is to introduce neural networks into the design process of fuzzy systems. Secondly, the MFs are designed completely at one stroke, rather than separately along each input axis. In this cooperative combination, an NN provides the MFs' parameters. Such a cooperative combination is shown in Figure 10.5. Adeli and Hung (1995) proposed an algorithm to determine MFs using a topology-and-weight-change classification with a two-layer NN. The number of input nodes equals the number of patterns (M) in each training instance and the number of output nodes equals the number of clusters.

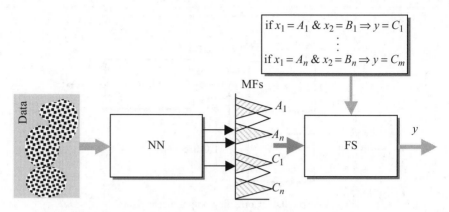

Figure 10.5 NN learning MFs' parameters from available data

The algorithm starts with an NN(M,1) network with M inputs and 1 output. As the training goes on with training instances, the algorithm adds output for a new cluster when the Euclidean distance (also called the degree of difference) between an instance X and a cluster C is greater than a predefined threshold. That is, the new instance is classified as a new cluster. Thus, it creates an NN(M,P) network with P outputs (or clusters) from N instances. The prototype for each cluster C_P is defined as the mean of all instances (n_P equals number of instances) belonging to cluster p, i.e., $C_P = 1/n_p \sum_{i=1}^{n_P} X_i^P$. The degree of membership of each instance in the cluster p is based on the similarity measure between X_i^P and C_P, which is defined as a weighted norm of the similarity function $D^w(X_i^P, C_P) = \|w_P(X_i^P, C_P)\|^w$. The weight parameters w_P and w are chosen arbitrarily depending on the application. A fuzzy MF can be defined for the ith instance belonging to the pth cluster as $\mu_P(X_i^P) = f[D^w(X_i^P, C_P)]$. A triangular MF would be defined as

$$\mu_P\left(X_i^P\right) = \begin{cases} 0 & \text{if } D^w\left(X_i^P, C_P\right) > \sigma \\ 1 - \dfrac{D^w\left(X_i^P, C_P\right)}{\sigma} & \text{if } D^w\left(X_i^P, C_P\right) \leq \sigma \end{cases} \tag{10.1}$$

Here, σ is the crossover value for overlapping MFs.

Though an FS can be constructed using expert knowledge and MFs defined in a qualitative manner, the precise definition of the parameters of the MFs from data clustering is not always possible as the data distribution may not be representative of the entire input space. As a first attempt, an NN is employed to acquire knowledge from the set of experiential data. A multi-dimensional function is then decomposed into single-dimensional functions. The error between the designed FS and the actual data depends on the parameters of the one-dimensional MFs. These MFs are tuned to minimize error in a manner similar to backpropagation learning. The model in Figure 10.6 uses an NN to optimize the parameters of the FS (i.e., the parameters of the MFs) by minimizing the error between the specification and the output of the FS. This type of combination is widely used in many applications and consumer products, such as washing machines, vacuum cleaners, rice cookers, dishwashers and photocopiers developed by Japanese companies.

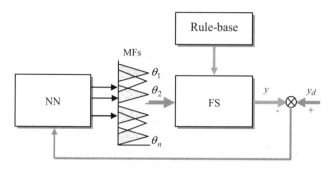

Figure 10.6   Developing tool-type combination of NN-FS

10.3.2.3   NN for Fuzzy Rules Learning

A neural network determines fuzzy rules from training data. A clustering approach is usually used by implementing self-organizing feature maps. SOMs are trained offline and then applied to the fuzzy system. The MFs of the fuzzy system are predetermined. Such a cooperative neuro-fuzzy system is shown in Figure 10.7. Pedrycz and Card (1992) used SOM to extract fuzzy rules from the data. Another way to create a fuzzy rule base is to use fuzzy associative memory (FAM), as proposed by Kosko (1992), where fuzzy rules are interpreted as an association between antecedent and consequents. If fuzzy sets are seen as points in the unit hypercube and rules are associations, then it is possible to use neural associative memory to represent fuzzy rules. A neural associative memory is also called a bidirectional associative memory (BAM), because creating its connection matrix corresponds to the Hebbian learning rule (Kosko, 1992). Kosko (1992) suggests a form of adaptive vector quantization (AVQ) to learn FAM from available data. AVQ is also known as learning vector quantization (LVQ). AVQ or LVQ learning is similar to SOM and realized using competitive learning. A detailed discussion on competitive learning is presented in Section 4.5.2.2 of Chapter 4.

If a fuzzy system uses multiple input and multiple output variables, it is difficult for an expert to formulate the fuzzy *if... then* rules. In that case it is desirable to extract the rules from

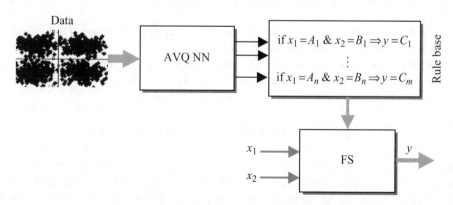

Figure 10.7   NN learning rules for FS

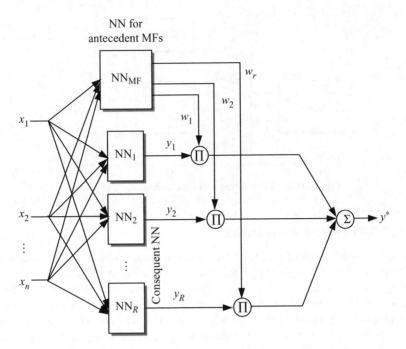

Figure 10.8   A block diagram of the Takagi–Hayashi method

available data from the physical systems to be modelled. Takagi and Hayashi (1991) proposed a neural network-driven fuzzy reasoning for a Sugeno-type fuzzy system of the form

$$\text{If } x_1 \text{ is } A_1 \text{ and } x_2 \text{ is } A_2, \dots, x_n \text{ is } A_n \text{ Then } y = f(x_1, x_2, \dots, x_n)$$

The Takagi–Hayshi method consists of three steps:

(i) Partition the input space into number of rules. This is done through clustering the available data.
(ii) Determine the MFs by identifying the given rule's antecedent values. This is done by employing an NN to derive the MF for each rule.
(iii) Determine the consequent value by identifying the consequent function. In the Takagi–Hayashi method, each rule's consequent function is replaced with an NN and the NN is trained with supervised learning, i.e., the rule would look like

$$\text{If } (x_1, x_2, \dots, x_n) \text{ is } A_s \text{ Then } y_s = \text{NN}_s(x_1, x_2, \dots, x_n)$$

Here, $x = (x_1, x_2, \dots, x_n)$ is the input vector, A_s is the antecedent MFs and $\text{NN}_s(x_1, x_2, \dots, x_n)$ is the consequent NN of the sth rule that generates the output. A generic block diagram of the Takagi–Hayashi method is given in Figure 10.8. The neural network NN_{MF} generates the MFs for the rule antecedent and neural networks $\text{NN}_1, \dots, \text{NN}_r$ generate the consequent function values. The final output y^* is the sum of all weighted outputs

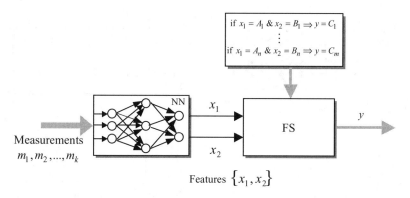

Figure 10.9 Extracting features as inputs to a fuzzy system

y_s, defined as $y^* = \sum_{s=1}^{R} w_s y_s$. Further details of the procedure can be found in Takagi and Hayashi (1991) and Tsoukalas and Uhrig (1997).

10.3.2.4 NN for Feature Selection

In many real-world systems, data obtained by measurements may have noisy, redundant and useless information that cannot be used effectively by the FS. The measurements are fed into an NN to extract useful information and provide the numerical values as inputs to the FS. Figure 10.9 shows a cooperative NN-FS system where the measurement space $\{m_1, m_2, \ldots, m_k\}$ is mapped to the feature space $\{x_1, x_2\}$ as inputs to the FS.

10.3.2.5 Cascade Combination

A similar configuration to that in Figure 10.9 can be thought of, where the NN does the pre-estimation of input parameters for the FS and the FS controls the system parameters. Toshiba applied such a combination for a range of toasters, as shown in Figure 10.10. The NN estimates the initial temperature and the number of pieces of bread from sensor information. Using this information and other sensor inputs, the FS determines the optimum time and energy required for the toasting process.

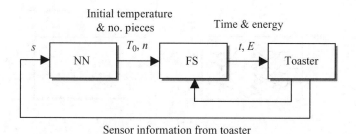

Figure 10.10 Cascade combination – NN collaborating with FS

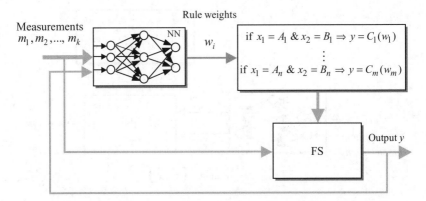

Figure 10.11   Learning fuzzy rule weights cooperatively from an NN

10.3.2.6   NN for Parameter Determination

A neural network determines parameters (scaling parameters) online (i.e., during use of the fuzzy system) to adapt the membership functions and it can also learn the weights of the rules online or offline. Figure 10.11 shows the cooperative combination of NN and FS where the NN determines the rule weights $\{w_1, w_2, \ldots, w_m\}$ of the FS. The NN can be trained offline using the error function derived from the difference between the desired output and the output y of the FS.

After designing an FS, it is sometimes necessary to adapt the MFs. Adapting the MFs can be done using the data distribution as shown in Figure 10.5. The other possibility for adapting the MFs is tuning or learning the scaling parameters of the FS. In many cases, tuning the scaling factors or adjusting the membership functions can lead to the same result. Adjustment of membership functions requires learning of several parameters, and hence scaling factor tuning is a much simpler task than adjusting the MF parameters (Chen and Linkens, 1998).

Figure 10.12 shows the cooperative combination of NN-FS where the NN determines the scaling factors for the MFs. The NN can be trained offline using the error function derived from the difference between the desired output and the output y of the FS.

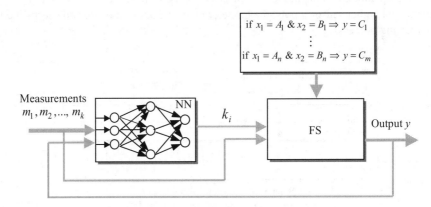

Figure 10.12   Learning scaling parameters from an NN

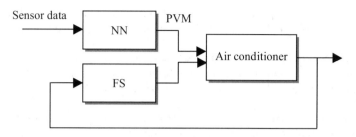

Figure 10.13 FS and NN working concurrently on the same plant

10.4 Concurrent Neuro-Fuzzy Systems

Fuzzy systems and neural networks can work in parallel for a plant without mutual cooperation among themselves. For example, some Japanese air conditioners use an FS to prevent a compressor from freezing in the winter and use an NN to estimate index parameters of comfort, known as predictive mean vote (PMV). PMV can be defined as a function of room temperature, mean radiant temperature, relative air velocity, humidity, thermal resistance of users' clothing, metabolic rate. Some of the PMV parameters cannot be measured using sensors, e.g., thermal resistance of clothing and metabolic rate.

An NN can be used to estimate the PMV index from a set of measured variables such as room temperature, time differential of room temperature, outdoor air temperature, air flow, setting temperature and direction of air flow (Saito *et al.*, 1990). Sensor data and PVM are used as the inputs and outputs of the NN and are defined as sextuples:

PVM = {room temperature, mean radiant temperature, relative air velocity, humidity, thermal resistance of users' clothing, metabolic rate}

Sensor data = {room temperature, time differential of room temperature, outdoor air temperature, air flow, setting temperature, direction of air flow}

The NN performs a nonlinear mapping from the measured sensor values onto the PMV index. The technique has been used by Matsushita Electric Company in their air-conditioning products as shown in Figure 10.13.

10.5 Hybrid Neuro-Fuzzy Systems

In any fuzzy system, inferencing using the rule base and defuzzification using different methods such as centre of gravity are the most time-consuming part. The idea of a hybrid approach is to interpret a fuzzy system in terms of a neural network. The strategy adopted here with a neuro-fuzzy system is, firstly, to replace the rule base with a neural network so that the inference processing is simplified and secondly, to find the parameters of a fuzzy system by means of learning methods obtained from neural networks. A common way to apply a learning algorithm to a fuzzy system is to represent it in a special neural-network-like architecture so that a learning algorithm, such as backpropagation, can be used to train the system. In the first kind of neuro-fuzzy system, there can be three types of fuzzy neural network where only the rule base is replaced and the input and output MFs are kept the same. Different pre-defuzzification methods are applied to minimize processing time. No learning algorithms are used to tune the MFs or parameters of the system. These are the three basic types of fuzzy system. Heuristic or trial-and-error methods are applied for tuning and adjusting parameters:

- Fuzzy neural networks with Mamdani-type fuzzy inference system,
- Fuzzy neural networks with Takagi–Sugeno–Kang-type fuzzy inference system,
- Fuzzy neural networks with Tsukamoto-type fuzzy inference system.

In the second kind of neuro-fuzzy system, learning algorithms such as backpropagation or hybrid training are applied to tune or adjust the parameters of the system. There are different types of neuro-fuzzy system reported in the literature of the 1990s. Some of these will be discussed in detail in the following sections:

- Fuzzy adaptive learning control network (FALCON),
- Approximate reasoning-based intelligent control (ARIC),
- Generalized approximate reasoning-based intelligent control (GARIC),
- Fuzzy basis function networks (FBFN),
- Fuzzy net (FUN),
- Adaptive neuro-fuzzy inferencing systems (ANFIS),
- Fuzzy inference and neural network in fuzzy inference software (FINEST),
- Neuro-fuzzy controller (NEFCON),
- MANFIS, CANFIS,
- Self-constructing neural fuzzy inference network (SONFIN),
- Fuzzy neural network (NFN).

10.5.1   Fuzzy Neural Networks with Mamdani-Type Fuzzy Inference System

The fuzzy-neural network discussed in this section is a Mamdani-type fuzzy system where the rule base is replaced with a neural network. A detailed description of Mamdani-type fuzzy inference system is provided in Chapter 3. For simplicity, a simple two-input single-output system is shown in Figure 10.14. The fuzzy-neural network shown in Figure 10.14 consists of five layers, described as follows.

> Layer 1: The nodes in this layer represent fuzzy MFs $\{A_j, B_j\}$ where x_1 and x_2 are two inputs. These nodes calculate the membership grade of the inputs by fuzzification operation:

$$\mu_{Aj}(x_1), \mu_{Bj}(x_2) \tag{10.2}$$

where $j = 1, 2$. There are two MFs for each input.

> Layer 2: Every node in this layer represents a rule of the fuzzy system. There are four rules, labelled r_1, r_2, r_3, r_4. Each node determines the firing strength of a rule, defined as

$$w_i = \Gamma\left\{\mu_{A_j}(x_1), \mu_{B_j}(x_2)\right\}, \quad i = 1, 2, 3, 4; j = 1, 2 \tag{10.3}$$

> The function $\Gamma(.)$ represents the inferencing operation using the product rule or min rule. For example, Equation (10.4) defines the firing strength using the product rule:

$$w_i = \mu_{A_j}(x_1) {}^* \mu_{B_j}(x_2), \quad i = 1, 2, 3, 4; j = 1, 2 \tag{10.4}$$

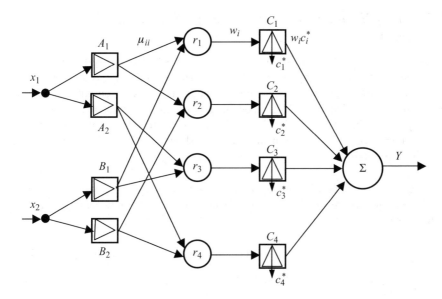

Figure 10.14 Hybrid Mamdani-type fuzzy-neural system

A normalization of the weights can be carried out which will provide the relative firing strength of the individual rules. The normalization is calculated according to Equation (10.5). The normalized weights are not used here in the above architecture.

$$\bar{w}_i = \frac{w_i}{\sum\limits_{i=1}^{4} w_i}, \quad i = 1, 2, \ldots, 4 \tag{10.5}$$

Layer 3: Every node in this layer represents the fuzzy MF C_k, $k = 1, 2, 3, 4$ for the output. The output MFs are pre-defuzzified and the defuzzification operation is denoted

$$c_i^* = \Psi(C_k) \tag{10.6}$$

$\Psi(.)$ is a defuzzification operation applied on the MFs C_k, i.e., c_i^*, $i = 1, 2, \ldots, 4$ are the defuzzified values of the consequent MFs of each rule. Different types of defuzzification operation, discussed in Chapter 2, can be applied here.

Layer 4: The single node in this layer produces the final output by aggregating all the fired rule values, defined as

$$Y = \sum_i w_i.c_i^*, \quad i = 1, 2, \ldots, 4 \tag{10.7}$$

As mentioned earlier, in a Mamdani-type fuzzy-neural system the rule base is replaced with a neural-network-like structure to simplify the inferencing mechanism. Therefore, the layer performing the normalization of the firing strength can be omitted without any significant performance degradation of the system.

10.5.2   Fuzzy Neural Networks with Takagi–Sugeno-type Fuzzy Inference System

The fuzzy-neural network discussed in this section is a Sugeno-type system (also known as Takagi-Sugeno-Kang-type fuzzy system). Current fuzzy-neural systems are mainly Sugeno-type fuzzy systems with the rule base replaced by a neural network and output MFs described by linear functions rather than fuzzy MFs. A detailed description of Sugeno-type fuzzy inference system is provided in Chapter 3. For simplicity, a simple two-input single-output system is shown in Figure 10.15. The fuzzy-neural network shown in Figure 10.15 consists of four layers, described as follows.

> Layer 1: Every node i in this layer is a node with fuzzy membership functions where x_1 and x_2 are two inputs. These nodes calculate the membership grade of the inputs:

$$\mu_{Aj}(x_1), \mu_{Bj}(x_2) \tag{10.8}$$

> where $j = 1, 2$.

> Layer 2: Every node in this layer is a fixed node representing the rules labelled $r_1, \ldots, r_4$. Each node determines the firing strength of a rule:

$$w_i = \Gamma\left\{\mu_{A_j}(x_1), \mu_{B_j}(x_2)\right\}, \quad i = 1, 2, 3, 4; j = 1, 2 \tag{10.9}$$

The function $\Gamma(.)$ represents the inferencing operation using the product rule or min rule. For example, $\Gamma(.)$ is a min operation: $\min\{\mu_{A_j}(x_1), \mu_{B_j}(x_2)\}$. A normalization of the weights can be carried out which will provide the relative firing strength of

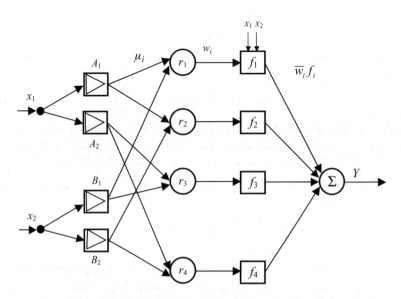

Figure 10.15   Hybrid Sugeno-type fuzzy-neural system

the individual rules. The normalization can be calculated according to Equation (10.5).

Layer 3: Every node in this layer is an output node representing a linear function, defined by

$$f_i = a_i.x_1 + b_i.x_2 + c_i, \quad i = 1, 2, \ldots, 4 \tag{10.10}$$

where a_i, b_i and c_i, $i = 1, 2, \ldots, 4$ are the parameters of the consequent part of the rule. Each node calculates the weighted value of the consequent part of each rule as

$$w_i.f_i = w_i(a_i x_1 + b_i x_2 + c_i), \quad i = 1, 2, \ldots, 4 \tag{10.11}$$

The normalized weights are not used in the architecture in Figure 10.15.
The parameters a_i, b_i and c_i are to be estimated using any heuristic or trial-and-error method. If the parameters $a_i = 0$ and $b_i = 0$, then $f_i = c_i$ is a constant value. Some researchers call it a zero-order Takagi–Sugeno-type system. c_i can be chosen arbitrarily or by trial and error. If c_i is chosen as c_i^* from the pre-defuzzified value of the output MFs of Mamdani-type, as shown in Figure 10.14, then the Takagi–Sugeno-type system is equivalent to a Mamdani-type system.

Layer 4: The single node in this layer produces the output by aggregating all the fired rule values:

$$Y = \sum_i w_i.f_i, \quad i = 1, 2, \ldots, 4 \tag{10.12}$$

If it is a zero-order Takagi–Sugeno-type system, the output is defined as

$$Y = \sum_i w_i.c_i, \quad i = 1, 2, \ldots, 4 \tag{10.13}$$

If c_i is chosen as c_i^*, then the output is defined as

$$Y = \sum_i w_i.c_i^*, \quad i = 1, 2, \ldots, 4 \tag{10.14}$$

Thus, a fuzzy-neural system has been created that is functionally equivalent to a Takagi-Sugeno-type fuzzy model. For a Mamdani-type inference system with max/min composition, a corresponding fuzzy-neural system can be constructed if discrete approximations are used to replace the integrals in a centroid (or other type) defuzzification scheme.

10.5.3 Fuzzy Neural Networks with Tsukamoto-Type Fuzzy Inference System

Tsukamoto-type fuzzy-neural systems are mainly Tsukamoto-type fuzzy systems with the rule base replaced by a feedforward neural network and output MFs described by monotonic MFs.

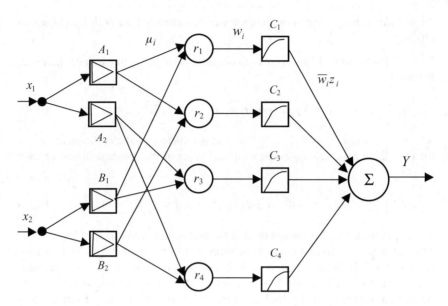

Figure 10.16 Hybrid Tsukamoto-type fuzzy-neural system

A detailed description of Tsukamoto-type fuzzy inference system is provided in Chapter 3. A Tsukamoto-type fuzzy-neural system with two inputs and one output is shown in Figure 10.16. The layers are described as follows.

Layer 1: Every node i in this layer is a node with fuzzy membership functions where x_1 and x_2 are two inputs. These nodes calculate the membership grade of the inputs:

$$\mu_{Aj}(x_1), \mu_{Bj}(x_2) \tag{10.15}$$

where $j = 1, 2$.

Layer 2: Every node in this layer is a fixed node representing the number of rules, labelled $r_1, \ldots, r_4$. Each node determines the firing strength of a rule as

$$w_i = \Gamma\left\{\mu_{A_j}(x_1), \mu_{B_j}(x_2)\right\}, \quad i = 1, 2, 3, 4; j = 1, 2 \tag{10.16}$$

The function $\Gamma(.)$ represents the inferencing operation using the product rule or min rule. For example, $\Gamma(.)$ defines a product rule as $\{\mu_{A_j}(x_1) * \mu_{B_j}(x_2)\}$.

Layer 3: Every node in this layer represents a monotone function $C_k, k = 1, 2, 3, 4$ for the output MFs. The output MFs' defuzzification operation is denoted as

$$z_i = \Psi(C_k) \quad i = 1, 2, \ldots, 4 \tag{10.17}$$

$\Psi(.)$ is a defuzzification operation applied to the MFs C_k, i.e., $z_i, i = 1, 2, \ldots, 4$ are the defuzzified values of the consequent MFs of each rule. The defuzzification operation on a monotone function is discussed in Chapter 2.

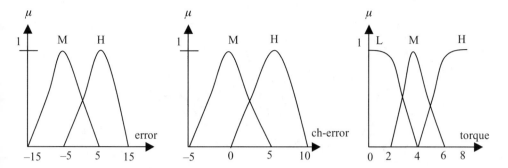

Figure 10.17   Membership functions for error, change of error and torque

Layer 4: The single node in this layer produces the output by aggregating all the fired rule values:

$$Y = \sum_i w_i . z_i, \quad i = 1, 2, \ldots, 4 \tag{10.18}$$

It is to be noted that the three models (Mamdani-type, Takagi–Sugeno-type and Tsukamoto-type) discussed above do not use any learning algorithms, rather they depend on heuristic or trial-and-error methods for the input and output MFs and other parameters. The fuzzy-neural system developed in this section is functionally equivalent to a Tsukamoto-type fuzzy model. For a Mamdani-type inference system with max/min composition, a corresponding system can be constructed if discrete approximations are used to replace the integrals in a centroid (or other type) defuzzification scheme.

Example 10.1   Construction of a zero-order Takagi–Sugeno-type fuzzy-neural system from the description of a Mamdani-type fuzzy system.

A Mamdani-type fuzzy system is described by two inputs, error and change of error, and a single output torque. There are two MFs for each input and three MFs for the output as shown in Figure 10.17. The rule base is shown in Table 10.2.

A zero-order Takagi–Sugeno-type fuzzy-neural system is to be developed from the above description of the Mamdani-type fuzzy system.

Table 10.2   Rule base for Mamdani-type fuzzy system

	Change of error	
Error	M	H
M	r_1: M	r_2: M
H	r_3: H	r_4: L

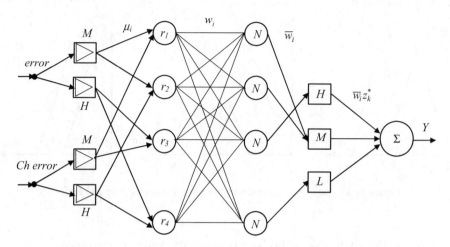

Figure 10.18   Mamdani-type FS converted into fuzzy-neural system

The fuzzy-neural system with two inputs and one output is shown in Figure 10.18. The description of the layers is as follows.

Layer 1: Every node i in this layer is a fixed node with triangular membership functions where x_1 and x_2 are error and change of error. These nodes calculate the membership grade of the inputs by fuzzification:

$$\mu_{Aj}(x_1),\ \mu_{Bj}(x_2) \tag{10.19}$$

where $j = 1, 2$.

Layer 2: Every node in this layer is a fixed node representing four rules, labelled $r_1, \ldots, r_4$. Each node determines the firing strength of a rule. $\Gamma\,[\cdot]$ defines the product or minimum operation of $\left\{\mu_{A_j}(x_1),\ \mu_{B_j}(x_2)\right\}$:

$$w_i = \Gamma\left[\mu_{A_j}(x_1),\ \mu_{B_j}(x_2)\right], \quad i = 1, 2, 3, 4;\ j = 1, 2 \tag{10.20}$$

Layer 3: Every node in this layer is a fixed node labelled N. Each node calculates the normalized firing strength:

$$\bar{w}_i = \frac{w_i}{\displaystyle\sum_{i=1}^{4} w_i}, \quad i = 1, 2, \ldots, 4 \tag{10.21}$$

Layer 4: Every node in this layer contains a pre-defuzzified constant value of the consequent MFs defined by

$$z_k = \Psi\,(MF_k), \quad k = 1, 2, 3 \tag{10.22}$$

where Ψ (MF$_k$) is the chosen defuzzification operation on MFs and z_k is the defuzzified values of MFs in the consequent part of the Mamdani-type fuzzy system. Each node calculates the weighted value of the consequent part of each rule as

$$\bar{w}_i.z_k, \tag{10.23}$$

Layer 5: The single node in this layer produces the output by aggregating all the fired rule values:

$$Y = \sum \bar{w}_i.z_k, \tag{10.24}$$

10.5.4   Neural Network-Based Fuzzy System (Pi–Sigma Network)

The fuzzy-neural systems described in Sections 10.4.1–10.4.3 usually replace the rule base with a neural network and apply a product or minimum rule for inferencing. The Takagi–Sugeno fuzzy system seems to be more flexible than the Mamdani-type fuzzy system. Nevertheless, there are still two shortcomings. Firstly, identification of the fuzzy system is not trivial, which makes it difficult to apply to real-time systems. Secondly, not only are the MFs limited to piecewise linear functions, but the consequent part is also assumed to be linear. This problem remains unresolved until neural networks are combined with fuzzy systems to incorporate suitable learning ability and nonlinear mapping capacity. Jin *et al.* (1995) proposed a hybrid neuro-fuzzy system where the rule firing strength is computed from the antecedent part of the Takagi–Sugeno fuzzy systems in the one part and the output of the consequent part is estimated from a pi–sigma neural network in the other part. In this architecture, a fuzzy neuron is used which performs some fundamental fuzzy operations (such as minimum and maximum operations). The fuzzy-neural architecture is shown in Figure 10.19. The architecture of the fuzzy-pi–sigma neural network is an extension of the Takagi–Sugeno fuzzy model (bottom) where a layered neural network model (upper) is used to estimate the consequent outputs and the two models are combined using a set of product nodes (pi-nodes).

The FS part of the model (lower part of Figure 10.19) represents the antecedent part of the Takagi–Sugeno fuzzy system and computes the rule firing strength w_i as follows.

Layer 1: Every node in this layer is a node with fuzzy linguistic term (MF) $A_k(x_j)$, where $k = 1, 2, \ldots, m$ and $x_j, j = 1, 2, \ldots, n$ are the inputs. These nodes calculate the membership grade of the inputs $\mu_{A_k}(x_j)$.

Layer 2: Every node in this layer represents a rule node and each node determines the firing strength of a rule as

$$w_i = \Gamma\left(\mu_{A_k}(x_1), \ldots, \mu_{A_k}(x_n)\right), \quad \text{where } i = 1, 2, \ldots, N \tag{10.25}$$

$\Gamma(.)$ represents a fuzzy neuron that performs a minimum or product operation. N is the number of rules of the fuzzy system.

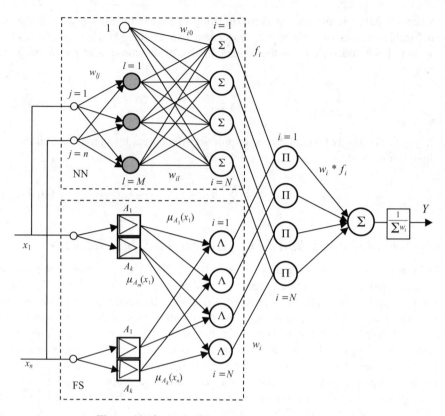

Figure 10.19   Hybrid fuzzy pi–sigma neural network

The NN part of the model represents the consequent part of the Takagi–Sugeno system and computes the output f_i as follows.

Layer 1: Every node in this layer is a neuron with nonlinear activation function. The outputs of this layer is calculated as

$$o_l = g \left(\sum_{j=1}^{n} w_{lj} x_j \right) \text{ where } j = 1, 2, \ldots, n, l = 1, 2, \ldots, M \qquad (10.26)$$

where w_{lj} are the connection weights between the inputs and the first layer and $g(.)$ is a sigmoidal-type nonlinear function.

Layer 2: Every node in this layer is a linear summation neuron (sigma-neuron). The outputs of this layer are calculated as

$$f_i = w_{i0} + \sum_{l=1}^{N} w_{il} o_l = w_{i0} + \sum_{l=1}^{N} \left\{ w_{il} g \left(\sum_{j=1}^{n} w_{lj} x_j \right) \right\} \text{ where } i = 1, 2, \ldots, N$$

$$(10.27)$$

where w_{il} are the connection weights between the first and second layer and w_{i0} are the biases to the second-layer neurons.

Layer 3: This is a common layer for both parts of the fuzzy neural system. Each node in this layer is a pi-neuron (Π) and calculates the product of w_i and f_i, i.e., $(w_i \cdot f_i)$.

Layer 4: This is also a common layer for both parts of the fuzzy neural system and calculates the sum of the products multiplied by the term $\frac{1}{\sum_{i=1}^{N} w_i}$ to yield the final output of the system Y as follows:

$$Y = \frac{\sum_{i=1}^{N} w_i f_i}{\sum_{i=1}^{N} w_i} = \sum_{i=1}^{N} \bar{w}_i f_i \tag{10.28}$$

This hybrid fuzzy-neural network is equivalent to a Takagi–Sugeno-type fuzzy system where the linear consequent functions have been extended to nonlinear functions and the parameters are estimated by the neural network shown in the upper part of the diagram in Figure 10.19.

To adjust the consequence parameters and the parameters of the MF, the error backpropagation algorithm should be extended as the gradient method requires differentiable functions. Therefore, the minimum operator will need to be transformed. Suppose the desired output of the pi–sigma network is Y_d. The error function is defined as follows:

$$E = \frac{1}{2}(Y - Y_d)^2 \tag{10.29}$$

According to the principle of error backpropagation, the generalized error of the final output node Σ is $(Y_d - Y)/\Sigma w_i$. Since the consequence f_i and the overall truth value of the premises w_i of the ith implication are multiplied in the multiplication node Π, the error cannot be backpropagated directly. However, considering w_i as the 'weight' connecting the consequence node Σ and the final output node Σ, the product node can be 'eliminated'. In this way, the generalized error of the ith consequence node Σ is obtained approximately as

$$\delta_i^1 = (Y_d - Y) w_i \bigg/ \sum_{i=1}^{N} w_i \tag{10.30}$$

Similarly, the generalized error of each fuzzy node is calculated as

$$\delta_i^2 = (Y_d - Y) f_i \bigg/ \sum_{i=1}^{N} w_i \tag{10.31}$$

Therefore, the consequence parameters are adjusted according to the following formulae:

$$\Delta w_{i0} = \eta \delta_i^1 \tag{10.32}$$

$$\Delta w_{lj} = \eta x_j \sum_{l=1}^{N} \left\{ w_{il} g \left(\sum_{j=1}^{n} w_{lj} x_j \right) \right\} \delta_i^1 \tag{10.33}$$

$$\Delta w_{il} = \eta x_j g \left(\sum_{j=1}^{n} w_{lj} x_j \right) \delta_i^1 \tag{10.34}$$

Here, η is a positive learning rate and the other parameters are defined as before. The MF used in the fuzzy part (lower) is a Gaussian MF defined by

$$\mu_{A_k}(x_j) = \exp\left[-\frac{\left(x_j - m_k^j\right)^2}{\sigma_k^j}\right] \tag{10.35}$$

where m_k^j is the centre and σ_k^j is the width of the membership function $A_k(x_j)$, respectively. The MF parameters can be adjusted according to the following rules:

$$\Delta m_k^j = \begin{cases} 2\eta\left(x_j - m_k^j\right) w_i \delta_i^2 / \sigma_k^j & \text{if } A_k(x_j) \text{ minimum} \\ 0 & \text{else} \end{cases} \tag{10.36}$$

$$\Delta\sigma_k^j = \begin{cases} \eta\left(x_j - m_k^j\right)^2 w_i \delta_i^2 / \sigma_k^j & \text{if } A_k(x_j) \text{ minimum} \\ 0 & \text{else} \end{cases} \tag{10.37}$$

Here, η is a positive learning rate and the other parameters are defined as before. A detailed derivation of the different update rules and terms can be found in Jin and Jiang (1999).

10.5.5 Fuzzy-Neural System Architecture with Ellipsoid Input Space

Traditional partitioning of the input space leads to exponential growth of fuzzy rules with increasing number of inputs. In order to control the exponential growth of the rules, Aoyama and Venkatasubramanian (1995) proposed a combination of grid partitioning and ellipsoidal partitioning. Grid partitioning is used for input dimensions where *a priori* knowledge is available and ellipsoidal partitioning is used for other input dimensions (Aoyama *et al.*, 1995). The fuzzy-neural network consists of four layers. The input dimensions are divided into two groups: a fuzzy grid partition denoted as g-dimensions and a fuzzy ellipsoid partition denoted as e-dimensions. Figure 10.20 shows the FS-NN architecture with ellipsoid inputs.

Layer 1: Nodes in this layer are g-nodes and e-nodes and linear transfer nodes responsible for inputs X^g and X_i^e, $i = 1, \ldots, n$ from the g-dimensions and e-dimensions, respectively. The number of nodes is equal to the number of input dimensions. They transfer the inputs to the fuzzification layer:

$$x^g = f\left(X^g\right) \tag{10.38}$$

$$x_i^e = f\left(X_i^e\right), \quad i = 1, 2, \ldots, n \tag{10.39}$$

Layer 2: This is the fuzzification layer consisting of bell-shaped MFs. Each node represents the antecedent part of the rule of the FS. The fuzzified outputs are

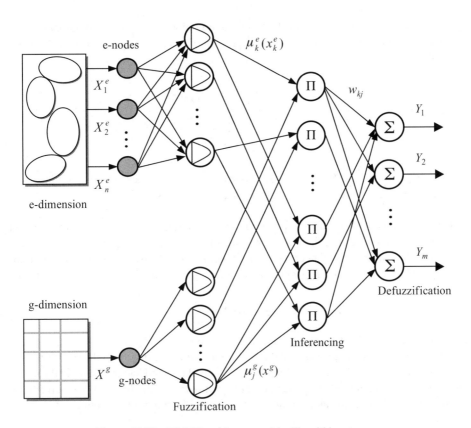

Figure 10.20 FS-NN architecture with ellipsoid input space

defined as

$$\mu_j^g = \exp\left(-\frac{\left(m_j^g - x^g\right)^2}{\sigma_j^g}\right), \quad j = 1, 2, \ldots, G \tag{10.40}$$

$$\mu_k^e = \prod_{i=1}^n \exp\left(-\frac{\left(m_{ki}^e - x_i^e\right)^2}{\sigma_{ki}^e}\right), \quad k = 1, 2, \ldots, E \tag{10.41}$$

where m_j^g and m_{ki}^e are the centres, σ_j^g and σ_{ki}^e are the variances of the g-dimensions and e-dimensions, respectively.

Layer 3: This is the fuzzy inference layer where each node corresponds to a rule. The nodes compute the rule firing strength w_{kj} defined as the product of μ_j^g and μ_k^e:

$$w_{kj} = \mu_k^e \times \mu_j^g \tag{10.42}$$

Layer 4: This is the defuzzification layer and computes outputs. The number of nodes in this layer is equal to the output dimension. The output is defined as follows:

$$Y_l = \frac{\sum\limits_{k=1}^{E} \sum\limits_{j=1}^{G} m_{kj} w_{kj}}{\sum\limits_{k=1}^{E} \sum\limits_{j=1}^{G} w_{kj}} \tag{10.43}$$

where G and E are the number of g-nodes and e-nodes in the fuzzification layer with $l = 1, 2, \ldots, m$ and m_{kj} are the centres of the MFs.

If the parameters of fuzzification operations and interference rules are determined by methods such as fuzzy clustering, the consequent parameters can be estimated by least-squares calculations. However, a gradient descent-based method like the backpropagation algorithm is normally applied to adaptive fuzzy systems where the goal is to minimize the error function

$$E = \frac{1}{2}(Y - Y_d)^2 \tag{10.44}$$

where Y_d is the desired output and Y is the actual output.

Each training data set is propagated through the network starting from the input nodes and the output Y is calculated. The error function E is calculated and backpropagated through the network starting from the output node in order to compute $\partial E / \partial w$ for all hidden-layer nodes. Assuming that w is an adjustable parameter, the general learning rule is defined as

$$w(t + 1) = w(t) + \eta \left(-\frac{\partial E}{\partial w} \right) \tag{10.45}$$

Here, η is a positive learning rate and the other parameters are defined as before. In most of the applications, a structure of the fuzzy neural network is provided, i.e., fuzzy rules are given as *a priori* knowledge and backpropagation learning is used to fine-tune the parameters. A detailed description of the derivation of the learning rule and different terms can be found in Aoyama *et al.* (1999).

The fuzzy neural network described in Equations (10.38)–(10.45) has been used to model physical systems. Chen and Teng (1995) used a fuzzy neural network with two membership functions for each dimension to model a SISO nonlinear system. The model was then used to identify a fuzzy neural network controller in a model reference control scheme.

10.5.6   Fuzzy Adaptive Learning Control Network (FALCON)

The learning rate in the backpropagation learning algorithm is limited due to the fact that the weights of the network are determined by the error function defined in terms of output subject to minimization. A substantial amount of computation time is spent in discovering the internal representation. To minimize the internal representation time, a fuzzy adaptive learning control network (FALCON) is proposed by Lin and Lee (1991), which is a connectionist model of a fuzzy logic controller and decision-making system. In this connectionist structure, the input and output nodes represent the input states and output control decision signals, respectively, and in the hidden layers there are nodes functioning as MFs and rules. The FALCON architecture consists of five layers of neurons (Lin and Lee, 1991, 1994) as shown in Figure 10.21. The individual layers are discussed in the following.

Layer 1: The nodes in this layer are input linguistic nodes which represent linguistic variables. The nodes in this layer transmit input values to the next layer directly.

Layer 2: The nodes in this layer act as MFs to represent the terms of the respective linguistic variable. In this layer, there can be either a single node that performs a simple MF (e.g., a triangle-shaped or bell-shaped function) or multilayer nodes (a subnet) to perform a complex MF (e.g., in an acoustic cue detector). In this case, the total number of layers in FALCON can be more than five. The typical MF used in FALCON is bell-shaped. A bell-shaped MF is defined using two parameters $A_j(m_{ij}, \sigma_{ij})$, such as the centre or mean m_{ij} (for the jth MF of the ith input x_i) and the width or variance σ_{ij} (for the jth MF of the ith input x_i). The link weights at this layer w_{ij} can be interpreted as m_{ij} when a bell-shaped MF is used. The nodes in this layer are fully connected between linguistic nodes and their corresponding term nodes:

$$\mu_j(x_i) = -\frac{(x_i - m_{ij})^2}{\sigma_{ij}}, \quad i = 1, \ldots, n, \quad j = 1, \ldots, m \qquad (10.46)$$

Here, n is the number of inputs and m is the number of MFs.

Layer 3: The nodes in this layer are rule nodes. The links in this layer are used to perform precondition matching of fuzzy logic rules. Each node represents one fuzzy rule performing a fuzzy AND operation, i.e., the minimum operation as defined in Equation (10.47).

$$w_k = \min\left[\mu_1^k(x_i), \ldots, \mu_m^k(x_i)\right], \quad k = 1, \ldots, M \qquad (10.47)$$

where $\mu_j(x_i)$ is the membership grade of the jth MF of the ith input, w_k is the rule firing strength of the kth rule and M is the number of rules. The rule base is represented by this layer.

Layer 4: The nodes in this layer are the output term nodes and they operate in two modes. The first mode is the down/up transmission mode, in which the links at layer four perform the fuzzy OR operation to integrate the fired rules. The second mode is the up/down transmission mode. The nodes in layer four and the links in layer five function exactly the same as those in layer two. Only a single node is used to perform a membership function for output linguistic variables.

Layer 5: This layer is the output layer. There are two types of node for each output variable. The first type of node performs the up/down transmission for the training data to feed the desired outputs y_k, $k = 1, \ldots, m$ into the network. The second type of node performs the down/up transmission for the decision signal outputs y_k^*, $k = 1, \ldots, m$. The arrow on the link indicates the normal signal flow direction when this network is in use after it has been built and trained. These nodes and the layer-five links attached to them act as the defuzzifier.

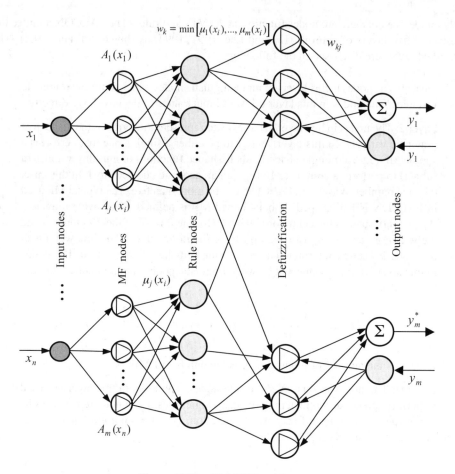

Figure 10.21   FALCON architecture

A two-phase hybrid learning scheme has been developed for FALCON combining a self-organizing and supervised learning algorithm. In the first phase, the self-organizing learning determines the initial MFs and builds the rule nodes. In the second phase, the supervised learning adjusts the MFs. Training data and the desired partitioning of the input/output MFs are provided. It has been shown that the hybrid learning algorithm outperforms the purely supervised learning algorithm due to the *a priori* classification of training data through an overlapping receptive field before the supervised learning.

10.5.7   Approximate Reasoning-Based Intelligent Control (ARIC)

The ARIC architecture, first proposed by Berenji (1992), is a fuzzy controller consisting of two specialized NNs: the action-state evaluation network (AEN) and the action selection network (ASN). The AEN is an adaptive critic that evaluates the ASN and provides advice to the main controller. The ASN is the direct representation of the fuzzy controller. The AEN and ASN modules are shown along with the ARIC architecture in Figure 10.22.

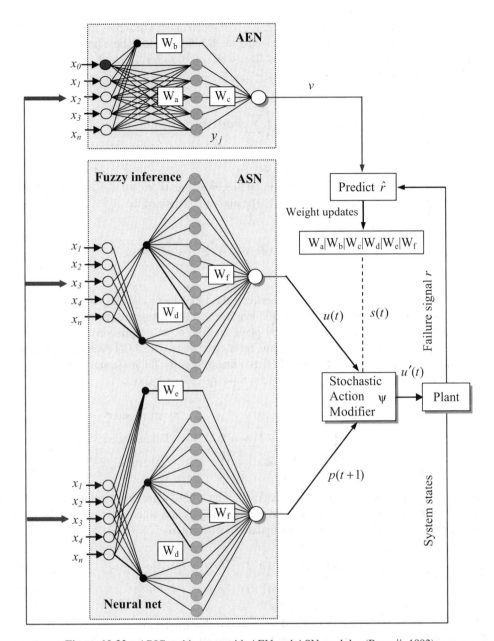

Figure 10.22   ARIC architecture with AEN and ASN modules (Berenji, 1992)

Action-State Evaluation Network

The AEN is a three-layer network, which assumes the role of an adaptive critic (Barto *et al.*, 1983). The input layer consists of n input units and receives information on the states of the physical system to be controlled in terms of its state variables x_i, $i = 1, 2, \ldots, n$ and a failure signal r.

The hidden layer consists of h_j, $j = 1, 2, \ldots, m$ hidden units. The inputs from the environment $(x_1, \ldots, x_n)$ and a bias (x_0) are connected to all hidden units. The connection weights between input and hidden units are denoted Wa_{ji}, with $i = 0, 1, 2, \ldots, n$ and $j = 1, 2, \ldots, m$. The hidden unit outputs h_j are defined as

$$h_j(t + 1) = g \left(\sum_{i=0}^{n} Wa_{ji}(t)x_i(t + 1) \right) \tag{10.48}$$

where $g(.)$ is a sigmoidal activation function.

The output layer consists of a single unit and receives inputs from $n + 1$ input-layer units (i.e., x_i) and m hidden-layer units (i.e., h_j). The output v is defined as

$$v(t + 1) = \sum_{i=0}^{n} Wb_i(t)x_i(t + 1) + \sum_{j=1}^{m} Wc_j(t)h_j(t + 1) \tag{10.49}$$

where Wb_i are the connection weights between the input layer and the output layer. Wc_i are the connection weights between the hidden layer and the output layer. The output v is the prediction of reinforcement. The AEN plays the role of an adaptive critic element and constantly predicts the future reinforcements for a given state. The AEN evaluates the action recommended by the action network (ASN) as a function of the failure signal and the change in state evaluation based on the system state at time $(t + 1)$:

$$\hat{r}(t + 1) = \begin{cases} 0 & \text{starting state} \\ r(t + 1) - v(t) & \text{failure state} \\ r(t + 1) + \alpha v(t + 1) - v(t) & \text{otherwise} \end{cases} \tag{10.50}$$

where $0 \leq \alpha \leq 1$ is the discount rate. In general, Sutton's AHC algorithm (Sutton, 1984, 1988) is used to train the weights Wb and Wc and the backpropagation algorithm is used to train the weights Wa. The weights Wb and Wc are updated according to

$$Wb_i(t + 1) = Wb_i(t) + \beta \hat{r}(t + 1)x_i(t) \tag{10.51}$$

$$Wc_j(t + 1) = Wc_j(t) + \beta \hat{r}(t + 1)h_j(t) \tag{10.52}$$

where $\beta > 0$ is the learning constant and $\hat{r}(t + 1)$ is the internal reinforcement at time $(t + 1)$. The weight update function for Wa is based on a modified version of the error backpropagation algorithm as there is no direct measurement of error possible. $\hat{r}$ is used as an error measure and the weights Wa are updated according to

$$Wa_{ij}(t + 1) = Wa_{ij}(t) + \beta \hat{r}(t + 1)h_j(t) \lfloor 1 - h_j(t) \rfloor \text{sgn} \lfloor Wc_j(t) \rfloor x_i(t) \tag{10.53}$$

A positive change in the state evaluations, i.e., a positive $\hat{r}$, results in an increase in weights and a negative $\hat{r}$ results in a decrease in weights.

Action Selection Network

The ASN consists of two networks. The first network is a fuzzy controller consisting of a fuzzifier, a rule base and decision-making logic, and a defuzzifier. The fuzzy controller is modelled by a two-layer neural network. The input layer performs fuzzification of the input variables $\{x_1, \ldots, x_n\}$ against the labels represented by the MFs and determines the degree of MFs $\{\mu_{i1}(x_i), \ldots, \mu_{im}(x_i)\}$, $i = 1, 2, \ldots, n$ and m is the number of MFs for each input. The hidden layer corresponds to the rules of the fuzzy controller and includes the decision-making logic. The antecedents are not visible in the structure in Figure 10.22. The inputs to the neuron are the preconditions of a rule and the output of the neuron is its conclusion. A multi-input single-output (MISO) control system is considered here. Let w_j represent the degree that a rule j is fired by an input state variable x_i in X, which means

$$w_j = \min \left\{ Wd_{j1}\mu_{1j}(x_1), \ldots, Wd_{ji}\mu_{ij}(x_i), \ldots, Wd_{jn}\mu_{nj}(x_n) \right\} \qquad (10.54)$$

where $\mu_{ij}(x_i)$ represents the degree of membership of the input x_i in a fuzzy set representing the label used in the first precondition of the rule j, Wd_{ji} is the weight from the j th rule node to the i th input node, i.e., the link between neuron i and neuron j (i.e., from input layer to hidden layer). Then, m_j represents the result of applying the w_j on the consequent part of rule j and is calculated as follows:

$$m_j = \mu_{C_j}^{-1}(w_j) \qquad (10.55)$$

where $j = 1, 2, \ldots, M$, M is the number of rules (or number of nodes in the hidden layer), μ_{C_j} represents the monotonic membership function of the consequent part of rule j. The inverse $\mu_{C_j}^{-1}$ of the MF μ_{C_j} is taken to mean a suitable defuzzification operation applicable to an individual rule. The output layer performs the defuzzification process and combines the consequent part of the individual rules by using the centre of area (COA) method. The amount of control action $u(t)$ is then calculated assuming discretized MFs according to

$$u(t) = \frac{\sum_{j=1}^{M} Wf_j \times m_j \times w_j}{\sum_{j=1}^{M} w_j \times Wf_j} \qquad (10.56)$$

where Wf_j are the connection weights from the hidden-layer nodes to the output-layer nodes. The architecture allows the rules in the control knowledge base to be simply translated into the action selection network. Adjusting the weights in the network represents fine-tuning of the control rules.

The second network of the ASN is a neural network that computes a probability value p to signify a measure of confidence associated with the selected action. The probability measure is used to modify the control action $u(t)$. The output of the hidden-layer units is

$$z_j(t) = g\left(\sum_{i=1}^{n} Wd_{ji}(t)x_i(t+1) \right), \quad j = 1, 2, \ldots, H \qquad (10.57)$$

where $g(.)$ is a sigmoidal function and H is the number of units in the hidden layer. The units in the output layer receive inputs from the hidden layer and input layer and compute the probability $p(t + 1)$ as follows:

$$p(t + 1) = \sum_{i=1}^{n} We_i(t)x_i(t + 1) + \sum_{j=1}^{H} Wf_j(t)z_j(t + 1) \tag{10.58}$$

We_i are the connection weights between the input layer and output layer, and $p(t + 1)$ is used to modify the action $u(t)$ of the fuzzy controller. The stochastic action modification is performed as follows:

$$u'(t) = \Psi [u(t), p(t + 1)] \tag{10.59}$$

$\Psi[.]$ is a stochastic modification function based on the probability $p(t + 1)$. Berenji (1992) uses a final measure $s(t)$ for stochastic action modification and it is computed based on the comparison of $u(t)$ and $u'(t)$ according to

$$s(t) = k \left[u(t), u'(t)\right] \tag{10.60}$$

The function $k[.]$ should be chosen depending on the application. $s(t)$ is then used to update the weights of ASN. AHC and the backpropagation algorithm are used to train the weights. The weight changes of We and Wf are proportional to $\hat{r}$, s and the corresponding output. The weights are updated as follows:

$$We_i(t + 1) = We_i(t) + \eta\hat{r}(t + 1)s(t)x_i(t) \tag{10.61}$$

$$Wf_j(t + 1) = Wf_j(t) + \eta\hat{r}(t + 1)s(t)z_j(t) \tag{10.62}$$

The weight update for Wd is based on a modified version of the backpropagation algorithm using $\hat{r}$ with $s(t)$ as an error measure. Backpropagating this error, the weights Wd are updated according to

$$Wd_{ji}(t + 1) = Wd_{ji}(t) + \eta\hat{r}(t + 1)z_j(t)\lfloor 1 - z_j(t)\rfloor \text{sgn}\lfloor Wf_j(t)\rfloor s(t)x_i(t) \tag{10.63}$$

where $\eta > 0$ is a learning constant. The AEN and ASN do not have the same number of nodes in the hidden layer or in the output layer. The ARIC architecture discussed above allows the rules in the control knowledge base to be translated into the ASN. It has also been demonstrated that changing the weights in this network represents fine-tuning the control rules.

10.5.8   Generalized ARIC (GARIC)

GARIC is an extension to ARIC developed by Berenji and Khedkar (1992, 1993). Like ARIC it consists of an evaluation network (AEN) and an action network (ASN). The architecture and learning algorithm for AEN of GARIC are exactly the same as ARIC, as described in the preceding section 10.5.7. The ASN is replaced with a modified five-layer feedforward network

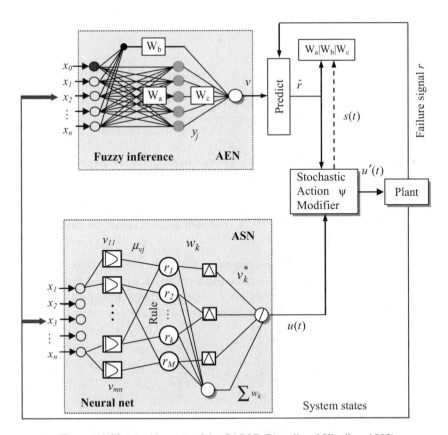

Figure 10.23 Architecture of the GARIC (Berenji and Khedkar, 1992)

where all the connection weights between the layers are unity. The five-layer GARIC model is shown in Figure 10.23 and the layers are discussed in the following.

Layer 1: This layer is the input layer consisting of real-valued input variables. Inputs are passed on to layer two.

Layer 2: The nodes in this layer represent m fuzzy MFs $\{A_{ji}(x_i)\}$, $j = 1, 2, \ldots, m$ for the inputs x_i, $i = 1, 2, \ldots, n$. These nodes calculate the membership grade of the inputs by fuzzification operation. GARIC uses asymmetric triangular MFs, defined as

$$
\mu_{v_{ji}}(x_i) = \begin{cases} 1 - \dfrac{|x_i - c_j|}{r_j} & \text{if } x_i \in [c_j, c_j + r_j] \\[2mm] 1 - \dfrac{|x_i - c_j|}{l_j} & \text{if } x_i \in [c_j - l_j, c_j] \\[2mm] 0 & \text{otherwise} \end{cases} \tag{10.64}
$$

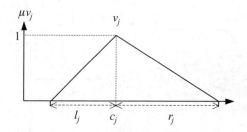

Figure 10.24   Asymmetric triangular MF used for the input and output variables

where c_j, r_j and l_j are the centre, right width and left width of the triangular MF, respectively as shown in Figure 10.24. The membership values are passed on to the second hidden layer.

Layer 3: Every node in this layer represents a rule of the fuzzy system. The rules are labelled $r_1, r_2, \ldots, r_M$ and M is the number of rules. Each node determines the degree of fulfilment (or firing strength) of a rule. A different minimum operation called softmin is used here, which is defined as

$$\widetilde{\min}\{\mu_1, \mu_2, \ldots, \mu_n\} = \frac{\sum\limits_{i=1}^{n} \mu_i e^{-\alpha \mu_i}}{\sum\limits_{i=1}^{n} e^{-\alpha \mu_i}} \tag{10.65}$$

Here, $\alpha > 0$ is a parameter which defines the behaviour of the function. When $\alpha = 0$, $\widetilde{\min}$ means an arithmetical mean. Each node in this layer calculates the $\widetilde{\min}$ operation for each rule as

$$w_k = \widetilde{\min}\left\{\mu_{v_j}^k(x_1), \mu_{v_j}^k(x_2), \ldots, \mu_{v_j}^k(x_n)\right\}, \quad j = 1, 2, \ldots, m \tag{10.66}$$

where $k = 1, 2, \ldots, M$. The rule firing strength w_k is passed on to the fourth layer and the fifth (output) layer.

Layer 4: Every node in this layer represents a consequent fuzzy MF. It is assumed that once again an asymmetric triangular MF is used. It is a similar MF used for the inputs shown in Figure 10.24. The defuzzification procedure used here is called the local mean-of-maximum and defined as follows:

$$v_{jk}^* = c_j + \frac{1}{2}(r_j - l_j)(1 - w_k) \tag{10.67}$$

Here, v_{jk}^* is the defuzzified value of the jth MF v_{jk} for the kth rule. c_j, r_j and l_j are the centre, right width and left width of the triangular MF v_j, $r_j \neq l_j$ and $j = 1, 2, \ldots, m$. m is the number of membership functions in the consequent part. The unusual feature of the nodes in the fourth layer is that they can produce multiple output values. The non-standard feature is eliminated by a single integrated value for the nodes in layer four (v_j^4) computed as follows:

$$v_{jk}^4 = c_j + \frac{1}{2}\left(r_j - l_j\right)\sum_k w_k - \frac{1}{2}\left(r_j - l_j\right)\sum_k (w_k)^2 \qquad (10.68)$$

The integrated output value in Equation (10.68) is necessary if strict compliance is required for a neural network model and the GARIC model is not influenced by the above integration of values.

Layer 5: The single node in the output layer produces the final output by aggregating all the fired rule values defined as

$$u = \frac{\displaystyle\sum_{k=1}^{M} w_k \cdot v_{jk}^*}{\displaystyle\sum_{k=1}^{M} w_k} = \sum_{k=1}^{M} \bar{w}_k \cdot v_{jk}^* \qquad (10.69)$$

Or, by taking advantage of the single integrated value computed in layer four, it can be written as

$$u = \frac{\displaystyle\sum_{k=1}^{M} v_{jk}^4}{\displaystyle\sum_{k=1}^{M} w_k} \qquad (10.70)$$

Thus, layer five produces a continuous value, which is the action selected by ASN. The action output is modified by the stochastic action modifier using the previous prediction $\hat{r}(t-1)$ from the AEN and the action selection $u(t)$ from the ASN. The stochastic action modification is performed as follows:

$$u'(t) = \Psi\left[u(t), \sigma[\hat{r}(t-1)]\right] \qquad (10.71)$$

$\Psi[.]$ is a stochastic modification function based on a Gaussian random variable with mean $u(t)$ and standard deviation $\sigma[\hat{r}(t-1)]$. $\sigma[.]$ is a non-negative monotonically decreasing function, e.g., $e^{-\hat{r}}$. Berenji and Khedkar (1992) use a final measure $s(t)$ for stochastic action modification using a perturbation at each time step computed based on the normalized deviation of $u(t)$ and $u'(t)$ according to

$$s(t) = \frac{u'(t) - u(t)}{\sigma[\hat{r}(t-1)]} \qquad (10.72)$$

This contributes as a learning factor in the weight updates of ASN.

Learning of the weights W_a, W_b and W_c in the AEN of GARIC is the same as the learning mechanism of ARIC. The weights W_a, W_b and W_c in the AEN of GARIC are updated according to Equations (10.53), (10.51) and (10.52), respectively. The connection weights of the neural net in ASN are unity. It is the parameters of the fuzzy MFs in layer two and four in the ASN of the GARIC that are to be learnt. Let p be the parameter vector that contains all the parameters (i.e., centre, left width and right width) of the antecedent and consequent MFs. v is the objective function to be maximized with respect to the parameter vector p. This can be done using a gradient descent learning algorithm

$$\Delta p = \eta \frac{\partial v}{\partial p} = \eta \frac{\partial v}{\partial u} \frac{\partial u}{\partial p} \tag{10.73}$$

Berenji and Khedkar used an approximation for the first term as

$$\frac{\partial v}{\partial u} \approx \frac{\Delta v}{\Delta u} \approx \frac{v(t) - v(t-1)}{u(t) - u(t-1)} \tag{10.74}$$

The sign of the quotient in Equation (10.74) is enough for updating the parameters. The second term $\frac{\partial u}{\partial p}$ can be determined easily for the consequent parameters by using Equations (10.67) and (10.69):

$$\frac{\partial u}{\partial p_j} = \sum_k \bar{w}_k \frac{\partial v_{jk}^*}{\partial p_j} \tag{10.75}$$

The differentiation of the term v_{jk}^* with respect to the three parameters $p_j = \{c_j, r_j, l_j\}$ of the consequent MF provides the following:

$$\frac{\partial v_{jk}^*}{\partial c_j} = \frac{\partial \left(c_j + \frac{1}{2}(r_j - l_j)(1 - w_k) \right)}{\partial c_j} = 1 \tag{10.76}$$

$$\frac{\partial v_{jk}^*}{\partial r_j} = \frac{\partial \left(c_j + \frac{1}{2}(r_j - l_j)(1 - w_k) \right)}{\partial r_j} = \frac{1}{2}(1 - w_k) \tag{10.77}$$

$$\frac{\partial v_{jk}^*}{\partial l_j} = \frac{\partial \left(c_j + \frac{1}{2}(r_j - l_j)(1 - w_k) \right)}{\partial l_j} = -\frac{1}{2}(1 - w_k) \tag{10.78}$$

The update rule for the consequent parameters $p_j = \{c_j, r_j, l_j\}$ is then defined as

$$\Delta p_j = \eta \operatorname{sgn} \left(\frac{\Delta v}{\Delta u} \right) s(t) \hat{r} \frac{\partial u}{\partial p_j} \tag{10.79}$$

where η is the learning rate. The use of $s(t)$ and $\hat{r}$ as factors in the learning rule is to give an extra reward to the large perturbation results in a good action. Updating only the consequent parameters may be sufficient in many applications.

The updating of the antecedent MF parameters can be done in a similar way. The action depends on the degrees of w_k, which in turn depends on the degrees of μ_j in layer two.

The updates can only be determined if the functions μ_j, defined in Equation (10.64), are differentiable with respect to p_j. To maximize u with respect to the parameters p_j of the antecedent MF, the gradient descent learning rule follows:

$$\frac{\partial u}{\partial p_j} = \frac{\partial u}{\partial \mu_j} \frac{\partial \mu_j}{\partial p_j} \tag{10.80}$$

The first term $\frac{\partial u}{\partial \mu_j}$ is computed as

$$\frac{\partial u}{\partial \mu_j} = \sum_k \frac{\partial u}{\partial w_k} \frac{\partial w_k}{\partial \mu_j} \tag{10.81}$$

The terms $\frac{\partial u}{\partial w_k}$ and $\frac{\partial w_k}{\partial \mu_j}$ are derived as follows:

$$\frac{\partial u}{\partial w_k} = \frac{c_j + \frac{1}{2}\left(r_j - l_j\right)(1 - 2w_k) - u}{\sum_k w_k} \tag{10.82}$$

$$\frac{\partial w_k}{\partial \mu_j} = \frac{e^{-\alpha\mu_j(x_i)}\left(1 + \alpha[w_k - \mu_j(x_i)]\right)}{\sum_i e^{-\alpha\mu_j(x_i)}} \tag{10.83}$$

The second term $\frac{\partial \mu_j}{\partial p_j}$ can be determined only if μ_j is differentiable at all points. An example of such a smooth function is given by

$$\mu_j(x_i) = \frac{1}{1 + \left|\frac{x_i - c_j}{s}\right|^b} \tag{10.84}$$

where $s = l_j$, if $x_i < c_j$ and $s = r_j$, if $x_i \geq c_j$ and b is a parameter which controls the shape of the function. An approximation according to Berenji and Khedkar (1992) for a triangular-shaped MF is given in Equation (10.64). Computation of the derivatives of μ_j with respect to the parameters $p_j = \{c_j, r_j, l_j\}$ is very straightforward. Details of the derivations for individual parameters can be found in Nauck et al. (1997). The updates of the antecedent parameters $p_j = \{c_j, r_j, l_j\}$ are computed using Equation (10.79). The derivatives can be computed locally using Equations (10.80)–(10.83). GARIC represents a general hybrid neuro-fuzzy architecture, which has been applied to various control problems with success. For example, Berenji and Khedkar (1992) demonstrated the application of GARIC to balance the well-known cart-pole system.

10.5.9   Fuzzy Basis Function Networks (FBFN)

The FBFN has a similar structure to the RBFN. The RBFN has been discussed in Chapter 4. The FBFN was first proposed by Wang and Mendel (1992) using the Stone–Weierstrass theorem, where it is shown that linear combinations of fuzzy basis functions are capable of approximating any real continuous function. In general, the backpropagation algorithm requires thousands of iterations in learning the fuzzy basis functions. Therefore, an orthogonal least squares (OLS) learning algorithm is proposed for designing fuzzy systems based on given input/output pairs. The algorithm selects significant fuzzy basis functions from an initial set of functions. These are then used to construct the final fuzzy system. The OLS learning algorithm is a one-pass procedure, which is much faster compared with the backpropagation algorithm.

The basic configuration of FBFN is similar to the fuzzy systems described in Section 3.4 in Chapter 3, where the main components such as fuzzification, rule base, inference engine and defuzzification are discussed in detail. An FBFN consists of five layers and uses Gaussian MF as inputs, a fuzzy inference system with product inference and singleton output MF with centroid defuzzification procedure. A multiple-input single-output (MISO) system is considered in this section. Consider an n-input single-output system with M rules of the form

$$R^k : If \ x_1 \ is \ A^k_{1j} \ and \ x_2 \ is \ A^k_{2j} \ and \cdots and \ x_n \ is \ A^k_{nj} \ Then \ y \ is \ B^k_j \qquad (10.85)$$

where A_{ij} are the Gaussian MFs for inputs and B_j are the singleton output MFs with $i = 1, 2, \ldots, n$ and $j = 1, 2, \ldots, m$. m is the number of MFs for each input and output assuming each input and output have the same number of MFs. $k = 1, 2, \ldots, M$ and M is the number of rules of the fuzzy system. The FBFN architecture with three inputs, single output and four rules is shown in Figure 10.25.

Layer 1: This layer is the input layer and responsible for transferring input values through to the second layer.

Layer 2: This layer is the fuzzification layer and consists of Gaussian MFs defined by

$$\mu_{A_{ji}}(x_i) = \exp\left[-\frac{1}{2}\left(\frac{x_i - m_{ji}}{\sigma_{ji}}\right)^2\right] \qquad (10.86)$$

where m_{ji} and σ_{ji} are the centre and width of the jth Gaussian MF for input x_i. The membership value $\mu_{A_{ji}}(x_i)$ is passed on to the rule layer.

Layer 3: This layer is the rule layer and uses the product rule for inferencing defined as follows:

$$p_k(x_i) = \prod_{i=1}^{n} \mu_{A_{ji}}(x_i), \quad k = 1, 2, \ldots, M \qquad (10.87)$$

Layer 4: This layer is the defuzzification layer and performs a defuzzification operation on the singleton output MF using a centroid defuzzifier. w_k are the connection weights between layer four and five.

Layer 5: This is the aggregation layer and defines the FBF as

$$y = f(x) = \frac{\sum\limits_{k=1}^{M} p_k(x)w_k}{\sum\limits_{k=1}^{M} p_k(x)}, \quad k = 1, 2, \ldots, M \qquad (10.88)$$

It is shown that the fuzzy inference system is equivalent to a FBF expansion, i.e., a linear combination of FBF or FBFN.

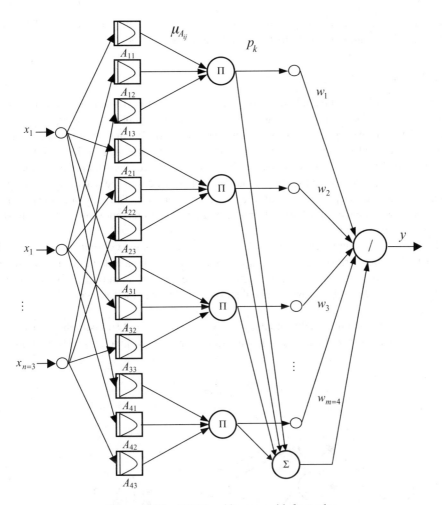

Figure 10.25   FBFN architecture with four rules

Assume an N-input/output data set $\{(x_1, d_1), (x_2, d_2), \ldots, (x_N, d_N)\}$ is available for training. The task of training the FBFN is to design an FBFN $f(x)$ such that

$$d(t) = f[x(t)] + e(t) = \sum_{k=1}^{M} \bar{p}_k(x)w_k + e(t) \tag{10.89}$$

where $\bar{p}_k(x) = \frac{p_k(x)}{\sum_{k=1}^{M} p_k(x)}$. Equation (10.89) can now be written in matrix notation form as

$$d = Pw + e \tag{10.90}$$

where $d = [d_1, \ldots, d_N]$, $P = \begin{bmatrix} p_{11} & \cdots & p_{1M} \\ \vdots & \ddots & \vdots \\ p_{N1} & \cdots & p_{NM} \end{bmatrix}$, $w = [w_1, \ldots, w_N]$ and $e = [e_1, \ldots, e_N]$.

The FBFN becomes a simple linear least-squares problem once the matrix P is known. The column vectors of P are the response vectors of the FBF nodes. In order to perform the OLS procedure, the learning of the parameters $\lambda = \{(m_1, \sigma_1), \ldots, (m_m, \sigma_m)\}$ of the FBF is to be performed on the input/output pairs (Shin and Xu, 2009).

10.5.10 Fuzzy Net (FUN)

Neural networks have been applied to fuzzy systems mainly to extract rules. Many researchers have applied NNs for learning fuzzy rules. NNs have also been applied to finding the parameters of MFs. There have been few attempts at learning rules and MFs in a system at the same time. Sulzberger *et al.* (1993) proposed a method for translating the fuzzy rules and MFs into a network called FUN (Fuzzy Net). The FUN uses special neurons which can evaluate logic expressions using their activation functions. The performances of the network and the quality of the rule base are improved by training the neural network. The network consists of five layers. Neurons of each layer have different activation functions representing different stages of the fuzzy inferencing.

The FUN architecture with inputs $\{x_1, x_2, \ldots, x_n\}$ and a single output y is shown in Figure 10.26. The network consists of an input, an output and three hidden layers. The neurons of each layer have different activation functions. The network is initialized with MFs and a fuzzy rule base. The architecture is described layer by layer in the following.

Layer 1: The neurons in this input layer are sensors and simply transfer the inputs to the next layer (fuzzification layer) without performing any kind of transformation:

$$x_i = f(x_i), \quad i = 1, 2, \ldots, n \qquad (10.91)$$

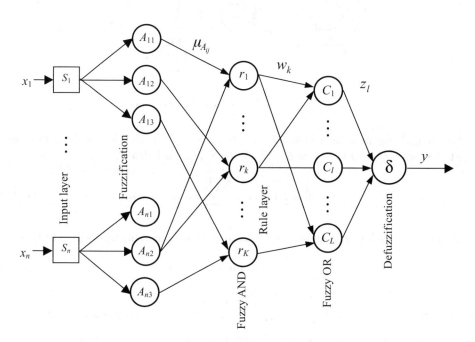

Figure 10.26 FUN architecture

Layer 2: The neurons in this layer (fuzzification layer) contain the MFs and perform fuzzification of the input values:

$$\mu_{A_k} = \Phi\lfloor A_{ij}(x_i)\rfloor, \quad j = 1, 2, \ldots, m, \quad k = 1, \ldots, \quad i \times j, \ldots, n \times m$$

$$(10.92)$$

where A_{ij} is the jth MF for the ith input (x_i). $\Phi\,[.]$ is the fuzzification operation. $\mu_{A_{ij}}$ is the fuzzified value for each MF. There are m membership functions for each input. There are only three MFs shown in Figure 10.26, i.e., $m = 3$.

Layer 3: In this layer (conjunction layer), the conjunctions (fuzzy-AND) are calculated:

$$w_k = \bigcap_{k=1}^{K} \mu_{A_k}, \quad k = 1, 2, \ldots, K \text{ and } K \subseteq n \times m \qquad (10.93)$$

Layer 4: The MFs of the output variables are stored in the third hidden layer. Their activation function is a fuzzy-OR.

Layer 5: Finally, the output neuron performs the defuzzification.

The network is initialized with a fuzzy rule base and the corresponding MFs and thereafter uses a stochastic learning technique that randomly changes parameters of MFs and connections within the network structure. The rules are represented by the connections between layers. To learn the rules, connections between the fuzzification and rule layer are changed. The learning algorithm for the MF is a combination of gradient descent and a stochastic search. To learn the MFs, data to the fuzzification layer (first hidden layer) and the fuzzy-OR layer (third hidden layer) are changed. The learning process is driven by a cost function, which is evaluated after the random modification. If the modification results in an improved performance the modification is kept, otherwise it is undone. The network can be trained with standard neural network training algorithms such as reinforcement or supervised learning (Sulzberger *et al.*, 1993).

Sulzberger *et al.* (1993) demonstrated, by applying FUN to different examples, that it has the ability to optimize a given rule base and the corresponding MFs.

10.5.11 Combination of Fuzzy Inference and Neural Network in Fuzzy Inference Software (FINEST)

The tuning of fuzzy inference mostly investigated fuzzy rules while other factors like parameters of aggregation operators, implication functions, combination functions and fuzzy predicates remained under-explored. The FINEST architecture was proposed by Tano *et al.* (1994, 1996) to develop a tuning mechanism for fuzzy inference and fuzzy predicates. It has two kinds of tuning process: tuning of fuzzy predicates, combination functions and the tuning of an implication function. The generalized *modus ponens* is improved in four ways: (i) aggregation operators that have synergy and cancellation nature; (ii) a parameterized implication function; (iii) a combination function that can reduce fuzziness; and (iv) backward chaining

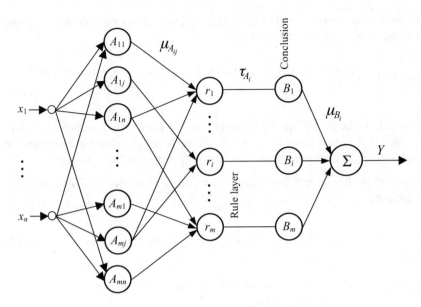

Figure 10.27   Four-layer FINEST architecture

based on generalized *modus ponens*. The backpropagation algorithm is used for fine-tuning the parameters. FINEST provides a framework to tune any parameter which appears in the nodes of the network representing the calculation process of the fuzzy data if the derivative function with respect to the parameters is given.

Consider the case of a fuzzy inference of the form given by the rule

$$\text{Rule } i: \text{if } x_1 \text{ is } A_{i1} \text{ and } \cdots \text{ and } x_n \text{ is } A_{in} \text{ then } y \text{ is } B_i \tag{10.94}$$

where $i = 1, \ldots, m$ and m is the number of rules, x_j with $j = 1, \ldots, n$ and n is the number of MFs for each input variable to the network, y is the conclusion of the fuzzy inference, and A_{ij} and B_i are the input and output membership functions, respectively. In order to carry out the tuning process, the fuzzy inferencing mechanism is converted into a neural network-like structure. There are four layers of neural structure in the FINEST architecture, which is shown in Figure 10.27. The calculation of the output of each layer is described layer by layer in the sequel.

Layer 1: This layer calculates the output of the neurons using a supremum operation as follows:

$$\tau_{A_{ij}}(a_{ij}) = \sup_{\mu_{A_{ij}}(x_j) = f(a_{ij})} \left\{ \mu_{A_{i1}}(x_1), \mu_{A_{i2}}(x_2), \ldots, \mu_{A_{in}}(x_n) \right\} \tag{10.95}$$

$\tau_{A_{ij}}(a_{ij})$ is the truth value from the sup(.) operation and a_{ij} is the set of parameters of the MFs.

Layer 2: This is the aggregation layer and calculates the antecedent part of the ith rule as follows:

$$\tau_{A_i}(a_i) = \sup_{and_i(a_{i1}, \cdots, a_{in})} \left\{ \tau_{A_{i1}}(a_{i1}) \wedge \cdots \wedge \tau_{A_{in}}(a_{in}) \right\} \tag{10.96}$$

where $i = 1, \ldots, m$. $\tau_{A_i}(a_i)$ is the truth value of the antecedent part of the ith rule. and_i is the parameterized aggregation function for rule i.

Layer 3: This layer performs inferencing, i.e., deduction of the conclusion of the ith rule according to

$$\mu_{B_i}(y) = \sup_{a_i} \left\{ \tau_{A_i}(a_i) \wedge I_i \left(a_i, \mu_{B_i'}(y) \right) \right\} \tag{10.97}$$

$\tau_{B_i}(y)$ is the conclusion of the consequent part of the ith rule. I_i is the parameterized implication function of rule i.

Layer 4: This layer performs combinations of the conclusions of all rules according to

$$\mu_B^*(y) = comb \left\{ \mu_{B_1}(y) \wedge \cdots \wedge \mu_{B_m}(y) \right\} \tag{10.98}$$

where *comb* is the parameterized combination function.

The important feature of FINEST is the parameterization of the inference procedure. There are many parameterized functions used in fuzzy inference. There are variants of the t-norm and t-conorm operations, such as max, min, average or product. A new parameterized aggregation function, denoted and_i, is defined in FINEST by adding a synergistic effect to an ordinary t-norm (Tano *et al.*, 1996). There are four parameters γ, (α, β) and p to be chosen arbitrarily by the user:

$$and_i(x_i, y_i) = w_i \times synergy(x_i, y_i) + (1 - w_i) \times basic(x_i, y_i) \tag{10.99}$$

with the following definitions of the terms:

$$w_i = equal(x_i, y_i) \times high(x_i, y_i) \times \gamma \tag{10.100}$$

$$equal(x_i, y_i) = almost(0, x_i - y_i, \alpha) \tag{10.101}$$

$$high(x_i, y_i) = almost(1, x_i, \beta) \times almost(1, y_i, \beta) \tag{10.102}$$

$$synergy(x_i, y_i) = 1 \tag{10.103}$$

$$almost(a, x_i, b) = \exp \left(\ln 0.5 \times (x_i - a)^2 / b^2 \right) \tag{10.104}$$

$$basic(x_i, y_i) = \frac{1}{1 + \sqrt[p]{((1 - x_i)/x_i)^p + ((1 - y_i)/y_i)^p}} \tag{10.105}$$

There are various implications used in fuzzy systems, but it is difficult to choose the appropriate function for a particular application. Therefore, a parameterized implication function I_i is defined in FINEST, which can easily be selected by changing the value of the parameters. Details of the parameterized implication function can be found in Tano *et al.* (1996).

A combination operation is the method of producing a combined-result fuzzy set from two fuzzy sets deduced by two inference processes. Mostly, the max operator is used as combination operator in applications, which causes a constant increase of the fuzziness. Tano *et al.* (1996) proposed two new parameters: the equilibrium E and dependence factors (α, β). A detailed description can be found in Arnould and Tano (1994), Oyama *et al.* (1994) and Tano *et al.* (1994, 1996).

10.5.12 Neuro-Fuzzy Controller (NEFCON)

A major problem encountered in designing a neuro-fuzzy controller is that it requires a set of input/output data for learning. A control problem cannot be solved by supervised learning as there may not be an input/output data set available. An alternative approach would be to use reinforcement learning. Reinforcement learning is employed when a teacher signal is not available and the learning is performed through continued interaction with the environment in order to minimize a scalar performance index (Barto *et al.*, 1983). The NEFCON is a Mamdani-type FIS implemented using reinforcement learning. The NEFCON consists of three layers of a neural network-like structure where an input layer represents input variables, a hidden layer represents fuzzy rules and an output layer consists of a single node. The NEFCON architecture can learn MFs as well as the rule base. The idea of NEFCON is to explore *a priori* knowledge such as known rules and measures of error. This architecture can be used to learn an initial rule base, if there is no *a priori* knowledge of the system available, or to optimize an initial manually defined rule base. The hidden layer represents the rules of the NEFCON of the form

$$Rule:\ if\ x_i\ is\ A_{ij}\ then\ u\ is\ B_k \tag{10.106}$$

where A_{ij} with $i = 1, \ldots, n$ and $j = 1, \ldots, p$. n is the number of inputs, p is the number of antecedent MFs, $k = 1, \ldots, q$ and q is the number of consequent MFs. The NEFCON architecture, shown in Figure 10.28, is a controller consisting of two inputs $\{x_1, x_2\}$ and a single control action u. Triangular MFs are used for the input layer. The input units perform the function of fuzzification. The fuzzified values are passed on to rule nodes. The firing strength is calculated for each rule represented by the nodes in the hidden layer. The output nodes are responsible for the defuzzification. The outputs are combined to a final output.

The process of learning in NEFCON is carried out in two stages: learning the structure (i.e., learning the rules) and learning the parameters (i.e., learning the MFs). When learning the parameters, it is assumed that the structure is already known. The learning is carried out using a backpropagation algorithm. If there is no known control strategy (i.e., rules), the neuro-fuzzy controller should be developing its own rules of operation. Barto *et al.* (1983) showed that neuro-controllers can learn the rules using reinforcement learning. A detailed description of the learning process can be found in Nauck *et al.* (1997).

There are two variants of NEFCON found in the literature: NEFPROX (Nauck and Kruse, 1997) and NEFCLASS (Nauck and Kruse, 1995; Nauck *et al.*, 1996). The NEFPROX model (for function approximation) is very similar to NEFCON, which can have more than one output and be trained using a supervised learning algorithm. The supervised training algorithm uses

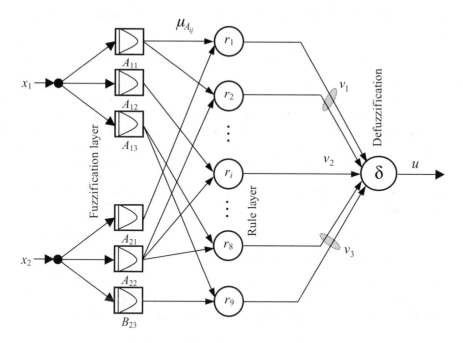

Figure 10.28   NEFCON architecture

an error defined by the difference between the desired and actual output. In NEFPROX, a fuzzy error measure can also be used. NEFCLASS (for classification tasks) was created to determine classes of given data. The classification problem is represented by an unknown function. The rule base of the NEFCLASS model approximates the unknown function by mapping the input pattern to an output class.

10.5.13   Self-constructing Neural Fuzzy Inference Network (SONFIN)

SONFIN is a modified Takagi–Sugeno-type fuzzy system which uses an NN learning mechanism. SONFIN starts with an empty rule base, then creates and adapts the rule base using an online learning mechanism. Both the structure and parameter identification are performed simultaneously to form fast learning. In the structure identification of the antecedent part, the input space is partitioned in a flexible way according to an aligned clustering-based algorithm. As to the structure identification of the consequent part, only a singleton value selected by a clustering method is assigned to each rule initially. Afterwards, some additional significant terms (input variables) selected via a projection-based correlation measure for each rule are added to the consequent part (forming a linear equation of input variables) incrementally as learning proceeds. For parameter identification, the consequent parameters are tuned optimally by either least mean squares or recursive least-squares algorithms and the antecedent parameters (i.e., MF parameters) are tuned by a backpropagation algorithm. To enhance the knowledge representation in the SONFIN, a linear transformation of the input variables can be incorporated into the network for further reduction of the rules.

The architecture of the SONFIN consists of six layers, which realize a TSK-type fuzzy system of the form

$$\text{Rule } k: \text{ if } x_i \text{ is } A_{ij} \text{ then } y_k = m_{0k} + a_{1k}x_1 + \cdots + a_{nk}x_n \tag{10.107}$$

where A_{ij} with $i = 1, \ldots, n$ and $j = 1, \ldots, p$. n is the number of inputs, p is the number of antecedent MFs, m_{0k} is the centre of a symmetric MF on y, a_{ik} are the consequent parameters with $k = 1, \ldots, m$. One important difference of SONFIN from TSK fuzzy systems is that not all of the parameters $\{a_{1k}, a_{2k}, \ldots, a_{nk}\}$ are used in the linear output function. The SONFIN architecture is shown in Figure 10.29. Only two inputs are shown in the figure, but can be extended to n inputs. The architecture is described layer by layer in the following.

Layer 1: Each node in this layer corresponds to one input variable. Nodes do not involve any processing and transmit the input values to the next layer, i.e.

$$x_i = f(x_i), \quad i = 1, 2, \ldots, n \tag{10.108}$$

Only two inputs are shown in Figure 10.29.

Layer 2: Each node in this layer corresponds to an MF (linguistic label) and fuzzifies the inputs. MFs can be of any type, such as triangular, Gaussian, bell-shaped, etc. In SONFIN, a Gaussian MF is used for the inputs $\mu_{ji}(x_i)$:

$$\mu_{ji}(x_i) = f\left(-\frac{\left(x_i - m_{ji}\right)^2}{\sigma_{ji}}\right) \tag{10.109}$$

where m_{ji} and σ_{ji} are the centre (or mean) and width of the jth Gaussian MF of the ith input and $i = 1, \ldots, n$, $j = 1, \ldots, p$. p is the number of MFs for the input x_i. $f(.)$ denotes the chosen activation function of the nodes. One important thing

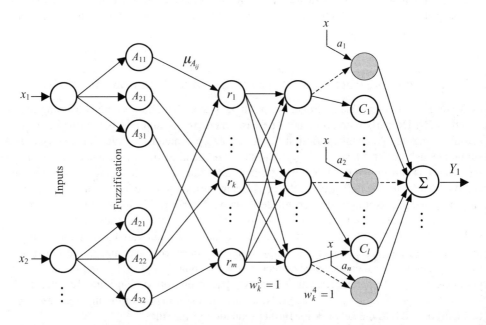

Figure 10.29 SONFIN architecture (Feng and Teng, 1998)

to note here is that the number of MFs for each input is not necessarily identical in SONFIN and the number of MFs for each input is also not the same. The maximum number of nodes in layer two is $m = n \times p$.

Layer 3: This is the rule layer, where each node represents the antecedent part of the rule:

$$\tau_k = \prod_{k=1}^{m} \mu_{k=i \times j} \tag{10.110}$$

where $k = 1, \ldots, m$ and m is the number of nodes from layer 2 contributing to precondition matching of a rule, i.e., the firing strength τ_k of the k th rule. The AND operation is used in SONFIN. The connection weights of layer 3 are unity, i.e., $w_k^3 = 1$.

Layer 4: The number of nodes in this layer is equal to the number of nodes in layer 3. The firing strength τ_k, $k = 1, \ldots, m$, is normalized in this layer and calculated as follows:

$$\bar{\tau}_k = \frac{\tau_k}{\sum\limits_{k=1}^{m} \tau_k} \tag{10.111}$$

The connection weights of layer 4 are also unity, i.e., $w_k^4 = 1$.

Layer 5: This is the consequent layer consisting of two types of node: blank and shaded nodes. The first type (blank) represent consequent MFs (i.e., MFs for output variable). The local mean of maximum defuzzification is applied to the Gaussian MFs and the width is used for output clustering. m_{0k} are the centre of the Gaussian MFs. This eventually delivers the centre of each Gaussian MF m_{0k} as the output to the next layer. Each blank node provides the output y_k^b as

$$y_k^b = m_{0k} \tag{10.112}$$

The other type (shaded) of node generates the consequent part of the rules. Each node receives one input from layer 4 outputs and the other input from layer 1 outputs. Each shaded node provides the output y_k^s as

$$y_k^s = m_{0k} + (a_{1k}x_1 + \cdots + a_{nk}x_n) = m_{0k} + \sum_{i=1}^{n} a_{ik}x_i \tag{10.113}$$

a_{ik} is the corresponding parameter of the input x_i. Combining these two types of node in layer 5, the whole function performed by this layer is given by

$$y_k = \left(y_k^s \right) y_k^b = \left(m_{0k} + \sum_{i=1}^{n} a_{ik}x_i \right) y_k^b \tag{10.114}$$

Layer 6: Each node in this layer corresponds to one output variable. The node in this layer performs defuzzification by integrating all actions from layer 5:

$$Y_1 = \sum_{k=1}^{m} y_k \tag{10.115}$$

Two types of learning occur in SONFIN: structure learning and parameter learning. The structure learning involves both the antecedent and consequent structure identification of the fuzzy if–then rule. The antecedent structure identification corresponds to the input space partitioning, which can be formulated as a combinatorial optimization problem. The objective of the optimization is to minimize the number of rules and MFs. The consequent structure identification corresponds to the creation of a new MF for output variables and inclusion of the input variables in the linear function of the consequent part. The parameter learning involves supervised algorithms such as backpropagation for the antecedent parameters and LMS or RLS algorithms for the consequent parameters. Details of the learning process are discussed in Feng and Teng (1998).

10.6 Adaptive Neuro-Fuzzy System

The neuro-fuzzy synergism comes from adaptive networks. An adaptive network is a network structure described by a set of modifiable parameters. Such an adaptive network combines both neural networks and fuzzy systems. Fuzzy systems under the framework of adaptive networks are known as ANFIS. The only constraint of an adaptive network is that the node transfer function should be piecewise differentiable. The only limitation of network configuration is that it should be of feedforward type. A decomposition of the antecedent and consequent parameters is required in order to apply a suitable learning rule. Besides, it is also shown how to apply the Stone–Weierstrass theorem to ANFIS with simplified fuzzy if–then rules and how the radial basis function network relates to this kind of simplified ANFIS (Jang and Sun, 1993). Different variants of ANFIS have appeared in the literature, as will be discussed in the following sections.

10.6.1 Adaptive Neuro-Fuzzy Inference System (ANFIS)

ANFIS, a class of adaptive networks that are functionally equivalent to a fuzzy inference system, was first proposed by Jang (Jang, 1993; Jang et al., 1997). The ANFIS architecture represents both the Sugeno and Tsukamoto fuzzy models (discussed in Chapter 2 and 3, respectively). It has also been shown that under minor constraints, the RBFN (discussed in Chapter 4) is functionally equivalent to the ANFIS architecture. For simplicity and ease of understanding of the ANFIS architecture discussed in this section, two inputs and a single output are used. The architecture is shown for a first-order Sugeno fuzzy model with four rules in Figure 10.30.

Layer 1: Every node in this layer is an adaptive node with bell-shaped MF, where x_1 and x_2 are the two input variables. Any parameterized MFs can be used for input variables. For simplicity, two MFs are used for each input variable here. These nodes calculate the membership grade of the inputs according to

$$\mu_{Aj}(x_1) = \frac{1}{1 + \left| \dfrac{x_1 - m_{Aj}}{\sigma_{Aj}} \right|^{2b_{Aj}}}, \quad \mu_{Bj}(x_2) = \frac{1}{1 + \left| \dfrac{x_2 - m_{Bj}}{\sigma_{Bj}} \right|^{2b_{Bj}}} \quad (10.116)$$

$\{m_{Aj}, \sigma_{Aj}, b_{Aj}\}$ and $\{m_{Bj}, \sigma_{Bj}, b_{Bj}\}$ with $j = 1, 2$ are the parameter set of the input MFs. Two MFs are used for each input in Figure 10.30.

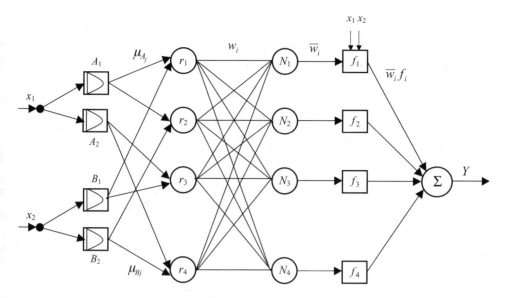

Figure 10.30 ANFIS architecture

Layer 2: Every node in this layer is a fixed node representing the rules labelled r_i, $i = 1, \ldots, 4$. Each node determines the firing strength of a rule, defined as

$$w_i = \mu_{A_j}(x_1).\mu_{B_j}(x_2), \quad j = 1, 2 \qquad (10.117)$$

Layer 3: Every node in this layer is a fixed node labelled N_i, $i = 1, \ldots, 4$. Each node calculates the normalized firing strength of the i th rule according to

$$\overline{w}_i = \frac{w_i}{\displaystyle\sum_{i=1}^{4} w_i}, \quad i = 1, 2, \ldots, 4 \qquad (10.118)$$

Layer 4: Every node in this layer is an adaptive node with a linear function defined by

$$f_i = a_i.x_1 + b_i.x_2 + c_i, \quad i = 1, 2, \ldots, 4 \qquad (10.119)$$

where $\{a_i, b_i, c_i\}$, $i = 1, \ldots, 4$ are the parameters of the consequent part of the rule r_i. Each node calculates the weighted value of the consequent part of each rule as

$$\overline{w}_i.f_i = \overline{w}_i(a_i x_1 + b_i x_2 + c_i), \quad i = 1, 2, \ldots, 4 \qquad (10.120)$$

Layer 5: The single node in this layer produces the overall output by aggregating all the fired rule values:

$$Y = \sum_i \overline{w}_i.f_i, \quad i = 1, 2, \ldots, 4 \qquad (10.121)$$

By using adaptive nodes in layer one and layer four, an adaptive network has been created that is functionally equivalent to a Sugeno-type fuzzy model. A Tsukamoto-type neuro-fuzzy inference system can be constructed by replacing the consequent function with a parameterized piecewise linear function with two parameters. For a Mamdani-type inference system with max/min composition, a corresponding adaptive system can be constructed if discrete approximations are used to replace the integrals in a centroid defuzzification scheme. The MFs to be used should reasonably be shaped so that they can be parameterized correctly to reflect the adequate constraints. It is also expected that the MFs should remain bell-shaped regardless of their parameter values. The fuzzy-neural systems discussed in Sections 10.5.1–10.5.3 are very similar to ANFIS. The only difference is that they don't have any adaptive nodes and no learning mechanism is applied for adjusting the antecedent and consequent parameters. If the MFs are kept fixed at the antecedent part and the consequent parameters are adapted, then ANFIS is seen as a functional-link network (Klassen *et al.*, 1988), where the enhanced representations of the input variables are obtained via the MFs. Similar adaptive fuzzy-neural structures have also been proposed independently by Lin and Lee (1991) and Wang and Mendel (1992). The adaptive capability of ANFIS makes it directly applicable to learning and control problems.

Learning in ANFIS can be separated into two learning schemes: learning of the antecedent MFs and learning of the consequent parameters. There are arguments that learning mechanisms should not be applied to MFs in Sugeno-type ANFIS since they represent a subjective description of the problem. MFs should be kept fixed if the available input/output data set is small. Learning of the MFs will not be useful in this case. If the input/output data set is large, fine-tuning of MFs is necessary as the MFs determined by an expert may not be optimal. If the antecedent parameters are fixed, the output can be expressed as a linear combination of the consequent parameters:

$$y = \overline{w}(ax_1 + bx_2 + c) = (\overline{w}x_1)a + (\overline{w}x_2)b + \overline{w}c \qquad (10.122)$$

For a training data set N, Equation (10.122) can be written in vector-matrix form

$$Ap = y \qquad (10.123)$$

where p is the unknown parameter vector and A is the coefficient matrix. Since the number of training pairs N is greater than $|p|$, there is no unique solution to Equation (10.123). Instead, a least-squares method of estimation (LSE) is suitable for identification of the parameter vector p which minimizes $\|Ap - y\|^2$:

$$\hat{p} = (A^T A)^{-1} A^T y \qquad (10.124)$$

$(A^T A)^{-1} A^T$ is the pseudo-inverse of A when $A^T A$ is non-singular. In general, the structure of ANFIS is assumed fixed and the parameter identification is solved through a hybrid learning rule. In the forward pass of the hybrid learning, node outputs are propagated through to layer four and the consequent parameters are estimated using a least-squares method. In the backward pass, the error (difference between the desired output and the actual output) is backpropagated through to layer one and the antecedent parameters are updated using gradient descent while the consequent parameters are fixed. Since the antecedent and consequent parameters are decoupled in a hybrid learning rule, further speedup of the learning is possible by using other variants of gradient methods or optimization techniques (Jang *et al.*, 1997).

10.6.2   Coactive Neuro-Fuzzy Inference System (CANFIS)

The problem encountered in ANFIS is that the model provides automatic tuning of a Sugeno-type fuzzy inferencing system generating a single output. As a result, the application of ANFIS is limited to multiple-input single-output (MISO) systems only and not multiple-input multiple-output (MIMO) systems. Attempts have been made to apply a multiple ANFIS (MANFIS), where each ANFIS has an independent set of adjustable parameters and fuzzy rules. In MANFIS, it is therefore difficult to find the correlation between inputs and outputs. As the number of outputs increases, the adjustable parameters also increase drastically. The problem is to find how multiple outputs can be generated from an ANFIS system. A generalized ANFIS is called CANFIS, where an NN and an FIS play an active role in the system (Mizutani *et al.*, 1994; Mizutani and Jang, 1995). The architecture of CANFIS is shown with two inputs and a single output in Figure 10.31. The CANFIS here consists of two parts: an FS model (upper part) that computes the normalized weights of the antecedent part of the rules and a layered NN model (lower part) that computes the consequent outputs using the weights from the FS of the MIMO system.

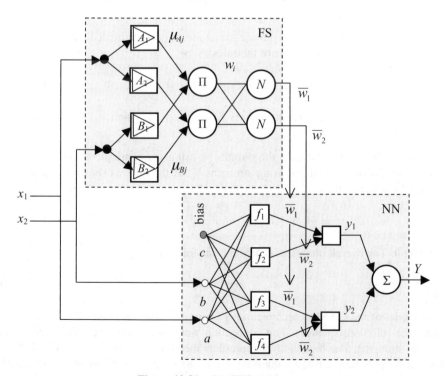

Figure 10.31   CANFIS architecture

The FS model consists of three layers, described as follows.

Layer 1: This is the fuzzification layer that calculates the membership grade $\mu_{Aj}(x_k)$ and $\mu_{Bj}(x_k)$, $k = 1, 2$ and $j = 1, 2$ for the two input variables $\{x_1, x_2\}$.

Layer 2: The second layer determines the firing strength of the individual rule according to

$$w_i = \Gamma\left(\mu_{A_j}(x_1), \mu_{B_j}(x_2)\right), \text{ where } i = 1, 2, \ldots, M(= 2 \times 2) \qquad (10.125)$$

Γ represents a fuzzy neuron that can perform a minimum or product operation. M is the number of rules in the fuzzy system.

Layer 3: The normalized firing strengths w_i are calculated as

$$\overline{w}_i = \frac{w_i}{\displaystyle\sum_{i=1}^{M} w_i} \qquad (10.126)$$

The NN model represents the consequent part of the rules of the Takagi–Sugeno system and computes the output $f_i = a_i x_1 + b_i x_2 + c_i$ as follows.

Layer 1: The nodes in this layer transfer the inputs $\{x_1, x_2\}$ to the next layer. The connection weights are the consequent parameters $\{a_i, b_i, c_i\}$ of the Takagi–Sugeno system.

Layer 2: The nodes in this layer are summation nodes and calculate the linear function of the consequent part of the rules:

$$f_i = a_i x_1 + b_i x_2 + c_i \quad \text{where } i = 1, 2, \ldots, 4 \qquad (10.127)$$

The nodes calculate the consequent parts $f_1 = a_1 x_1 + b_1 x_2 + c_1, \ldots, f_4 = a_4 x_1 + b_4 x_2 + c_4$.

Layer 3: The nodes compute the outputs y_i using the consequent functions f_i weighted by the normalized firing strengths $\overline{w}_i$ of the rules as shown in the NN model of Figure 10.31:

$$y_1 = \overline{w}_1 f_1 + \overline{w}_2 f_2 \quad \text{and} \quad y_2 = \overline{w}_1 f_3 + \overline{w}_2 f_4 \qquad (10.128)$$

Layer 4: The overall output y is calculated as follows:

$$Y = y_1 + y_2 \qquad (10.129)$$

The NN model of the CANFIS architecture is a simple backpropagation MLP network representing the rule consequences. The connection weights are numeric values representing connection strengths. The hidden layer represents the number of rules in the CANFIS. The connection weights between the hidden layer and the output layer correspond to membership values between the consequent layer and the fuzzy association layer. Membership values are dynamically changing, depending on the input patterns. That is, the CANFIS is locally tuned. This is where the powerful capability of CANFIS comes from. Performance may be improved without increasing the rules or MFs by using a nonlinear consequent function such as a sigmoidal function, defined as

$$f = \frac{1}{1 + \exp(ax_1 + bx_2 + c)} \tag{10.130}$$

CANFIS can be extended to multiple outputs by putting ANFIS in juxtaposition for the required number of outputs. The combination is called multiple ANFIS (MANFIS). The problem with MANFIS is that each ANFIS has its own fuzzy rules and no modifiable parameters are shared by the ANFIS. The modifiable parameters increase drastically as the number of outputs increases in MANFIS. A clever way of designing multiple outputs would be to share parameters by using the same antecedent fuzzy rules. That is, fuzzy rules are constructed with shared membership values to establish possible correlations between outputs. Further precision can be obtained for the consequent NN by entwining multiple neural modules (also called local experts) for each neural rule. The advantage of neural modules is that they help in reducing modifiable parameters. The architecture is equivalent to modular networks. The architecture of a MIMO CANFIS, i.e., two inputs, two outputs and two neural rules, is shown in Figure 10.32.

10.7 Fuzzy Neurons

The neurons in neural networks and neural systems, discussed in Chapters 4 and 5, respectively, consist of processing units that handle numeric inputs (mostly sensor measurements) and outputs. There are many examples from real-world applications where such numerical measurements are not available or are corrupt with noise and involve uncertainties. Therefore, some researchers have suggested that neurons should be able to handle such real-world situations and attempted to incorporate them into a fuzzy neuron. A fuzzy neuron has the same basic structure as the artificial neuron except the inputs, processing and outputs are described through fuzzy logic. Therefore, a variety of fuzzy neurons are devised and found in the literature and a few of them are shown in Figure 10.33. The architecture of a fuzzy-neuron network is shown in Figure 10.34.

Figure 10.33(a) shows a fuzzy Min neuron, which performs the aggregation operation that selects the minimum of the weighted inputs defined by

$$O = f \left(\bigwedge_{i=1}^{n} x_i w_i + b \right) \tag{10.131}$$

Similarly, Figure 10.33(b) shows a fuzzy Max neuron, which performs the aggregation operation that selects the maximum of the weighted inputs defined by

$$O = f \left(\bigvee_{i=1}^{n} x_i w_i + b \right) \tag{10.132}$$

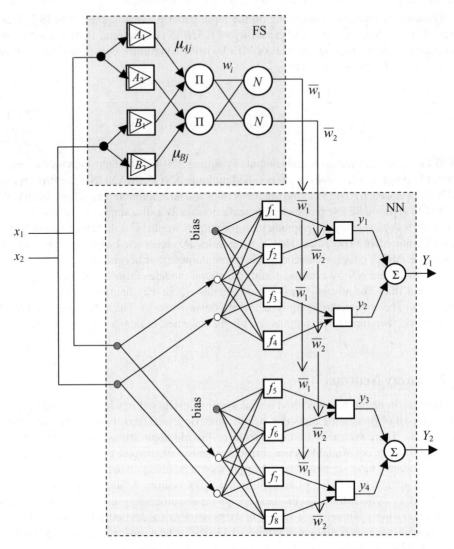

Figure 10.32   MIMO CANFIS (MANFIS) architecture

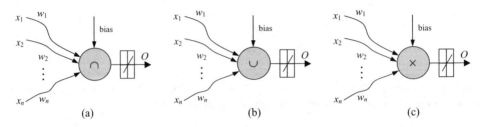

Figure 10.33   Different fuzzy neurons. (a) AND (Min) fuzzy neuron; (b) OR (Max) fuzzy neuron; (c) Product (x) fuzzy neuron

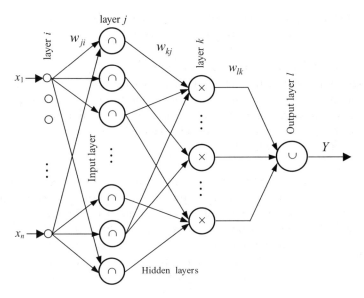

Figure 10.34 A simple fuzzy-neuron network

Figure 10.33(c) shows a fuzzy product neuron, which performs the product operation of all the weighted inputs defined by

$$O = f \left(\overset{n}{\underset{i=1}{\times}} x_i w_i + b \right) \tag{10.133}$$

$f(.)$ is the activation function. Different types of activation function are discussed in Chapter 3. Figure 10.34 shows a multilayer fuzzy neural network. The first layer is an input layer which transfers the inputs x_i weighted by w_{ji} to the second layer. The second layer consists of AND (Min) neurons and outputs the maximum of its input values. The third layer consists of product neurons and outputs the product of all its inputs. The final layer is the output layer, which aggregates all the inputs using the OR (Max) neuron. It is obvious that training algorithms like backpropagation can be used for training with a set of input/output data. There have been a number of applications of fuzzy neural networks reported in the literature (Pedrycz and Rocha, 1993; Tsoukalas and Uhrig, 1997; Zhang and Kandel, 1998).

10.8 MATLAB® Programs

MATLAB® provides ANFIS tools to use with the Fuzzy Logic Toolbox. ANFIS is a method of fuzzy modelling to learn the MF parameters, inference system and rules. The learning method works similarly to that of neural networks. Using a given input/output data set, the ANFIS constructs a fuzzy inference system whose membership function parameters are tuned (adjusted) using either a backpropagation algorithm alone, or in combination with a least-squares type method. This allows the fuzzy systems to learn from the data. ANFIS is much more complex than the fuzzy inference systems discussed so far, and is not available for all of the fuzzy inference system options. Specifically, ANFIS only supports Sugeno-type fuzzy systems, and these must have the following properties:

- First- or zeroth-order Sugeno-type fuzzy system.
- Have a single output, obtained using weighted average defuzzification. All output membership functions must be either linear or constant and must be of the same type.
- Different rules cannot share the same output membership function, namely the number of output membership functions must be equal to the number of rules.
- Have unity weight for each rule.

An error occurs if the ANFIS structure does not comply with these constraints. Moreover, ANFIS cannot accept all the customization options which the basic fuzzy inference allows. That is, ANFIS does not accept customized membership and defuzzification functions. ANFIS can be accessed either from the command line or through the ANFIS GUI editor. Since the functionality of the command-line function and GUI editor is similar, further description of the GUI editor will not be provided in this section. Appendix G presents a description of ANFIS tools and demonstrates examples of modelling and control using these tools.

References

Adeli, H. and Hung, S.-L. (1995) *Machine Learning – Neural Networks, Genetic Algorithms and Fuzzy Systems*, John Wiley & Sons, New York.

Aoyama, A. and Venkatasubramanian, V. (1995) Internal model control using neural networks for modelling and control of a bio-reactor, *Engineering Application and Artificial Intelligence*, 8, 689–701.

Aoyama, A., Doyle, F.J. and Venkatasubramanian, V. (1995) Fuzzy neural network approach for nonlinear process control, *Engineering Application and Artificial Intelligence*, 8, 483–493.

Aoyama, A., Doyle III, F.J. and Venkatasubramanian, V. (1999) Fuzzy neural network systems techniques and their applications to nonlinear chemical process control systems. In *Fuzzy Theory Systems: Techniques and Applications*, C. Leondes (ed.), Academic Press, New York, Vol. II, pp. 485–526.

Arnould, T. and Tano, S. (1994) Definition and formulation of backward-reasoning with fuzzy if... then... rules. Proceedings of the 3rd IEEE Conference on Fuzzy Systems (FUZZ-IEEE'94), IEEE World Congress on Computational Intelligence, Vol. 2, pp. 864–869.

Barto, A.G., Sutton, R.S. and Anderson, C.W. (1983) Neuronlike adaptive elements that can solve difficult learning control problems. *IEEE Transactions on Systems, Man, and Cybernetics*, 13, 834–846.

Berenji, H.R. (1992) A reinforcement learning-based architecture for fuzzy logic control, *International Journal of Approximate Reasoning*, 6, 267–292.

Berenji, H.R. and Khedkar, P. (1992) Learning and tuning fuzzy logic controllers through reinforcements, *IEEE Transactions on Neural Networks*, 3(5), 724–740.

Berenji, H.R. and Khedkar, P. (1993) Clustering for product space for fuzzy inference. IEEE International Conference on Neural Networks, San Francisco, pp. 1402–1407.

Chen, M. and Linkens, D.A. (1998) A hybrid neuro-fuzzy PID controller, *Fuzzy Sets and Systems*, 99, 27–36.

Chen, Y.-C. and Teng, C.-C. (1995) A model reference control structure using a fuzzy neural network, *Fuzzy Sets and Systems*, 73, 291–312.

Feng, J.C. and Teng, L.C. (1998) An online self constructing neural fuzzy inference network and its applications, *IEEE Transactions on Fuzzy Systems*, 6(1), 12–32.

Halgamuge, S.K., Mari, A. and Glesner, M. (1994) Fast perceptron learning by fuzzy controlled dynamic adaptation of network parameters. In *Fuzzy Systems in Computer Science*, R. Kruse, J. Gebhardt and R. Palm (eds), Vieweg, Braunschweig.

Haykin, S. (1999) *Neural Networks – A Comprehensive Foundation*, Prentice-Hall, Upper Saddle River, NJ.

Hu, Q. and Hertz, D.B. (1994) Fuzzy logic controlled neural network learning, *Information Sciences Applications*, 2(1), 15–33.

Ishibuchi, H., Fujioka, R. and Tanaka, H. (1993) Neural networks that learn from fuzzy if–then rules, *IEEE Transactions on Fuzzy Systems*, 1(2), 85–97.

Ishibuchi, H., Morioka, K. and Turksen, I. (1995) Learning by fuzzified neural networks, *International Journal of Approximate Reasoning*, 13(4), 327–358.

Jacobs, R.A. (1988) Increased rates of convergence through learning rate adaptation, *Neural Networks*, 1, 295–307.

Jang, J.-S.R. (1993) ANFIS: adaptive-network-based fuzzy inference system, *IEEE Transactions on Systems, Man and Cybernetics*, 23(3), 665–685.

Jang, J.-S.R. and Sun, C.-T. (1993) Functional equivalence between radial basis function networks and fuzzy inference systems, *IEEE Transactions on Neural Networks*, 4(1), 156–159.

Jang, J.-S.R., Sun, C.-T. and Mizutani, E. (1997) *Neuro-fuzzy and Soft Computing*, Prentice-Hall, Englewood Cliffs, NJ.

Jin, Y. and Jiang, J. (1999) Techniques in neural network based fuzzy systems identification and their application to complex systems. In *Fuzzy Theory Systems: Techniques and Applications*, C. Leondes (ed.), Academic Press, New York, Vol. I, pp. 111–128.

Jin, Y.C., Jian, J.P. and Zhu, J. (1995) Neural network-based fuzzy identification and its application to modelling and control of complex systems, *IEEE Transactions on Systems, Man and Cybernetics*, 25(6), 990–997.

Klassen, M.S. and Pao, Y.-H. (1988) Characteristics of the functional link net: a higher order delta rule net. IEEE Proceedings of the International Conference on Neural Networks, San Diego, pp. 507–513.

Kosko, B. (1992) *Neural Networks and Fuzzy Systems: A Dynamical Systems Approach to Machine Learning*, Prentice-Hall, Englewood Cliffs, NJ.

Lin, T.C. and Lee, C.S. (1991) Neural network based fuzzy logic control and decision system, *IEEE Transactions on Computers*, 40(12), 1320–1336.

Lin, T.C. and Lee, C.S. (1994) *Neural Fuzzy Control Systems with Structure and Parameter Learning*, World Scientific, Singapore.

Mizutani, E. and Jang, J.-S.R. (1995) Coactive neural fuzzy modelling. Proceedings of the International Conference on Neural Networks, pp. 760–765.

Mizutani, E., Jang, J.-S.R., Nishio, K., Takagi, H. and Auslander, D.M. (1994) Coactive neural networks with adjustable fuzzy membership functions and their applications. Proceedings of the International Conference on Fuzzy Logic and Neural Networks, Japan, pp. 581–582.

Nauck, D. and Kruse, R. (1995) NEFCLASS – a neuro-fuzzy approach for the classification of data. Proceedings of ACM Symposium on Applied Computing, K. George, J.H. Carrol, E. Deaton, D. Oppenheim and J. Hightower (eds), ACM Press, New York, pp. 461–465.

Nauck, D. and Kruse, R. (1997) Neuro-fuzzy systems for function approximation. 4th International Workshop on Fuzzy-Neuro Systems, A. Gruel, W. Becker and F. Belli (eds), pp. 316–323.

Nauck, D., Nauck, U. and Kruse, R. (1996) Neuro-fuzzy classification with NEFCLASS. Operation Research Proceedings 1995, P. Kleinschmidt, A. Bachem, U. Derigs, D. Fischer, U. Leopold-Wildburger and R. Moehring (eds), Springer-Verlag, Berlin, pp. 294–299.

Nauck, D., Klawonn, F. and Kruse, R. (1997) *Foundations of Neuro-Fuzzy Systems*, John Wiley & Sons, Chichester.

Oyama, T., Tano, S. and Arnould, T. (1994) A tuning method for fuzzy inferencing with fuzzy input and fuzzy output. Proceedings of the 3rd IEEE Conference on Fuzzy Systems (FUZZ-IEEE'94), IEEE World Congress on Computational Intelligence, Vol. 2, pp. 876–881.

Pedrycz, W. and Card, H.C. (1992) Linguistic interpretation of self-organising maps. Proceedings of the IEEE International Conference on Fuzzy Systems, San Diego, CA, pp. 371–378.

Pedrycz, W. and Rocha, R.A. (1993) Fuzzy set based models of neurons and knowledge-based networks, *IEEE Transactions on Fuzzy Systems*, 1(4), 254–266.

Saito, M., Naka, M., Yoshida, K. and Akamine, I. (1990) Estimation of thermal comfort by neural network. Japanese Association of Refrigeration Annual Conference, pp. 125–128.

Shin, Y.C. and Xu, C. (2009) *Intelligent Systems: Modelling, Optimisation and Control*, CRC Press, Boca Raton, FL.

Siddique, N.H. (2002) Intelligent control of flexible-link manipulator system, PhD Thesis, Department of Automatic Control and Systems Engineering, The University of Sheffield, UK.

Sulzberger, S.M., Tschicholg-Gurman, N.N. and Vestli, S.J. (1993) FUN: optimization of fuzzy rule based systems using neural networks. Proceedings of the IEEE Conference on Neural Networks, San Francisco, pp. 312–316.

Sutton, R.S. (1984) Temporal credit assignment in reinforcement learning. PhD Thesis, University of Massachusetts.

Sutton, R.S. (1988) Learning to predict by method of temporal differences, *Machine Learning*, 3, 9–44.

Takagi, H. (1992) Applications of neural networks and fuzzy logic to consumer products. The First International Workshop on Industrial Applications of Fuzzy Control and Intelligent Systems, Texas, pp. 1629–1633.

Takagi, H. (1995) Applications of neural networks and fuzzy logic to consumer products. *Industrial Applications of Fuzzy Control and Intelligent Systems*, J. Yen, R. Langari and L. Zadeh (eds), IEEE Press, Piscataway, NJ, pp. 93–106.

Takagi, H. (1997) Introduction to fuzzy systems, neural networks and genetic algorithms. In *Intelligent Hybrid Systems*, Da Ruan (ed.), Kluwer Academic, Dordrecht, pp. 3–33.

Takagi, H. and Hayashi, I. (1991b) NN-driven fuzzy reasoning, *International Journal of Approximate Reasoning*, 5(3), 191–212.

Takagi, T. and Sugeno, M. (1985) Fuzzy identification of systems and its applications to modeling and control, IEEE Transactions on System, Man and Cybernetics, 15, pp. 116–132.

Tano, S., Oyama, T., Arnould, T. and Bastian, A. (1994) Definition and tuning of unit-based fuzzy systems in FINEST. FUZZ-IEEE'94, pp. 436–441.

Tano, S., Oyama, T. and Arnould, T. (1996) Deep combination of fuzzy inference and neural network in fuzzy inference, *Fuzzy Sets and Systems*, 82(2), 151–160.

Tsoukalas, L.H. and Uhrig, R.E. (1997) *Fuzzy and Neural Approaches in Engineering*, John Wiley & Sons, New York.

Tsukamoto, Y. (1979) An approach to a fuzzy reasoning method. In *Advances in Fuzzy Set Theory*, M. Gupta, R. Ragade and R. Yager (eds), North-Holland, Amsterdam.

Wang, L.X. and Mendel, J.M. (1992) Fuzzy basis functions, universal approximation, and orthogonal least-squares learning, *IEEE Transactions on Neural Networks*, 3(5), 807–814.

Yager, R.R. (1994) Modelling and formulating fuzzy knowledge bases using neural networks, *Neural Networks*, 7(8), 1273–1283.

Yea, B., Konishi, R., Osaki, T. and Sugahara, K. (1994) Discrimination of many kinds of odor species using fuzzy reasoning and neural networks, Sensors and Actuators, *A: Physical*, 45(2), 159–165.

Zhang, Y.-Q. and Kandel, A. (1998) Compensatory neuro-fuzzy systems with fast learning algorithms, *IEEE Transactions on Neural Networks*, 9(1), 83–105.

Appendix A

MATLAB® Basics

A.1.1 Variables

The prompt (>>) in the command window indicates that MATLAB® is ready to accept input. A variable (x or a) is entered at the prompt and MATLAB® responds in the following way:

```
>> x=2.66
x =
    2.6600

>> a=x+0.0099
a =
    2.6699
```

Numerical output is suppressed by putting a semicolon (;) at the end of the line. For example:

```
>> x=2.66;
>> x=x+0.33
x =
    2.9900
```

A.1.2 Input and Output Commands

MATLAB® supports simple input/output commands. The command 'input' can be used for assigning values to variables. The general form of use is shown as follows:

```
variable = input('prompt')
```

The command displays the prompt as a message to the user on the screen, waits for input from the keyboard and returns the value entered in the variable. The response to the input

Computational Intelligence: Synergies of Fuzzy Logic, Neural Networks and Evolutionary Computing, First Edition.
Nazmul Siddique and Hojjat Adeli.
© 2013 John Wiley & Sons, Ltd. Published 2013 by John Wiley & Sons, Ltd.

prompt can be any MATLAB® expression, which is evaluated using the variables in the current workspace. For example:

```
>> x = input('Enter x: ')
Enter x: 5
x =
    5
>> y = input('Enter y: ')
Enter y: x+2
y =
    7
```

The input command can also be used to assign a string to a variable. The general form of use is shown as follows:

```
variable = input('prompt', 's')
```

The command returns the entered string as a text variable rather than as a variable name or numerical value. For example:

```
>> val = input('Enter y/n: ', 's')
Enter y/n: y
val =
    y
```

MATLAB® automatically generates a display when commands are executed. In addition to this automatic display, MATLAB® has several commands that can be used to generate displays or outputs. Two commands that are frequently used to generate output are 'disp' and 'fprintf'. The command 'disp' outputs (displays) text or array. The general form is as follows:

```
disp(x)
```

disp(x) displays an array. It does not print the array name. If x contains a text string, the string is displayed. For example:

```
>> x=5;
>> disp(x)
    5

>> x='yes';
>> disp(x)
Yes
```

The command 'fprintf' is slightly complicated. More detail about the command can be found in MATLAB® documentation.

A.1.3   Vectors

A vector in MATLAB® is represented by any variable, e.g., x, y. A row vector x can be initialized as

```
>> x = [6 5 4 8]
x =
    6     5     4     8
```

If elements are separated by a semicolon, a column vector y is initialized as

```
>> y = [1; 2; 3; 4]
y =
    1
    2
    3
    4
```

Individual elements of a vector can be accessed by the vector name followed by an index running from 1 to n (max number of elements) to point to the elements. For example:

```
>> x(4)
ans =
    8
```

returning the fourth element of vector x, which is 8.

The transpose of a column vector results in a row vector, and vice versa. For example:

```
>> z=y'
z =
    1    2    3    4
```

Vectors of the same size can be added or subtracted, where addition is performed component-wise. However, for multiplication, specific rules must be followed in order to obtain the correct resulting values. The operation of multiplying a vector x by a scalar k is performed component-wise. For example:

```
>> x=[1  3  -5];
>> k=5;
>> z=k*x
z =
    5    15    -25
```

The operator '.*' performs element-by-element operation. For example:

```
>> x.*z    % x and z are both row vectors
ans =
    5    45    125
```

The inner product or dot product of two vectors x and z (both row vectors) is a scalar quantity. The inner product is given by

```
>> s=x*z'
s =
 175
```

Various norms (measures of size) of a vector can be obtained. For example, the Euclidean norm is the square root of the inner product of the vector and itself. For example:

```
>> N=norm(z)
N =
 29.5804
```

The angle between two vectors x and y is defined by $\cos\theta = \frac{x^*y}{\|x\|^*\|y\|}$, where x^*y is the inner product and $\|x\|$, $\|y\|$ are the norms of the vectors. The angle between the vectors is calculated as follows:

```
>> x=[1 6 9];
>> y=[2.5 3.1 7.0];
>> theta=acos(x'*y/(norm(x)*norm(y)))
theta =
    1.5422    1.5354    1.4907
    1.3985    1.3566    1.0700
    1.3107    1.2462    0.7668
```

The zero vector is a vector with all components equal to zero. To generate a zero vector of size 4, use

```
>> Z=zeros(1,4)
Z =
    0    0    0    0
```

The sum vector is a vector with each component equal to one. To generate a sum vector of size 4, use

```
>> E=ones(1,4)
E =
    1    1    1    1
```

In MATLAB®, the colon (:) can be used to generate a row vector. For example, $x = 1:5$ generates a row vector of integers from 1 to 5:

```
>> x=1:5
x =
    1    2    3    4    5
```

For increments other than unity, any value can be used as follows:

```
>> z=0:pi/3:pi
z =
        0    1.0472    2.0944    3.1416
```

For negative increments:

```
>> x=5.5:-1.5:0
x =
    5.5000    4.0000    2.5000    1.0000
```

A.1.4   Matrices

A matrix is represented by a variable name (e.g., w), elements in each row are separated by blanks or commas. A semicolon must be used to indicate the end of a row. If a semicolon is

not used, each row must be entered in a separate line. Matrix elements can be any MATLAB®
expression. For example, a 3×3 matrix can be initialized as

```
>> w = [4 5 6; 0 4 7; 3 5 1]
w =
     4      5      6
     0      4      7
     3      5      1
>> w=[2*2 5 3*2; 0 2*2 7; 3 5 1]
w =
     4      5      6
     0      4      7
     3      5      1
```

Individual elements of a matrix can be accessed by the matrix name followed by two indices
running from 1 to n (max row) and 1 to m (max column) to point to the elements. For example,
$w(2, 3)$ returns the element on row 2 column 3, which is 7:

```
>> w(2,3)
ans =
     7
```

The entire row of a matrix can be addressed by means of the symbol (,:). For example, $w(2, :)$
returns the second row of the matrix w:

```
>> r2w=w(2,:)
r2w =
     0      4      7
```

Similarly, the entire column of the matrix can be accessed by means of (:,). For example,
$w(:, 2)$ returns the second column of the matrix w:

```
>> c2w=w(:,2)
c2w =
     5
     4
     5
```

MATLAB® has dozens of functions that create different kinds of matrices. Pascal(n) creates
an $n \times n$ symmetric matrix. For example:

```
>> A=pascal(3)
A =
     1      1      1
     1      2      3
     1      3      6
```

Magic(n) creates an $n \times n$ non-symmetric matrix. For example:

```
>> B=magic(3)
B =
     8      1      6
     3      5      7
     4      9      2
```

Addition and subtraction of matrices is defined, just as it is for vectors, element by element. Addition and subtraction require both matrices to have the same dimension or one of them to be a scalar. For example:

```
>> X=A+B
X =
     9     2     7
     4     7    10
     5    12     8

>> Y=B-A
Y =
     7     0     5
     2     3     4
     3     6    -4
```

Matrix B can be transposed (rows will convert to columns) by using the transpose operator. For example:

```
>> B'
ans =
     8     3     4
     1     5     9
     6     7     2
```

The matrix product $C = A \times B$ is defined when the column dimension of A is equal to the row dimension of B. If A is $m \times p$ and B is $p \times n$, their product C is $m \times n$. MATLAB® uses a single asterisk to denote matrix multiplication. For example:

```
>> C=A*B
C =
    15    15    15
    26    38    26
    41    70    39

>> D=B*A
D =
    15    28    47
    15    34    60
    15    28    43
```

It can be seen from the above two products that $A^*B \neq B^*A$.

An identity matrix has ones in the diagonal and zeros elsewhere. Generally, I is used to denote the identity matrix. These matrices have the property that $AI = IA = A$. The function eye(n) returns an $n \times n$ identity matrix. For example:

```
>> I=eye(3)
I =
     1     0     0
     0     1     0
     0     0     1
```

The determinant of a square matrix is computed by the function det(). For example, the determinant of $A = \begin{bmatrix} 1 & 1 & 1 \\ 1 & 2 & 3 \\ 1 & 3 & 6 \end{bmatrix}$ is computed by

```
>> d=det(A)
d =
    1
```

If the determinant of a matrix is not equal to zero, the matrix is called non-singular. If A is a square and non-singular matrix, then the equations $Ax = I$ and $xA = I$ have the same solution x. This solution is called the inverse of A and denoted A^{-1}. The inverse of a matrix is computed by the function inv(). For example:

```
>> X=inv(A)
X =
     3    -3     1
    -3     5    -2
     1    -2     1
```

Rectangular matrices do not have determinants or inverses. That means at least one of the equations $Ax = I$ and $xA = I$ does not have a solution. In that case, the Moore–Penrose pseudo-inverse is to be computed for rectangular matrices. MATLAB® provides the function pinv() for computing pseudo-inverses. For example, the pseudo-inverse of $M = \begin{bmatrix} 1 & 2 & 3 \\ 2 & 3 & 4 \\ 5 & 1 & 3 \\ 1 & 2 & 6 \end{bmatrix}$

is computed by

```
>> Z=pinv(M)
Z =
   -0.0214    0.0129    0.2208   -0.1083
    0.2390    0.4566   -0.1618   -0.3430
   -0.0600   -0.1640    0.0182    0.2969
```

A.1.5 Polynomials

MATLAB® provides functions for standard polynomial operations such as polynomial roots, evaluation and differentiation. Advanced operations such as curve fitting and partial fraction expansion are also supported by MATLAB®. Polynomials are represented by row vectors containing coefficients ordered by descending powers. For example, the polynomial $p(x) = x^3 - 2x - 5$ is represented by the row vector $p = [1 \quad 0 \quad -2 \quad -5]$.

The solutions to the polynomial (roots) can be found using the function roots(). There is one real solution and two imaginary solutions to the polynomial. For example:

```
>> r=roots(p)
r =
    2.0946
   -1.0473 + 1.1359i
   -1.0473 - 1.1359i
```

The polynomial can be evaluated at a specific value. The function polyval() evaluates the polynomial at $x = 5.9$. For example:

```
>> p1=polyval(p,5.9)
p1 =
   188.5790
```

The derivative of the polynomial can also be found using the function polyder(). For example:

```
>> p_dash=polyder(p)
p_dash =
     3      0     -2
```

The derivative of the polynomial function can be reconstructed from the vector as $\dot{p}(x) = 3x^2 - 2$.

A.1.6   Control Structures

MATLAB$^{®}$ has four control structures:

- if statement,
- for loop and
- while loop.

A.1.6.1   The if ... end structure

MATLAB$^{®}$ supports the following variants of the 'if' construct:

```
if ... end
if ... else ... end
if ... elseif ... else ... end
```

The general form of the 'if' statement is as follows:

```
if expression
   statements
elseif
   statements
else
   statements
end
```

The following examples of solutions to the well-known quadratic equation will make the usage of the three constructs clear.

Example A.1.1

```
%Example A.1.1
%Control construct if ... end
%Quadratic equation - discr<0
```

```
a=1;b=2;c=3;
discr = b*b - 4*a*c;
if discr < 0
  disp('Warning: Discriminant is negative.')
  disp('Solutions are imaginary');
end
```

Example A.1.2

```
%Example A.1.2
%Control construct if ... else ... end
%Quadratic equation - discr<0 or discr>0
discr = b*b - 4*a*c;
if discr < 0
  disp('Warning: Discriminant is negative.')
  disp('Solutions are imaginary');
else
  disp('Discriminant is positive.')
  disp('There are two real solutions')
end
```

Example A.1.3

```
%Example A.1.3
%Control construct if ... elseif ... else ... end
%Quadratic equation - discr<0, discr==0 or discr>0
discr = b*b - 4*a*c;
if discr < 0
  disp('Warning: Discriminant is negative.')
  disp('Solutions are imaginary');
elseif discr == 0
  disp('Discriminant is zero.');
  disp('There are two identical solutions')
else
  disp('Discriminant is positive.')
  disp('There are two real solutions')
end
```

A.1.6.2   The for ... end structure

In the 'for ... end' structure a sequence of commands (or statements) is executed repeatedly, a fixed and predetermined number of times. The general form of the 'for ... end' structure is as follows:

```
for variable = expression
    statements
end
```

Usually, the expression is a vector of the form

```
initial_value : step : end_value.
```

A simple example of the 'for' loop is

```
for i = 1: 0.5: 5
  x=i*i
end
```

It is a good idea to indent the loops for readability, especially when they are nested. The following example demonstrates a nested loop that creates a 5×5 symmetric matrix A with (i, j) element i/j for $j \geq i$.

Example A.1.4

```
%Example A.1.4
n = 5;
A = eye(n);
for j=2:n
  for i=1:j-1
    A(i,j)=i/j
    A(j,i)=i/j
  end
end
```

A.1.6.3   The while ... end structure

The 'while ... end' structure is used when the number of iterations is not specified. The looping continues until the expression is satisfied. The general form of the 'while' loop is as follows:

```
while expression
  statements
end
```

The statements are executed as long as the expression is true. For example:

```
x = 1
while x <= 10
 x = 3*x
end
```

Care must be taken when defining the condition for the loop. If it is not well-defined, it will enter into an infinite loop, i.e., the looping will continue indefinitely. Other control statements include 'return', 'continue', 'switch', etc. More detailed information about these commands can be found in MATLAB® documentation.

A.1.7   Reading Data Files

MATLAB® provides many ways to load data from disk files into workspace (the process of importing data) and to save workspace variables to disk files (the process of exporting data).

The command 'load' reads workspace variables from disk. For example, loading variables from the disk file (e.g., 'signal.dat') can be done in the following way:

```
>> load d:\signal.dat;
```

$d : \backslash$specifies the path and the semicolon stops displaying values on the screen. The loaded data can be assigned to any other variable. For example:

```
>>s = signal;
```

Only specified variables from the data file can be read. For example:

```
>> load signal x y z
```

The command above just loads the specified variables x, y and z.

The load command can be used in functional form, such as load('file name'). For example:

```
>> s=load(' d:\signal.dat');
```

$s = $ load ('...') returns the contents of the data file in variable s. If the data file is a MAT-file, s is a structure containing fields matching the variables. If the data file is an ASCII file, s is an array.

A.1.8 Plotting Functions

MATLAB® has an excellent set of graphic tools for representing experimental results in 2-D and 3-D plots. Plotting a given data set or the results of a computation is possible with very few commands. The function plot() is used to plot 2-D figures; it has different forms depending on the input arguments. The general form is as follows:

```
plot(y)
plot(x,y)
plot(x,y,LineSpec, ... )
```

plot(y) plots the columns of y versus their index, plot (x, y) plots vector x versus vector y, plot $(x, y, $ LineSpec, ...) plots all lines defined by $x, y,$ LineSpec. If x or y is a matrix, then the vector is plotted versus the rows or columns of the matrix, whichever line up. If x is a scalar and y is a vector, length(y) disconnected points are plotted.

Various line types, plot symbols and colours may be used with plot $(x, y, 's')$, where s is a character string made up of one element from any or all the columns in Table A.1.1. For example:

```
>>plot(x, y, 'c+:') %plots a cyan dotted line with a plus at each
                    data point;
>>plot(x, y, 'bd') %plots a blue diamond at each data point
                    but does not draw any line.
```

The title, labels for x-axis and y-axis and legend of a figure can also be created by

```
xlabel('x-axis')
ylabel('y-axis')
title('Title of plot')
legend('Fist plot', 'Second plot')
```

Table A.1.1 Plot symbols for colour, marker and line types

Symbol	Colour	Symbol	Marker	Symbol	Marker	Symbol	Line type
b	blue	.	point	v	triangle(down)	–	solid
g	green	o	circle	^	triangle(up)	:	dotted
r	red	x	x-mark	<	triangle(left)	-.	dashdot
c	cyan	+	plus	>	triangle(right)	–	dashed
m	magenta	*	star	p	pentagon		
y	yellow	s	square	h	hexagon		
k	black	d	diamond				

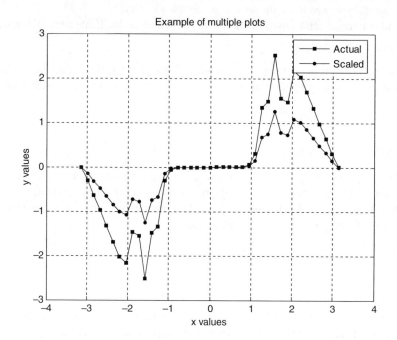

Figure A.1.1   2-D multiple plots of trigonometric function

A sample code of a 2-D plot is given in Example A.1.5, which creates a 2-D plot shown in Figure A.1.1.

Example A.1.5

```
%Example A.1.5
%Example of plot functions
x = -pi:pi/20:pi;
y1 = tan(sin(x)) - sin(tan(x));
y2 = 0.5*(tan(sin(x)) - sin(tan(x)));
plot(x,y1,'-ks',x,y2,'-ko','LineWidth',1, ...
            'MarkerEdgeColor','k', ...
            'MarkerFaceColor','k', ...
            'MarkerSize',3)
```

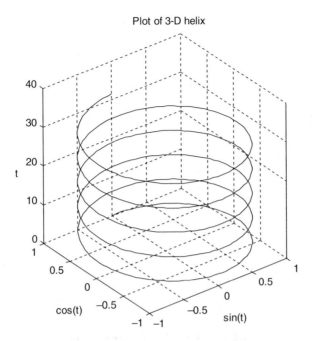

Figure A.1.2   3-D plot in MATLAB®

```
xlabel('x values')
ylabel('y values')
legend('Actual','Scaled')
grid
title('Example of multiple plots')
```

MATLAB® also provides functions for displaying a 3-D plot of a set of data points. The function plot3() is used to plot 3-D figures. The general form is as follows:

```
plot3(X, Y, Z, LineSpec, ...)
```

where *X, Y, Z* are vectors or matrices. It plots one or more lines in 3-D space through the points whose coordinates are the elements of *X, Y, Z*. A sample code of a 3-D plot is given in Example A.1.6, which creates a 3-D plot shown in Figure A.1.2.

Example A.1.6

```
%Example A.1.6
%Example of 3-D plot
t = 0:pi/50:10*pi;
plot3(sin(t),cos(t),t,'-k')
grid on
axis square
xlabel('sin(t)')
```

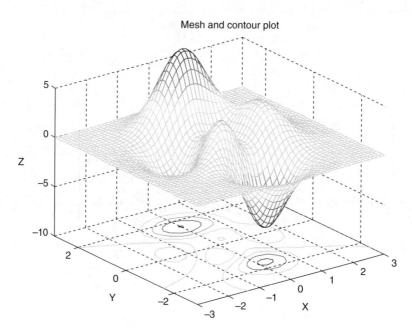

Figure A.1.3   Mesh and contour plot

```
ylabel('cos(t)')
zlabel('t')
title('Plot of 3-D helix')
```

MATLAB® also provides other powerful 3-D plot functions such as mesh, surface and contour plots. The functions mesh(.), meshc(.) and meshz(.) create parametric surfaces specified by X, Y and Z, with colour specified by C. mesh (X, Y, Z) draws a mesh with colour, which is proportional to surface height and determined by Z. If X and Y are vectors of dimension n and m, respectively, the dimension of Z is $[m, n]$. If $X(j)$ and $Y(i)$ are matrices, then $X(j)$, $Y(i)$ and $Z(i, j)$ are the intersections of the grid lines. meshc(.) draws a contour plot beneath the mesh. meshz(.) draws a curtain plot (i.e., a reference plane) around the mesh. A sample code for a mesh and contour plot is given in Example A.1.7, which creates a 3-D mesh and contour plot shown in Figure A.1.3.

Example A.1.7

```
%Example A.1.7
%Example of mesh and contour plot
[X,Y] = meshgrid(-3:.125:3);
Z = peaks(X,Y);
meshc(X,Y,Z);
axis([-3 3 -3 3 -10 5])
grid on
xlabel('X')
ylabel('Y')
zlabel('Z')
title('Mesh and contour plot')
```

A.1.9   Mathematical Functions

MATLAB® offers a wide range of built-in mathematical functions supporting advanced technical computing. Trigonometric and exponential functions such as sin(), cos(), tan(), log() and exp() can be used in MATLAB®. For example:

```
sin(x) is the sine of the elements of x,
cos(x) is the cosine of the elements of x,
tan(x) is the tangent of the elements of x,
log(x) is the natural logarithm, i.e., log base e.
```

To find out how many bits are required to represent the decimal number 128 as a binary number, use the log base 2 function as follows:

```
>> log2(128)
ans =
    7
```

To evaluate $\sin(x^2 + 5)$ where $x = [1 \quad 2 \quad 3]$, use the sin() function in the following way:

```
>> x=[1 2 3];
>> sin(x.^2+5)
ans =
 -0.2794     0.4121     0.9906
```

Here, (.^) is the array exponential operator. This enables the function to accept x as an array.

Table A.1.2 lists some commonly used mathematical functions, where variables x and y can be numbers, vectors or matrices.

A.1.10   User-Defined Functions

The first line in a function should begin with the function definition, with a list of inputs and outputs. This line distinguishes a function M-file from a script M-file. The general format is as follows:

```
function [output variables] = name (input variables)
            body of function
```

Table A.1.2   MATLAB® mathematical functions

Function	Description	Function	Description
$\cos(x)$	cosine	$abs(x)$	absolute value
$\sin(x)$	sine	$sign(x)$	signum function
$\tan(x)$	tangent	$max(x)$	maximum value
$acos(x)$	arc cosine	$min(x)$	minimum value
$asin(x)$	arc sine	$ceil(x)$	round towards $+\infty$
$atan(x)$	arc tangent	$floor(x)$	round towards $-\infty$
$exp(x)$	exponential	$round(x)$	round to nearest integer
$sqrt(x)$	square root	$rem(x)$	remainder after division
$\log(x)$	natural logarithm	$angle(x)$	phase angle
$\log 10(x)$	common logarithm	$conj(x)$	complex conjugate

'name' should be a valid function name, input variables are the function arguments and output variables are the values to be returned by the function. For example:

Example A.1.8

```
%Example A.1.8
%Definition of a function
function z = fun (x,y)
u = 5*x;
z = u+6*y.^2;
```

The function can be called from the command prompt with actual arguments. For example:

```
>> z=fun(3.33,7.5)
z =
354.1500
```

Mathematical functions can be represented by expressing them as MATLAB® functions in M-files or as inline objects. For example, consider the function described by

$$f(x) = \frac{1}{(x-0.3)^2 + 0.01} + \frac{1}{(x-0.9)^2 + 0.04} - 6$$

The inline object can be created and evaluated as follows:

```
>> f=1./((x-0.3).^2+0.01)+1./((x-0.9).^2+0.04)-6;
>> f(2.0)
ans =
 -5.8261
```

This function can be used as input to any function as well as being defined in an M-file named humps.m. For example:

Example A.1.9

```
%Example A.1.9
%Example of humps function
function y=humps(x)
  y=1./((x-0.3).^2+0.01)+1./((x-0.9).^2+0.04)-6;
```

It is also common to use a handle for functions defined using @ as follows:

```
fh=@humps;
```

The function can be evaluated at any point. For example:

```
>> feval(fh,2.0)
ans =
 -4.8552
```

MATLAB® also supports plotting mathematical functions between a given set of values using the function fplot(). In Example A.1.10, the 'humps' function is plotted within the interval of [–5 5]. The plot is shown in Figure A.1.4.

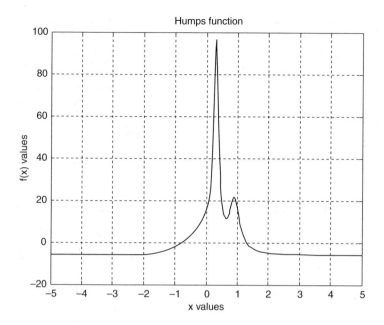

Figure A.1.4   Plot of humps function

Example A.1.10

```
%Example A.1.10
%Plot of Humps function
fplot(fh,[-5 5],'k')
grid
xlabel('x values')
ylabel('f(x) values')
title('Humps function')
```

A.1.11   M-file Scripting

MATLAB® allows writing lines of code using MATLAB® functions and statements in a file and saving it as an M-file. The M-file can be run from the command prompt. M-file scripts operate on existing data in the workspace, or they can create new data on which to operate. Any variables that the script creates remain in the workspace. For example, the following code calculates the radius of several trigonometric functions for angles, then creates a series of polar plots. The code is saved in an M-file named flower.m to be run from the command prompt.

Example A.1.11

```
%Example A.1.11: flower.m
%Example of M-file script for flower petal
theta=-pi:0.01:pi
r(1,:)=2*sin(5*theta).^2;
r(2,:)=cos(10*theta).^3;
```

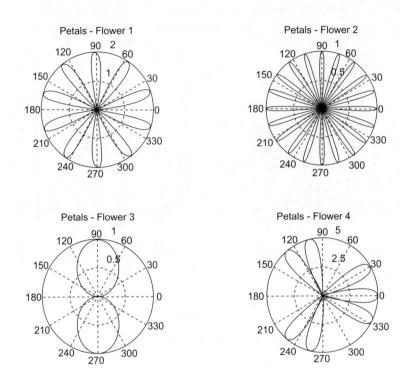

Figure A.1.5   Output of the M-file

```
r(3,:)=sin(theta).^2;
r(4,:)=5*cos(3.5*theta).^3;
%Polar makes a polar plot using angle and radius
subplot(221)
    polar(theta, r(1,:),'k')
    title('Petals - Flower 1')
subplot(222)
    polar(theta, r(2,:),'k')
    title('Petals - Flower 2')
subplot(223)
    polar(theta, r(3,:),'k')
    title('Petals - Flower 3')
subplot(224)
    polar(theta, r(4,:),'k')
    title('Petals - Flower 4')
```

The M-file flower.m can now be run from the command prompt, which will generate the petals of flowers as shown in Figure A.1.5.

```
>> flower
```

The remaining appendices will illustrate and demonstrate different ways of writing M-files for a wide range of applications.

Appendix B

MATLAB® Programs for Fuzzy Logic

B.1.1   Membership Functions

The Fuzzy Logic Toolbox provides a number of membership functions. The most widely used MFs are triangular, Gaussian, bell-shaped and trapezoidal. trimf(), trapmf(), gaussmf() and gbellmf() are built-in functions for triangular, trapezoidal, Gaussian and bell-shaped MFs. The general forms of use with parameters are described below:

```
y = trimf(x,[a b c])
y = trapmf(x,[a b c d ])
y = gaussmf(x,[a c]),
y = gbellmf(x,[a b c])
```

Different parametric membership functions are discussed in Section 2.4 of Chapter 2. The triangular membership function trimf() depends on three parameters a, b and c. The parameters a, b and c locate the 'feet' of the triangle and the parameter c locates the peak. The trapezoidal function trapmf() depends on four scalar parameters a, b, c and d. The parameters a and d locate the 'feet' of the trapezoid and the parameters b and c locate the 'shoulders'. The symmetric Gaussian membership function gaussmf() depends on two parameters a and c. The parameter c is the centre of the function and the width is determined by the parameter a of the Gaussian function. The generalized bell function gbellmf() depends on three parameters a, b and c. The parameter b is usually positive. The parameter c locates the centre and the parameter a determines the width of the membership function. Sample plots of these MFs are shown in Figure B.1.1.

A second set of built-in MFs sigmf(), dsigmf(), psigmf(), pimf(), zmf() and smf() are sigmoidal, difference sigmoidal, product sigmoidal, Π-shaped, Z-shaped and S-shaped functions,

Computational Intelligence: Synergies of Fuzzy Logic, Neural Networks and Evolutionary Computing, First Edition.
Nazmul Siddique and Hojjat Adeli.
© 2013 John Wiley & Sons, Ltd. Published 2013 by John Wiley & Sons, Ltd.

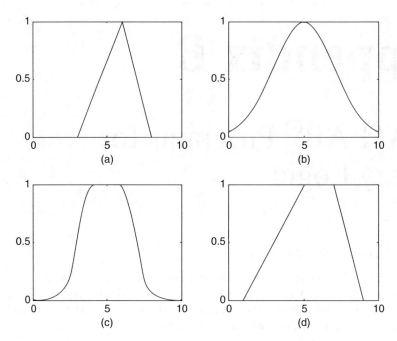

Figure B.1.1 (a) Triangular MF = [3 6 8]; (b) Gaussian MF = [2 5]; (c) Bell-shaped MF = [2 3 5] and (d) Trapezoidal MF = [1 5 7 9]

respectively. The parameterized general forms are described below:

```
y = sigmf(x,[a c])
y = dsigmf(x,[a1 c1 a2 c2])
y = psigmf(x,[a1 c1 a2 c2]),
y = pimf(x,[a b c d])
y = zmf(x,[a b])
y =smf(x,[a b])
```

The sigmoidal membership function sigmf() depends on the two parameters a and c and is given by $f(x, a, c) = \dfrac{1}{1 + e^{-a(x-c)}}$. The parameter c is the centre of the sigmoidal function. The sign of the parameter a determines the spread of the sigmoidal membership function, i.e., whether it inherently open to the right or to the left. Thus, the parameter a is appropriate for representing concepts of linguistic hedges such as 'very large' or 'more or less small'.

The difference sigmoidal function dsigmf() depends on four parameters a_1, c_1, a_2 and c_2 and is the difference between two sigmoidal functions defined by $f_1(x, a_1, c_1) - f_2(x, a_2, c_2) = \dfrac{1}{1 + e^{-a_1(x-c_1)}} - \dfrac{1}{1 + e^{-a_2(x-c_2)}}$. The product sigmoidal function psigmf() is simply the product of two sigmoidal curves defined by

$$f_1(x, a_1, c_1) * f_2(x, a_2, c_2) = \frac{1}{1 + e^{-a_1(x-c_1)}} * \frac{1}{1 + e^{-a_2(x-c_2)}}.$$

The parameters of dsigmf() and psigmf() are listed in the order $[a_1\ c_1\ a_2\ c_2]$. The Π-shaped membership function pimf() is a spline-based curve. It is named Π-shape because of its shape.

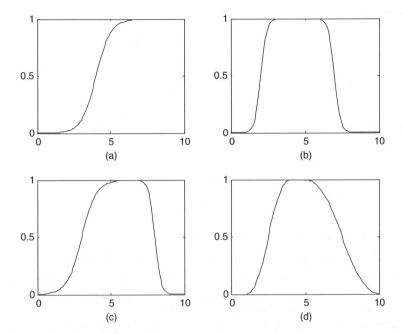

Figure B.1.2 (a) Sigmoidal MF = [2 4]; (b) Difference sigmoidal MF = [5 2 5 7]; (c) Product sigmoidal MF = [2 3 −5 8] and (d) Π-shaped MF = [1 4 5 10]

The parameters a and d locate the 'feet' of the curve, while b and c locate its 'shoulders'. Plots of sigmf(), dsigmf(), psigmf() and pimf() MFs on a vector x are shown in Figure B.1.2.

There are also two spline-based functions zmf() and smf(). They are so named because of their Z-shape and S-shape (also called Zadeh's S-function) (Driankov *et al.*, 1993). The parameters a and b locate the extremes of the sloped portion of the curve. Plots of S-shaped and Z-shaped functions are shown in Figure B.1.3. A sample of MATLAB® code is given in Example B.1.1 and Example B.1.2 for difference sigmoidal and Π-shaped functions below.

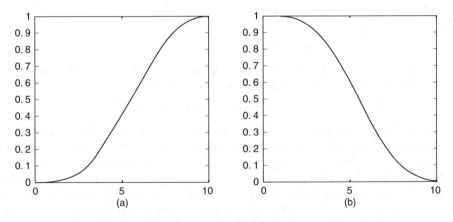

Figure B.1.3 (a) S-shaped MF = [1 10]; (b) Z-shaped MF = [1 10]

Example B.1.1

```
x=0:0.1:10;
y=dsigmf(x,[5 2 5 7]);
plot(x,y)
xlabel('dsigmf, P=[5 2 5 7]')
```

Example B.1.2

```
x=0:0.1:10;
y=pimf(x,[1 4 5 10]);
plot(x,y)
xlabel('pimf, P=[1 4 5 10]')
```

B.1.2   Fuzzy Inference

To build a fuzzy system entirely from the command line, the following commands are used: newfis(), addvar(), addmf(), addrule(). MATLAB® supports only Mamdani- and Sugeno-type inference systems. To perform a fuzzy inference calculation, the evalfis() command is used. To view the control surface, the gensurf() command is used.

There are two types of fuzzy inference system that can be implemented in the Fuzzy Logic Toolbox: Mamdani-type and Sugeno-type. newfis() creates a new fuzzy inference system. newfis() can be used to create a Mamdani-type or Sugeno-type FIS structure and can be used in the following general format:

```
A = newfis(FIS_Name, [FIS_Type], [AND_Method], [OR_Method],
[Imp_Method], ...
[Agg_Method], [Defuzz_Method]) creates a FIS structure for a
Mamdani or Sugeno-style system with the name FIS_Name
```

The other six optional arguments are as follows: FIS_Type, which specifies the FIS structure of 'Mamdani' or 'Sugeno' type and by default is Mamdani-type; AND_Method, OR_Method, Imp_Method, Agg_Method and Defuzz_Method, which specify the methods for AND, OR, implication, aggregation and defuzzification, respectively.

For example:

```
A=newfis ('NewSys', 'mamdani');
A=newfis ('NewSys', 'sugeno');
```

Once an FIS is created, addvar() defines input or output variables for an FIS structure with respective input and output ranges. The general form of the command is

```
A = addvar (A, 'Var_Type', 'Var_Name', Var_Bounds)
```

addvar() has four arguments in this order. A is the name of a FIS structure created by newfis() in the MATLAB® workspace. A string representing the type of the variable is specified by Var_Type and can be 'input' or 'output'. Var_Name is a string representing the name of the variable, e.g., 'X1', 'Y2'. Var_Bounds is a vector describing the range values for the variables, e.g., [−5 11]. Indices are applied to variables in the order in which they are added, so the first input variable added to a system will always be known as input variable number one for that system. Input and output variables are numbered independently.

For example:

```
A=newfis ('NewSys');
A=addvar (A,'input','X1',[10 20]);
A=addvar (A,'input','X2',[-7 9]);
A=addvar(A,'output','Y1',[-5 5]);
A=addvar(A,'output','Y2',[-15 10]);
```

Once the input/output variables are defined, addmf() adds membership functions to the FIS structure. The general format of the command is as follows:

```
A = addmf (A, 'Var_Type', Var_Index, 'mfName', 'mfType', mfParams)
```

A membership function can only be added to a variable in an existing FIS structure in the MATLAB® workspace. Indices are assigned to membership functions in the order in which they are added, so the first membership function added to a variable will always be known as membership function number one for that variable. A membership function cannot be added to input variable number two of a system if only one input has been defined. The function requires six input arguments in this order. A is the FIS structure name in the workspace. Var_Type is a string 'input' or 'output' representing the type of variable to be added to the membership function. Var_Index is the index of the variable. String 'mfName' represents the name of the membership function described in Section 2.14.1 of Chapter 2. String 'mfType' is the type of the new membership function, such as 'gaussmf', 'trimf', etc. mfParam is the vector of parameters that specify the membership function, e.g., a triangular membership function has three parameters [–3, 1, 5], a Gaussian membership function has two parameters [0.5, 5]. For details of different parameters of membership functions, see Sections 2.4 and 2.15.1 in Chapter 2.

For example:

```
A = newfis ('NewSys');
A = addvar (A,'input','X1',[-10 10]);
A = addvar (A,'input','X2',[-5 15]);
A = addvar(A,'output','Y',[-5 5]);
A = addmf(A,'input',1,'nagative','gaussmf',[1.5 -5]);
A = addmf(A,'input',1,'zero','gaussmf',[1.5 5]);
A = addmf(A,'input',2,'small','gaussmf',[1.5 0]);
A = addmf(A,'input',2,'medium','gaussmf',[1.5 10]);
A = addmf(A,'output',1,'small','gaussmf',[1.5 0]);
A = addmf(A,'output',1,'medium','gaussmf',[1.5 2.5]);
```

B.1.3   Fuzzy Rule Base

Once the FIS structure is defined with appropriate variables and membership functions, a rule base has to be defined. addrule() adds a list of rules to an FIS structure. The following general form is used:

```
A = addrule(A, ruleList)
```

addrule() has two arguments. The first argument is the name of the FIS structure. The second argument is a matrix of one or more rows, each of which represents a given rule. The format

Table B.1.1   Rule base of a simple fuzzy system

x_1	x_2	
	mf_1	mf_2
mf_1	mf_1	mf_2
mf_2	mf_2	mf_1

that the rule list matrix must take is very specific. If there are m inputs to a system and n outputs, there must be exactly $m + n + 2$ columns to the rule list. The entries in the first m columns refer to the inputs of the system. Each column contains a number that refers to the index of the membership function for that variable. The entries in the next n columns refer to the outputs of the system. Each column contains a number that refers to the index of the membership function for that variable. The $m + n + 1$ column contains the weight that is to be applied to the rule. The weight must be a number between 0 and 1, and is generally left as 1. The $m + n + 2$ column contains a 1 if the fuzzy operator for the rule's antecedent is AND. It contains a 2 if the fuzzy operator is OR. For example, if the FIS structure A has two inputs x_1, x_2 and one output y with each input/output having two membership functions mf_1 and mf_2, the rule base will look like Table B.1.1.

The first two rules (first row of rule table) can be described as:

```
If x1 is mf1 and x2 is mf1 Then y is mf1 (rule weight 1)
If x1 is mf1 and x2 is mf2 Then y is mf2 (rule weight 1)
```

The above rule list can be written as a rule matrix below:

```
ruleList = [1 1 1 1 1
            1 2 2 1 1];
```

The rule matrix is then added to the FIS system using the addrule() function:

```
A = addrule(A, ruleList)
```

The fuzzy inference diagram for an FIS is stored in a file; 'A.FIS' can be viewed by invoking the rule viewer function ruleview(). This is used to view the entire implication process from beginning to end. You can move around the line indices that correspond to the inputs and then watch the system readjust and compute the new output.

For example:

```
ruleview('A')
```

B.1.4   Defuzzification

The Fuzzy Logic Toolbox provides a number of defuzzification methods. The most widely used methods are centroid of area, bisector of area, mean value of maximum, smallest (absolute) value of maximum and largest (absolute) value of maximum. The following general form is used for the defuzz() function in MATLAB®:

```
df = defuzz (x, mf, type)
```

defuzz $(x$, mf, type) returns a defuzzified value df of a membership function mf positioned at the associated variable value x, using one of several defuzzification strategies, according to the specified type. The variable type can be one of the following:

- 'centroid' for centroid of area
- 'bisector' for bisector of area
- 'mom' for mean value of maximum
- 'som' for smallest (absolute) value of maximum
- 'lom' for largest (absolute) value of maximum.

If type is not specified, the Fuzzy Logic Toolbox assumes it to be a user-defined function. x and mf are passed to this function to generate the defuzzified output value.

Example B.1.3

```
x = -10:0.1:10;
%Trapezoidal MF is defined for defuzzification
mf = trapmf(x,[-10 -8 -4 7]);
%Defuzzification of trapezoidal MF is carried out using different
methods supported by
%MATLAB toolbox
dfc = defuzz(x,mf,'centroid');
dfb = defuzz(x,mf,'bisector ');
dfm = defuzz(x,mf,'mom');
dfs = defuzz(x,mf,'som');
dfl = defuzz(x,mf,'lom');
```

B.1.5   Simulation of FIS

evalfis() simulates the FIS for the input data and returns the output data. The general form is

```
Y = evalfis (X, A)
```

X is the input data matrix of $M \times N$ dimension. Each row of the matrix is a particular input vector. Y is the return output data, an $M \times L$ matrix, each row being a particular output vector. For example, if an FIS structure A has two inputs x_1 and x_2 with $x_1 = 4$ and $x_2 = 9$, then the FIS A can be simulated using the function evalfis().
   For example:

```
Y = evalfis ([4 9], A)
```

Example B.1.4   The membership functions for x_1, x_2 and y of a Mamdani-type fuzzy system are defined within the universes of discourse [2 11], [4 14] and [1 9], respectively and are shown in Figure B.1.4. For each of the variables, the MFs are taken to be {A1, A2}, {B1, B2} and {C1, C2}.

The rule base of the Mamdani-type fuzzy inference system is shown in Table B.1.2.

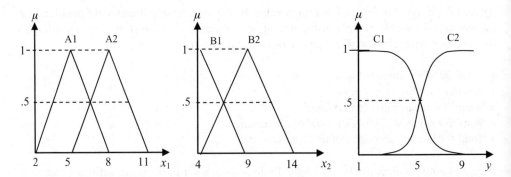

Figure B.1.4   MFs for x_1, x_2 and y

Table B.1.2   Rule base for Mamdani-type FLC

	X_2	
X_1	B1	B2
A1	C1	C2
A2	C2	C1

Simulate the Mamdani-type fuzzy system for initial conditions $x_1(0) = 4$ and $x_2(0) = 8$:

```
%Mamdani-type Fuzzy Inferencing
clear all;
close all;
sys=newfis('ExampleB.1.4');
%Inputs -------------------------------------------------
%Define input variable x1 to FIS within interval [2 11]
sys=addvar(sys,'input', 'x1', [2 11]);
%Define input variable x2 to FIS within interval [4 14]
sys=addvar(sys,'input', 'x2', [4 14]);
%Define Triangular MFs A1 and A2 for input x1
sys=addmf(sys,'input',1,'A1','trimf',[2 5 8]);
sys=addmf(sys,'input',1,'A2','trimf',[5 8 11]);
%Define Trinagular MFs B1 and B2 for input x2
sys=addmf(sys,'input',2,'B1','trimf',[4 4 9]);
sys=addmf(sys,'input',2,'B2','trimf',[4 9 14]);
%outputs------------------------------------------------
%Define output variable y to FIS within interval [1 9]
sys=addvar(sys,'output', 'y', [1 9]);
%Define MFs C1 and C2 for output y
sys=addmf(sys,'output',1,'C1','zmf',[1 9]);
sys=addmf(sys,'output',1,'C2','smf',[1 9]);
```

```
%Rules-----------------------------------------------------
%Define rules and add to FIS
rule=[1 1 1 1 1;
      1 2 2 1 1;
      1 1 2 1 1;
      2 2 1 1 1];
sys=addrule(sys,rule);
%Plot------------------------------------------------------
figure(1);
plotfis(sys);
figure(2)
ruleview(sys)
%Perform fuzzy inference for x1=4 and x2=8
y=evalfis([4 8], sys)
%------end of program--------------------------------------
```

Example B.1.5 The membership functions A_1, A_2, B_1, B_2 for the inputs x_1 and x_2 of a Takagi–Sugeno-type fuzzy system are defined by Gaussian functions:

$$\mu_{A_1}(x_1) = \exp\left[-\frac{1}{2}\left(\frac{x_1 - m_1}{\sigma_1}\right)^2\right], \mu_{A_2}(x_1) = \exp\left[-\frac{1}{2}\left(\frac{x_1 - m_2}{\sigma_2}\right)^2\right],$$

$$\mu_{B_1}(x_2) = \exp\left[-\frac{1}{2}\left(\frac{x_2 - m_3}{\sigma_3}\right)^2\right], \mu_{B_2}(x_2) = \exp\left[-\frac{1}{2}\left(\frac{x_2 - m_4}{\sigma_4}\right)^2\right],$$

Assume $m_1 = 2, m_2 = 3, m_3 = 3, m_4 = 4$ and $\sigma_1 = \sigma_2 = \sigma_3 = \sigma_4 = 2$. Consider the following rules for the Takagi–Sugeno fuzzy system:

> If x_1 is A_1 and x_2 is B_1 Then $z_1 = x_1 + x_2 + 1$
> If x_1 is A_2 and x_2 is B_1 Then $z_2 = 2x_1 + x_2 + 1$
> If x_1 is A_1 and x_2 is B_2 Then $z_3 = 2x_1 + 3x_2$
> If x_1 is A_2 and x_2 is B_2 Then $z_4 = 2x_1 + 5$

Compute the value of the output z for $x_1 = 1$ and $x_2 = 4$:

```
%Sugeno-type Fuzzy System
%Chapter 2 Example B.1.5
clear all;close all;
sys=newfis('Example2.14.4','sugeno');
%Inputs -----------------------------------------------
%Define input variable X1 within interval [1 4]
sys=addvar(sys,'input', 'X1', [1 4]);
%Define MFs A1 and A2 for input X1
sys=addmf(sys,'input',1,'A1','gaussmf',[1 2]);
sys=addmf(sys,'input',1,'A2','gaussmf',[1 3]);
```

```
%Define input variable X2 within interval [2 5]
sys=addvar(sys,'input', 'X2', [2 5]);
%Define MFs B1 and B2 for input X2
sys=addmf(sys,'input',2,'B1','gaussmf',[1 3]);
sys=addmf(sys,'input',2,'B2','gaussmf',[1 4]);

%outputs----------------------------------------
%Define output variable Z within interval [-9 9]
sys=addvar(sys, 'output', 'Z', [-9 9]);
sys=addmf(sys,'output',1,'Z1','linear',[1 1 1]);
sys=addmf(sys,'output',1,'Z2','linear',[2 1 1]);
sys=addmf(sys,'output',1,'Z3','linear',[2 3 0]);
sys=addmf(sys,'output',1,'Z4','linear',[2 0 5]);

%Rules-------------------------------------------
%Define rules and add to FIS
%1If (x1 is A1) and (x2 is B1) then (z is z1) (1)
%2If (x1 is A2) and (x2 is B1) then (z is z2) (1)
%3If (x1 is A1) and (x2 is B2) then (z is z3) (1)
%4If (x1 is A2) and (x2 is B2) then (z is z4) (1)

% A1 B1 C1 W &=1
rule=[1 1 1 1 1;
      2 1 2 1 1;
      1 2 3 1 1;
      2 2 4 1 1];

sys=addrule(sys,rule);

figure(1);
plotfis(sys);  %Figure is not shown here
figure(2)
ruleview(sys)  %Figure is not shown here
%Perform fuzzy inference for x1=1.5 and x2=4
y=evalfis([1.5 4], sys)
%------end of program------------------------
```

Appendix C

MATLAB® Programs for Fuzzy Systems

Example C.1.1　A system is described by two inputs (X_1, X_2) and a single output (Y) within the universes of discourse $-10 \le X_1 \le 10$, $-30 \le X_2 \le 30$ and $-20 \le Y \le 20$, respectively. In order to develop a Mamdani-type fuzzy system, the membership functions for (X_1, X_2) and (Y) are defined in Figure C.1.1. The rule base for the Mamdani-type fuzzy system is given in Table C.1.1.

A Mamdani-type fuzzy inference system for the above MFs and rule base is developed using MATLAB® and the Fuzzy Logic Toolbox.

```
%Fuzzy system - Example C.1.1
clear all;close all;
sys=newfis('ExampleFS_1');
%----------------Inputs definition --------------------
%Define input variable X1 to FIS within interval [-10 10]
sys=addvar(sys,'input', 'X1', [-10 10]);
%Define MFs N, Z and P for input X1
sys=addmf(sys,'input',1,'N','trimf',[-10 -10 0]);
sys=addmf(sys,'input',1,'Z','trimf',[-10 0 10]);
sys=addmf(sys,'input',1,'P','trimf',[0 10 10]);

%Define input variable X2 to FIS within interval [-30 30]
sys=addvar(sys,'input', 'X2', [-30 30]);
%Define MFs N, Z and P for input X2
sys=addmf(sys,'input',2,'N','trimf',[-30 -30 0]);
sys=addmf(sys,'input',2,'Z','trimf',[-30 0 30]);
sys=addmf(sys,'input',2,'P','trimf',[0 30 30]);
%----------------output definitions -------------------
```

Computational Intelligence: Synergies of Fuzzy Logic, Neural Networks and Evolutionary Computing, First Edition.
Nazmul Siddique and Hojjat Adeli.
© 2013 John Wiley & Sons, Ltd. Published 2013 by John Wiley & Sons, Ltd.

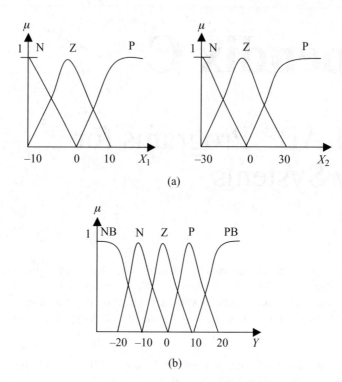

Figure C.1.1   Input and output membership functions. (a) Input membership functions; (b) Output membership function

```
%Define output variable Y to FIS within interval [-20 20]
sys=addvar(sys,'output', 'Y', [-20 20]);

%Define MFs NB, N, Z, P, PB for output Y
sys=addmf(sys,'output',1,'NB','trimf',[-20 -20 -10]);
sys=addmf(sys,'output',1,'N','trimf',[-20 -10 0]);
sys=addmf(sys,'output',1,'Z','trimf',[-10 0 10]);
sys=addmf(sys,'output',1,'P','trimf',[0 10 20]);
sys=addmf(sys,'output',1,'PB','trimf',[10 20 20]);
```

Table C.1.1   Rule base for a Mamdani-type system

	X_2		
X_1	N	Z	P
N	PB	P	N
Z	P	Z	N
P	Z	Z	NB

```
%----------------Rules--------------------------
%Define rules and add to FIS
%    N   Z   P
%---|-------------------
%N  |  PB  P   N
%Z  |  P   Z   N
%P  |  Z   Z   NB
%The rule-base is interpreted as If Input_1 is Mf_1 and Input_2 is
Mf_1 Then %Output_1 is Mf_5, followed by weight of rule and
connective used i.e.
% N N PB W=1 &=1

rule =[1 1 5 1 1;
       1 2 4 1 1;
       1 3 2 1 1;
       2 1 4 1 1;
       2 2 3 1 1;
       2 3 2 1 1;
       3 1 3 1 1;
       3 2 3 1 1;
       3 3 1 1 1];
sys=addrule(sys,rule);
```

Example C.1.2   A system with single input and single output is shown in Figure C.1.2. The mathematical description of the system is given by the difference equation (C.1). An incremental PI-like fuzzy controller is to be developed so that the system's output follows a reference signal.

$$y(k+1) = 0.0237u(k) + 0.0175u(k-1) + 1.407y(k) - 0.407y(k-1) \qquad (C.1)$$

The incremental PI-like fuzzy controller is described in Section 3.5.4 of Chapter 3. The control output is defined, and can be rewritten as $\Delta u = k_P \cdot \Delta e + k_i \cdot e$ where k_p and k_i are the proportional and the integral gain coefficients to be adjusted.

In this case, to obtain the value of the control output $u(k)$, the change of control output $\Delta u(k)$ is added to $u(k-1)$ such that

$$u(k) = \Delta u(k) + u(k-1) \qquad (C.2)$$

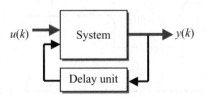

Figure C.1.2   System described by equation (C.1)

Table C.1.2   Rule base for Mamdani-type fuzzy controller.

	CE	
E	B_1	B_2
A_1	C_1	C_2
A_2	C_2	C_3

The PI-like Mamdani-type fuzzy controller's rule base accordingly consists of rules of the form

$$\text{If } e \text{ is } A_i \text{ and } \Delta e \text{ is } B_j \text{ Then } \Delta u \text{ is } C_k$$

The rule base is shown in Table C.1.2.

Triangular membership functions are chosen for all inputs and outputs and are defined as $A_1 = \{-1, -1, 1\}$, $A_2 = \{-1, 1, 1\}$, $B_1 = \{-1, -1, 1\}$, $B_2 = \{-1, 1, 1\}$, $C_1 = \{-1, -1, 0\}$, $C_2 = \{-1, 0, -1\}$ and $C_3 = \{0, 1, 1\}$. The final fuzzy controller with inputs $\{e(k), \Delta e(k)\}$ and output $\Delta u(k)$ is shown in Figure C.1.3.

Note that the incremental PI-type fuzzy controller in Figure C.1.3 looks like a PD-type fuzzy controller but is different from a PD-type controller. For details, see Section 3.5.4 of Chapter 3.

The MATLAB® codes for implementation of a PI-like fuzzy controller are given in the following:

```
%Fuzzy system -- Example C.1.2
clear all;close all;
sys=newfis('ExampleFS_2');
%Inputs ----------------------------------------------
%Define input variable E to FIS within interval [-1 1]
sys=addvar(sys,'input', 'E', [-1 1]);
%Define MFs A1 and A2 for input E
sys=addmf(sys,'input',1,'A1','trimf',[-1 -1 1]);
```

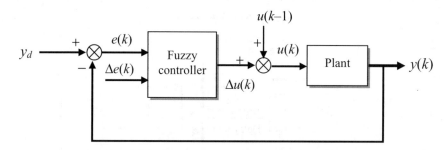

Figure C.1.3   Incremental PI-like fuzzy controller

```
sys=addmf(sys,'input',1,'A2','trimf',[-1 1 1]);
%Define input variable CE to FIS within interval [-1 1]
sys=addvar(sys,'input', 'CE', [-1 1]);
%Define MFs B1 and B2 for input CE
sys=addmf(sys,'input',2,'B1','trimf',[-1 -1 1]);
sys=addmf(sys,'input',2,'B2','trimf',[-1 1 1]);
%outputs---------------------------------------------
%Define output variable CU to FIS within interval [-1 1]
sys=addvar(sys,'output', 'CU', [-1 1]);
%Define MFs C1 and C2 for output CU
sys=addmf(sys,'output',1,'C1','trimf',[-1 -1 0]);
sys=addmf(sys,'output',1,'C2','trimf',[-1 0 1]);
sys=addmf(sys,'output',1,'C3','trimf',[0 1 1]);
%---------------Rules--------------------------------
%Define rules and add to FIS
%    B1 B2
%A1 C1 C2
%A2 C2 C3
%The rule-base is interpreted as If Input_1 is Mf_1 and Input_2 is
Mf_1 Then %Output_1 is Mf_1, followed by weight of rule and
connective used i.e.
% A1 B1 C1 W=1 &=1
rule=[1 1 1 1 1;
      1 2 2 1 1;
      2 1 2 1 1;
      2 2 3 1 1];
sys=addrule(sys,rule);
%Define constants
kp=0.029;
ki=0.049;
kc=60;
%Define reference signal
for k=1:500
  if(k<=250)
    yd(k)=15;
  else yd(k)=25;
  end
end
%Simulation of the FLC
y(1)=0;
y(2)=0;
u(1)=0;
v(1)=0;
for k=1:499
    e(k)=yd(k)-y(k);
  if(k==1)
    v(1)=0;
```

```
    else
       v(k) = (e(k)-e(k-1))/0.25;
    end; %end if
    einp(k) =e(k)*kp;
    vinp(k) =v(k)*ki;
    %deltau calculated by FLC
    delu(k) =evalfis([vinp(k) einp(k)],sys);
    u(k+1) =u(k) +delu(k)*kc;
    if(k==1)
      y(2)=0;
    else
      y(k+1) =0.0237*u(k) +0.0175*u(k-1) +1.407*y(k) -0.407*y(k-1);
    end;%end if
end; %end for
%Plot------------------------------------------------------------
k=1:500;
plot(k,yd,'- k',k,y,'-. k');
grid
title('Incremental PI-like FLC')
xlabel('Time index')
ylabel('Desired output & response')
figure
gensurf(sys)
ruleview(sys)
```

The performance of the Mamdani-type fuzzy controller with triangular membership functions is shown in Figure C.1.4, where the dotted line represents the desired output and the solid line shows the response of the system. The control surface is shown in Figure C.1.5. As mentioned earlier, the shape of the membership function is arbitrary from the point of view of simplicity, convenience, speed and efficiency. The fuzzy controller shows better performance if the membership functions for A_1 and A_2 are chosen as Z- and S-shaped, respectively. See Figures C.1.6 and C.1.7. A_1 and A_2 are defined as $A_1 = \{-1, -1, 1\}$, $A_2 = \{-1, 1, 1\}$, $B_1 = \{-1, -1, 1\}$, $B_2 = \{-1, 1, 1\}$, $C_1 = \{-1, -1, 0\}$, $C_2 = \{-1, 0, -1\}$ and $C_3 = \{0, 1, 1\}$.

Example C.1.3   An incremental PI-like Sugeno-type fuzzy controller is to be developed for the system given by the difference equation (C.1) described in Example C.1.2. The membership functions for the inputs are the same as in Example C.1.2. The output membership functions in a Sugeno-type fuzzy controller are linear functions defined as $Z_1 = a_1e + b_1\Delta e + c_1$, $Z_2 = a_2e + b_2\Delta e + c_2$ and $Z_3 = a_3e + b_3\Delta e + c_3$ with $a_1 = 0.01, b_1 = 0.0, c_1 = -1.001, a_2 = 0.0, b_2 = 0.0, c_2 = 0.0, a_3 = 0.01, b_3 = 0.0$ and $c_3 = 0.999$. The rule base of the Sugeno-type fuzzy controller consists of rules of the form

$$\text{If } e \text{ is } A_i \text{ and } \Delta e \text{ is } B_j \text{ Then } \Delta u \text{ is } Z_k = a.e + b.De + c$$

The rule base is shown in Table C.1.3.

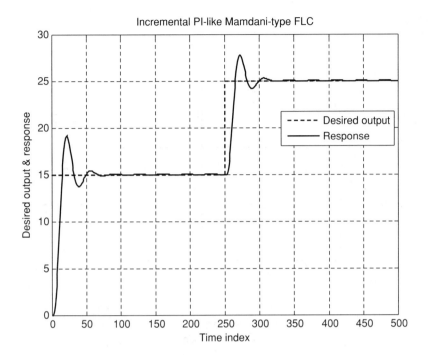

Figure C.1.4   Performance of the Mamdani-type fuzzy controller with triangular membership functions

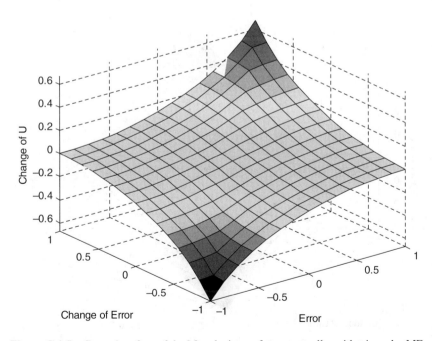

Figure C.1.5   Control surface of the Mamdani-type fuzzy controller with triangular MFs

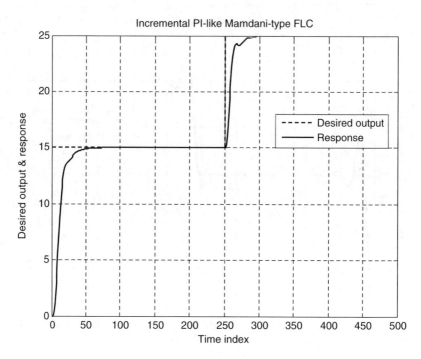

Figure C.1.6 Performance of the Mamdani-type fuzzy controller with Z- and S-shaped membership functions

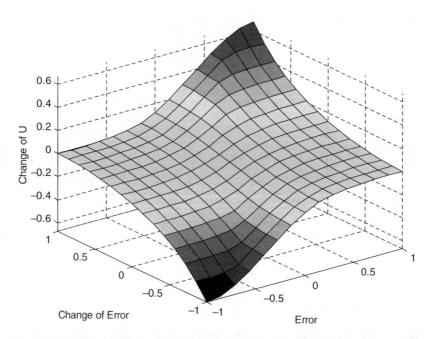

Figure C.1.7 Control surface of the Mamdani-type fuzzy controller with Z,S-shaped MFs

Table C.1.3 Rule base for Sugeno-type fuzzy controller.

		CE
E	B_1	B_2
A_1	Z_1	Z_2
A_2	Z_2	Z_3

The MATLAB® code for implementation of an incremental PI-like Sugeno-type fuzzy controller is given in the following:

```
%Sugeno-type Fuzzy Controller
%Chapter 3 Example C.1.3
clear all;close all;
sys=newfis('ExampleFS_3','sugeno');
%Inputs -------------------------------------------------
%Define input variable x1 to FIS within iterval [-1 1]
sys=addvar(sys,'input', 'Error', [-1 1]);
%Define MFs A1 and A2 for input x1
sys=addmf(sys,'input',1,'A1','gbellmf',[1 0.5 -1]);
sys=addmf(sys,'input',1,'A2','gbellmf',[1 0.5 1]);
%Define input variable x2 to FIS within iterval [-1 1]
sys=addvar(sys,'input', 'Change of Error', [-1 1]);
%Define MFs B1 and B2 for input x2
sys=addmf(sys,'input',2,'B1','gbellmf',[1 0.5 -1]);
sys=addmf(sys,'input',2,'B2','gbellmf',[1 0.5 1]);

%outputs------------------------------------------------
%Define output variable y to FIS within iterval [-1 1]
sys=addvar(sys,'output', 'Change of U', [-1 1]);

%Define linear MFs Z1, Z2 and Z3 for output y
%Z1=a1x1+b1x2+c1 with a1=0.01, b1=0.0, c1=-1.001
%Z2=a2x1+b2x2+c2 with a2=0.0, b2=0.0, c2=0.0
%Z3=a3x1+b3x2+c3 with a3=0.01, b3=0.0, c3=0.999

sys=addmf(sys,'output',1,'Z1','linear',[0.01 0.0 -1.001]);
sys=addmf(sys,'output',1,'Z2','linear',[0.0 0.0 0.0]);
sys=addmf(sys,'output',1,'Z3','linear',[0.01 0.0 0.999]);
%Rules--------------------------------------------------
%Define rules and add to FIS
%   B1 B2
%A1 Z1 Z2
%A2 Z2 Z3
%   A1 B1 Z1 W &=1
```

```
rule=[1 1 1 1 1;
      1 2 2 1 1;
      2 1 2 1 1;
      2 2 3 1 1];
sys=addrule(sys,rule);
%Define constants
kp=0.029;
kd=0.049;
kc=60;
%Define referece signal
for k=1:500
   if(k<=250)
       yd(k)=15;
   else yd(k)=25;
   end
end
%Simulation of the FLC
y(1)=0;
y(2)=0;
u(1)=0;
v(1)=0;
for k=1:499
    e(k)=yd(k)-y(k);
    if(k==1)
        v(1)=0;
    else
        v(k)=(e(k)-e(k-1))/0.25;
    end;%end if
    einp(k)=e(k)*kp;
    vinp(k)=v(k)*kd;
    %deltau calculated by FLC
    delu(k)=evalfis([vinp(k) einp(k)],sys);
    u(k+1)=u(k)+delu(k)*kc;
    if(k==1)
        y(2)=0;
    else
        y(k+1)=0.0237*u(k)+0.0175*u(k-1)+1.407*y(k)-0.407*y(k-1);
    end;%end if
end;%end for
%Plot-------------------------------------------------------
k=1:500;
plot(k,yd,'-. b',k,y,'- b');
grid
title('Incremental PI-like Sugeno-type FLC')
xlabel('Time index')
ylabel('Desired output & response')
figure
gensurf(sys)
```

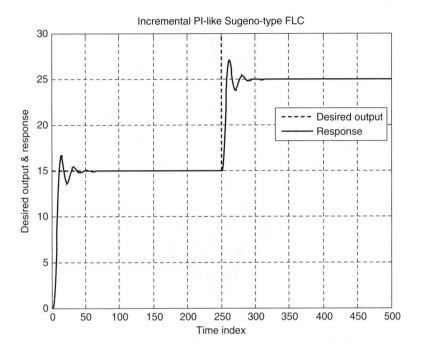

Figure C.1.8   Performance of the Sugeno-type fuzzy controller

The performance of the Sugeo-type fuzzy controller is shown in Figure C.1.8, where the dotted line represents the desired output and the solid line shows the response of the system. The control surface is shown in Figure C.1.9.

Example C.1.4   The Khepera is a miniature robot that has diameter 55 mm, height 30 mm and weight 70g, as shown in Figure C.1.10. The robot is supported by two wheels and two small Teflon balls placed under its platform. The wheels are controlled by two DC motors with an incremental encoder (12 pulses per millimetre of robot advancement) and can rotate in both directions. The geometrical shape and the motor layout of the Khepera make the robot able to navigate in a sophisticated environment even when its control system is immature. It is provided with eight infrared proximity sensors placed around its body, which are based on emission and reception of infrared light, as shown in Figure C.1.11. Each receptor can measure both the ambient infrared light and the reflected infrared light emitted by the robot itself. The Khepera robot includes eight infrared sensors, allowing it to detect by reflection (small rectangles) the proximity of objects in front of it, behind it and to the right and left of it. Each sensor returns a value ranging between 0 and 1023 represented in gradual colour levels. 0 means that no object is perceived whereas 1023 means that an object is very close to the sensor (almost touching the sensor). Intermediate values may give an approximate idea of the distance between the sensor and the object. Each motor can take a speed value ranging between −10 and +10. A Sugeno-type fuzzy controller has to be developed for the Khepera robot to navigate in a cluttered environment by avoiding obstacles.

To develop a fuzzy controller for the Khepera robot, a reasonable number of inputs and outputs of the system are to be chosen. The eight infrared sensors are used to detect objects

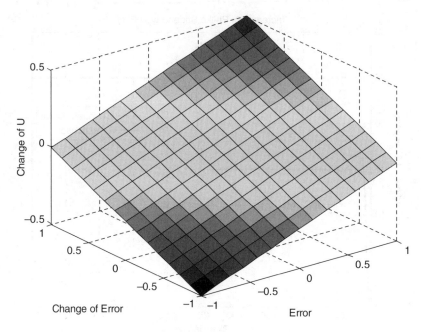

Figure C.1.9   Control surface of the Sugeno-type fuzzy controller

Figure C.1.10   Khepera robot

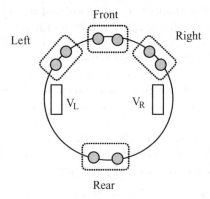

Figure C.1.11   Layout of infrared sensors and wheels of Khepera robot

in different directions of the Khepera. The two wheel speeds are the outputs of the system. Design of a fuzzy controller for a system with eight inputs and two outputs will be complicated. Therefore, to reduce the number of inputs of the Khepera, the eight sensors are grouped into four pairs $\{s_1, s_2\}$, $\{s_3, s_4\}$, $\{s_5, s_6\}$ and $\{s_7, s_8\}$. Each pair of sensors can be fused to obtain an accurate measure of distances to obstacles in the four directions left, front, right and rear, respectively as shown in Figure C.1.11.

Since the robot will move forward only avoiding obstacles to the left or right, we will not use the two rear sensors. Thus, the number of inputs to the system reduces to only three: left, forward and right. The two membership functions are chosen for each input. Simple triangular membership functions are chosen for all inputs and are defined as $small = \{-1023, 0, 1023\}$, $big = \{0, 1023, 2046\}$. A Sugeno-type fuzzy system uses a linear polynomial as output membership function (see Section 3.4.2 of Chapter 3). When the linear polynomial is constant, it is a zero-order Sugeno fuzzy system, which can be considered as a special case of a Mamdani fuzzy system, in which the consequent of each rule is specified by a fuzzy singleton or by a pre-defuzzified consequent (see Section 2.13.2 of Chapter 2). Two constant values are chosen for all outputs and are defined as $C_1 = \{10\}$ and $C_2 = \{-10\}$. The final fuzzy controller with three inputs (left, forward, right) and two outputs V_L (v_l = velocity of left wheel) and V_R (v_r = velocity of right wheel) has the following rule base:

If (left is small) and (forward is small) and (right is small)
then (v_l is C2)(v_r is C2) (1)
If (left is small) and (forward is small) and (right is big)
then (v_l is C1)(v_r is C2) (1)
If (left is small) and (forward is big) and (right is small)
then (v_l is C1)(v_r is C2) (1)
If (left is small) and (forward is big) and (right is big)
then (v_l is C1)(v_r is C2) (1)
If (left is big) and (forward is small) and (right is small)
then (v_l is C2)(v_r is C1) (1)
If (left is big) and (forward is small) and (right is big)
then (v_l is C2)(v_r is C2) (1)
If (left is big) and (forward is big) and (right is small)
then (v_l is C2)(v_r is C1) (1)
If (left is big) and (forward is big) and (right is big)
then (v_l is C1)(v_r is C2) (1)

The MATLAB® code for implementation of a Sugeno-type fuzzy controller for the Khepera robot is given in the following:

```
%Sugeno-type Fuzzy Controller for Khepera Robot
%Chapter 3 Example C.1.4
clear all;close all;
sys=newfis('Khepera','sugeno');
%Inputs -------------------------------------------------------------
%Define input variable Left to FIS within interval [-1023 2046]
sys=addvar(sys,'input', 'left', [-1023 2046]);
%Define MFs small and big for input Left
```

```
sys=addmf(sys,'input',1,'small','trimf',[-1023 0 1023]);
sys=addmf(sys,'input',1,'big','trimf',[0 1023 2046]);
%Define input variable Forward to FIS within iterval [-1023 2046]
sys=addvar(sys,'input', ' forward ', [-1023 2046]);
%Define MFs small and big for input Front
sys=addmf(sys,'input',2,'small','trimf',[-1023 0 1023]);
sys=addmf(sys,'input',2,'big','trimf',[0 1023 2046]);
%Define input variable Right to FIS within interval [-1023 2046]
sys=addvar(sys,'input', 'right', [-1023 2046]);
%Define MFs small and big for input Right
sys=addmf(sys,'input',3,'small','trimf',[-1023 0 1023]);
sys=addmf(sys,'input',3,'big','trimf',[0 1023 2046]);

%outputs--------------------------------------------------------
%Define output variable y to FIS within interval [-10 10]
sys=addvar(sys, 'output', 'v_l', [-10 10]);
sys=addmf(sys,'output',1,'C1','linear',[0 0 0 10]);
sys=addmf(sys,'output',1,'C2','linear',[0 0 0 -10]);

%Define linear MFs C1, C2 for outputs VL and VR
%C1=a1x1+b1x2+c1x3+d1 with a1=0, b1=0, c1=0, d1=10
%C2=a2x1+b2x2+c2x3+d2 with a2=0, b2=0, c2=0, d2=-10
sys=addvar(sys, 'output', 'v_r', [-10 10]);
sys=addmf(sys,'output',2,'C1','linear',[0 0 0 10]);
sys=addmf(sys,'output',2,'C2','linear',[0 0 0 -10]);

%Rules----------------------------------------------------------
%Define rules and add to FIS
%If (left is small) and (forward is small) and (right is small)
    then (v_l is C1)(v_r is C1)
%If (left is small) and (forward is small) and (right is big)
    then (v_l is C2)(v_r is C1)
%If (left is small) and (forward is big) and (right is small)
    then (v_l is C2)(v_r is C1)
%If (left is small) and (forward is big) and (right is big)
    then (v_l is C2)(v_r is C1)
%If (left is big) and (forward is small) and (right is small)
    then (v_l is C1)(v_r is C2)
%If (left is big) and (forward is small) and (right is big)
    then (v_l is C1)(v_r is C1)
%If (left is big) and (forward is big) and (right is small)
    then (v_l is C1)(v_r is C2)
%If (left is big) and (forward is big) and (right is big)
    then (v_l is C2)(v_r is C1)

% 1st rules is coded as small small small C1 C1 W &=1
% indexed as 1 1 1 1 1 1 1;
```

```
rule=[1 1 1 1 1 1 1 ;
      1 1 2 2 1 1 1 ;
      1 2 1 2 1 1 1 ;
      1 2 2 2 1 1 1 ;
      2 1 1 1 2 1 1 ;
      2 1 2 1 1 1 1 ;
      2 2 1 1 2 1 1 ;
      2 2 2 2 1 1 1 ];

sys=addrule(sys,rule);
%An alternative will be to read the FIS file created using GUI
%e.g. sys=readfis('khepera.fis');

max_time=700;
i=1;
ref= kopen([0,9600,1])
sen=zeros(max_time,8);
vel=zeros(max_time,2);

while i<max_time,
    s= kProximity(ref);
    left=max(s(1),s(2));
    front=max(s(3),s(4));
    right=max(s(5),s(6));
    v = evalfis([left, front, right], sys);
    kSetSpeed(ref,v(1),v(2));
    vel(i,:)=v;
    sen(i,:)=s;
    i=i+1;
end
kStop(ref)
kclose(ref)
v1=vel(:,1);
v2=vel(:,2);
t=1:max_time;
plot(t, v1, '.-k',t,v2,'-k')
ylabel('Wheel speed')
xlabel('Time instance')
legend('Left wheel speed', 'Right wheel speed')
```

kopen([0,9600,1]), kclose(ref), kProximity(ref), kSetSpeed (ref, left, right), kStop(ref) are MATLAB® routines that permit the user to interact with Khepera over a serial connection. Windows DLLs are included to perform the system-level serial-port communication, and a library of useful MATLAB® M-files to read proximity sensors, set wheel speeds, stop moving, etc. The first thing is to open the serial port com1 at band rate 9600 with 1-s timeout using the command kopen([0, 9600, 1]). kopen() returns a reference value to be used for subsequent commands. We used 'ref' as the return variable for the above example and will use

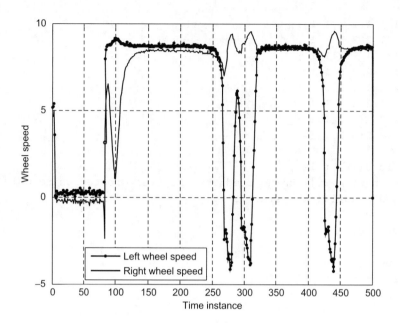

Figure C.1.12   Left and right wheel speed

this variable in all examples in subsequent chapters. The serial port must be closed using the command kclose(ref). kProximity(ref) returns the proximity sensor readings as an 8-element vector. kSetSpeed (ref, left, right) sets the motor speed of the left and right wheel. To stop the Khepera, use kStop(ref).

For an explanation of the actual Khepera commands sent by these MATLAB® routines, please consult Appendix A of the Khepera User Manual to be found at http://www.k-team.com.

The performance of the Sugeo-type fuzzy controller for the Khepera robot can be illustrated by the left and right wheel speeds as shown in Figure C.1.12. The rise of the right wheel speed and the fall of the left wheel speed down to a negative value within a time interval represent a left turn to avoid an obstacle on the right. Otherwise the Khepera moves straight forward.

C.1.1   GUI Interface Tools

Section 3.6.1 of Chapter 3 describes how to develop a control application by working from the command line. It is much easier to build a system using GUI tools. There are five GUI tools for developing an application: FIS Editor, MF Editor, Rule Editor, Rule Viewer and Surface Viewer. The tools are dynamically linked to each other, meaning change of any tool will automatically update other tools. Figure C.1.13 shows the different editors of the GUI tools and the links between them.

FIS Editor

The FIS Editor handles the high-level issues of the system (e.g., number of input and output variables and their names). The Fuzzy Logic Toolbox does not limit the number of inputs.

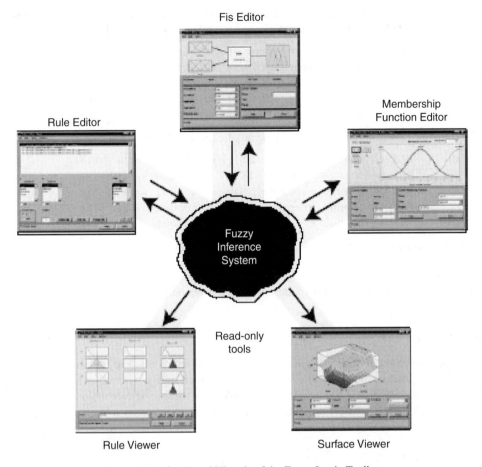

Figure C.1.13   Five GUI tools of the Fuzzy Logic Toolbox

However, the number of inputs may be limited by the complexity of the system (i.e., if the number of inputs and outputs is too large or if the number of MFs is too big, then it may be difficult to analyse the system using the GUI tools).

MF Editor

The MF Editor is used to define the shapes of the MFs by changing the parameters and ranges of each MF associated with each input and output variable. The names of each MF can also be edited here.

Rule Editor

The Rule Editor is for editing the list of rules that defines the behaviour of the system. Creation of the rule base is very straightforward, simply by clicking the appropriate MFs, connection and rule weight for each rule. There are three ways to express the rules: verbose, symbolic and indexed. The rule base appearing in the window in verbose form looks like

```
If (error is NB) and (change_error is NS) then (torque is PB) (1)
If (error is NB) and (change_error is ZO) then (torque is PB) (1)
If (error is NB) and (change_error is PS) then (torque is PS) (1)
If (error is NB) and (change_error is PB) then (torque is ZO) (1)
If (error is NS) and (change_error is NB) then (torque is PB) (1)
```

The rule base appearing in the window in symbolic form looks like

```
(error == NB) & (change_error == NS) => (torque == PB) (1)
(error == NB) & (change_error == ZO) => (torque == PB) (1)
(error == NB) & (change_error == PS) => (torque == PS) (1)
(error == NB) & (change_error == PB) => (torque == ZO) (1)
(error == NS) & (change_error == NB) => (torque == PB) (1)
```

The rule base appearing in the window in indexed form looks like

```
1 3, 3 (1) : 1
```

where error and change_error are inputs and torque is the output.

Rule Viewer

The Rule Viewer and the Surface Viewer are strictly read-only tools. They are used for looking at the FIS as a diagnostic. It can show (for example) which rules are active, or how individual MF shapes are influencing the results. The Rule Viewer shows one calculation at a time and in great detail. In this sense, it presents a sort of overview of the fuzzy inference system.

Surface Viewer

The Surface Viewer is used to display the dependency of one of the outputs on any one or two of the inputs – that is, it generates and plots an output surface map for the system. If an entire output surface of the system is to be examined based on the entire span of the output set over the entire span of the input set, then Surface Viewer is useful for such an analysis.

C.1.2   Simulink® Blocks

The Fuzzy Logic Toolbox is designed to work with Simulink®, the simulation software available from MathWorks. Once we have created a fuzzy system using the GUI tools or some other method, it can be embedded directly into a simulation system. The Fuzzy Logic Toolbox provides different Simulink® blocks. These are called Fuzzy Logic Controller blocks.

Appendix D

MATLAB® Programs for Neural Systems

D.1.1 Defining Feedforward Network Architecture

Feedforward networks often have one or more hidden layers of sigmoid neurons followed by an output layer of linear neurons. Multiple layers of neurons with nonlinear transfer functions allow the network to learn nonlinear and linear relationships between input and output vectors. The function newff() creates a feedforward backpropagation network architecture with desired number of layers and neurons. The general form of use of the function is given below, which returns an N-layer feedforward backpropagation network object:

```
net = newff([PN],[S1 S2 ... SN],{TF1 TF2 ... TFN},BTF,LF,PF);
```

where the first input PN is an $N \times 2$ matrix of minimum and maximum values for N input elements. S1 S2 . . . SN are the sizes (number of neurons) of the layers of the network architecture. TFi is the transfer function of the ith layer; the default is 'tansig'. The transfer functions TFi can be any differentiable transfer function such as tansig, logsig or purelin. BTF is the backpropagation network training function; the default is 'trainlm'. Different training functions with their features are described in Section 4.7.2 of Chapter 4. LF is the backpropagation weight/bias learning function with gradient descent, such as 'learngd', 'learngdm'. The default is 'learngdm'. The function 'learngdm' is used to calculate the weight change dW for a given neuron from the neuron's input P and error E. Learning occurs according to learngdm's learning parameters such as the weight (or bias) W, learning rate and momentum constant, according to gradient descent with momentum and returns weight change and new learning states. PF is the performance function such as mse (mean squared error), mae (mean absolute error) and msereg (mean squared error with regularization). The default is 'mse'. For example:

```
net=newff([-1 2; 0 5],[3,1],{'tansig','purelin'},'traingd',
          'learngdm', 'mae');
```

The function creates a two-input single-output feedforward network with single hidden layer. The first input [−1 2; 0 5] specifies the minimum and maximum values for each of the input vectors. The second input is an array containing the sizes of each layer, i.e., the network

---

Computational Intelligence: Synergies of Fuzzy Logic, Neural Networks and Evolutionary Computing, First Edition.
Nazmul Siddique and Hojjat Adeli.
© 2013 John Wiley & Sons, Ltd. Published 2013 by John Wiley & Sons, Ltd.

has 3 neurons in the hidden layer and 1 neuron in the output layer. The third input is a cell array containing the names of the transfer functions to be used in each layer, i.e., tansig for hidden layer and purelin (linear) activation function for output layer. There are other different activation functions with distinct features, such as logsig, hardlim. The final input contains the name of the training function to be used. 'traingd' is one of the training functions used by the network. newff() will also automatically initialize the weights and biases of the network.

D.1.1.1   Creating RBF Network Architecture

In an RBF network, there can be a maximum of M inputs and a maximum of N radial basis neurons in the hidden layer. There are no weights between inputs and hidden neurons. Each radial basis neuron is connected to the output neuron through the weight matrix W, which has to be learned. Using the MATLAB® functions, the architecture of an RBF network with $m = 1, 2, 3, \ldots, M$ input elements and $n = 1, 2, 3, \ldots, N$ radial basis neurons (in the hidden layer) can be created. All details of designing a radial basis function network are built into the design functions newrbe() and newrb(), and their outputs can be obtained with sim(). The functions are called in the following way:

```
net = newrbe(P, T, Spread);
```

The function newrbe() takes matrices of input vectors P and target vectors T, and a spread for the radial basis layer, and returns a network with weights and biases such that the outputs are exactly T when the inputs are P. The value for the spread constant should be larger than the distance between adjacent input vectors, so as to get a good generalization, but smaller than the distance across the whole input space. The function newrbe() creates as many radial basis neurons as there are input vectors in P. The drawback to newrbe() is that it produces a network with as many hidden neurons as there are input vectors. For this reason, newrbe() does not return an acceptable solution when many input vectors are needed to properly define a network, as is typically the case (Demuth and Beale, 2000). newrb() is a more efficient design function, which creates a radial basis network one neuron at a time. Neurons are added to the network until the sum-squared error falls beneath an error goal or a maximum number of neurons have been reached. The call for this function is

```
net = newrb(P, T, Goal, Spread);
```

The function newrb() takes matrices of input vectors P and target vectors T, and design parameters goal and spread for radial basis layer, and returns the desired network with weights and biases such that the outputs are exactly T when the inputs are P. The design method of newrb() is similar to that of newrbe(). The difference is that newrb() creates neurons one at a time. The error of the new network is checked, and if low enough newrb() is finished. Otherwise the next neuron is added. This procedure is repeated until the error goal is met or the maximum number of neurons is reached.

Thus, newrbe() creates a network with zero error on training vectors. The only condition required is to make sure that SPREAD is large enough so that the active input regions of the radbas neurons overlap enough so that several radbas neurons always have fairly large outputs at any given moment. This makes the network function smoother and results in better generalization for new input vectors occurring between input vectors used in the design. (However, SPREAD should not be so large that each neuron is effectively responding in the same large area of the input space.)

RBF networks, even when designed effectively with newrbe(), tend to have many times more neurons than a comparable MLP network with tansig or logsig neurons in the hidden layer. This is because sigmoid neurons can have outputs over a large region of the input space, while RBF neurons only respond to relatively small regions of the input space. The result is that the larger the input space (in terms of number of inputs, and the ranges those inputs vary over) the more RBF neurons are required. On the other hand, designing an RBF network often takes much less time than training a sigmoid/linear network, and can sometimes result in fewer neurons being used.

D.1.1.2   Creating GRNN Network Architecture

A generalized regression neural network is often used for function approximation. A GRNN network with $m = 1, 2, 3, \ldots, M$ input elements and $n = 1, 2, 3, \ldots, N$ radial basis neurons (in the hidden layer) can be created using the function newgrnn() where the first layer is just like that of newrbe() or newrb() but has a slightly different second layer. It has as many neurons as there are input/target vectors in P. Specifically, the first-layer weights are set to P'. The bias b1 is set to a column vector of 0.8326/Spread. The user chooses 'Spread', the distance an input vector must be from a neuron's weight vector to be 0.5. Each neuron's weighted input is the distance between the input vector and its weight vector. The second layer also has as many neurons as input/target vectors, but there the weights are set to target T. The function is called in the following way:

```
net = newgrnn(P, T, Spread);
```

The function newgrnn() takes matrices of input vectors P and target vectors T, and a spread for the radial basis layer, and returns a network with weights and biases such that the outputs are exactly T when the inputs are P. The value for spread constant should be larger than the distance between adjacent input vectors, so as to get good generalization, but smaller than the distance across the whole input space. To fit data closely, a smaller spread is suggested, i.e., smaller than the typical distance between input vectors. To fit the data more smoothly, a larger spread is to be chosen.

A larger spread leads to a large area around the input vector where layer 1 neurons will respond with significant outputs. Therefore, if the spread is small the radial basis function is very steep, so that the neuron with weight vector closest to the input will have a much larger output than other neurons. The network tends to respond with the target vector associated with the nearest design input vector.

As the spread becomes larger the radial basis function's slope becomes smoother and several neurons can respond to an input vector. The network then acts as if it is taking a weighted average between target vectors whose design input vectors are closest to the new input vector. As the spread becomes larger, more and more neurons contribute to the average, with the result that the network function becomes smoother.

D.1.1.3   Creating PRNN Network Architecture

A PNN network can be created by calling the function in the following way:

```
net = newpnn(P, T, Spread);
```

The function newpnn() takes matrices of input vectors P and target vectors T, and a spread for the radial basis layer, and returns a network with weights and biases such that the outputs are exactly T when the inputs are P. If the spread is near zero, the network will act as a nearest-neighbour classifier. As the spread becomes larger, the designed network will take into account several nearby design vectors.

Although the PNN was derived from the same mathematical merits and similarities to those of RBF and GRNN networks, after defining the architecture it is found to be more appropriate for classification problems rather than prediction or approximation problems.

Probabilistic neural networks can be used for classification problems. When an input is presented, the first layer computes distances from the input vector to the training input vectors and produces a vector whose elements indicate how close the input is to a training input. The second layer sums these contributions for each class of inputs to produce as its net output a vector of probabilities. Finally, a complete transfer function on the output of the second layer picks the maximum of these probabilities, and produces a 1 for that class and a 0 for the other classes.

D.1.2   Training Networks

Different backpropagation training algorithms are available as functions in MATLAB®. They have their own features and advantages. Some of the most widely used functions are discussed briefly.

> traingd – basic gradient descent learning algorithm. It has slow response but can be used in incremental mode training.

> traingdm – gradient descent with momentum. It is generally faster than traingd and can be used in incremental mode training.

> traingdx – adaptive learning rate. It has faster training time than traingd but can only be used in batch mode training.

> trainrp – resilient backpropagation. It is a simple batch mode training algorithm with fast convergence and minimal storage requirements.

> trainlm – Levenberg–Marquart algorithm. It is a faster training algorithm for networks of moderate size. It has a memory reduction feature for use when the training set is large.

There are several parameters associated with training algorithms. The parameters are learning rate, error goal, epochs and show. These parameters are defined as:

```
net.trainParam.lr - specifies learning rate
net.trainParam.goal - specifies error goal
net.trainParam.epochs - specifies the number of iterations
net.trainParam.show - displays status for every show.
```

Once the network has been defined and the parameters are set, the network can be trained using the function train() as

```
[net, tr] = train(net, P, T)
```

where net is the network object, tr contains information about the progress of training, P and T are the input and output vectors, respectively. Typically, one epoch of training is defined as a single presentation of all input vectors to the network. The network is then updated according to the results of all those presentations. Training occurs until a maximum number of epochs occur, the performance goal is met or any other stopping condition of the function is met. For example:

```
net.trainParam.lr =0.05;
net.trainParam.goal = 0.01;
net.trainParam.epochs =100;
net.trainParam.show = 25;

[net,tr]=train(net, P, T)
```

The network will be trained using the input and output data P and T, respectively for up to 100 epochs or when an error goal of 0.01 is reached.

D.1.3   Simulating Networks

The function sim() simulates a network. It takes the network input P and the network objects net and returns the network output $\hat{y}$. A single matrix of concurrent vector is presented to the network and the network produces a single matrix of concurrent vector as output:

$\hat{y}$= sim(net, P)

D.1.4   Creating Neural Network Subsystem

Once the network has been trained and tested with training and checking data, a Simulink® model can be created using the MATLAB® function gensim(). The function gensim() generates block descriptions of networks so that it can simulate the neural network in Simulink®. The function is called in the following way:

```
gensim(net, st)
```

gensim() takes these inputs net $\triangleq$ neural network defined either in an M-file or NN Toolbox and st $\triangleq$ sample time and creates a Simulink® system containing a block which simulates a neural network with a specified sampling time:

```
gensim(net,st)
```

The second argument to gensim() determines the sample time, which is normally chosen to be some positive real value. If the network has no delays associated with its input weights or layer weights, this value is set to –1. For example:

```
gensim(net,-1)
```

The value of the parameter st is –1, i.e., –1 tells gensim to generate a network with continuous sampling.

Example D.1.1: Define a feedforward network, train and simulate with input data   In this example, a two-layer feedforward network is created. The network's input ranges from 0 to 10. The first layer has five neurons with tansig function; the second layer has one neuron with linear function. The training function 'traingd' is used to train the network.

```
%Chapter 4 Example D.1.1
%Create a feedforward NN and train with data set [P    T]
%Training set
P = [0 1 2 3 4 5 6 7 8 9 10];
T = [0 1 2 3 4 3 2 1 2 3 4];

net = newff([0 10],[5 1],{'tansig' 'purelin'},'traingd');

%Set network parameters as follows
net.trainParam.lr = 0.01              %Learning rate
net.trainParam.goal = 0.1             %Performance goal
net.trainParam.epochs = 50            %This sets maximum number of epochs
                                       in a training
net.trainParam.show = 25              %This displays training status
                                       after each 25 epoch
net.trainParam.time = inf             %Maximum time to train in seconds
net.trainParam.min_grad=1e-10         %Minimum performance gradient

%Here the network is simulated and its output plotted against
 the targets.
net = train(net,P,T);
Y = sim(net,P);
plot(P,T, '-k', P,Y, 'ok')
```

Example D.1.2: Updating weights of a two-layer neural net   A two-layer neural network with two inputs $x = [x_1, x_2]$ and one output y is given by $y = W_2^T [f (W_1^T x + b_1)] + b_2$, where $W_1^T = \begin{bmatrix} -2.69 & -2.78 \\ -2.39 & -3.56 \end{bmatrix}, b_1 = \begin{bmatrix} -2.29 \\ 3.67 \end{bmatrix}, W_2^T = [-3.91 \ 5.95]$ and $b_2 = [-2.81]$. Update the weights and biases of the network, simulate the network and plot the output surface over the grid $[-2, 2] \times [-2, 2]$:

```
%Chapter 4 Example D.1.2
%A two layer NN is given by y=W2'[f(W1'x+b1)]+b2
%with W1=[-2.69 -2.78; -2.39 -3.56]; b1=[-2.29; 3.67]; W2=[-3.91
 5.95]; b2=[-2.81];
%Update NN with W1, W2, b1 and b2
%Plot the NN output surface y as a function of x over grid [-2,2]
 x[-2,2]
%Weights and bias
W1=[-2.69 -2.78; -2.39 -3.56];
b1=[-2.29; 3.67];
```

```
W2=[-3.91 5.95];
b2=[-2.81];
%Output surface
[x1, x2]=meshgrid(-2:0.1:2);
%Compute NN output
p1=x1(:);
p2=x2(:);
p=[p1';p2'];
%NN weights and bias
%nnt2ff () updates NN with specified weights and biases
net=nnt2ff(minmax(p),{W1,W2},{b1,b2},{'tansig', 'purelin'});
%Simulate
a=sim(net,p);
%Arrange results for mesh plot
a1=eye(41);
a1(:)=a';
mesh(x1,x2,a1);
AZ=60, EL=30;
view(AZ,EL);
xlabel('x1');
ylabel('x2');
title('NN output surface for tansigmoid function')
```

See Figure D.1.1 for the result.

Example D.1.3: Approximation of output surface   In this example, an output surface of a nonlinear function is approximated. The nonlinear function is defined by $f(x, y) =$

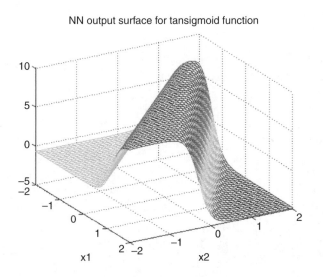

Figure D.1.1 Output surface of a two-layer network

$\sin(\pi x)^* \cos(\pi y)$ with $x \in [-2, 2]$ and $y \in [-2, 2]$. An MLP with 20 tansigmoidal neurons and 1 linear neuron can approximate the function after training the network for 500 epochs:

```
%Chapter 4 - Example D.1.3
%NN Function approximation
[x, y]=meshgrid(-2:0.1:2);
%Nonlinear function defined by
z=sin(pi*x).*cos(pi*y);
%Generate Input & Target data
for i=1:2000
  p(:,i)=4*(rand(2,1)-0.5);
  T(:,i)=sin(pi*p(2*i-1))*cos(pi*p(2*i));
end
%----------------------------------------------------------------
%Two-layer NN created with 20 tansig
%and one purelin neuron
net=newff(minmax(p), [20,1], {'tansig', 'purelin'}, 'trainlm');
net.trainParam.show=50;
net.trainParam.epochs=500;
net.trainParam.goal=1e-6;
[net,tr]=train(net,p,T)
%Simulate the net
a=zeros(41,41);
[x1, y1]=meshgrid(-2:0.1:2);
for i=1:1681
  a(i)=sim(net,[x1(i);y1(i)]);
end
figure(1)
%Original nonlinear function
subplot(1,2,1)
mesh(x,y,z);
title('Original function graphics');
xlabel('<--x-->')
ylabel('<--y-->')
zlabel('<--z-->')
AZ=151; EL=59;
view(AZ,EL)
subplot(1,2,2)
mesh(x1,y1,a);
title('NN approximated graphics')
xlabel('<--x-->')
ylabel('<--y-->')
zlabel('<--z-->')
AZ=151; EL=59;
view(AZ,EL)
```

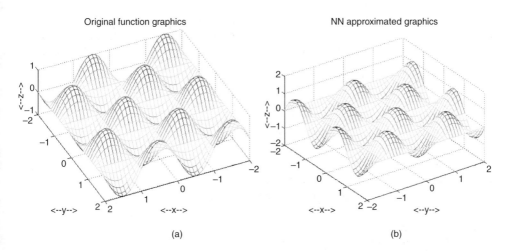

Figure D.1.2 Approximation of a nonlinear function. (a) Nonlinear function; (b) NN approximated function

See Figure D.1.2.

Example D.1.4: Approximation of a nonlinear function   In this example, a nonlinear function defined by the input/output data is approximated using a two-layer feedforward network. The network's weights and biases are shown after training. The plot will also show how good the approximation is:

```
%Chapter 4 Example D.1.4
%Function approximation
clear all; close all;

%Training data:examplar input pattern and target output vector
x=-1:0.1:1;
y=[-0.96,-0.577,-0.073, 0.377, 0.641, 0.66, 0.461, 0.134,...
    -0.201, -0.434, -0.5, -0.393, -0.165, 0.099, 0.307, 0.396,...
    0.345, 0.182, -0.031, -0.219, -0.32];

%Define a NN and initialise weights
net=newff(minmax(x), [7 1], {'tansig', 'purelin'},'trainlm');

%Output of NN with initial weights
ycap1=sim(net,x);

%Train the NN
net.trainParam.epochs = 5000        %Maximum number of epochs to train
net.trainParam.goal = 0.001         %Performance goal
net.trainParam.lr = 0.01            %Learning rate
%net.trainParam.min_grad=1e-10      %Minimum performance gradient
```

```
net.trainParam.show = 100          %Epochs between displays
net.trainParam.time = inf          %Maximum time to train in seconds
[net,tr]=train(net,x,y);

%Output of NN
figure(1)
%Generalisation: input vector is different
%from the one used for training
x2=-1:0.01:1;
ycap2=sim(net,x2);
plot(x1,ycap1,x2,ycap2,'-',x,y,'o');
plot(x,ycap1,'--k', x2,ycap2,'-.k',x,y,'ok');
title('Function approximation');
xlabel('x-values');
ylabel('y-values');
legend('Before Training', 'After Training','Function');

%Show weights and biases of NN
w=net.IW{1,1}
bw=net.b{1}
v=net.LW{2,1}
bv=net.b{2}
```

See Figure D.1.3.

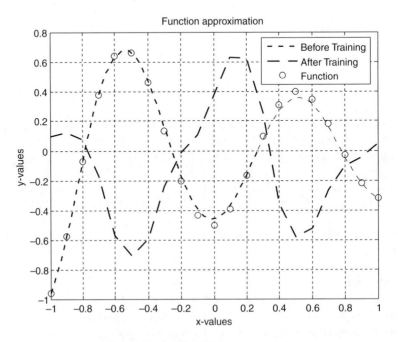

Figure D.1.3   Function approximation

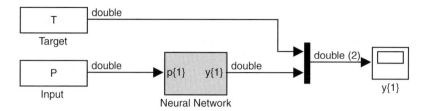

Figure D.1.4   NN block description for Simulink®

Example D.1.5: Creating an NN subsystem for simulation   In this example, a neural network block description is created for simulation to be used under Simulink®. To do this, an NN is trained with a set of input/output data and simulated. Once this is done, an NN block description is created using the gensim() function:

```
%Chapter 4 Example D.1.5
%Training data set
p=[-1:0.05:1];
%noisy sine wave
t=sin(2*pi*p)+0.1*randn(size(p));

net=newff([-1 1],[20,1],{'tansig','purelin'},'traingdx');
net.trainParam.show=50;
net.trainParam.epochs=1300;

pt=p*0.979;
y=sim(net,pt)
plot(p,t,'-',p,y,'o')

%generate NN block description
gensim(net,-1)
```

The generated NN Simulink® block description is shown in Figure D.1.4.

The Neural Network Toolbox provides three popular neural network Simulink® blocks for prediction and control that have been applied to many applications: Model Predictive Control, NARMA-L2 (or Feedback Linearization) Control and Model Reference Control. These Simulink® control blocks are discussed in detail in Chapter 5.

Appendix E

MATLAB® Programs for Neural Control Design

Example E.1.1: Neural network for systems modelling   The universal approximation capabilities of the multilayer perceptron make it a popular choice for modelling nonlinear systems. It is desired to design a two-layer feedforward neural network to model the nonlinear system described by the function $y = f(x)$. The input/output data given below describes the input/output behaviour of the system.

$$x = [-1 : 0.1 : 1]$$
$$y = [-0.96, -0.577, -0.073, 0.377, 0.641, 0.66, 0.461, 0.134, -0.201, -0.434, -0.5, \ldots$$
$$-0.393, -0.165, 0.099, 0.307, 0.396, 0.345, 0.182, -0.031, -0.219, -0.320]$$

A neural network model has to be developed using the input/output data.

The hidden layer with five neurons has tansigmoidal activation functions and the output layer is linear. Using batch gradient descent with momentum, train the network so that the mean square error (mse) < 0.005. The learning rate $\eta = 0.01$. Repeat training three times with different initial weights. What is the conclusion of the experiment?

```
%Chapter 5 Example E.1.1
%Function approximation
clear all;close all;
%Training data:examplar input pattern and htarget output vector
x=-1:0.1:1;
y=[-0.96,-0.577,-0.073, 0.377, 0.641, 0.66, 0.461, 0.134,...
   -0.201, -0.434, -0.5, -0.393, -0.165, 0.099, 0.307, 0.396,...
   0.345, 0.182, -0.031, -0.219, -0.32];
%Define a NN and initialise weights
net=newff(minmax(x), [10 1], {'tansig', 'purelin'},'traingd');
%Output of NN with initial weights
```

Computational Intelligence: Synergies of Fuzzy Logic, Neural Networks and Evolutionary Computing, First Edition.
Nazmul Siddique and Hojjat Adeli.
© 2013 John Wiley & Sons, Ltd. Published 2013 by John Wiley & Sons, Ltd.

```
ycap1=sim(net,x);
figure(1)
plot(x,ycap1,'-',x,y,'o');
%Train the NN
figure(2)
net.trainParam.epochs = 5000      %Maximum number of epochs to train
net.trainParam.goal = 0.001       %Performance goal
net.trainParam.lr = 0.01          %Learning rate
%net.trainParam.min_grad=1e-10    %Minimum performance gradient
net.trainParam.show = 100         %Epochs between displays
net.trainParam.time = inf         %Maximum time to train in seconds
[net,tr]=train(net,x,y);
%Output of NN
figure(3)
%Generalisation: input vecto is different
%from the one used for training
x2=-1:0.01:1;
ycap2=sim(net,x2);
plot(x2,ycap2,'-',x,y,'o');
%Weights and biases of NN
w=net.IW{1,1}
bw=net.b{1}
v=net.LW{2,1}
bv=net.b{2}
```

Example E.1.2 Neural network for system identification   The nonlinear system is described by the following equation:

$$y(k+1) = \frac{y(k)\,[y(k-1)+2]\,[y(k)+2.5]}{8.5+[y(k)]^2+[y(k-1)]^2} + u(k)$$

where $y(k)$ is the output of the system at the kth time step and $u(k)$ is the input. The system is stable for $u(k) \in [-2, 2]$ and $u(k)$ is a uniformly bounded function of time. The above equation can be written in the following form for identification purposes:

$$\hat{y}(k+1) = f(y(k), y(k-1)) + u(k)$$

where $f(y(k), y(k-1))$ is to be represented by a feedforard neural network. A neural network-based system for identification of the above model is to be developed:

```
%Chapter 5 Example E.1.2
%System Identification
clear all;close all;
%Define initial values
y(1)=0;y(2)=0;
%Create 500 random values within [-2,2]
%and obtain training pattern
N=500;
```

```
for k=2:N
  f(k)=y(k)*(y(k-1)+2)*(y(k)+2.5)/(8.5+...
      y(k)*y(k)+y(k-1)*y(k-1));
  u(k)=(rand-0.5)*4;
  y(k+1)=f(k)+u(k);
end
for k=1:(N-1)
  input(1,k)=y(k+1);
  input(2,k)=y(k);
  target(k)=f(k+1);
end
%Define a NN and initialise weights
net=newff(minmax(input), [10 1], {'tansig', 'purelin'});
net.trainParam.goal = 0.005      %Performance goal
net=train(net,input,target);
%Test NN
for k=1:200, u(k)=2*cos(2*pi*k/100);end
for k=201:500, u(k)=1.2*sin(2*pi*k/20);end
yp(1)=0;yp(2)=0;ycap(1)=0;ycap(2)=0;
%yp - from direct calculation
%ycap - NN simulation
for k=2:500
  yp(k+1)=yp(k)*(yp(k-1)+2)*(yp(k)+2.5)/(8.5+...
      yp(k)*yp(k)+yp(k-1)*yp(k-1))+u(k);
  ycap(k+1)=sim(net,[ycap(k); ycap(k-1)])+u(k);
end
figure(2)
plot(1:501,yp,'r',1:501,ycap,'g');
```

Example E.1.3: Neural network for adaptive identification A linear system is described by the following equation:

$$y(k) = u(k) + 0.5u(k-1) - 1.5u(k-2)$$

The input signal is defined by $u(k) = 0.6\sin(2\pi k/10) + 1.2\cos(2\pi k/10)$ with $k \in \{1, 2, \ldots\}$. A neural network-based prediction system is to be developed for the system output y for given current and two previous inputs. The parameters of the linear system are not robust and hence should be taken care of by implementing an adaptive identification method and training the network incrementally. Assume that the measurements are $y_m = y + 0.1*\text{randn}(\text{size}(y))$.

```
%Chapter 5 Example E.1.3
%Adaptive Identification
clear all;close all;
%Input signal
N=100; T=10;
index=1:N;
u=0.6*sin(2*pi*index/T)+11.2*cos(2*pi*index/T);
%Measured values of output
```

```
y(1)=u(1);y(2)=u(2)+0.5*u(1);
for k=3:N
  y(k)=u(k)+0.5*u(k-1)-1.5*u(k-2);
end
randn('state',0);
ym=y+0.1*randn(size(y));
%Training data
for k1=3:N
  k=k1-2;
  P(1,k)=u(k1);
  P(2,k)=u(k1-1);
  P(3,k)=u(k1-2);
  t(k)=ym(k1);
end
%Define a NN and initialise weights
a=min(u);b=max(u);
net=newff([a b;a b;a b], [20 1], {'tansig', 'purelin'},'traingdm');
%Incremental training
net.trainParam.epochs = 200;
net.trainParam.goal = 1.0e-4;
net.trainParam.lr = 0.5;
%Convert the input and output from numeric array to
%cell array for incremental training
P1=num2cell(P,1);
t1=num2cell(t,1);
disp('Starting adaptation. Please wait....');
for i=1:10
  [net, netOut, netError]=adapt(net,P1,t1);
  disp(strcat('Pass',num2str(i),'Complete'));
end;
%Display mean squared adaptation error
figure(1)
%Combine a cell array into one matrix
plot(cell2mat(netError));
disp(strcat('Mean squared adaptation error=',num2str(mse(netError))));
%Plot measured output and NN estimate
figure(2)
plot(1:N-2,t,'b',1:N-2,sim(net,P),'r');
```

E.1.1 GUI Interface of Neural Network Toolbox

This section introduces three popular neural network architectures for prediction and control that have been implemented in the Neural Network Toolbox:

- Model Predictive Control,
- NARMA-L2 (or Feedback Linearization) Control and
- Model Reference Control.

In the system identification stage, a neural network model of the plant has to be developed before it can be controlled. The prediction error between the plant output and the neural network output is used as the neural network training signal. The process is described in Section 5.3.1 and illustrated in Figures 5.1 and 5.2 of Chapter 5. This network can then be trained offline in batch mode using data collected from the operation of the plant. In the control design stage, the neural network plant model is used to design (or train) the controller. In each of the three control architectures described in this chapter, the system identification stage is the same. It is the control design stage that is different for each of the architectures:

- For model predictive control, the plant model is used to predict future behaviour of the plant, and an optimization algorithm is used to select the control input that optimizes future performance.
- For NARMA-L2 control, the controller is simply a rearrangement of the plant model.
- For model reference control, the controller is a neural network that is trained to control a plant so that it follows a reference model. The neural network plant model is used to assist in the controller training.

These three controllers are implemented as Simulink® blocks, which are contained in the Neural Network Toolbox block set. Each controller has its own advantages and pitfalls for any application. The choice of controller is eventually problem-dependent.

Example E.1.4: NN predictive control of continuous stirred tank reactor   The CSTR, also known as a vat reactor or backmix reactor, is a common ideal reactor type in chemical engineering. A CSTR often refers to a model that is used to estimate the key unit operation variables when using a continuous agitated-tank reactor to reach a specified output. From a control point of view, product concentration at the output of the process is an interesting control problem. The mathematical model works for all fluids: liquids, gases and slurries. A CSTR system for mixing two liquids of concentration C_{b1} and C_{b2} with flow rates w_1 and w_2 is shown in Figure E.1.1. The problem here is to control the concentration C_b of the mixture in the reactor by adjusting the flow w_1 while the flow w_2 is

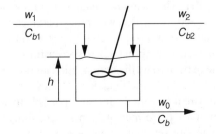

Figure E.1.1   Continuous stirred tank reactor

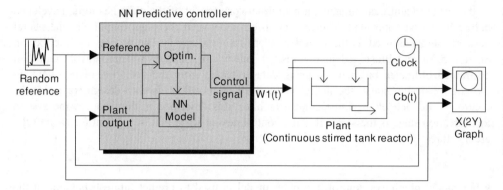

Figure E.1.2   NN Predictive Control of a CSTR system

fixed. The dynamic model of a CSTR system is described by the two differential equations below:

$$\frac{dh(t)}{dt} = w_1(t) + w_2(t) - 0.2\sqrt{h(t)} \tag{E.1}$$

$$\frac{dC_b h(t)}{dt} = (C_{b1} - C_b(t))\frac{w_1(t)}{h(t)} + (C_{b2} - C_b(t))\frac{w_2(t)}{h(t)} - \frac{k_1 C_b(t)}{(1 + k_2 C_b(t))^2} \tag{E.2}$$

where $h(t)$ is the liquid level, $C_b(t)$ is the product concentration at the output of the process, $w_1(t)$ is the flow rate of the concentrated feed $C_{b1}(t)$ and $w_2(t)$ is the flow rate of the diluted feed C_{b2}. The input concentrations are set to $C_{b1} = 24.9$ and $C_{b2} = 0.1$. The constants associated with the rate of consumption are $k_1 = 1$ and $k_2 = 1$. The objective of the controller is to maintain the product concentration by adjusting the flow $w_1(t)$. To simplify the demonstration, set $w_2(t) = 0.1$. The level of the tank $h(t)$ is not controlled for this experiment. A Simulink® model of a CSTR, shown in Figure E.1.2, has been developed from the differential equations (E.1) and (E.2).

The Neural Network Toolbox provides a blockset for an NN Predictive Controller. The first step is to copy the NN Predictive Controller block from the blockset to the Simulink® model window. Connect both the NN Predictive Controller block and the CSTR model as shown in Figure E.1.1. A random reference signal is used from the Simulink® source.

Double-clicking on the NN Predictive Controller block brings up the design window for the model predictive controller. This window enables us to set the controller horizons N_2 and N_u where N_1 is kept fixed at 1. The weighting parameter p is also defined in this window. The search parameter α is used to control the optimization. It determines how much reduction in performance is required for a successful optimization step. The linear minimization search routine to be used by the optimization algorithm is also selected here. The linear minimization routines are slight modifications of those discussed in backpropagation, for example backtracking or Brent's methods. It is also possible to decide how many iterations of the optimization algorithm are performed at each sample time. Once the Neural Network Predictive Control parameters are set, plant identification can be started by clicking on the Plant Identification button. This opens up the Plant Identification window.

Plant identification is carried out in three steps: defining the neural network architecture, generating the training data and training the neural network model. The defined neural network will be used as the plant model and hence will be called the NN model. The NN model predicts the future plant outputs. The optimization algorithm uses these predictions to determine the control inputs that optimize future performance. The plant model neural network has one hidden layer. The size of the hidden layer and the number of delayed inputs and delayed outputs are set in this window. A set of input/output data is required to train the neural network plant model.

The next step is to generate a set of input/output data using the Simulink® model for the CSTR described above. Training samples, maximum and minimum of plant input/output data and minimum interval value are set and then the data-generation process started by clicking on the Generate Training Data button. The program generates training data by applying a series of random step inputs to the CSTR Simulink® model. The potential training data is then generated and displayed in a separate window. Once the data generation is complete, accept the data by clicking on the Accept Data button in this window.

The final step is to train the NN model using the accepted data. Any of the training functions described in the backpropagation algorithm can be used to train the NN model. To do this, training epochs and training function are selected and then, by selecting Train Network from the Plant Identification window, the plant model training is started. The training proceeds according to the chosen training algorithm (`trainlm`, `traingdx`, etc.). This is a straightforward application of batch training, as described in backpropagation. After the training is complete, the responses of the resulting plant model and NN model are displayed. The random input data to the plant model and its response are shown in Figure E.1.3(a,b). The training error after training and the NN model output after training are shown in Figure E.1.3(c,d). There are also separate plots for validation and testing data, if they exist. Training can be continued with the same data set by selecting Train Network again or data can be deleted by selecting Erase Generated Data. Once the NN model has been trained, accept the current NN plant model and begin simulating the closed-loop system. For this example, follow the steps below to begin the simulation.

Select OK in the Plant Identification window. This loads the trained neural network plant model into the NN Predictive Controller block. Select OK in the Neural Network Predictive Control window. This loads the controller parameters into the NN Predictive Controller block. Return to the Simulink® model and start the simulation by choosing the Start command from the Simulation menu. As the simulation runs, the plant output and the reference signal are displayed as in Figure E.1.4.

Example E.1.5: NARMA-L2 control of magnetic levitation system A magnetic levitation system (MagLev) is a system that uses magnetic fields as a means to levitate an object in a certain position. If an object is placed far away from the magnetic source, the magnetic force is too weak to support the weight of the object. If placed too close to the magnetic source, the magnetic field becomes too strong and causes the object to move towards the source and come into contact with the magnet. Thus, the MagLev device is inherently an unstable system and thus poses a very interesting control problem. In this example, the objective is to control the position $y(t)$ of a magnet suspended above an electromagnet, where the magnet is constrained so that it can only move in the vertical direction. A simple MagLev system is shown in Figure E.1.5. The equation of motion for the system is given by

$$\ddot{y}(t) = -g + \frac{\alpha}{m}\frac{i^2(t)}{y(t)} - \frac{\beta}{m}\dot{y}(t) \tag{E.3}$$

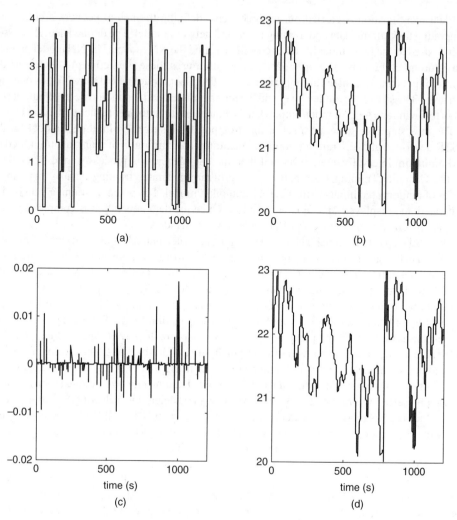

Figure E.1.3 Plant identification. (a) random input data; (b) CSTR plant output data; (c) error after training; (d) NN model output after training

where $y(t)$ is the distance of the magnet above the electromagnet, $i(t)$ is the current flowing in the electromagnet, m is the mass of the magnet and g is the gravitational constant. The parameter β is a viscous friction coefficient that is determined by the material in which the magnet moves, and α is a field strength constant that is determined by the number of turns of wire on the electromagnet and the strength of the magnet.

A Simulink® model of the MagLev system, shown in Figure E.1.6, has been developed from the differential equation (E.3). The Neural Network Toolbox provides a blockset for the NARMA-L2 Controller. The NARMA-L2 Controller block is copied from the blockset to the Simulink® model window. Connect both the NARMA-L2 Controller block and the MagLev model as shown in Figure E.1.6. A random reference signal is used from the Simulink® source.

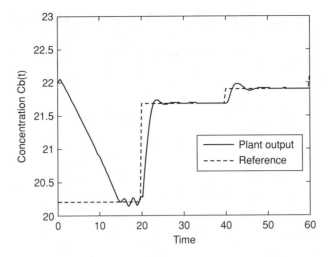

Figure E.1.4   Performance of the NN Predictive Controller showing CSTR plant output compared with reference signal

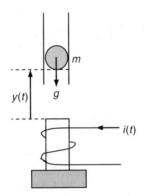

Figure E.1.5   Magnetic levitation system

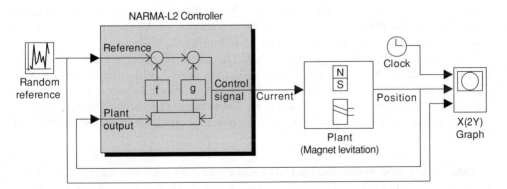

Figure E.1.6   Simulink® model of NARMA-L2 Controller

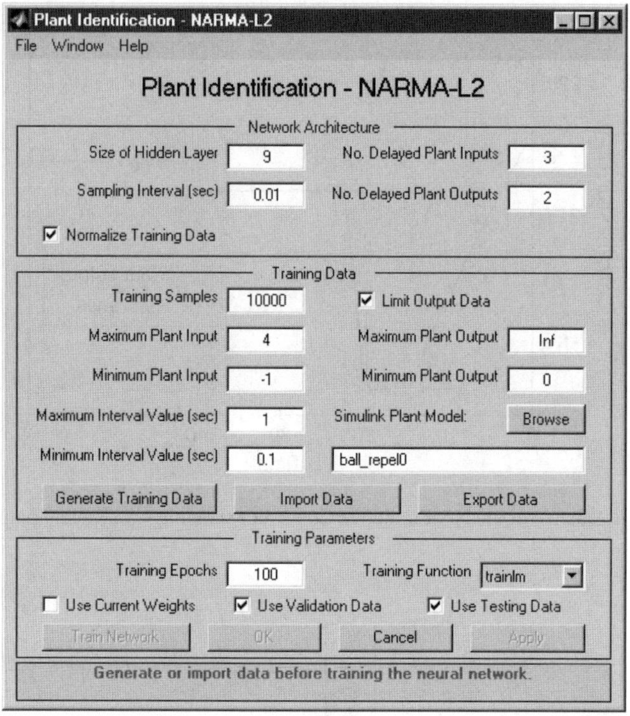

Figure E.1.7　Plant identification window of NARMA-L2 Controller

Plant identification of the MagLev system is carried out first. It is the same procedure as the plant identification in Model Predictive Control. Double-clicking on the NARMA-L2 Controller block brings up the plant identification window of NARMA-L2 shown in Figure E.1.7. The network architecture and training data parameters are specified to generate training data. The random input data to the MagLev plant model and its output data are shown in Figure E.1.8(a,b). The training parameters and algorithm are chosen for training the network. The training error after training and the NN model output after training are shown in Figure E.1.8(c,d). There is no separate window for the controller design, because the controller is determined directly from the model, unlike the Model Predictive Controller. This window works the same as the other Plant Identification windows, so the training process is not repeated. Instead, simulation of the NARMA-L2 Controller can be started.

Select OK in the Plant Identification window. This loads the trained neural network plant model into the NARMA-L2 Controller block. Select OK in the NARMA-L2 Control window. This loads the controller parameters into the NARMA-L2 block. Return to the Simulink® model and start the simulation by choosing the Start command from the Simulation menu. As the simulation runs, the plant output and the reference signal are displayed as shown in Figure E.1.9.

Example E.1.6: Model reference control of robot arm　　Online computation of the Model Reference Controller, like NARMA-L2, is also minimal. However, one difference is that the model reference architecture requires a separate neural network controller to be trained

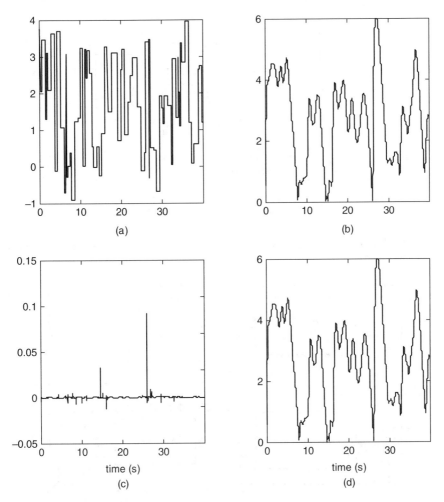

Figure E.1.8   Plant identification. (a) random input data; (b) MagLev plant output data; (c) error after training; (d) NN model output after training

offline, in addition to the neural network plant model identification. The controller training is computationally expensive, because it requires the use of dynamic backpropagation (Narendra and Parthasarathy, 1991). The advantage is that the model reference control applies to a larger class of plant than that of NARMA-L2 control. The neural network MRC architecture uses two neural networks: a controller network and a plant model network. The architecture is shown in the block diagram of Figure E.1.10. The plant model is identified first, and then the controller is trained so that the plant output follows the reference model output.

In this example, the objective is to control the movement of a rigid single-link robot arm, as shown in Figure E.1.11. The equation of motion for the arm is given by the differential equation

$$\frac{d^2\theta}{dt^2} = -10\sin\theta - 2\frac{d\theta}{dt} + u \tag{E.4}$$

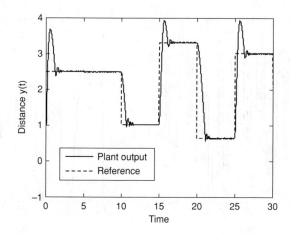

Figure E.1.9 Performance of the NARMA-L2 Controller showing MagLev plant output compared with reference signal

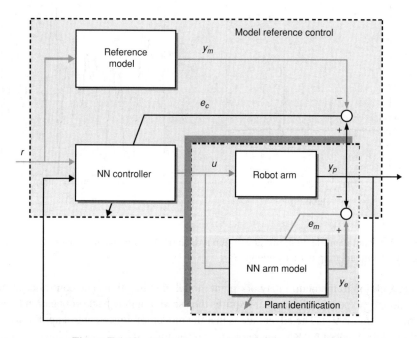

Figure E.1.10 Block diagram of model reference control

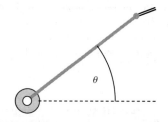

Figure E.1.11 Single-link robot arm

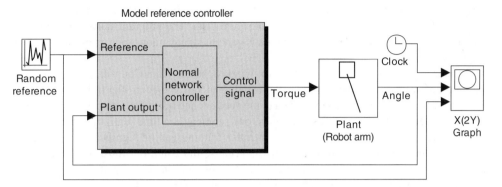

Figure E.1.12   Model reference control of single-link robot arm

where θ is the angle of the arm and u is the torque supplied by the DC motor. The objective is to train the controller so that the arm tracks the reference model. The reference model is described by the differential equation

$$\frac{d^2 y_r}{dt^2} = -9y_r - 6\frac{dy_r}{dt} + 9u_r \tag{E.5}$$

where y_r is the output of the reference model and u_r is the input reference signal. A detailed description of the model can be found in Kim and Lewis (1998).

A Simulink® model of the robot arm system has been developed from the differential equation (E.4). The Neural Network Toolbox provides a blockset for the MRC. The MRC block is copied from the blockset to the Simulink® model window. Connect both the MRC block and the robot arm model. Figure E.1.12 shows details of the MRC as implemented in the Neural Network Toolbox, and the Simulink® model of the robot arm. A reference model is also required which will generate the reference signal to be traced by the controller. A Simulink® model of the reference model has also been developed using the differential equation (E.5). A random reference signal generator from the Simulink® source is used here to provide a reference signal to both the MRC and the reference model.

Double-clicking on the MRC block brings up the Model Reference Control window. The first step would normally be to select Plant Identification, which opens the Plant Identification window. A neural network plant model is defined and trained from the generated data. A Simulink® plant model for the robot arm defined in (E.4) is used for plant identification. Because the Plant Identification window is identical to the one used with the previous controllers in Model Predictive or NARMA-L2 Control, that process is repeated here.

Once the plant identification is complete, the MRC architecture is defined. The architecture is defined by the number of neurons to use in the hidden layer and three other input parameters. The three sets of controller parameters are: delayed reference inputs, delayed controller outputs and delayed plant outputs. For each of these inputs, select the number of delayed values to use. Typically, the number of delays increases with the order of the plant. In this example, the architecture of the controller is a 5–13–1 neural network. The inputs to the controller consist of two delayed reference inputs, two delayed plant outputs and one delayed controller output. A sampling interval of 0.05 s is used. The Simulink® model of the reference model defined in (E.5) is used for this experiment. After entering the maximum/minimum reference value, interval value and training samples, select Generate Data. The program starts generating the data for training the controller. After the data is generated, a window appears prompting us to accept the data.

On return to the Model Reference Control window, select Train Controller. The controller training requires two parameters: controller training epochs and controller training segments. The program presents one segment of data to the network and trains the network for a specified number of iterations. This process continues, one segment at a time, until the entire training set has been presented to the network. Controller training can be significantly more time-consuming than plant model training. This is because the controller must be trained using *dynamic backpropagation* (see Hagan *et al.*, 1999). After training is complete, the response of the resulting closed-loop system is displayed, as shown in Figure E.1.13.

If the performance of the controller is not satisfactory, then train the controller again, which continues the controller training with the same data set. If there is a problem with the generated data, generate or import a new data set to continue training. Be sure that Use Current Weights is selected if you want to continue training with the same weights. It might also be necessary to retrain the plant model. If the plant model is not accurate, it can affect the controller training.

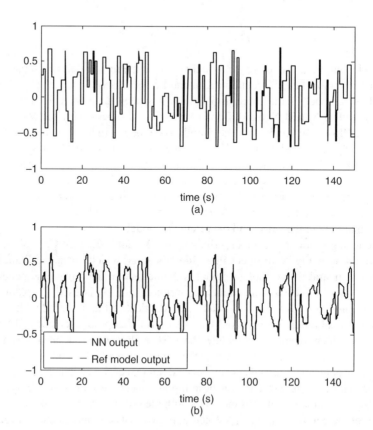

Figure E.1.13 Controller training. (a) reference model input; (b) reference model output and neural network output after training the controller for 10 epochs in 10 segments

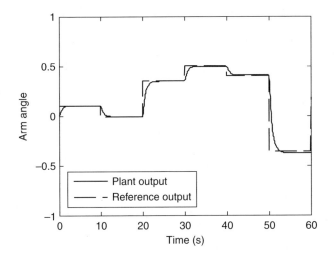

Figure E.1.14   Performance of the Model Reference Controller showing robot arm plant output compared with reference signal

For this example, make sure that the controller is accurate enough. If so, select OK. This loads the controller details and weights into the Simulink® model.

Return to the Simulink® model and start the simulation by selecting the Start command from the Simulation menu. As the simulation runs, the plant output and reference signal are displayed, as shown in Figure E.1.14.

Appendix F

MATLAB® Programs for Evolutionary Algorithms

F.1.1 Using the Genetic Algorithm Toolbox

The toolbox provides the function ga() to implement the genetic algorithm at the command line or in M-files to minimize an objective function. The general form of the function ga() is

```
x = ga(fitness_function, no_variables, options)
[x, fval, reason, output, population, scores] = ga( ... )
```

where fitness_function, no_variables and options are the input arguments. For standard optimization algorithms, fitness_function is known as the objective function. The objective function is written as an M-file and used as a function handle input argument in the ga() function. no_variables is the number of independent variables of the fitness function, and options is the set of parameters of the genetic algorithm defined with gaoptimset.

After running the algorithms for the maximum number of generations, ga(...) returns x, fval, reason, output, population and scores. x is the final solution, fval is the final value of the fitness function at x, reason is a string containing the reason for termination of the algorithm, output is a structure that contains the following fields:

randstate – the state of the MATLAB® random number generator
randnstate – the state of the MATLAB® normal random number generator
generations – the number of generations computed
funccount – the number of evaluations of the fitness function
message – the reason the algorithm terminated (same as reason).

ga(...) also returns a population of solutions in matrix form, whose rows are the solution to the optimization problem. Finally, ga() returns the scores of the final population.

F.1.1.1 Defining the Fitness Function

The ga() function actually minimizes the objective function $f(x)$, i.e., it solves a problem of the form $\min_x f(x)$. To maximize $f(x)$, it finds $-f(x)$. Firstly an objective function is

Computational Intelligence: Synergies of Fuzzy Logic, Neural Networks and Evolutionary Computing, First Edition.
Nazmul Siddique and Hojjat Adeli.
© 2013 John Wiley & Sons, Ltd. Published 2013 by John Wiley & Sons, Ltd.

defined for the optimization problem and then an M-file is written. For example, the function $f(x_1, x_2) = x_1^2 + x_2^2 - 6x_1 + 6x_2$ represents an objective function to be optimized. The M-file is written as follows:

```
function z = objfcn(x)
z=x(1)^2+x(2)^2-6*x(1)+6*x(2);
```

A function handle is used for the fitness (objective) function as @objfcn that computes the fitness value.

F.1.1.2 Setting the Options for GA

Option is the set of parameters of the genetic algorithm defined with gaoptimset, which creates the genetic algorithm options structure. The general form is as follows:

```
options = gaoptimset('param1',value1,'param2',value2,...)
```

gaoptimset() creates a structure of options and sets the value of 'param1' to value1, 'param2' to value2, and so on. Any unspecified parameters are set to their default values. The parameters, for example, are population size, selection function, crossover function, mutation function, plot functions. It is sufficient to type enough leading characters to define the parameter name uniquely. Case is ignored for parameter names. For example:

```
options = gaoptimset('PopulationSize', 100, 'PlotFcns', @gaplotbestf)
```

This creates an options structure with field values set to their default except for populationSize and PlotFcns, which are set to 100 and @gaplotbestf, respectively. A new set of parameters can be set or changed. For example:

```
options = gaoptimset(options, 'SelectionFcn', @selectionstochunif)
```

This preserves the current values of all fields of the options and sets the selection function to @selectionstochunif.

The toolbox provides a variety of selection, crossover and mutation functions.

Selection functions
SelectionFcn is a handle to functions that selects parents for crossover and mutation. The toolbox provides a number of functions: @selectionremainder, @selectionuniform, @selectionstochunif, @selectionroulette, @selectiontournament.

Crossover functions
CrossoverFcn is a handle to crossover functions. There are a number of such functions: @crossoverheuristic, @crossoverintermediate, @crossovertwopoint, @crossoverarithmatic, @crossoversinglepoint, @crossoverscattered.

Mutation functions
MutationFcn is a handle to mutation functions. There are a number of mutation functions provided by the toolbox: @mutationuniform, @mutationadaptfeasible, @mutationgaussian.

The GA Toolbox also supports a number of plot functions using the handle PlotFcns. Some of these are @gaplotbestf, @gaplotbestindiv, @gaplotdistance.

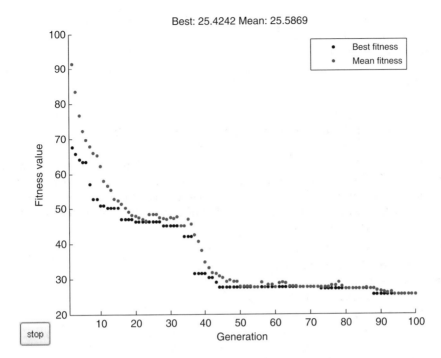

Figure F.1.1 Best and mean fitness over generations

Example F.1.1 This example demonstrates the optimization of Rastrigin's function, defined by

$$f(x_1, x_2) = 20 + x_1^2 + x_2^2 - 10(\cos 2\pi x_1 + \cos 2\pi x_2)$$

The handle to the objective function is @rastriginsFcn. The number of variables is set to 10. The options parameters are single point crossover and uniform mutation with all other parameters set to default. After running the GA for maximum generations (default 100), it will plot the best and mean fitness over the generations as shown in Figure F.1.1. Finally, the following code (M-file) will output the solution vector x, final fitness at x and the reason for termination of the GA:

```
clear all; close all; format compact
%Finding the Minimum of Rastrigins Function using GA
%Refer to Example F.1.1
FitnessFcn=@rastriginsFcn;
No_Variables=10;
options=gaoptimset('CrossoverFcn',@crossoversinglepoint, ...
    'MutationFcn',@mutationuniform, 'PlotFcns',@gaplotbestf);
%Run GA with options
[x,fval,reason]=ga(FitnessFcn,No_Variables,options)
x =
Columns 1 through 7
```

```
 0.8008   0.0672   0.0767   0.0383   0.2520   0.1295   0.0211
Columns 8 through 10
 0.9975   0.0475   0.0612
fval =
 25.4242
reason =
```

Optimization terminated: maximum number of generations exceeded.

Example F.1.2   This example demonstrates the effect of different settings for crossover fraction. The following code runs the GA 21 times, varying the crossover fraction (probability) from 0 to 1 in increments of 0.05 then records the result and plots the fitness value over the crossover fractions as shown in Figure F.1.2. The plot shows that the GA provides the best result for a crossover setting between 0.65 and 0.95.

```
clear all; close all; format compact
%Finding the Minimum of Rastrigins Function using GA
%Refer to Example F.1.2
FitnessFcn=@rastriginsFcn;
No_Variables=10;
record=[];
for n=0:0.05:1
  options=gaoptimset('Generations', 300, 'CrossoverFraction',n);
```

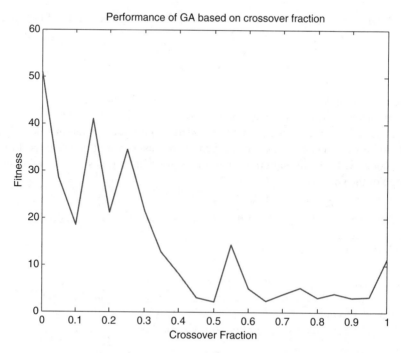

Figure F.1.2   Performance of GA based on crossover fraction (probability)

```
%Run GA with options
[x,fval]=ga(FitnessFcn, No_Variables, options);
record=[record; fval];
end
%Plot values of fval against the crossover fraction
plot(0:0.05:1,record);
xlabel('Crossover Fraction');
ylabel('Fitness')
title('Performance of GA based on crossover fraction')
```

Example F.1.3   This example demonstrates the GA plot functions. The following code (M-file) generates the four plots showing the best and mean fitness, current best individual, fitness scaling and diversity in Figure F.1.3:

```
clear all; close all; format compact
%Finding the Minimum using GA
%Refer to Example F.1.3
FitnessFcn=@rastriginsFcn;
No_Variables=3;
options=gaoptimset('PlotFcns',{@gaplotbestf,@gaplotbestindiv, ...
                @gaplotexpectation,@gaplotscorediversity});
```

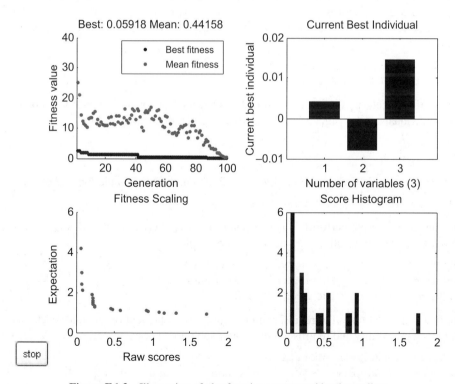

Figure F.1.3   Illustration of plot functions supported by the toolbox

```
%Run GA with options
[x,fval,reason]=ga(FitnessFcn,No_Variables,options)
```

The University of Sheffield developed a GA Toolbox under MATLAB®. It has many powerful functions. A few functions are discussed in this section, and a simple GA example is
demonstrated in Example F.1.4. For these exercises, the user needs to install or write a few
functions and put them in their work directory. These functions are crtbp(), crtrp(), bs2rv(),
select(), xovrsp() and mut(). These functions are a simple demonstration of the concepts of
evolutionary computing. To create a binary or real-valued population, functions such as crtbp()
or crtrp() are used. To convert a binary population into a real-valued population, use bs2rv().
For genetic operators such as selection, crossover and mutation, use the functions select(),
xovrsp() and mut(), respectively.

   The function crtrp() creates a population of given size of random real values. Nind is a
scalar containing the number of individuals in the new population and FieldDR is a matrix of
size 2 by number of variables describing the boundaries of each variable. For example:

```
Nind = 6;
FieldDR = [-10 -5 -3 -2;   % lower bound
            10  5  3  2]   % upper bound
Chrom = crtrp(Nind, FieldDR)
```

The function bs2rv() decodes binary chromosomes into vectors of real values. The chromosomes are seen as the concatenation of binary strings of given length, and decoded into
real numbers in a specified interval using either standard binary or Gray decoding. FieldD
is a matrix describing the length and how to decode each substring in the chromosome. For
example:

```
FieldD = [ 8; -1; 25; 1; 0; 1; 1];
Phen = bs2rv(Chrom,FieldD)
```

The function select() performs a universal selection from the population. Consider a population
of six individuals with assigned fitness values and generation gap as follows:

```
FitnV = [0.5; 1.16; 0.83; 0.723; 1.6; 0.19];
GGap = 0.9;
```

Using the Stochastic Universal Sampling 'sus' function, the new population will be

```
SelCh= select('sus', Chrom, FitnV, GGap)
```

The function xovrsp() performs a single-point crossover between pairs of individuals and
returns the current generation after mating. For example:

```
Xovrp = 0.87;      %cross over probability
OldChrom = Chrom; %Chrom created in example 1
NewChrom = xovrsp(OldChrom, Xovrp)
```

The function mut() takes the representation of the current population, mutates each element
with given probability and returns the resulting population. For example:

```
pm=0.12;   %mutation probability
OldChrom = NewChrom;    %Chrom created in example 6
NewChrom = mut(OldChrom, pm)
```

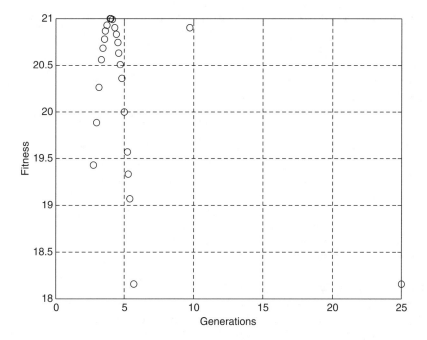

Figure F.1.4   GA achieves the maximum at 4

Example F.1.4   The function given by $f(x) = x^2 + 8x + 15$ is to be optimized using GA. The function has a maximum at 4. The following M-file generates an initial binary population of 20 individuals of length 8. It runs the GA for 50 generations using the functions explained above. The GA achieves the maximum of the function in only 5 generations, as shown in Figure F.1.4.

```
%Refer to Example F.1.4
%--GA parameter settings
Nind = 20;       % No of chromosomes
Lind = 8;        % Length of string
xovpr = 0.87;    % Crossover probability
mutpr = 0.12;    % Mutation probability
maxgen = 50;     % Maximum generation
%create a binary population of 20 chrom of string length 8
Chrom = crtbp(Nind, Lind);
%Field description for convertion
FieldD=[8; 0; 25; 1; 0; 1; 1];

gen = 1;
while gen < maxgen
  %convert the binary strings into real value
  Phen = bs2rv(Chrom, FieldD);
  x = Phen;
```

```
%Calculate Fitness value FitnV - scale the fitness value between 0 and 2
  ObjV = -x.^2+8*x+5;
  mx = max(abs(ObjV));
  FitnV = ObjV./mx+2;
  %Show the max value at each gen
  mxx = ObjV(1);
  for i = 2:length(x)
    if mxx < ObjV(i)
        mxx = ObjV(i);
        ix = i;
    end
  end
  mxx;
  p = x(ix);
  %Select using 'sus' selection
  SelCh = select('sus', Chrom, FitnV);

  % Single point Cross over is performed
  % here with a cross over probability of 0.87
  % that means 87% of the individuals undergoe crossover
  OldChrom = SelCh;
  NewChrom = xovsp(OldChrom, xovpr);

  % Mutation is performed here with a mutation probability of 0.12
  % that means 12% of the individuals undergoe mutation
  Chrom = mut(NewChrom, mutpr);

  %Plot the max value and x at each generation
  plot(p, mxx, 'ok')
  xlabel('Generations')
  ylabel('Fitness')
  hold on
  grid
  gen = gen +1
end
```

Appendix G

MATLAB® Programs for Neuro-Fuzzy Systems

The modelling approach used by ANFIS is similar to many system identification techniques and can be broken down into the following steps:

- Set of input/output data,
- Parameterized model structure relating to input/output MFs and rules.

In some cases, data is collected using noisy measurements, and the training data cannot be representative of all the features of the data that will be presented to the model. This is where model validation and testing come into play. The whole model-building process is divided into three steps:

- Model building,
- Model validation and
- Model testing.

Model validation is the process by which the input vectors from testing the I/O data set are presented to the trained ANFIS model to see how well the ANFIS model predicts the corresponding data set output values. To perform the above tasks, the whole data set is divided into three sets of data:

- Training data,
- Testing data and
- Checking data.

To create a training set from the available historical sequence first requires the choice of how many and which delayed outputs affect the next output. Each item in the training data set should have a value, because that is what the classifier uses to learn how to predict. In the checking data, each item may or may not have a correct value specified for the class value.

Computational Intelligence: Synergies of Fuzzy Logic, Neural Networks and Evolutionary Computing, First Edition.
Nazmul Siddique and Hojjat Adeli.

To evaluate how accurate the classifier is, the true class values or checking data are needed. The classifier won't use them when making predictions, but they will be used to calculate how accurate the predictions are. If there are about 1000 I/O data points, the first 500 data points are used for ANFIS training (called the training data set) while the others are used as checking data for validating the identified model. To perform the above tasks, the whole data set is divided into two sets of data:

- Training data and
- Checking data.

The following simple code can be used to generate the data sets:

```
train_Data=data(1:500,:);
check_Data=data(501:end,:);
```

In order to achieve good generalization capability, it is important to ensure that the number of training data points is several times larger than the number of parameters being estimated.

G.1.1 Creating ANFIS Structure

The first thing in ANFIS modelling is to define an ANFIS structure and set the initial parameters for learning. The tools provide two functions, genfis1() and genfis2(), to generate the ANFIS structure. The general form of the two functions is as follows:

```
fismat = genfis1(data, numMFs, inmftype, outmftype)
fismat = genfis2(Xin, Xout, radii, xBounds, options)
```

genfis1() generates a Sugeno-type FIS structure from a training data set using a grid partition on the data without applying clustering. The number of MFs (numMFs), type of input and output MFs (inmftype, outmftype) can also be specified. genfis1 uses the following arguments: 'data' is the training data matrix, which must be provided with all columns representing input data, and the last column should represent the single output. numMFs specifies the number of membership functions associated with each input. If numMFs is omitted, the default value of 2 is used. If the same number of membership functions is desired for each input, it suffices to make a single number. inmftype is a string that specifies the type of membership function for each input. If the same membership type is desired for each input, it suffices to name a single type. If the membership function type is omitted, the default input membership function type 'gbellmf' is used. outmftype is a string that specifies the membership function type for the output. There can only be one output, since this is a Sugeno-type system. The output membership function type must be either linear or constant. The default output membership function type is 'linear'.

The number of membership functions associated with the output is the same as the number of rules generated by genfis1. Default settings are used whenever genfis1 is invoked without numMFs, inmftype, outmftype.

Example G.1.1

```
%Example G.1.1 - use of genfis1 function
data = [rand(100,1) 10*rand(100,1)-5 rand(100,1)];
numMFs = [3 3];
mfType = str2mat('pimf','trimf');
```

```
fismat = genfis1(data, numMFs, mfType);
[x,mf] = plotmf(fismat, 'input',1);
subplot(2,1,1), plot(x, mf);
xlabel('input 1 (pimf)');
[x,mf] = plotmf(fismat, 'input',2);
subplot(2,1,2), plot(x, mf);
xlabel('input 2 (trimf)');
```

The genfis1 function generates initial MFs that are equally spaced and cover the whole input space. The genfis1-generated MFs are shown in Figure G.1.1.

Given separate sets of input and output data, genfis2 generates an initial ANFIS structure for training by first implementing subtractive clustering on the data set. genfis2 does this by extracting a set of rules that models the behaviour of the system represented by the data set. The rule extraction method determines the number of rules and antecedent membership functions and then uses linear least-squares estimation to determine each rule's consequent equations. genfis2 uses the following arguments: Xin is a matrix in which each row contains the input values of a data point. Xout is a matrix in which each row contains the output values of a data point. radii is a vector that specifies a cluster centre's range of influence in each of the data dimensions. For example, if the data dimension is 3 (i.e., two columns of Xin and one column of Xout), radii = [0.5 0.4 0.3] specifies the ranges of influence in the first, second and third data dimensions. xBounds is a 2-by-N optional matrix that specifies how to normalize the data in Xin and Xout into values in the range [0 1] for processing. N is the data (or row) dimension of the matrix. The first row of xBounds contains the minimum and the second row contains the maximum range values for scaling the data in each dimension. For example, xBounds = [−10 0 −2; 10 30 1] specifies that the first, second and third data dimension values are to be

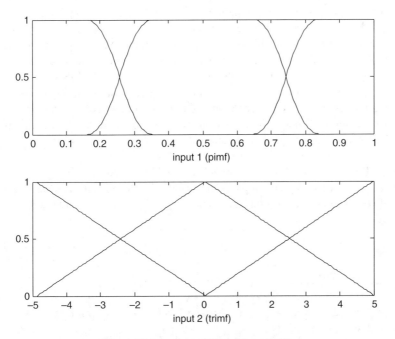

Figure G.1.1   Input MFs of the ANFIS

scaled from the range [–10 +10], [0 30] and [–2 1] into values in the range [0 1], respectively. If xBounds is not provided, then xBounds uses the minimum and maximum data values found in each data dimension. 'options' is an optional vector for specifying algorithm parameters to override the default values. There are four options available. The parameters are explained in the help text for subclust. Default values are in place when this argument is not specified.

Example G.1.2

```
%Example G.1.2 - use of genfis2 function
data = [rand(100,1) 10*rand(100,1)-5 rand(100,1)];
Xin=data(:,1:2);  % column 1 and 2 of data
Xout=data(:,3);    % column 3 of data
fismat = genfis2(Xin,Xout,[0.5 0.25 0.3]);
[x,mf] = plotmf(fismat, 'input',1);
subplot(2,1,1), plot(x, mf,'k');
xlabel('input 1');
[x,mf] = plotmf(fismat, 'input',2);
subplot(2,1,2), plot(x, mf, 'k');
xlabel('input 2');
```

G.1.2 Training ANFIS

Once a Sugeno-type FIS is generated, an ANFIS can be trained using the function anfis(). The general form is as follows:

```
[fismat, error, stepsize] = anfis(trnData, fismat)
```

This is the major training function for Sugeno-type fuzzy inference systems. anfis() uses a hybrid learning algorithm to identify parameters of Sugeno-type fuzzy inference systems. It applies a combination of the least-squares method and the backpropagation gradient descent method for training FIS membership function parameters to emulate a given training data set. anfis() can also be invoked using an optional argument for model validation. trnData is the training data used in defining the ANFIS structure. fismat is the ANFIS structure created by the function genfis1(). chkData is the optional checking data for overfitting model validation. The training process stops whenever the maximum epoch number is reached or the training error goal is achieved. When anfis is invoked with two or more arguments, the rest of the optional arguments will take on their default values.

Example G.1.3   A re-vibration system is described by the equation $y = \frac{\sin(2x)}{\exp(-x/5)}$, where x is the input to the system. A prediction model has been developed using ANFIS. The following segment of code shows how to cerate and train the ANFIS model and predict the output for a noisy input:

```
clear all;
clc;
%ANFIS for prediction of a re-vibration system
%-------------------------------------------------------
%The re-vibration system is described by the equation
%y = sin(2*x)./exp(-x/5);
```

```
%To identify the re-vibration system, I/O data is generated first
x = (0:0.1:10)';
y = sin(2*x)./exp(-x/5);
trnData=[x y];
%--Setting ANFIS parameters
nummfs = 5;
mftype = 'gbellmf';
epoch_n = 20;


%--ANFIS structure is created using the training data
in_fis  = genfis1(trnData,nummfs,mftype);


%--ANFIS is then trained using  training data
out_fis = anfis(trnData,in_fis,epoch_n);


%--Prediction of y with additive noise using the model
x=x+0.01;
yp=evalfis(x,out_fis)


%--Plot the plant input and output and reference output.
plot(x,y,'--k',x,yp,'.-k');
grid;
legend('Desired output','Predicted Output');
xlabel('Input x');
ylabel('Model output');
disp('Sum of absolute error: ')
SAE=sum(abs(y-yp))
```

See Figure G.1.2.

Example G.1.4: Developing ANFIS controller for water bath system   The temperature control of a water bath plant is an interesting problem and used as a benchmark problem for much control system research. The water bath temperature plant is described by

$$y(k+1) = a(T)y(k) + \frac{b(T)u(k)}{1 + \exp(0.5y(k) - r)} + (1 - a(T))Y_0$$

where $a(T) = \exp(-\alpha T)$ and $b(T) = \frac{\beta(1-\exp(-\alpha T))}{\alpha}$. The parameters of the plant are $\alpha = 1 \times 10^{-4}$, $\beta = 8.7 \times 10^{-3}$, $r = 40$ and $Y_0 = 25\,°C$. The plant input $u(k)$ is defined as $0 \le u(k) \le 5$ V. The sampling period is $T_s = 25$ s. This reference signal is defined by

$$\text{ref}(k) = \begin{cases} 35\,°C & \text{for } 0 \le t \le 40\,\text{min} \\ 50\,°C & \text{for } 40 < t \le 80\,\text{min} \\ 65\,°C & \text{for } 80 < t \le 120\,\text{min} \\ 80\,°C & \text{for } 120 < t \le 180\,\text{min} \end{cases}$$

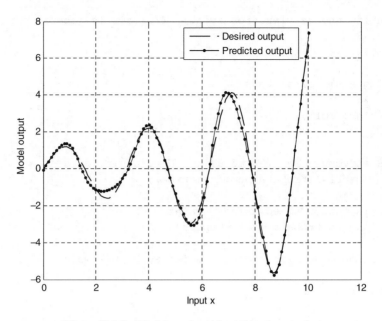

Figure G.1.2   Model output with additive input noise

A neuro-fuzzy (ANFIS) controller has to be developed to control the water temperature to follow a reference signal as closely as possible. One of the uses of ANFIS in a control system is to learn the inverse of the plant model, so that it can be used as a controller after training. The general procedure is to generate a random input signal $u(k)$, apply it to the plant and measure the output $y(k + 1)$. Using the output and delayed output signal $\{y(k + 1), y(k)\}$ and input $u(k)$, an inverse ANFIS model of the plant is trained. Figure G.1.3 shows the learning stage of the inverse ANFIS model. Notice that the learning of ANFIS is based on error backpropagation.

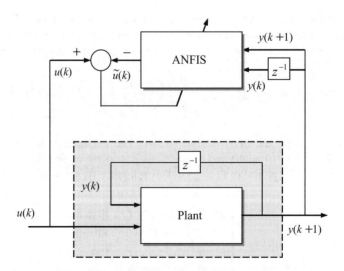

Figure G.1.3   Training stage of the ANFIS model

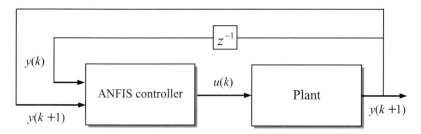

Figure G.1.4   ANFIS controller deployed for the plant

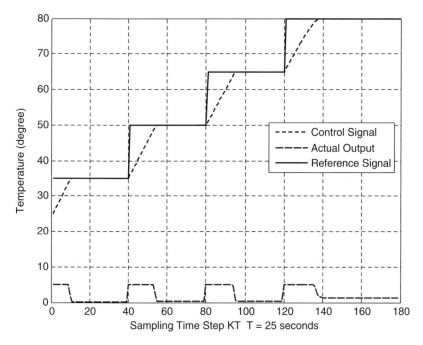

Figure G.1.5   Plant's response using the ANFIS controller

Once the ANFIS model is trained with sufficient training data, it is used as a controller for the plant. The ANFIS controller deployed for the water bath plant is shown in Figure G.1.4. The following segment of MATLAB® code shows how to create an ANFIS model, train the ANFIS model and apply it as a controller for the water bath plant. Figure G.1.5 shows the control signal produced by the ANFIS controller, which is able to control the plant's output following the reference signal closely.

```
clear all;
clc;
%ANFIS Controller for the Water Bath System
%-----------------------------------------------
```

```
%The water bath system is described by the equation
%y(k+1) = a(T)*y(k)+b(T)/(1+exp(0.5*y(k)-r))*u(k)+(1-a(T))*yo;
%with 0<= u(k)<=5 volts.
p = 1*10^(-4);
q = 8.7*10^(-3);
yo = 25;
y(1) = yo;
%--Sampling period Ts=25 sec
Ts = 25;
%--Parameters of plant a(T),b(T)& r are defined as
a = exp(-p*Ts);
b = (q/p)*(1-exp(-p*Ts));
r = 40.0;
%--The reference signal is defined as
for k = 1:180
  if (k <= 40)
      ref(k) = 35;
  elseif (k > 40 & k <= 80)
      ref(k) = 50;
  elseif (k > 80 & k <= 120)
      ref(k) = 65;
  elseif (k > 120)
      ref(k)=80;
  end;
end;
%To identify the water bath plant, I/O data set is required
%The simplest wat to do it is to input random signal to
%the plant and measure the output.
%Random input signal is generated and ouput signal is collected
for k = 1:120
  u(k) = rand(1,1)*5;
  y(k+1) = a*y(k)+b/(1+exp(0.5*y(k)-r))*u(k)+(1-a)*yo;
end;
%--Training data created
trndata = [y(2:101);y(1:100);u(1:100)]';

%To start the ANFIS training, we need an FIS structure
%and initial parameters of the FIS for learning.
%The user can use the command "genfis1" to generate an FIS matrix
%from the training data using the grid-type partition
%according to the given number and types of membership functions.
%In this demonstration, five 'gaussmf'-type MFs are used.

%--Setting ANFIS parameters: 5 Gaussian MFs are chosen
nummfs = 3;
mftype = 'gaussmf';
```

```
%--ANFIS structure is created using the training data
fismat = genfis1(trndata,nummfs,mftype);

%--MFs before training
figure
[x,mf] = plotmf(fismat, 'input',1);
subplot(2,1,1), plot(x, mf,'k');
xlabel('input 1 (gaussmf)');
[x,mf] = plotmf(fismat, 'input',2);
subplot(2,1,2), plot(x, mf, 'k');
xlabel('input 2 (gaussmf)');

%--The ANFIS is then trained to identify the inverse model
%--of the water bath system using the gathered training data.
[outfismat, error,stepsize] = anfis(trndata, fismat,5);

%     Save ANFIS data
save myanfis.mat outfismat
disp('----------------------------------------------------')

%--The ANFIS outfismat is now used as a controller for the
%--water bath system

controller=outfismat;

%--Test 180 time-steps.
for k=1:179
  %Controller is simulation using ref signal & plant output
  u(k) = evalfis([ref(k+1) y(k)],controller);
  if (u(k) >= 5)
      u(k)=5;
  elseif (u(k) <= 0)
      u(k)=0;
  else
      u(k)=u(k);
  end;
  %Plant simulation using the control input
  y(k+1)=a*y(k)+b/(1+exp(0.5*y(k)-r))*u(k)+(1-a)*yo;
end;
figure
%--Plot the plant input and output and reference output.
hold on;
grid;
plot(u(1:179),'--k');
plot(y(1:180),'-.k');
plot(ref(1:180),'-k');
```

```
xlabel('Sampling Time Step KT  T = 25 seconds');
ylabel('Temperature(degree)');
legend('Control Signal','Actual Output','Reference Signal');
disp('Sum of absolute error: ')
AE=sum(abs(ref-y));
```

Index

Computational Intelligence: Synergies of Fuzzy Logic, Neural Networks and Evolutionary Computing, First Edition.
Nazmul Siddique and Hojjat Adeli.
© 2013 John Wiley & Sons, Ltd. Published 2013 by John Wiley & Sons, Ltd.